SFRC87

Proceedings of the
International Symposium on

FIBRE REINFORCED CONCRETE

Madras, India
December 16-19, 1987

ORGANISED BY

STRUCTURAL ENGINEERING RESEARCH CENTRE, MADRAS

(COUNCIL OF SCIENTIFIC & INDUSTRIAL RESEARCH)

VOLUME I

CRC Press
Taylor & Francis Group
Boca Raton London New York

CRC Press is an imprint of the
Taylor & Francis Group, an informa business
A BALKEMA BOOK

GENERAL EDITORS:
V.S. Parameswaran & T.S. Krishnamoorthy

Published by:
CRCPress/Balkema
P.O. Box 447, 2300 AK Leiden, The Netherlands
e-mail: Pub.NL@taylorandfrancis.com
www.crcpress.com – www.taylorandfrancis.com

ISBN 13: 978-90-6191-796-0 (set)
ISBN 13: 978-90-6191-797-7 (hbk) (vol I)

Visit the Taylor & Francis Web site at
http://www.taylorandfrancis.com

and the CRC Press Web site at
http://www.crcpress.com

ORGANISING COMMITTEE

Chairman : M. Ramaiah, Structural Engineering Research Centre, Madras

Secretary : V.S. Parameswaran, Structural Engineering Research Centre, Madras
Members :
P.V. Naik, Gammon India Limited, Bombay
V. Murahari Reddy, Public Works Department (Andhra Pradesh) Hyderabad
Chief Engineer, Tamil Nadu Housing Board, Madras
S. Durvasula, Indian Institute of Science, Bangalore
A.M. Rao, Engineers India Limited, New Delhi
R.D. John, Department of Space, Bangalore
K.R. Puttanna, Bharat Heavy Electricals Limited, Hyderabad
K.A. Ramabhadran, Neyveli Lignite Corporation Limited, Neyveli
L.V. Ramakrishna, DRDO, Ministry of Defence, New Delhi
K.V. Rangaswami, L & T Limited, (ECC Construction Group), Madras
C.R. Alimchandani, STUP Consultants Limited, Bombay
R.N. Raiker, Institution of Engineers (India), Calcutta
N.G. Joshi, Indian Hume Pipe Company Limited, Bombay
R.K. Bhandari, Central Building Research Institute, Roorkee
S.R. Kulkarni, M.N. Dastur & Co. (P) Ltd., Calcutta
N. Bhattacharya, FGP Limited, Bombay
J.L. Ajmani, Tata Iron & Steel Company Limited, Jamshedpur
Zacharia George, Indian Concrete Institute, Madras
R.K. Sundaram, Central Public Works Department, New Delhi
S. Chockalingam, Annamalai University, Annamalainagar

TECHNICAL COMMITTEE

Chairman : A. Ramakrishna, L & T Limited, ECC Construction Group, Madras
Secretary : V.S. Parameswaran, Structural Engineering Research Centre, Madras
Members :
N. Bhattacharya, FGP Limited, Bombay
K. Ranga Rao, Department of Space, Bangalore
S. Subramanian, L & T Limited, ECC Construction Group, Madras
S.R. Kulkarni, M.N. Dastur & Company (P) Limited, Calcutta
Mohan Rai, Central Building Research Institute, Roorkee

INTERNATIONAL STEERING COMMITTEE

Chairman : M. Ramaiah, Structural Engineering Research Centre, Madras, India
Secretary : V.S. Parameswaran, Structural Engineering Research Centre, Madras, India
Members :
S.P. Shah, North-Western University, Evanston, Illinois, U.S.A.
V. Ramakrishnan, South Dakota School of Mines & Technology, Rapid City, South Dakota, U.S.A.
R.N. Swamy, University of Sheffield, Sheffield, England
Y. Ohama, Nihon University, Japan
Herbert Krenchel, Technical University of Denmark, Lyngby, Denmark
V.M. Malhotra, CANMET. Department of Energy, Mines & Resources, Ontario, Canada
R.K. Bhandari, Central Building Research Institute, Roorkee, India
K.N. Johry, Council of Scientific & Industrial Research, New Delhi, India
R.G. Kumble, Department of Science & Technology, New Delhi, India

PREFACE

Fibre reinforced concrete is a relatively new construction material developed through extensive research and developmental work during the last two decades. It has already found a wide range of practical applications and has proved as a reliable construction material having superior performance characteristics compared to conventional concrete. Incorporation of steel or other fibres in concrete has been found to improve several of its properties : cracking resistance, impact and wear resistance, ductility, and fatigue resistance. It is for this reason that FRC is now being increasingly used in structures such as airport pavements, bridge decks, machine foundations, blast-resistant structures, piles, pipes, sea-protective structures, ship hulls, and storage tanks.

Even though interest in this new material was generated about two decades ago, large-scale use of FRC has been on the increase only over the past ten years. Several types and configurations of fibres have now been developed and significant improvements have been made in the mixing and placing of fibre reinforced concrete. Steel, glass, and polypropylene fibres as well as natural fibres such as sisal, coir, and jute have been used in certain applications.

Several organisations in India have been carrying out research and development work on fibre reinforced concrete composites from the early 1970s. The applications so far have been in the construction of industrial floors and airport apron slabs and in the production of precast concrete components such as manhole covers. Most of these applications were with steel fibres. Research efforts are also in progress in the use of vegetable fibres and polymer-impregnated glass fibres. There is vast scope for application of fibre reinforced concrete in precast construction and in in situ repair works.

This International Symposium on Fibre Reinforced Concrete is being organised to focus attention of the Indian engineers and their counterparts in other countries on the present state of technology and the vast potential of this new and promising construction material. The symposium is being co-sponsored by several institutions and organisations in the country. It is gratifying to note that our call for papers has received an overwhelming response and over 80 papers have been finally chosen after review and published in the two volumes. It is hoped that these two publications will serve as a reliable and authoritative source of information on the latest developments and applications in FRC composites.

As the Chairman of the Organising Committee, I wish to record my sincere thanks to all the members of the Organising and Technical Committees who carried a heavy burden in bringing out these publications and in the organisation of the symposium. I am grateful to Mr. A. Ramakrishna, Chairman of the Technical Committee for his valuable contribution in guiding the deliberations of the Technical Committee. My special thanks and appreciation go to Mr. V.S. Parameswaran, Secretary of the Organising Committee, and his colleagues Mr. T.S. Krishnamoorthy and Mr. K.S. Balasubramanian who were responsible for coordinating the various activities and arrangements for the entire symposium.

I also thank all those who reviewed the papers and the staff of SERC and the co-sponsoring organisations who helped in several technical and organisational matters connected with the symposium. I also thank the authors of papers.

The Organising Committee is also grateful to all the co-sponsoring organisations for their keen cooperation and participation in all the activities of the symposium. I also wish to thank M/s Oxford & IBH Publishing Company Private Limited, New Delhi, who have brought out these publications in an excellent manner.

M. RAMAIAH
Chairman, Organising Committee

Madras
November 1987

CONTENTS

CHAPTER II : PROPERTIES AND BEHAVIOUR OF STEEL FIBRE REINFORCED CONCRETE

VOLUME II: CHAPTERS IV TO VIII

CHAPTER IV : PRODUCTION, TESTING METHODS AND SPECIFICATIONS

CHAPTER V : DURABILITY AND LONG-TERM PERFORMANCE OF FIBRE REINFORCED CONCRETE COMPOSITES

CHAPTER VI : APPLICATION OF FIBRE REINFORCED CONCRETE COMPOSITES TO IN SITU AND PRECAST CONCRETE CONSTRUCTION

CHAPTER VII : OTHER RELATED TOPICS

THE CONTRIBUTORS

Achyutha H.
Department of Civil Engineering,
Indian Institute of Technology,
Madras,
India.

Agrawal T.P.
Department of Civil Engineering,
Banaras Hindu University,
Varanasi,
India.

Andrzej Burakiewicz
Institute of Fundamental Technological Research,
Polish Academy of Sciences,
Warsaw,
Poland.

Antoine E. Naaman
University of Michigan,
Ann Arbor,
USA.

Arockiasamy M.
Department of Ocean Engineering,
Florida Atlantic University,
Boca Raton,
USA.

Asili M.
Department of Civil Engineering,
The Polytechnic of Wales,
Pontypridd, Mid Glamorgan,
England.

Backer S.
Department of Mechanical Engineering,
MIT, Cambridge,
USA.

Balaguru P.
Department of Civil and Environmental Engineering,
The State University of New Jersey,
Rutgers, Piscataway,
USA.

Balasubramanian K.
Structural Engineering Research Centre,
Madras,
India.

Beat Schaffner
SKAT, St, Gallen,
Switzerland.

Bentur A.
Department of Civil Engineering,
Technion,
Israel.

Bhatt R.D.
Civil Engineering Department,
M.S. University of Baroda,
Baroda,
India.

Cha-Don Lee
Michigan State University,
East Lansing, Michigan,
USA.

Charles Alguire
Sargent Lundy,
Chicago, Illinois,
USA.

Chiam T.P.
University of Sheffield,
England.

Chitharanjan N.
Structures Division,
Anna University,
Madras,
India.

Chockalingam S.
Department of Structural Engineering,
Annamalai University,
Annamalainagar,
India.

Corradini G.
Dipartimento di Chimica,
Ingegneria Chimica e Materiali,
Universita dell' Aquila, L'Aquila,
Italy.

David R. Lankard
Lankard Materials Laboratory,
Columbus, Ohio,
USA.

Dellaripa F.
Seafloor Engineering Division,
Naval Civil Engineering Laboratory,
Port Hueneme, California,
USA.

Deshpande S.K.
Maharashtra Engineering Research Institute,
Nasik,
India.

Devadas Manoharan P.
Structures Division,
Anna University,
Madras,
India.

Dwarakanath H.V.
Department of Civil Engineering,
M.S. Ramaiah Institute of Technology,
Bangalore,
India.

Ferry R.
Pilkington Reinforcements Limited,
St. Helens,
England.

Fwa Tien Fang
Department of Civil Engineering,
National University of Singapore,
Singapore.

Ganesan N.
Civil Engineering Department,
Regional Engineering College,
Calicut,
India.

Garg S.K.
Central Building Research Institute,
Roorkee,
India.

Geetha Rao
Department of Ocean Engineering,
Massachusetts Institute of Technology,
Cambridge,
USA.

George C. Hoff
Mobil Research and Development Corporation,
Texas,
USA.

Godbole P.N.
Civil Engineering Department,
University of Roorkee,
Roorkee,
India.

Gopalaratnam V.S.
Department of Civil Engineering,
University of Missouri-Columbia,
Columbia,
USA.

Gordon B. Batson
Clarkson University,
Potsdam, New York,
USA.

Govinda Rao V.V.
Consulting Engineer,
New Delhi,
India.

Guan Liqiu
Dalian Institute of Technology,
Dalian,
China.

Guofan Zhao
Dalian Institute of Technology,
Dalian,
China.

Gupta V.K.
Civil Engineering Department,
University of Roorkee,
Roorkee,
India.

Gupte K.A.
Mukand Iron and Steel Works Limited,
Bombay,
India.

Hafeez Khan T.A.
University of Constantine,
Constantine,
Algeria.

Halim J. Abu-Mathkour
Department of Civil Engineering,
University of Missouri-Columbia,
Columbia,
USA.

Hans Cederqvist
EKEBRO International AB
Vasteras,
Sweden.

Hans Erik Gram
SKAT—Swiss Centre for Appropriate Technology at ILE
St. Gallen,
Switzerland.

Hariraj P.
Highways and Rural Works Department,
Tamil Nadu,
India.

Heinz W. Poesch
Technical University of Berlin,
Berlin,
Federal Republic of Germany.

Herbert Krenchel
Technical University of Denmark,
Lyngby,
Denmark.

Indradatta I. Pandya
Department of Applied Mechanics,
M.S. University of Baroda,
Baroda,
India.

Irshaid A.
Department of Civil Engineering,
North Carolina State University,
Raleigh,
USA.

Jain O.P.
Consulting Structural Engineer,
New Delhi,
India.

Jakob Šušteršič,
ZRMK—Institute for Material and Structure Research
Ljubljana,
Yugoslavia.

Jeanfreau
Department of Ocean Engineering,
Florida Atlantic University,
Boca Raton,
USA.

Jerzy Piasta
Technical University of Kieice,
Kieice,
Poland.

Johar P.S.
Regional Research Laboratory,
Jammu,
India.

Jones R.
University of Sheffield,
England.

Josifek C.W.
Ribbon Technology Corporation,
Gahanna, Ohio,
USA.

Kadir M.R.A.
University of Salahaddin,
Arbil,
Iraq.

Kameswara Rao C.V.S.
Department of Civil Engineering,
Harcourt Butler Technological Institute,
Kanpur,
India.

Kanchi M.B.
Bechtel Energy Corporation,
Houston, Texas,
USA.

Karol Komloš
Institute of Construction and Architecture of S A V,
Bratislava,
Czechoslovakia.

Kashyap P.L.
Central Soil and Materials Research Station,
New Delhi,
India.

Katsunori Demura
College of Engineering,
Nihon University,
Koriyama, Fukushima-Ken,
Japan.

Kaushik S.K.
Civil Engineering Department,
University of Roorkee,
Roorkee,
India.

Kawa Taha Abu Al Awffa
University of Salahaddin,
Arbil,
Iraq.

Khalid M.
Central Building Research Institute,
Roorkee,
India.

Kharat B.R.
Research and Development Establishment (Engineers),
Dighi, Pune,
India.

Khorshed Alam A.K.M.
Housing and Building Research Institute,
Mirpur, Dhaka,
Bangladesh.

Krishnamoorthy T.S.
Structural Engineering Research Centre,
Madras,
India.

Kukreja C.B.
Thapar Institute of Engineering and Technology,
Deemed University,
Patiala,
India.

Kulshrestha H.K.
International Airports Authority of India,
Chanakyapuri,
New Delhi,
India.

Kumar V.
Department of Civil Engineering,
Banaras Hindu University,
Varanasi,
India.

Ladislav Brüll
Liko Grt, Bratislava,
Czechoslovakia.

Lakshmanan N.
Structural Engineering Research Centre,
Madras,
India.

Lakshmipathy M.
Structures Division,
Anna University,
Madras,
India.

Lee S.L.
Department of Civil Engineering,
National University of Singapore,
Singapore.

Leo Van De Loock
Department of Construction,
Katholieke Universitiet Te Leuven,
Heverlee,
Belgium.

Les Law Hebda
Technical University of Kieice,
Kieice,
Poland.

Leung C.
Department of Civil Engineering,
MIT, Cambridge,
USA.

Lim T.Y.
Department of Civil Engineering,
National University of Singapore,
Singapore.

Londhe B.A.
Research and Development
Establishment (Engineers),
Dighi, Pune,
India.

Mangat P.S.
Department of Engineering,
University of Aberdeen,
Aberdeen, Scotland,
England.

Marc Vandewalle
N.V. Bekaert S.A.,
Zwevegem,
Belgium.

Marian Grondziel
Harex,
Herne 1,
Federal Republic of Germany.

Md. Rafiqul Islam
Housing and Building Research Institute,
Mirpur, Dhaka,
Bangladesh.

Meek J.L.
Department of Civil Engineering,
University of Queensland,
St. Lucia,
Australia.

Mehta S.K.
Regional Research Laboratory,
Jammu,
India.

Motamedi Azari M.
Department of Engineering,
University of Aberdeen,
Aberdeen, Scotland,
England.

Munaswamy K.
Faculty of Engineering and Applied Science
Memorial University of New Foundland,
New Foundland,
Canada.

Murty D.S.R.
Andhra University,
Waltair,
India.

Muthumani K.
Structural Engineering Research Centre,
Madras,
India.

Nagaraj T.S.
Department of Civil Engineering,
Indian Institute of Science,
Bangalore,
India.

Nagaraja R.
J.N. Technological University,
Hyderabad,
India.

Nagarkar P.K.
Maharashtra Engineering Research Institute,
Nasik,
India.

Nalin J. Dave
Department of Civil Engineering,
University of Salford,
Salford,
England.

Neelamegam M.
Structural Engineering Research Centre,
Madras,
India.

Nemkumar Banthia
Department of Civil Engineering,
University of British Columbia,
Vancouver,
Canada.

Otter D.E.
University of Michigan,
Ann Arbor,
USA.

Parmar N.B.
Civil Engineering Department,
M.S. University of Baroda,
Baroda,
India.

Paramasivam P.
Civil Engineering Department,
National University of Singapore,
Singapore.

Parameswaran V.S.
Structural Engineering Research Centre,
Madras,
India.

Parviz Soroushian
Department of Civil and Environmental Engineering,
Michigan State University,
East Lansing,
USA.

Pazare D.G.
Maharashtra Engineering Research Institute,
Nasik,
India.

Prakash Desayi
Civil Engineering Department,
Indian Institute of Science,
Bangalore,
India.

Proctor B.A.
Pilkington Brothers plc.,
Research and Development Laboratories,
Lathom Ormskirk, Lancashire,
England.

Radomski W.
Warsaw University of Technology,
Warsaw,
Poland.

Ramachandra Rao R.
Urban Development Authority,
Visakhapatnam,
India.

Ramakrishna B.
J.N. Technological University,
Hyderabad,
India.

Ramakrishnan V.
South Dakota School of Mines and Technology,
Rapid City,
USA.

Ramamurthy K.N.
Department of Civil Engineering,
University of West Indies,
St. Augustine, Trinidad,
West Indies.

Ramanathan R.
Structural Engineering Department,
Annamalai University,
Annamalainagar,
India.

Rao D.L.N.
Andhra University,
Visakhapatnam,
India.

Rao S.V.
Urban Development Authority,
Visakhapatnam,
India.

Rau R.H.G.
Mukand Iron and Steel Works Limited,
Bombay,
India.

Reddy D.V.
Department of Ocean Engineering,
Florida Atlantic University,
Boca Raton,
USA.

Rehsi S.S.
Central Building Research Institute,
Roorkee,
India.

Sabapathi P.
Department of Civil Engineering,
Thiagarajar College of Engineering,
Madurai,
India.

Sabir B.B.
Department of Civil Engineering,
The Polytechnic of Wales,
Pontypridd, Mid Glamorgan,
England.

Sachan A.K.
Department of Civil Engineering,
Harcourt Butler Technological Institute,
Kanpur,
India.

Samai Mohamed Laid
University of Constantine,
Constantine,
Algeria.

Santhakumar A.R.
Structures Division,
Anna University,
Madras,
India.

Sayanam R.A.
Regional Research Laboratory,
Jammu,
India.

Scoccia G.
Dipartimento di Chimica,
Ingegneria Chimica e Materiali,
Universita dell'Aquila,
L'Aquila,
Italy.

Sethunarayanan Rm.
Annamalai University,
Annamalainagar,
India.

Senthil Kumar
South Dakota School of Mines and Technology,
Rapid City,
USA.

Shah S.P.
North Western University,
Evanston, Illinois,
USA.

Sharief M.F.
Jawaharlal Nehru Technological University,
Hyderabad,
India.

Sharma D.K.
Small Industries Service Institute,
Kanpur,
India.

Shreeram K. Damle
Department of Applied Mechanics,
M.S. University of Baroda,
Baroda,
India.

Shuaib H. Ahmad
Department of Civil Engineering,
North Carolina State University,
Raleigh,
USA.

Sidney Mindess
Department of Civil Engineering,
University of British Columbia,
Vancouver,
Canada.

Singh R.N.
Department of Civil Engineering,
Bihar Institute of Technology,
Sindri,
India.

Singh R.P.
Department of Civil Engineering,
Banaras Hindu University,
Varanasi,
India.

Sinha A.N.
Department of Civil Engineering,
Banaras Hindu University,
Varanasi,
India.

Siraskar K.A.
Research and Development
Establishment (Engineers),
Dighi, Pune,
India.

Sivarama Sarma B.
Structural Engineering Research Centre,
Madras,
India.

Srinivas Y.M.
Mukand Iron and Steel Works Limited,
Bombay,
India.

Srinivasulu P.
Structural Engineering Research Centre,
Madras,
India.

Stanislav Urbančič,
ZRMK—Institute for Material and Structure Research,
Ljubljana,
Yugoslavia.

Stone D.
Cellate Industries,
Carole Park, Queensland,
Australia.

Sudhindra C.
Central Soil and Materials Research Station,
New Delhi,
India.

Summerhayes S.D.
Civil Engineering Department,
The Polytechnic of Wales,
Pontypridd, Mid Glamorgan,
England.

Sundararajan R.
Structures Division,
Anna University,
Madras,
India.

Suresh Chand
Central Building Research Institute,
Roorkee,
India.

Suri S.B.
Central Soil and Materials Research Station,
New Delhi,
India.

Swami B.L.P.
Civil Engineering Department,
Jawaharlal Nehru Technological University,
College of Engineering,
Anantapur,
India.

Swamidas A.S.J.
Faculty of Engineering and Applied Science,
Memorial University of New Foundland,
New Foundland,
Canada.

Swamy R.N.
University of Sheffield,
England.

Taketo Uomoto
Institute of Industrial Science,
University of Tokyo,
Tokyo,
Japan.

Tambe S.K.
Maharashtra Engineering Research Institute,
Nasik,
India.

Tarafdar N.K.
Civil Engineering Department,
Bihar Engineering College,
Patna,
India.

Tiwari V.K.
Metallurgical Engineering Department,
University of Roorkee,
Roorkee,
India.

Tor Arne Hammer
SINTEF div FCB—Cement and Concrete Research
Institute,
The Norwegian Institute of Technology,
Trondheim,
Norway.

Vasan R.M.
QIP Centre,
University of Roorkee,
Roorkee,
India.

Venkatacharyulu T.
Regional Engineering College,
Warangal,
India.

Venkateswarlu B.
Structural Engineering Research Centre,
Madras,
India.

Victor C. Li
Department of Civil Engineering,
MIT, Cambridge,
USA.

Vilijem Kuhar
ZRMK—Institute for Material
and Structure Research,
Ljubljana,
Yugoslavia.

Volpe R.
Dipartimento di Chimica,
Ingegneria Chimica e Materiali,
Universita dell' Aquila,
L' Aquila,
Italy.

Wang Y.
Department of Mechanical Engineering,
MIT, Cambridge,
USA.

Yamanobe K.
Department of Civil Engineering,
MIT, Cambridge,
USA.

Yasuhiko Sato
College of Engineering,
Nihon University,
Koriyama, Fukushima-Ken,
Japan.

Yoshihiko Ohama
College of Engineering,
Nihon University,
Koriyama, Fukushima-Ken,
Japan.

Yogendra Joshi
Civil Engineering Department,
M.S. University of Baroda,
Baroda,
India.

CHAPTER I

THEORY OF FIBRE REINFORCED CONCRETE COMPOSITES

Proceedings of the International Symposium on Fibre Reinforced Concrete, December 16-19, 1987, Madras, India

STRENGTH EVALUATION AND FAILURE MECHANISMS OF FIBER REINFORCED CONCRETE

Dr. Surendra P. Shah
Professor of Civil Engineering
Director, Center for Concrete and Geomaterials
Northwestern University
Evanston, IL 60201 U.S.A.

ABSTRACT

The most widely used types of man made fibers are steel, glass and polymeric fibers. They increasingly are being used in a variety of construction and building projects. New types of fibers, new methods of producing composites and different applications are continuously evolving. To evaluate new developments, to efficiently design composites and to guide future developments, it is helpful to have a theoretical model to predict performance of these new composites.

Uniaxial tensile stress-strain response of fibers can be divided into three states: elastic, pre-peak nonlinear and post-peak response. Various theoretical models proposed to predict the response during each of these stages are described in this keynote paper. The most critical and also most difficult to analyze is the pre-peak nonlinearity. Progressive debonding, multiple matrix cracking and friction slip occur during this stage. The author and his coworkers at Northwestern University have examined the inelastic response of fiber reinforced composites using (1) a micromechanical model, (2) a linear elastic fracture mechanics model for the pullout fracture and (3) a nonlinear, fracture mechanics model for predicting the post-peak response. These studies are summarized in this paper.

INTRODUCTION

The current interest in fiber reinforced cement-based composite materials can be traced to three different sets of events. The activity with steel fiber reinforced concrete stems largely from research reported by Romualdi and Batson published in 1963 [1]. Our interest in glass fiber reinforced concrete can be traced to research conducted at Building Research Station in England by Dr. Majumdar and his colleagues [2]. Professor Krenchel has pioneered the use of polypropylene fibers in cement composites (see for example Ref. 3).

The most widely used types of man made fibers are steel, glass and polymeric. They increasingly are being used in a variety of construction projects. New types of fibers, new methods of producing the composites and different applications are continuously evolving. For example, now steel fibers can be chopped and sprayed, in a manner similar to that for glass fibers [4]. Thin sheets (10 ft. x 30 ft. and 1/2 in. thick) of glass fiber reinforced cement as well as thin panels reinforced with polymeric fibers (to simulate asbestos cement sheets) are being increasingly used in the U. S. and Europe [5, 6, 7]. A method of incorporating a large volume of steel fibers for defense related purposes (SIFCON) also has been developed [4, 5].

To evaluate the new developments, efficiently design fiber reinforced cement composites, and to guide the future developments, it is helpful to have a theoretical model to predict performance of these new composites. A model based on classical linear elastic fracture mechanics was used by Romualdi and Batson in the 1960s [1].

Shah and Rangan proposed a model based on composite materials concept to explain reinforcing mechanism of fiber reinforced concrete also in the 1960's [8]. Since then a large number of articles have been published which deal with various aspects of failure mechanisms [4, 5].

In this paper, various analytical models developed to predict the tensile response of fiber reinforced concrete (FRC) are described. When FRC specimens are subjected to uniaxial tension, the response is linearly elastic until the initial matrix cracking. During the inelastic range (between the first cracking and the peak) progressive debonding and frictional slip at the fiber matrix interface occur. Many have attempted to predict the stresses and strains at the initial matrix cracking. Most of these models predict an enhancement of the first cracking strain resulting from the fiber reinforcement, although the experimentally observed value for FRC is not significantly different from that observed for unreinforced matrix.

The more important and critical aspect of modeling the mechanical behavior of FRC is the post matrix cracking response. Three theoretical models developed recently by the author and his co-workers to predict this nonlinear response are described in this paper.

TENSILE STRESS-STRAIN CURVE

One of the most important properties for fiber reinforced concrete is the tensile stress-strain curve. A possible stress-strain response is shown in Fig. 1. Three stages are distinguished: (1) elastic response which is terminated at the initiation of matrix cracking, (2) nonlinear response which results from multiple matrix cracking, fiber-matrix debonding and frictional slip, and (3) the post-peak strain-softening response. It should be noted that as a result of strain localization post-peak stress-strain curve is not unique [9, 10]. After the peak, it is more accurate to refer to stress-crack width relationship. In fact it is likely that the prepeak nonlinearity may also be substantially controlled by widening at a single crack. This aspect is currently being studied at Northwestern University.

In addition to the method of testing, the extent of prepeak nonlinear zone can vary with the type, amount, and spatial distribution of fibers, matrix material, and time under accelerated aging environment. This stage is the most critical for fiber reinforced concrete and most difficult to predict theoretically.

STRESS AND STRAIN AT FIRST CRACKING

Several theoretical models have been proposed to predict the influence of fibers on the stresses and strain at first cracking of matrix. Romualdi and Batson [1] applied the concepts of linear elastic fracture mechanics (LEFM) to predict the stress at which matrix will crack in fiber reinforced cement composites. They assumed that fiber matrix interface has a perfect bond and that the matrix cracks when a critical stress intensity factor is reached. Based on these assumptions they noted that the stress required for cracking to occur is an inverse function of the square-root of the fiber spacing. The applicability of LEFM using a single parameter fracture criteria for small crack lengths (corresponding to fiber spacing) has been questioned by Shah and Rangan [8] and Jenq and Shah [11]. Also the assumption of perfect bond is arguable.

Aveston, Cooper and Kelly [12], assuming a uniform distribution of bond stresses, have proposed the following equation:

$$\varepsilon_{mu} = \left\{\frac{12\tau\gamma_m E_f V_f^2}{E_c E_m^2 r V_m}\right\}^{1/3} \qquad (1)$$

where ε_{mu} is the first cracking strain, r is the radius of fiber and γ_m is the surface work of fracture for matrix. This equation was derived by considering the changes in energy that occur when a crack is formed across a tensile specimen. It was assumed that a crack will form when the sum of the energy required to create two new surfaces in the matrix, the energy absorbed in frictional sliding at the fiber-matrix interface, and the increase in elastic energy in fibers is exceeded by the sum of the work done by the applied stresses and the elastic energy released by the matrix. Note that only the specimen length equal to twice the minimum crack spacing was considered in the derivation of Eq. (1).

More recently several so-called "strain-relief" models have been proposed to predict the stresses and strains at first cracking [13, 14, 15]. These models assume an elliptical zone around a crack at a given length in the matrix. Outside the elliptical zone the material is assumed to be unaffected by the presence of the crack. Energy changes associated with the initiation of the crack are considered only within the elliptical zone.

A crack is assumed to propagate when the rate of release of elastic strain energy is larger than the rate of energy absorbed in the processes of creating new surfaces and frictional sliding along the fiber-matrix interface. A purely frictional type of bond is assumed and the cracked matrix is assumed to transfer no traction. The results of these models are sensitive to the assumed initial flaw size [13].

In regard to the various models proposed to predict the first cracking stress, there are two points worth noting. First, they all make some assumptions which are not strictly valid for FRC composites. Second, the experimentally observed value of the first crack stress for FRC composites is not significantly different from that observed for unreinforced matrix.

The more important and critical aspect of modeling the mechanical behavior of FRC composites is the post matrix cracking response. The models which attempt to predict this nonlinear response are described next.

CLASSICAL COMPOSITE MATERIAL MODEL

Design of conventionally reinforced concrete can be viewed as if based on the classical composite material model. A composite stress-strain curve for concrete reinforced with a continuous steel reinforcing bar subjected to uniaxial tension is shown in Fig. 2. Before matrix cracking, the stress-strain curve of composite can be predicted using the composite modulus of elasticity (E_c) given by

$$E_c = E_f V_f + E_m V_m \qquad (2)$$

where E_f, V_f are the modulus of elasticity and volume fraction of fiber (steel rod in this case) and the subscript m refers to the corresponding value on the matrix (concrete, in this case).

After the matrix cracking, if it is assumed that concrete does not contribute any tensile reinforcement, then the composite modulus is given by

$$E_c = E_f V_f \qquad (3)$$

The maximum load is assumed to be reached when the fiber reaches its ultimate strength (σ_{fu}) , which in the case of reinforced concrete is taken as yield strength of steel. The maximum composite stress is given by

$$\sigma_{cu} = \sigma_{fu} V_f \qquad (4)$$

For reinforced concrete design and practice, application of Eqs. (1), (2) and (3) is implied when one deals with the transformed elastic, uncracked and cracked section.

It is recognized that the assumption that concrete does not resist any tension after cracking is erroneous as shown by the actual response in dotted line in Fig. 2.

For fiber reinforced concrete, fibers are discontinuous, not always unidirectional, and fibers generally do not fracture (or yield) at maximum load. These aspects can be taken empirically by (1) multiplying E_f in Eq. (1) by a factor to account

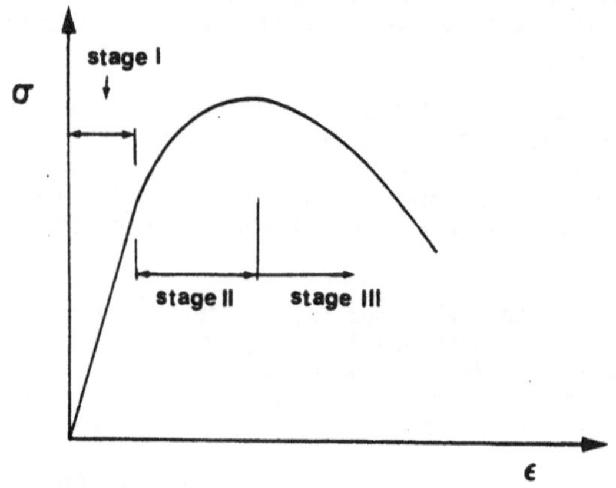

Fig. 1. Tensile stress-strain curve of GFRC (current work at Northwestern University).

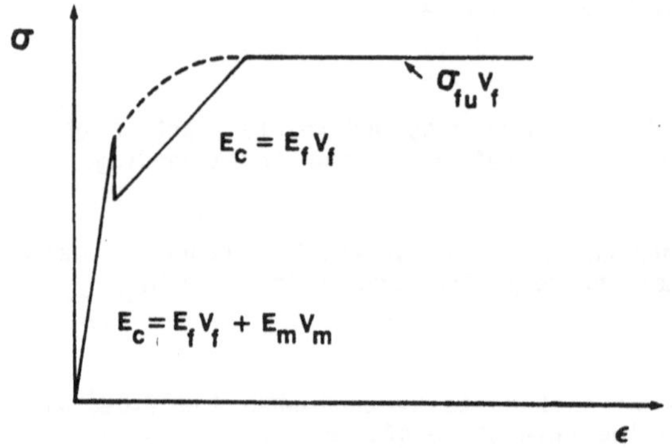

Fig. 2. Stress-strain curve for concrete reinforced with a continuous reinforcing bar subjected to uniaxial tension.

for discontinuity and random orientation and (2) multiplying σ_{fu} by a factor to account for pullout of fibers at the ultimate. With such modifications attempts have been made to predict the initial modulus of elasticity and the ultimate strength [16]. However, the above mentioned factors can only be obtained from curve fitting the test results and can not be applied to other types of tests or design in general. In addition, there has been no successful attempt, using such an approach, to predict displacement or strains at the peak load since the extent of stage 2 in terms of strains is one of the most important contribution of fibers. Therefore, we need to accurately and rationally predict the peak strains.

A MICROMECHANICAL MODEL FOR THE SINGLE FIBER PULLOUT PROBLEM

To analyze the response of FRC (where fibers are short and discontinuous) to loads beyond that corresponding to matrix cracking, it is helpful to understand the mechanics of bond failure at the fiber-matrix interface. What happens to the fiber-matrix interface after matrix cracking can be better understood by examining the case of the single fiber pullout test. Consider a fiber pulling out from a matrix as shown in Fig. 3.

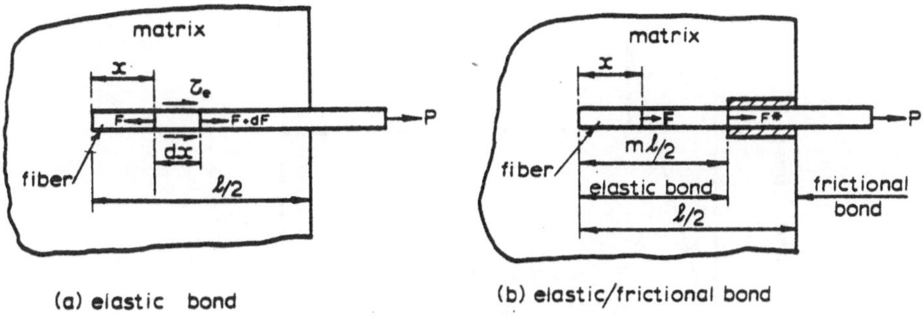

Fig. 3. Schematic of the fiber pull-out problem.

The calculated distribution of bond stress along the length of the fiber depends upon, among other things, the assumption made regarding the shear stress vs. shear strain relationship at the interface. For the small load P, it is reasonable to assume that the bond is elastic, that is, there is no relative movement between the fiber and the matrix at the interface (Fig. 3). Under this condition, shear-stress distribution can be calculated from the theory of elasticity using compatibility, equilibrium and boundary conditions. If it is assumed that the influence of the fiber is felt only within a concentric cylinder (the radius of the cylinder = half the spacing of fibers), then using the method developed by Cox [17], Lawrence [18], and Laws, Lawrence and Nurse [19], Gopalaratnam and Shah [10] have shown that the elastic bond stresses τ_e are given by:

$$\tau_e(x) = -\frac{\beta}{2\pi r}(C_1 \sinh\beta x + C_2 \cosh\beta x) \tag{5}$$

where $\beta^2 = \dfrac{2\pi G_m}{\ell n(\frac{1}{2}\sqrt{\pi/V_f})A_f E_f}$; $\alpha = \dfrac{A_f E_f}{A_m E_m}$

$$C_1 = -\alpha P$$

$$C_2 = \frac{P(1 - \alpha(1 - \cosh\frac{\beta\ell}{2}))}{\sinh\frac{\beta\ell}{2}}$$

and where G_m = shear modulus of the matrix, A_f and A_m are the cross-sectional areas of fiber and matrix and x is measured from the embedded length of the fiber (Fig. 3).

Note that the shear stress is maximum at $x = \ell/2$, that is, at the cracked surface of the matrix. When the maximum shear stress exceeds the shear strength of the interface (τ_s), debonding of fiber is initiated. The debonded portion is assumed to have frictional resistance τ_i (Fig. 4).

The partially debonded case is shown in Fig. 3. The fiber is assumed to be elastically bonded up to a distance $\frac{m\ell}{2}$ from the embedded end. The interface stress for $\frac{m\ell}{2} \leq x \leq \frac{\ell}{2}$, τ_i is considered a constant and purely frictional. The relation between the pullout load P and the debonded length $\ell(1-m)$ can be approximated as [18]:

$$P \approx \left[\frac{2\pi r \tau_s}{\beta(1-\alpha)} \tanh \frac{\beta m \ell}{2} + \frac{\pi r \ell \rho_i (1-m)}{(1-\alpha)}\right] \qquad (6)$$

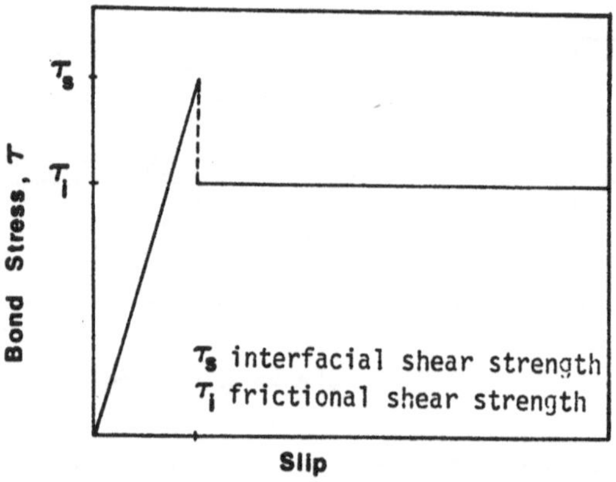

Fig. 4. Schematic of the interface property assumed.

This equation is plotted in Fig. 5. It can be seen that for a reasonable value of τ_i, the maximum pullout load ($\frac{dP}{dm} = 0$) is reached only after a stable debonding process. The average composite stress-strain curve will be nonlinear due to this stable debonding process prior to pullout of the fiber.

Typically in most experimental studies on the fiber-matrix bond behavior in pullout, the average interface strength τ is computed from the maximum pullout load assuming a uniform shear stress distribution. This value, however, depends on the length of the fiber embedment, as well as on the radius of the fiber, and thus is not likely to be a material property. Gopalaratnam and Shah [10] have shown that if a two parameter model

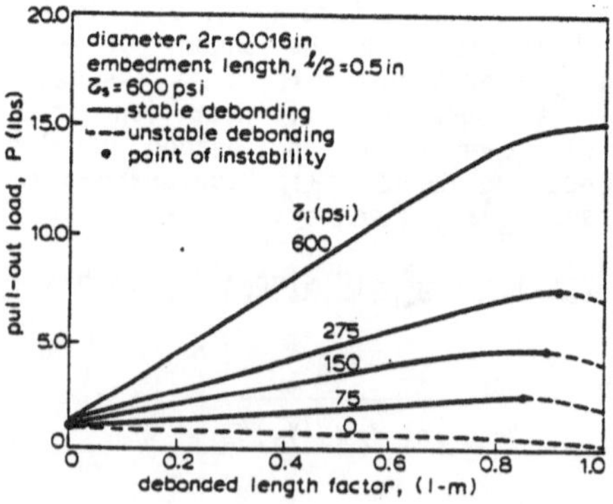

Fig. 5. Plot of model predicted pullout load vs. debonded length factor showing stable and unstable debonded processes.

(τ_s, τ_i) of interface behavior is assumed, then one can demonstrate that the calculated τ will depend on the length of the fiber. The value of τ_s and τ_i were calibrated by them from the test results of Naaman and Shah [20] on steel fibers of one diameter as shown in Fig. 6. From these values of τ_s and τ_i, the maximum pullout load was theoretically calculated for different fiber diameters. The theoretical results are compared with test data in Fig. 7. The theoretically predicted trend for various embedment length and diameter is shown in Fig. 8.

Fig. 6.
Comparison of the model predicted pull-out load versus displacement behavior with experimental results.

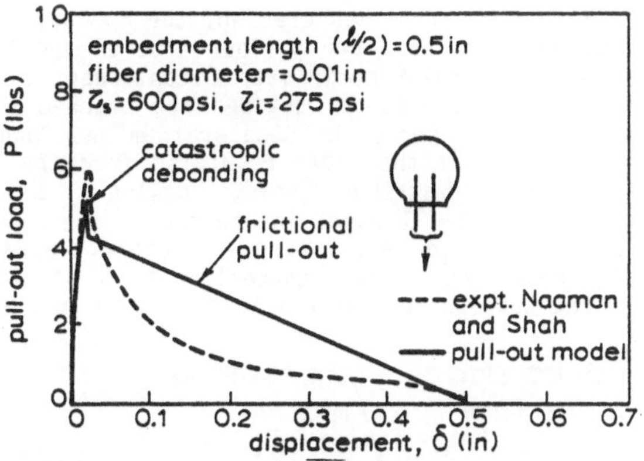

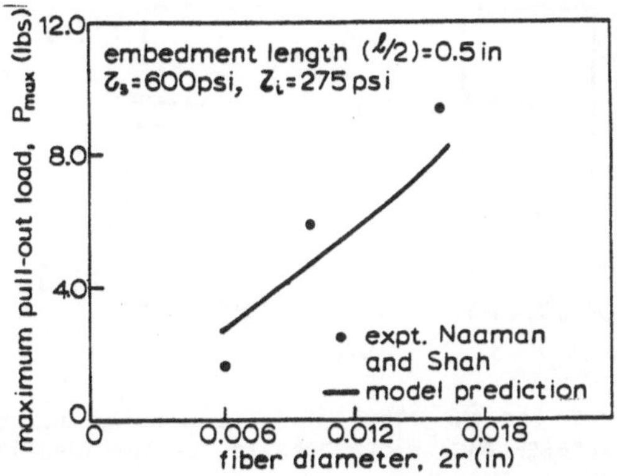

Fig. 7.
Experimental and theoretical plots of variation of pull-out load with respect to the fiber diameter.

Fig. 8.
Parametric study of the model predictions.

1 in. = 25.4 mm
1 lb. = 0.45 Kg
1 psi = 6.9 KN/m^2

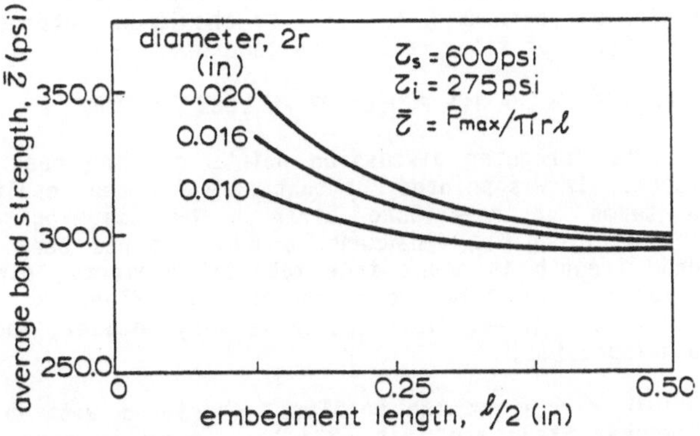

Idealized Uniaxial Tensile Response of FRC Composites

A calculation of the average stress-strain curve of a FRC composite can be obtained from the pullout behavior described above, if it is assumed that the pullout occurs only at one critical, cracked section. At that section, up to the start of matrix cracking, Eq. (2) can be applied, provided that it is assumed that the matrix, fiber and interface behave elastically up to that point. This point is taken to be the experimentally observed limit of proportionality (LOP). The matrix is assumed to crack when the tensile strength of the unreinforced matrix is exceeded.

After the matrix has cracked, the composite response is obtained by superimposing the results for the fiber pullout problem and the matrix softening problem (Fig. 9). The cracked (unreinforced) matrix continues to transfer a decreasing load, depending upon the average distance between the cracked surfaces (crack width). This has been experimentally observed by Gopalaratnam and Shah [10]. The total crack-width is the geometric slip of the fiber resulting from the pullout (which can be calculated from the procedure described above, knowing τ_s and τ_i) minus the displacement of the matrix due to the stresses introduced by the traction. Based on these considerations, it is possible to calculate the average composite stress and average strain during stable debonding. The composite peak stress is reached when the fiber-matrix interface subsequently starts to debond in an unstable manner.

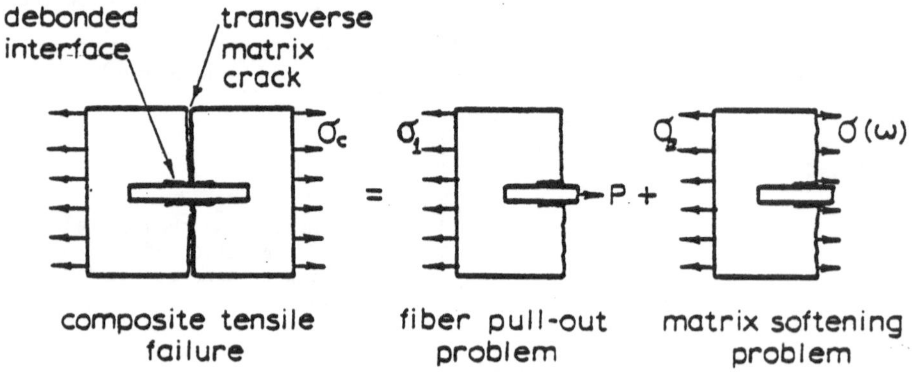

Fig. 9. Coupled superimposition of solutions for the fiber pull-out problem and the matrix softening problems yields stress-crack width behavior of the idealized composite.

This calculation procedure has led to a satisfactory prediction for steel fiber reinforced concrete [10, 13]. It should be interesting to see its applicability for other types of fiber reinforced concrete.

A FRACTURE MECHANICS MODEL FOR PULLOUT FRACTURE

The foregoing discussion points out the need to study the pullout behavior of fibers. It was pointed out that normally the results of a pullout test are expressed in terms of shear-bond strength by assuming a uniform distribution of bond stresses along the embedment length. It has been shown, however, that this average bond strength is not a true material property, since it varies with the length and diameter of the fiber, even for the same fiber and matrix (Fig. 8, and Ref. 10). The actual bond stress distribution is very complex, and difficult to estimate as well as to measure [21].

To circumvent the problems associated with defining bond behavior in terms of stresses, Stang and Shah [22] have recently proposed a fracture mechanics model to analyze the classical pullout test. No assumptions regarding the distribution of shear stresses are necessary for this approach. In the analytical model, the entire zone of debonding is treated as an interfacial crack.

A pullout specimen is shown in Fig. 10. For simplicity of presentation, let us assume that the debonded zone is stress free. Then the debonded zone can be treated as a crack with a length b ($b < \ell$) and a total crack surface of $2(2\pi rb)$ where r is the radius of the fiber.

Griffith type energy balance equation

$$\frac{1}{2}\frac{\partial C}{\partial b}(P_{cr})^2 = 2\pi\ r\ 2\gamma_i$$

P_{cr} = maximum pullout load

C = compliance = C(b)

b = crack length

r = radius of fiber

γ_i = specific interfacial work of fracture

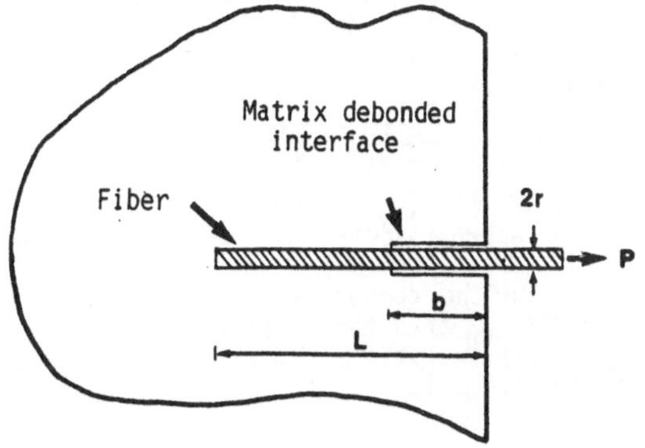

Fig. 10. Single Fiber Pull-out Test.

The maximum pullout load (P_{cr}) is reached when the Griffith energy criteria is satisfied; the rate of increase in energy required to create new surface is just equal to the rate of elastic energy release as a result of incremental crack growth. This can be expressed as:

$$\frac{1}{2}\frac{dC}{db}(P_{cr})^2 = 2\pi r 2\gamma_i \tag{7}$$

where C is the compliance of the specimen and γ_i is the specific internal work of fracture. If the fiber can be considered long ($\ell >> r$) and elastic $[\frac{E_f}{E_m}(\frac{r}{\ell})^2 \ell n\frac{2\ell}{r} << 1]$, then Stang and Shah [22] have theoretically estimated the compliance of the pullout specimen and shown that Eq. (6) reduces to:

$$\frac{1}{2}\frac{(P_{cr})^2}{E_f\pi r^2} = 2\pi r 2\gamma_i \tag{8}$$

Thus knowing the maximum load one can calculate γ_i which is the material property characterizing the interface. The accuracy of the above equation was demonstrated by measuring changes in compliance resulting from artificially introduced debonded zones of different lengths [23]. In addition, how to include frictional energy and how to calibrate it from the pullout test was also discussed in Ref. 24. Although additional research is necessary, this approach provides an alternative method of analyzing the results of the pullout test in terms of fracture energy rather than bond strength.

Uniaxial Tensile Specimen

To predict the tensile strength of fiber-reinforced cement-based composites whose ultimate strength is dictated by the strength of the relatively weak fiber matrix interface, the analyses described for the single-fiber pull-out case have been applied. The specimen considered here is shown in Fig. 11. It is assumed that (a) one macroscopic, transverse matrix crack is formed; (b) interfacial cracks have initiated; and (c) the fibers are continuous (that is, sufficiently long in the sense of Eqs. (6) and (7)) and aligned in the direction of the applied tensile stress. For the case of elastic fibers, Stang and Shah [23] have shown that using the Griffith energy criteria for critical interfacial crack propagation, the maximum tensile stress can be given by:

$$\sigma_{cr} = 2V_f\left(\frac{E_cE_f}{E_c - V_fE_f}\,\frac{2\gamma_i}{a}\right)^{1/2} \qquad (9)$$

Note that if σ_{cr} is less than the tensile strength σ_u of the unreinforced matrix (which is approximately the same as the limit of proportionality for the commonly employed FRC composites), then the stress-strain curve labelled (a) in Fig. 12 is observed for the composites. On the other hand, if σ_{cr} calculated from Eq. (9) is greater than σ_u then the stress-strain curve labelled (b) in Fig. 12 is predicted from the proposed analysis. Both of these types of curves have been reported for FRC composites (Fig. 12).

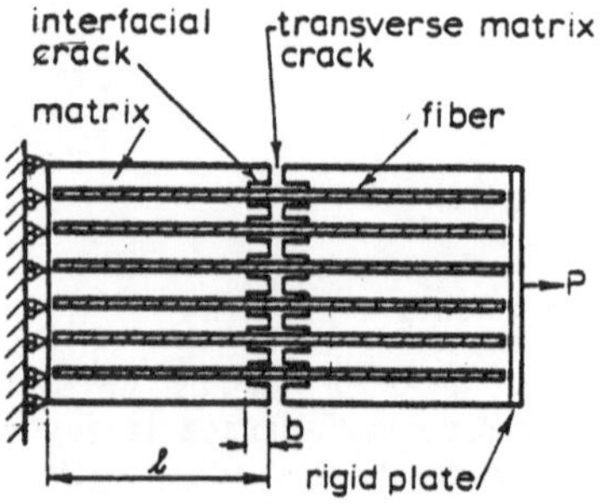

Fig. 11. Idealized tensile specimen considered for the composite model.

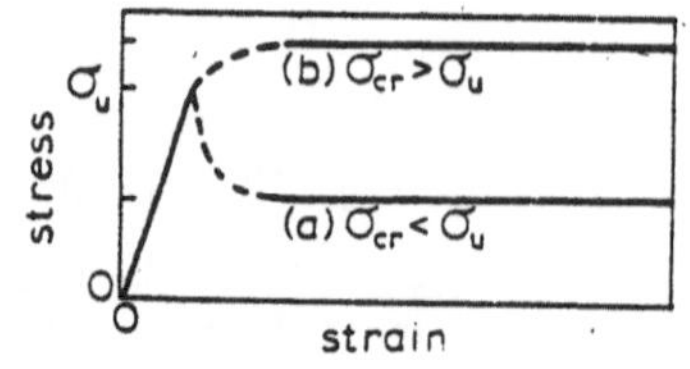

Fig. 12. Possible model predicted solutions for composite tensile stress-strain behavior.

FRACTURE MECHANICS MODEL FOR PREDICTING POST-PEAK RESPONSE

When a concrete structure is subjected to an external tensile loading, microcracks will occur at the cement-aggregate interfaces, around entrapped air voids or along the matrix-fiber interfaces. The existence of the microcracks have been investigated by Shah and Slate [25] among others. As the load starts increasing, some of the microcracks will grow and connect with each other and form a larger macrocrack. However, some of the microcracks will be arrested by aggregate or air void. The connecting and arresting processes of microcracks is complex and time-dependent and will not be discussed in this paper. When the applied load reaches the critical load, one of the macrocracks (the one which has the highest stress intensity factor) starts propagating at constant value of critical stress intensity factor (K_{Ic}^{s} in the global sense). Depending upon the loading configuration, the critical load may not be the peak load and a second crack may be generated. Assuming that the critical load is the peak load, then only one main crack will propagate throughout the structure (for all the other microcracks will be in unloading condition) and the complete loading history will be dominated by this single crack. (See also Ref. 9 for experimental verification of this localization phenomenon.) To characterize the behavior of this main crack, critical stress intensity factors (K_{Ic}^{s}) and critical crack tip opening displacement ($CTOD_c$) are proposed to be the fracture criteria as detailed later. Though only tensile failure (mode I) will be discussed in the present paper, the basic principle can be extended to compressive failure for local tensile stress will occur (due to the heterogeneous property of concrete) in the structure even under compressive loading.

Experimental Determination of Two Fracture Parameters

To determine the fracture parameters of the proposed model, notched beams were tested in a three-point bending arrangement. Three different beam sizes and five different mix-proportions were employed. For all beam sizes, the span to depth ratio was equal to 4 and the initial notch length (a_0) to depth ratio was equal to 0.3. Beams were tested in a closed-loop testing machine so as to maintain a constant rate of CMOD (crack mouth opening displacement). The details of testing and detailed results are reported in Ref. 28, 29.

From the measured maximum load and the measured corresponding elastic component of the CMOD, the value of the effective crack (a) was determined such that the calculated CMOD (see Ref. 28 for the LEFM based equation used) was equal to the measured value. For that value of a, K_{Ic}^{s} and $CTOD_c$ were determined using equations given in Ref. 28. The average values of K_{Ic}^{s} and $CTOD_c$ are given in Table I where it can be seen that K_{Ic}^{s} and $CTOD_c^{*}$ are essentially independent of the size of the beams.

From the proposed model it is possible to calculate the peak load of structures with different geometries. It is assumed that a crack will initiate at the point and perpendicular to the direction where the principle tensile stress is maximum. The proposed model requires three material properties: E (the Young's modulus of elasticity), K_{Ic}^{s} and $CTOD_c$. These three quantities can be combined into a single parameter Q such that:

$$Q = \left\{\frac{E \cdot CTOD_c}{K_{Ic}^{s}}\right\}^2 \qquad (10)$$

The parameter Q has dimension of length and can be used as material length to normalize the model predictions to facilitate parametric study. The values of Q for the five materials composition tested by the authors are given in Table I. Note that parameter Q is similar to the characteristic length defined in Ref. 28.

Using the proposed model it was possible to explain some of the experimentally observed size and geometry effects for notched and unnotched tensile and flexural tests of plain concrete [27].

Crack Resistant Mechanisms of Fiber Reinforced Concrete

The prediction of the response of fiber reinforced composites (FRC) containing an initial notch of length a_0 can be facilitated by dividing the response into three stages (Fig. 13).

* $CTOD_c$ for the larger concrete beam is an exception to this general observation.

(a) Linear Range: Within the linear range of the unreinforced matrix the FRC also behaves linearly. The initial Young's modulus for FRC may be different than that for matrix depending upon the amount of fibers.

(b) Nonlinear Crack Growth: When the stress intensity factor becomes greater than approximately half the critical stress intensity factor (K_{Ic}^{s}) nonlinear crack growth commences and the fibers will tend to resist this growth. The fiber bridging forces will depend on the total crack opening displacement while the calculations of stress intensity factor will depend on the effective crack growth and the elastic crack opening displacement. It is assumed that the crack profile and the critical effective crack extension remain the same for FRC as those for unreinforced matrix.

The total load P acting on the composite structure can be divided into three parts

$$P = P^{M} + P_{k}^{f} + P_{s}^{f} \tag{11}$$

where P^M is the contribution due to matrix and is related to K_I, P_k^f is related to the singularity effect due to fiber bridging, and P_s^f satisfies global equilibrium due to fiber bridging forces.

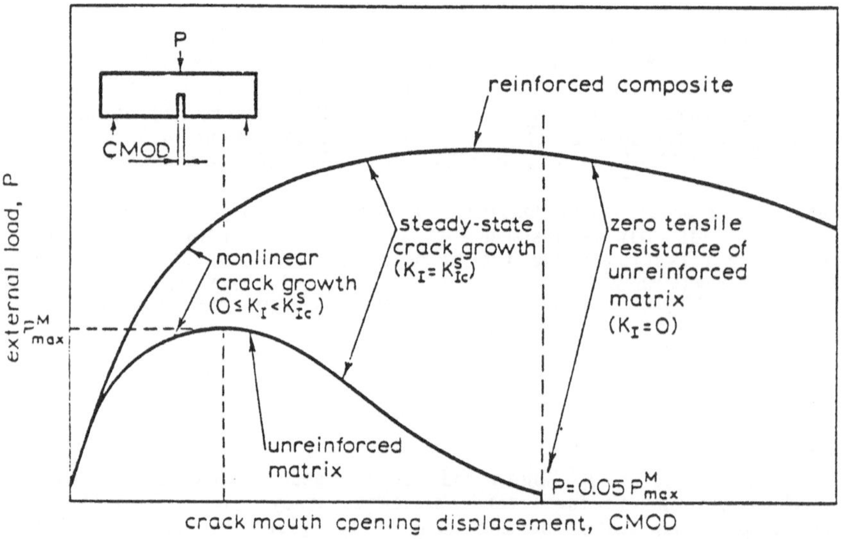

Fig. 13. Fracture Resistance Mechanisms of Fiber Reinforced Concrete

Note that the value of P for FRC does not necessarily attain a maximum when K_I just reaches K_{Ic}^{s}. Depending upon the volume of fibers, the maximum load for FRC may occur for a larger crack length than that corresponding to the peak load in the unreinforced matrix (Fig. 13).

(c) Completely cracked matrix: When the crack opening displacement (CMOD in Fig. 13) becomes very large, the resistance offered by matrix becomes negligible and eventually the stress intensity factor (K_I) becomes zero. Further crack separation is now mainly resisted by fibers. At this stage the load (P) and the corresponding CMOD can be calculated from only the global equilibrium consideration. That is

Table I. Relative Material Properties of Concrete, Mortar and Cement Paste

SERIES	COMPRESSIVE STRENGTH f'_c (psi)	YOUNG'S MODULUS $E (\times 10^6 psi)$	$K^s_{Ic} (psi \sqrt{in.})$				$CTOD_c$ ($\times 10^{-3}$ in.)				Q
			Large	Medium	Small	Average	Large	Medium	Small	Average	(in.)
C1	3650	4.87	930.8	813.5	904.5	882.9	0.3845	0.7035	0.792	0.63	12.08
M1	3942	3.68	631.1	654.0	644.1	643.1	0.324	0.463	0.302	0.363	4.15
M2	5718	4.71	918.9	879.3	816.3	871.5	0.475	0.336	0.332	0.381	4.24
M3	7950	5.41	--	--	963.5	963.5	--	--	0.394	0.394	4.89
P1	4013	3.01	595.5	544.8	547.1	562.1	0.267	0.244	0.301	0.271	2.11

Beam Dimensions (s x b x t x a):

Large (36" x 9" x 3.375" x 3"), Medium (24" x 6" x 2.25" x 2"), Small (12" x 2" x 1.125" x 1")

(1 in. = 2.54 cm : 145 psi = 1 MN / m^2)

$$P = P_s^f \qquad (12)$$

Resistance Provided by Fibers

The resistance offered by fibers depends primarily on the interfacial bond between the fibers and the matrix. This is because of the short length (of the order of 25mm) of the fibers and rather weak zone in the matrix that is observed in surrounding fibers. It is assumed that the fiber-bridging forces can be calculated from the pull-out test results of a single, aligned fiber.

From the experimental results of the pull-out tests it was observed that the pull-out load vs. slip relationship can be expressed only as a function of the maximum pull-out stress and the length of the fiber [30]. The maximum pull-out stress will depend on the number of fibers crossing the cracked surfaces, that is, on the volume fraction of fibers.

An analytical equation was derived to express pull-out load vs. slip relationship. This relationship was used to calculate the stress intensity factor due to the fiber-bridging pressure. The detailed procedure for calculating the net stress intensity factor - K_I for FRC is given in Ref. 11.

Calculations of Load-Deflection Response of FRC

To relate the load with the load-point deflection analytically, the concept of global energy balance was used. The total strain energy release rate (termed G_R) for the critical section can be derived and expressed as elastic and inelastic energy consumption during new crack formation.

$$G_R = C \cdot \frac{K_I^2}{E} + \int_0^{CTOD} \sigma(w)dw \qquad (13)$$

in which K_I = net stress intensity factor due to applied load and closing pressure, CTOD = crack opening displacement at the original crack tip, and C = energy correction factor to account for the energy absorbed by microcracks at the noncritical section. To obtain the loading history of the main crack in the ascending part a power law relationship of CTOD and K_I was assumed, i.e. $(CTOD/CTOC_c) = (K_I/K_{Ic}^s)^n$.

For a certain amount of increment of effective crack extension (ℓ_e), the total energy absorbed determined from G_R vs. ℓ_e is equated to that obtained from the load vs. load-point deflection curve. From this equality and knowing the R-curve, deflections can be determined. It was assumed that unloading was elastic prior to peak whereas for load-deflection curve beyond the peak, it was assumed that the elastic deflection after unloading remained constant and was equal to that at the peak. More details about the derivation of load-deflection and load-CMOD relationship are given in Ref. 10.

Comparison with Experimental Results of Steel Fibers

The experimental results of load-deflection curves for beams made with different fiber volume fraction are plotted in Fig. 14 and compared with theoretical prediction. The theoretical prediction is judged to be quite satisfactory. Fig. 15 shows a plot of the theoretically calculated relative peak load values vs. effective fiber volume fractions. The strength of SFRC beams with effective fiber volume fraction of 2.5% is about twice the strength of unreinforced matrix. The calculated value of energy release rate (G_R values at ℓ_e equals 1.8 in.) are also plotted in Fig. 15 for different values of V_{ef}. It can be seen that the energy absorption ability for beams with V_{ef} = 2.5% is about 30 times that of unreinforced matrix. In comparison to the improvement of energy absorption, the strength improvement due to the addition of fibers is less significant.

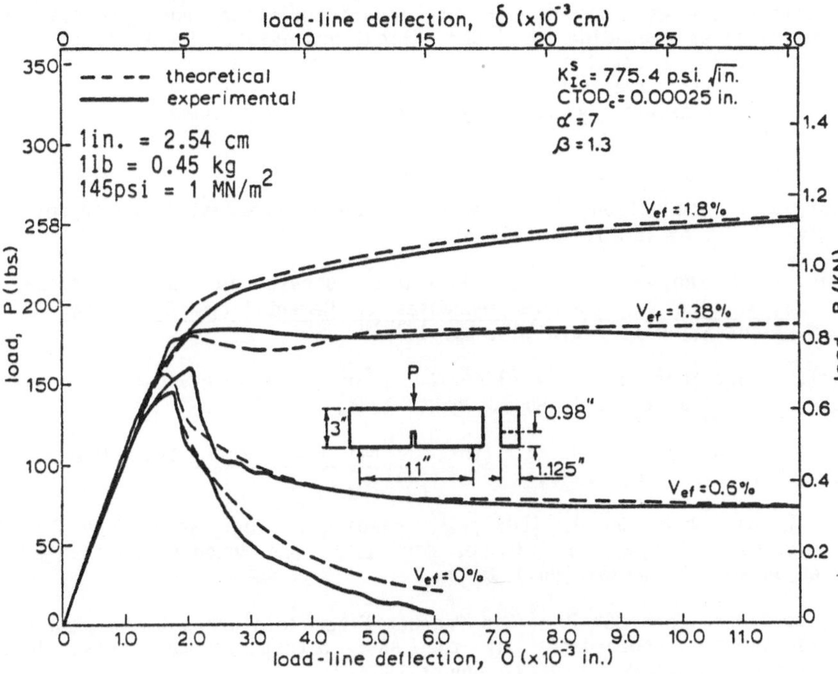

Fig. 14. Theoretical and Experimental plots of Load-Defln. curve.

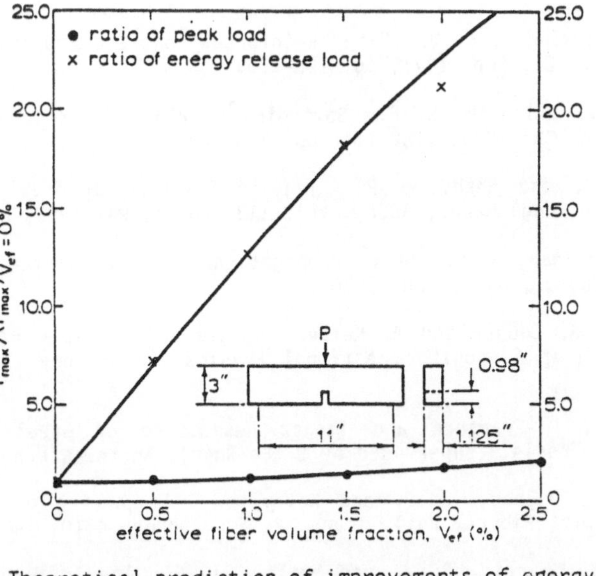

Fig. 15. Theoretical prediction of improvements of energy absorption and strength for different effective fiber volume fractions.

SUMMARY

Enhanced fracture toughness is likely to be the most important property for cement composites made with fibers. Fibers impart additional energy during fracture of composites by fiber debonding and fiber pullout processes. These processes must be accurately modeled to select the optimum compositions. The three theoretical models described in the paper are aimed at accurately describing the fracture toughness mechanisms of composites which result from these inelastic phenomena.

REFERENCES

1. Romualdi, J. P. and Batson, G. B. Mechanics of Crack Arrest in Concrete, J. Engr. Mech. Div., ASCE, 89 (EM3), June 1963.

2. Majumdar, A. J. and Walton, P. L. Fracture Processes in Fibre Reinforced Cement Sheets. Application of Fracture Mechanics to Cementitious Composites S. P.Shah (ed.), Martinus Nijhoff Pub., 1986, pp. 157-186.

3. Krenchel, H. and Shah, S. P. Applications of Polypropylene Fibers in Scandinavia, Concrete International, V. 7, No. 3, March 1985, pp. 32-35.

4. Shah, S. P. and Skarendahl, A. (Eds.) Steel Fiber Concrete, Elsevier Applied Science Publishers, Ltd., 1986.

5. Batson, G. and Shah, S. P. (Eds.), Proceedings, Fiber Reinforced Concrete in Infrastructure; ACI Symp., Baltimore, Nov. 1986; San Antonio, March 1987; to be published as an ACI Special Publication.

6. Daniel, J. I. and Shah, S. P. Thin Precast Fiber Reinforced Cement Panels, To be published in the Proceedings of the ASCE Structures Concgress, Innovative Materials for Concrete Structures, August 1987.

7. Shah, S. P., Ludirdja, D., and Daniel, J. Toughness of Glass Fiber Reinforced Concrete Panels Subjected to Accelerated Ageing, Accepted for publication, PCI Journal, June 1987.

8. Shah, S. P. and Rangan, B. V. Fiber Reinforced Concrete Properties, J. Am. Conc. Inst., Proc., 68 (2), Feb. 1971, pp. 126-135.

9. Gopalaratnam, V. and Shah, S. P. Softening of Plain Concrete in Direct Tension, ACI J., May-June 1985, Proc., 82 (3), pp. 310-323.

10. Gopalaratnam, V. and Shah, S. P. Tensile Fracture of Steel Fiber Reinforced Concrete, Jnl. of Eng. Mech., ASCE., Vol. 113, No. 5, May 1987, pp. 635-653.

11. Jenq, Y. S. and Shah, S. P. Crack Propagation in Fiber Reinforced Concrete, Jnl. of Structural Engrg., ASCE, Jan. 1986.

12. J. Aveston, G. A. Cooper and A. Kelly. Single and Multiple Fracture, Proc. The Properties of Fiber Composites, National Physical Laboratory, Guildford, pp. 15-26.

13. Gopalaratnam, V. Fracture and Impact Resistance of Steel Fiber Reinforced Concrete, Ph.D. Thesis, (supervised by S. P. Shah), Northwestern University, 1985.

14. Hannant, D. J., Hughes, D. C. and Kelly, A. Toughening of Cement and Other Brittle Solids with Fibers, Phil. Trans. R. Soc. London A310, 1983, pp. 175-196.

15. Korezynsloyj, Y., Harris, S. J. and Moxley, J. G. The Influence of Reinforcing Fibers on the Growth of Cracks in Brittle Matrix Composites, Journal of Materials Science, 16, 1981, pp. 1533-1547.

16. Swamy, R. N. (Ed.) Testing and Test Methods of Fibre Cement Composites, RILEM Symposium, 1978.

17. Cox, H. L. The Elasticity and Strength of Paper and Other Fibrous Materials, British J. of App. Physics, 3, Mar. 1952, pp. 72-79.

18. Lawrence, P. Some Theoretical Considerations of Fiber Pull-out From an Elastic Matrix, J. of Mat. Science, 7, 1972, pp. 1-6.

19. Laws, V., Lawrence, P. and Nurse, R. W. Reinforcement of Brittle Materials By Glass Fibers, J. of Physics, D. Applied Physics, 6, 1973, pp. 523-537.

20. Naaman, A. E. and Shah, S. P.. Pullout Mechanism in Steel Fiber Reinforced Concrete, Jnl. of the Structural Div., ASCE, Vol. 102, No. ST8, Aug. 1976, pp. 1537-1548.

21. Jiang, D. H., Shah, S. P. and Andonian, A. Transfer of Tensile Forces By Bond, ACI Journal, Proc. V. 81, May-June 1984, pp. 251-259.

22. Stang, H. and Shah, S. P. Failure of Fiber Reinforced Composites by Pull-Out Fracture, Journal of Materials Science, 21, 1986, pp. 953-957.

23. Stang, H. and Shah, S. P. Fracture Mechanical Interpretation of the Fiber/Matrix Debonding Process in Cementitious Composites, Proc. Intl. Conference on Fracture Mechanics of Concrete, Laussane, Switzerland, Oct. 1985.

24. Morrison, J. K., Shah, S. P. and Jenq, Y.-S. Analysis of Fiber Debonding and Pull-out in Composites, Accepted for publication, Journal of Engineering Mechanics (3258-2 EMM) 1987.

25. Shah, S. P. and Slate, F. O., in The Structure of Concrete, Cement and Concrete Association, London, 1968 pp. 82-92.

26. Jenq, Y. S. and Shah, S. P. A Fracture Toughness Criterion for Concrete, Eng. Fracture Mechanics, 21 (5), 1985, pp. 1055-1069.

27. Jenq, Y.-S. and Shah, S. P. Nonlinear Fracture Parameters for Cement Based Composites: Theory and Experiments, in Application of Fract. Mech. to Cementitious Composites (ed. S. P. Shah), Martinus Nijhoff Publ., 1986, pp. 319-359.

28. Jenq, Y.-S. and Shah, S. P. A Two Parameter Fracture Model for Concrete, J. of Eng. Mech., ASCE, 111 (10), Oct. 1985, pp. 1227-1241.

29. Hillerborg, A., Modeer, M., and Petersson, P. E. Analysis of Crack Formation and Crack Growth in Concrete by Means of Fracture Mechanics and Finite Elements, Cement and Concrete Research, 6, 1976, pp. 773-782.

30. Wecharatna, M. and Shah, S. P. A Model for Predicting Fracture Resistance of Fiber Reinforced Concrete, Cem. and Conc. Res. 13, 1983, pp 819.

Proceedings of the International Symposium on Fibre Reinforced Concrete, December 16-19, 1987, Madras, India

CONSTITUTIVE RELATIONS AND MODELLING FOR CONCRETE FIBRE COMPOSITES

A STATE-OF-THE-ART-REPORT

V. Ramakrishnan, Professor of Civil Engineering, and

M. Senthil Kumar, Graduate Student,

South Dakota School of Mines and Technology
Rapid City, South Dakota, 57701, U.S.A

SYNOPSIS :

Random oriented fibre reinforced concrete is one of the most promising composites used in construction. The addition of steel fibres to concrete has considerably improved the flexural strength, direct tensile strength, shock resistance, ductility, and failure toughness. While a substantial progress has been made in the design and use of concrete fibre composites, there are still many questions yet to be answered and a great deal of research work is needed to find these answers. Rational design methods are not available for several applications of fibre reinforced concrete because of lack of adequate experimental results and analytical models to predict closely the experimentally observed behavior.

This report presents a review of all the important analytical models published so far (June 1987). The merits and limitations of all these models in their application to concrete fibre composites are discussed. For both design and research purposes, mathematical models and constitutive relations have been proposed to predict the mechanical behavior of concrete fiber composites. Many types of models have been proposed ranging from the Law of Mixture type models, which consider only the mechanical properties of the matrix and fibres, and the percent by volume of fibres, to more complex models which consider the internal structure of the composite. As the model becomes more complex requiring more experimental data, it generally yields a more accurate prediction of the behavior. There is also a serious controversy over the application of some of these models to concrete fibre composites.

A fully applicable and accurate analytical model to predict the entire range of behaviour of concrete fiber composites and the variability of their mechanical properties has not yet been formulated. However, some models are available for design applications involving cracking, impact, flexural strength and flexural toughness.

INTRODUCTION :

All cement based materials and composites are essentially discontinuous, anisotropic and heterogeneous. These materials contain bond microcracks and interfacial discontinuities which are the root cause for unstable crack propagation and low tensile strength. Fibre reinforced cement and concrete were developed to overcome these problems associated with cement based materials. The incorporation of short, discrete fibres in a relatively brittle cement matrix transforms uncontrolled and unstable tensile crack propagation into a slow controlled growth. This gives the cement based materials a maximum ductility overcoming it's low tensile strength properties.

Considerable information regarding the behavior of various types of fibres in fibre reinforced concrete is known (1 to 3). The analytical methods proposed so far, range from Composite Models to Non-Linear Fracture Mechanics Models.More recently Endochronic Theory has also been used in developing constitutive relations. However,there is a serious controversy over the applicability of some models to the concrete fibre composites. There are also many difficulties in adopting some of the more basic theoretical models to concrete because of the non-homogeneous and complex micro-structure of concrete. The applicability of the existing models to fibre reinforced concrete is critically reviewed.

Constitutive Relations For Fibre Reinforced Concrete:

The addition of fibre considerably increases the area under the stress strain curve and improves the tensile strength of the member under consideration as shown in fig 1. This is due to the the crack controlling property of the fibre reinforcement with the following effects : (4)

1) Inclusion of fibre delays the occurence of first tensile cracking. This increases the tensile strain capability of the matrix.

2) It gives the member a well defined post-cracking behavior resulting in an increase of post-maximum load ductility.

3) It improves the energy absorption characteristics of the member by enhancing the crack resistance.

Thus the effect of inclusion of fiber is to improve the tensile fracture of the material. Therefore,in 1960 fracture mechanics was used as the starting point for investigation of fiber reinforced concrete. This was followed by several other concepts.

FRACTURE MECHANICS MODEL:

In 1961, Kaplan proposed the application of fracture mechanics to concrete (5). After a series of experimental investigations he confirmed that G_c was approximately a constant for different notches and that Linear Elastic Fracture Mechanics (LEFM) provided an approximate fracture criterion for concrete. This conclusion created a raging controversy over the application of fracture mechanics to concrete even as early as in 1962 which is going on unabated till today.

Blakey in 1962, rejected Kaplan's interpretation and felt that LEFM cannot be applied to concrete (6). He suggested that Kaplan's data could be better explained by "simpler" ideas such as the size effect, variation in the type of loading and the energy dissipated as heat. Glucklich and Irwin supported Kaplan's approach with the former pointing out three limitations in the approach and the later suggesting usage of sophisticated measuring devices for the crack toughness measurements in concrete (7,8).

In 1963. Romualdi et al were the first to incorporate steel fibres to serve as crack arrest mechanism (9). They added about 7% by volume of continuous aligned fibres and found that with appropriate wire spacings, flaws could be prevented from propagating. In the same year

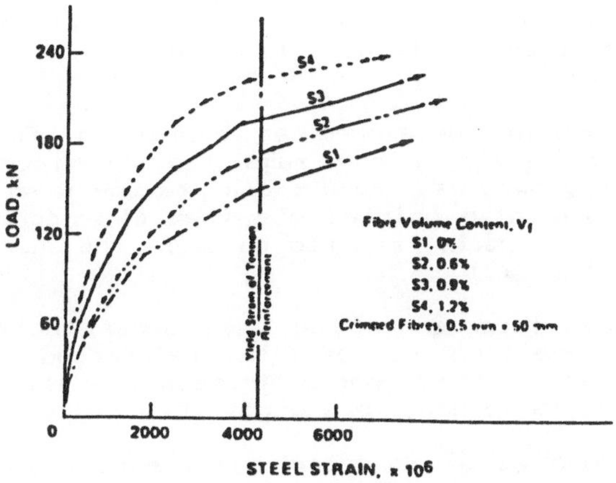

Fig 1 : Influence of steel fibers on tension steel strain near mid span. Ref. 4.

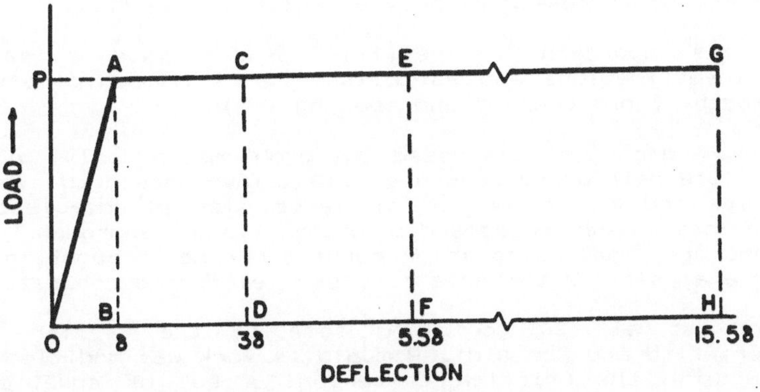

Fig 2 : Definition of Toughness Indices in terms of Multiples of First-Crack Deflection and Elastic-Plastic Material Behavior - Ref. 33.

they reinforced a series of beams with varying amounts of aligned steel fibres and tested in four point bending (10). The results were analysed in terms of a fracture arrest concept and confirmed that cracks could be contained between adjacent reinforcing elements.

In 1964 Romualdi and Mandel applied fracture arrest concepts to wire reinforced concrete (11). They found that the closely spaced fibres increased the post-cracking strength resulting in an increased tensile strength of the member.

In 1965, Romualdi developed an expression for G_c in fibre reinforced concrete (12). In the same year he found out that the resistance to crack growth was proportional to the inverse square of the fibre spacing. It was also pointed out that G_c increased with fibre content because of the energy required to cause the short fibres to slip ahead of the crack tip (13).

In 1971, Shah recommended the area under the load -deflection curve for the fracture toughness of fibre reinforced concrete (14). It was also found that this toughness increased with increasing fibre volume, fibre aspect ratio and fibre ductility.

In 1972, Brown et al rejected the fracture toughness model of Shah (15). They felt as Shah had used a soft testing machine for determining the stress-strain relationship the area under the load deflection curve included a considerable portion of the energy released by the machine on unloading. In the same year Parimi and Rao, assuming a random, uniform distribution of fiber, investigated the increase in fracture toughness of steel fibre reinforced concrete (16). They derived a formula for strain energy release rate as follows:

$$G_c = U^2 L^3 P \,/\, 314 \; E_f d^2 \qquad \ldots \quad 1$$

It should be noted that G_c derived using this equation is different from the one derived by Romualdi (12). Harris et al studied the effect of percentage of various fibres on the work of fracture, γ_f and K_c. They found that while γ_f ,increased by two orders of magnitude with 2% by volume of fibres, K_c increased by only a factor of two (17).

Using the approach in ref(11), Synder made a series of experimental investigations and established a relationship between the flexural strength, fibre content and spacing (18).

In 1975, Argon et al discussed the problems in applying fracture mechanics to fibre reinforced concrete (19). They showed the importance of size effects and the formation of large size pseudo-plastic zone ahead of the crack and recommended a non-linear approach to fibre reinforced concrete. Their suggestion perhaps has now brought in a spate of non-linear analysis, in the area of fiber reinforced concrete.

Spurrier et al (20) combined some of the earlier theories proposed by Griffith and Romualdi. Romualdi's work was modified to make it compatible with the Griffith criterion. A set of equations were developed based on this modification to give an indication of the fibre size and concentration to improve the "first crack strength" of the reinforced brittle matrix.

In 1976, Mangat improved the accuracy of the spacing concept of Romualdi and Mandel (11) for predicting the tensile strength of fiber reinforced concrete (21). His spacing concept was not only based on the geometric distribution of fibers as in earlier theories but also based on fiber-matrix interaction and the mode of failure. The experimental data reported (21) showed that the tensile strength tends to decrease as the effective spacing increased.

In 1977, Gustaffson studied the condition for stable crack growth of brittle materials in a 3 point bending test (22). When the total energy in the system divided by the fracture surface area was less than G_c, crack growth was stable. The area under the load deflection curve divided by the fracture surface area gave the value G_c. This value was determined for a 2-dimensionally oriented steel fiber reinforced concrete and was found not proportional to the fiber volume.

In 1979, Bunsell examined the load deflection curve pertaining to a compact asbestos cement tension specimen and studied the slow crack growth (23). He defined the stress intensity that was required to cause propagation using the following relation:

$$K = Ko + Kr \qquad \ldots \qquad 2$$

where Ko is the stress intensity factor of the matrix in the presence of fibers and Kr the inhibiting factor due to fibers. He concluded that LEFM could be used to characterize fracture in asbestos cement. Djabarav, noted that the increase in K_c with 'natural' cracks was greater than with sawn notches, since in the former case fiber bridged the gap (24). It appeared that the matrix failure occured at about the same load for both plain and fiber reinforced concrete. It was also observed that the function of fibers were to inhibit the microcrack growth in the composite.

Javan, found that cracks at the cement-aggregate interface existed even before the specimens were loaded (25). With increasing crack depth and fiber content, the value G_c increased. The increase was twice as effective in case of steel fibers than in case of polypropylene fibres.

Stroeven, using the fracture mechanics approach of Romualdi (12,13) derived the direct tensile strength of fiber reinforced materials taking into consideration the effect of matrix strength, bond strength, the age and the orientation of fibers and the concrete mix parameters (26). Though G_c was expressed in terms of the crack opening, no method for determining the crack opening was presented.

In 1980, Bazant pointed out that because of large microcracked zone, LEFM can be applied only to structures whose cross sectional dimensions are atleast 100 times the maximum aggregate particles (27). He further showed that since most specimens were much smaller, K_I & G_I, were not unique and depended upon the experimental conditions. In another work he disapproved the method in which the crack band was extended on the basis of a strength criteria (28) and recommended the energy criteria taking into account the bond forces into consideration instead of the bond slip between the steel and the concrete in the finite element analysis.

In the same year, Swamy studied the influence of slow crack growth on the fracture resistance of fiber cement composites (29). The fracture toughness vs crack growth curves gave a clear picture of the influence of slow crack growth on the fracture behavior of fiber composites and emphasized that slow crack growth must not be neglected in fracture toughness tests. During investigation they observed considerable branch cracking and even low fiber volumes imparted some stable growth to the system. The apparent K_c at first increased linearly with the crack growth, and then much more rapidly .

Saouma et al reviewed the application of fracture toughness of concrete (30). They modified the stress intensity factor (K_I)

equation to account for the finite height of the specimen as suggested in (31). Their finite element solution showed that toughness was independent of the specimen geometry for atleast five of the eight mixes tested. The second correction on load records produced drastic improvements in the consistency of the test results. Based on these corrections, they recommended a single parameter K_{IC} approach to crack propagation in concrete structures.

LIMITATIONS :

The concept of LEFM assumes infinite stresses and fracture zone of zero length in a material. But in a real material there is always a finite stress and fracture zone of finite length in front of a crack with tensile stresses perpendicular to the crack plane. Therefore, LEFM can never describe the behaviour of a material like concrete correctly (32).

Two fracture parameters measured by LEFM are the critical strain energy release rate and critical stress intensity factor. Several experimental data show that both these parameters are strongly dependent on the specimen geometry, the type of test and the method of measurement (4).

In G_C measurements it is essential to account for slow crack propagation which precedes rapid crack propagation. Neglecting the velocity of crack propagation will result in an underestimation of values (4).

When the fracture energy and toughness calculations are involved in using the load-displacement curve; it is important to recognize the effect of the testing machine stiffness on the energy absorbed by the material during fracture. The machine interaction can also produce extraneous fracture patterns and mask the true behavior of a material. Therefore, it is important to recognize this fact and avoid any discrepancies between the measured and computed values (15).

Determination of K_{IC} requires the formation of a natural crack at the root of the notch. It has been reported that the largest source of error in K_C and G_C measurement is the uncertainity about the exact crack length at the moment of instability. This requires usage of sophisticated measurement equipments like that of ultrasonic method or high speed photography (4). However this problem has been overcome in the compliance technique.

As of now, there appears to be no experimental or theoretical method to quantify the various energy dissipating mechanisms for concrete fracture. The results obtained by the different investigators show a wide range of values which confirms that other than surface energy,several other forms of energy dissipation are also involved which have not been accounted so far in the calculated values (4).

ACI Committee 544 recommends a method for determination of the Toughness Index. In this; the toughness index is defined as the measure of amount of energy required to deflect a fiber concrete beam used in the modulus of rupture test, by a given amount compared to the energy required to bring the fiber beam to the point of first crack. Barr et al have recommended a modification to this toughness index definition (4). In their recommended method, the specimen is loaded until the load-deflection curve extends to twice the deflection at first crack. The value of toughness index calculated using the modified method ranges from 0.25 for a plain concrete to 1.0 for very tough fiber reinforced concrete materials. It will be of interest to note that this modified method does not impose any restriction on specimen type.

This concept was extended and refined by Johnston (33). He proposed a new sytem of defining toughness based on a series of indices dependent on material behavior up to specified multiples of the first-crack deflection (Fig 2). This definition has been incorporated in the ASTM recommended test procedure (34) for the determination of flexural toughness.

The application of fracture mechanics concept to fiber reinforced concrete has been questioned by many. Hillerborg in his paper observes that when the fiber content reaches a certain level the material begins to behave in a plastic manner and therefore the term fracture toughness in the ordinary fracture mechanics sense becomes irrelevant (32). He explains that in the case of real bending, the transition from a nearly elastic to a nearly plastic state does not occur so suddenly; due to the formation of damage zones which increase with the increase in fiber content. Therefore, the behavior of such members will change towards a more plastic behavior. Thus he feels the structure may be analysed as if the material were plastic even with much lower fiber contents than those which cause a plastic behavior under tension.

LAW OF MIXTURE MODEL :

Law of Mixture models have been used extensively to predict the mechanical properties of composites (35). They have been more successfully used in composites with high percentages of fibres. The model assumes alignment of fibres in the direction of the load. It further assumes that the fibres are bonded to the matrix and that the poisson's ratio for the fibres and matrix as zero. The model computes the average stress in the matrix (σ_m & σ_f) and the effective modulus of the composite (E_c) using the principle of statistics and compatibility of strain at fibre matrix interface. The calculations are as shown below:

$$\sigma_m = \sigma_c \,/\, 1 + V_f\,(E_f / E_m - 1)$$

$$\sigma_f = \sigma_m \, E_f \,/\, E_m \qquad \ldots\ 3$$

$$E_c = E_f \, V_f + E_m\,(1 - V_f) \qquad \ldots\ 4$$

where V_f - fibre fraction of the volume, E_f and E_m - Young's Modulus of the fibres and matrix, and σ_c - applied stress on the composite. Correction factors for the 1-D alignment of fibres was proposed by Cox (36) which was later extended for 2D and 3D alignment by Krenchel (37). Allen proposed another correction factor for the fibre length (38). Both these corrections were applied to the volume fraction V_f by multiplying the correction factors with V_f to account for fibre orientation and length .

OTHER COMPOSITE MODELS :

The composite material approach to FRC based on the properties of individual components was applied by Shah et al (39). In 1973, Naaman proposed a fracture model for fibre reinforced cementitious materials (40). They developed a statistical model to predict the tensile properties of FRC. Based on the mechanics and statistics of composite material they simulated the ductile failure of FRC. They incorporated the concept of fracture mechanics in the second part of the model to cover the brittle failure. Both parts of the model were combined using the "Weakest Link " Theory which also introduced the size effect.

In 1973, Swamy et al studied the mechanics of fibre reinforcement in cement matrices (41). They observed that fibre-matrix stiffness, the interfacial bond stress and the fibre-matrix strain compatibility determined the role of fibres in cement matrices. As the crack arrest mechanism and geometrical spacing equations did not adequately explain the behavior of steel fibre concrete, they proposed a crack control composite approach. For randomly oriented short discontinuous fibres, they considered the interfacial bond stress distribution due to load transfer and also the presence of a crack. This approach enabled them to develop a relationship between fibre spacing and flexural strength.

In 1974, Swamy accounted for the partial effectiveness of matrix at first crack and ultimate tensile strength of fibrous composites (42). The matrix strength was multiplied by a coefficient γ whose value was smaller than 1.0 but greater or equal to 0. He also proposed the average interfacial bond stress τ at first crack and ultimate tensile strength of the composite. A comparison of these values with the values in reference (21) is shown in the following table.

Comparison of τ and γ values

Values	First Crack		Ultimate	
	Ref 42	Ref 21	Ref 42	Ref 21
τ psi	1035.5	-	1203.7	117.5
γ	0.843	-	0.97	1.1

The values of reference 42 were from flexural tests whereas values of reference 21 was from direct tensile test. The table shows a large variation in the values corresponding to the ultimate strength in flexure and in direct tension. The values of γ derived by Swamy appears to indicate the effectiveness of matrix at the ultimate stage where the matrix would have severely cracked rather than in the first crack condition. As this may not be the actual case several investigators have questioned the reliability of the composite material model (43).

In 1975, Aveston extended the composite material approach to long continuous fibres parallel to load and generated a stress-strain curve for it (44). The crack width and spacing were related to the volume percentage of fibres, the elastic moduli and the ultimate stresses of the matrix and fibres. He then extended his model for brittle matrices with discontinuous fibres.

LIMITATIONS :

In composite material model, fibers are assumed to contribute to the first crack and ultimate tensile strengths by mobilising their pull-out resistance. This contradicts with many of the recent findings including that of Soroushian who, based on the fiber pull-out test data, concluded that the crack openings upto the peak tensile strength in steel fiber reinforced concrete were not sufficient to mobilise any reasonable pull-out resistance of fibers (45).

INTERFACIAL BOND AND ORIENTATION EFFECTS :

The effect of fibres in a cementitious material is principally to cause relief of tensile stress at the crack tip and prevent unstable crack propagation. Papers published till 1970 mostly dealt with the stress-strain behavior of FRC and mortar under static loading in flexure, splitting tension and compression. In 1970, Kelly investigated the mechanism of fibre pullout (46). The typical load extension curve presented by him is shown in fig.3.

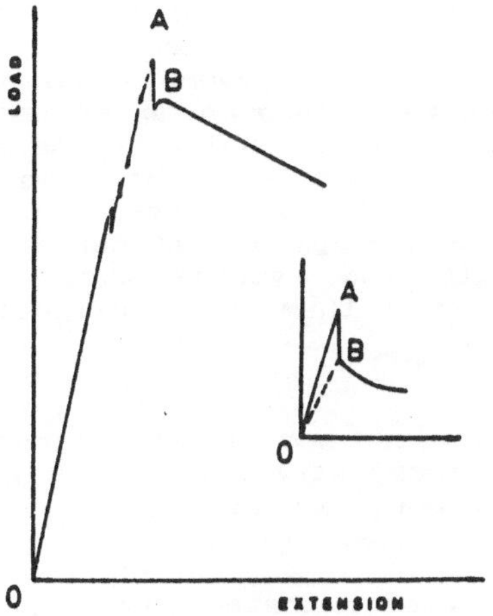

Fig 3 : Typical load extension curve for fiber pull-out test - Ref. 46.

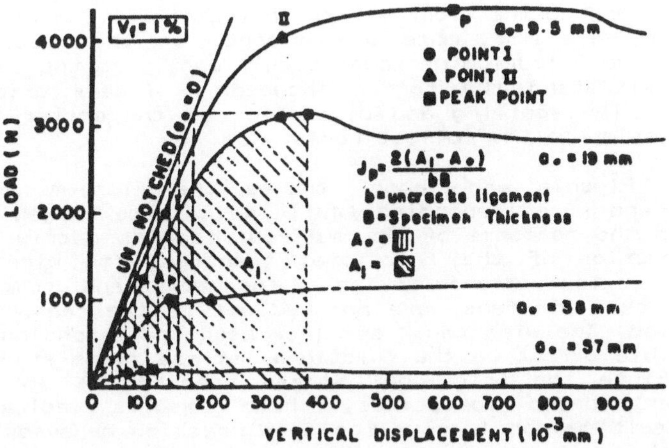

Fig 4 : Load-deflection curves for 1% fiber - Ref. 54.

Debonding of fibres characterizes the straight line portion of the curve OA. In the case of short fibres, the debonding occurs at the maximum load. The debonding energy per unit area is obtained by dividing the area OAB under the stress strain curve by the surface area of the fiber. This however does not apply to long fibres, because complete debonding does not take place at the peak load. The OA portion of the curve has a sawtooth form and the debonding energy is calculated using each of the sawtooth curves. Through experimental investigation for fiber pullout it was found that the value of the shear stress associated with the critical fiber length was the same as the shear stress developed during sliding friction. Based on this, equations were developed to estimate the average stress carried by the fiber at the ultimate tensile strength and average work of pullout fiber. The critical fiber length given by them is as follows:

$$L_c = d\,\sigma_f \,|\, 2.0 \quad \ldots \quad 5$$

where d and σ_f are the diameter and ultimate strength of the fiber. They further assumed that the shear stress distribution was uniform over the fiber length. This model does not adequately cover inclined fibers or the interface action with a crack in it.

In 1974, Johnson et al reported that the percentage increase in uniaxial strength and strain at failure were 1) independent of the shape of the fiber cross section, 2) independent of fiber size at constant aspect ratio, and 3) directly proportional to fiber concentration for a constant aspect ratio (47). Although they based their work on mortar, they felt that the same conclusion could be extended for concrete but with less increase in strength and other properties.

In 1974, Naaman and Shah reported that the fibers were effective in a brittle matrix only after the matrix has cracked (48). They realised that the pullout resistance calculated for a single fiber may not always correspond to the fiber action in a composite where the group action of fibers were much more predominant. In the case of group fibers, the fiber contribution depended significantly on the capacity of the matrix to withstand the forces induced by fibers bridging the cracked surfaces. The spalling and disruption of the mortar lead to a substantial reduction in the pullout resistance.

In 1978, Pinchin and Tabor studied the fibre-matrix stress transfer before and after debonding (49). Their tests showed inelastic behavior both in the response of the pull-out load to matrix shrinkage and in the reduction of the frictional transfer with wire movement during the pull out test. Microcracking of the matrix and compaction of the cement near the interface were pointed out as the reasons for the inelastic behavior. The wire roughness increased the debonding load but had very negligible effect on the frictional stress transfer subsequent to debonding. Since the frictional stress transfer is an important aspect of cement based composites, their results indicated that roughness by itself maynot improve the post cracking behavior of fibre reinforced cement.

Morton, using the pullout behavior of individual inclined wires estimated the work of fracture and post-cracking stress of FRC (50). He observed that where the interfacial shear stress was low as in case of short fibers, much of the work of fracture was due to plastic deformation of fibers inclined to the matrix crack face. Therefore, he concluded that aligned fibers were not effective as random fibers.

J-INTEGRAL :

In 1977, Mindess et al applied the concept of J-integral as a fracture criterion for fiber reinforced concrete (51). Beams were tested in a 4-point bending, using hardened cement paste, plain concrete and eight different steel and fiber reinforced concretes. Kc, Gc, and Jc values were evaluated. Experimental values for $J_{I}c$ and $G_{I}c$ were substantially identical for cement paste, but the former were greater for plain concrete and significantly greater for fiber reinforced concrete.

While $J_{I}c$ indicated the effectiveness of fiber; neither $G_{I}c$ nor $K_{I}c$ adequately reflected the changes in concrete behaviour due to fiber additions in it. It was observed that the fiber addition of less than 0.75% by volume did very little to improve the fracture behavior of the composites. He suggested that J-integral might be an appropriate model to describe the fracture criteria of FRC. Brandt found that the J integral approach appeared to be useful in characterising FRC (52). As a true brittle fracture could not be achieved, a great deal of scatter was found in their data. Carrato stated that scatter in J-integral determination could be reduced if the J-integral was calculated using a displacement criterion rather than a maximum load criterion (53).

Velazco et al investigated FRC beams with varying notch depths and different volume fractions of steel fibers (54) for a set of specimen shown in fig. 4. They calculated J-integral at the maximum load made with 1% volume of fibers but with different initial notch depth. The value J_p was not constant but varied substantially with the initial notch depth. Therefore, J_p cannot be considered as a geometry independent material fracture toughness parameter.

J_p values varied considerably for an identical initial notch depth (19 mm) but with different volume fractions of fibers . J-integral values were evaluated at two additional points besides the maximum load to confirm the J-integral as a geometry independent value. Fig. 5 shows the relation between CMOD and vertical displacement with three distinct regions in it indicating the elastic deformation, a slow crack growth and crack mouth opening by fiber pullout. The values of J were evaluated at I and II. While J_{I}, is independent of notch depth it is not sensitive to fiber concentration unlike J_{II} and J_p.

LIMITATIONS :

If there is no difference between loading and unloading curves then the stress-strain diagram for the material is said to be unique. This is assumed to be formally fulfilled for materials as long as no unloading occurs. Most metals conform to this assumption and the stress (Dugdale model) is fairly a constant within the fracture zone. Therefore, in metals where the amount of unloading before crack advance is negligible and where the crack is well defined, J approach yields fairly good results. But, in the case of concrete unloading of material occurs within the fracture zone due to the descending curve. The amount of unloading of material is predominant as the size of fracture zone in concrete is also appreciable. Moreover, the position of the crack tip is not well defined as in the case of metals. Therefore, J approach to concrete or FRC may not yield correct results.

The value of fracture toughness of similar concretes is a variable and not a constant. As stated by Swamy, this is due to the differing fracture processes and fracture paths arising out of the relative stiffness and toughness of the matrix and aggregates (4). Therefore, it appears that the energy dissipation mechanisms in heterogenous materials like concrete composites are not quite related to new fracture surface alone but also to other factors such as strength and factors influencing the strength.

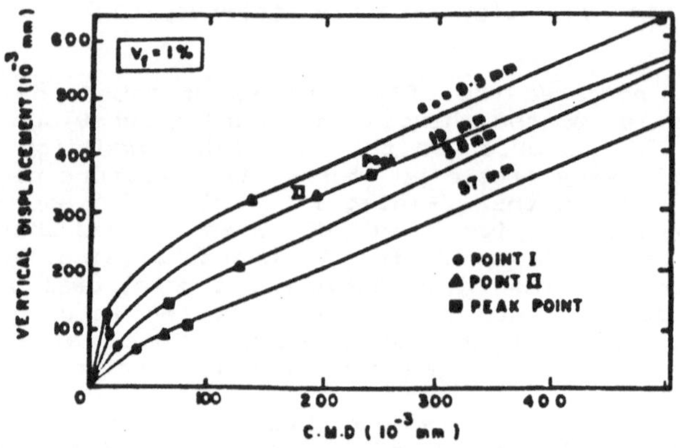

Fig 5 : Deflection - CMD curves for 1% fiber reinforced mortar - Ref. 54.

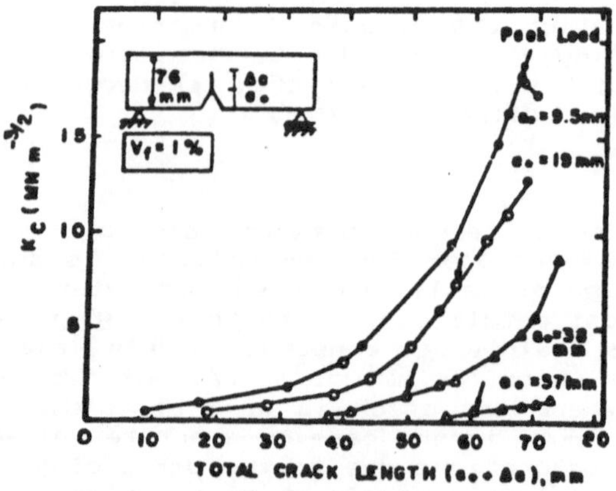

Fig 6 : Typical variation of stress intensity factor - Ref. 54.

The experimental results obtained by Velazco and Visalvanich indicated that the value J_p is not a constant but varied substantially with the initial notch depth (54). Therefore, J_p cannot be considered as a geometry-independent material fracture toughness parameter. This directly contradicts with several other earlier conclusions in which J_p ,parameter was shown independent of the specimen geometry. Mindess in his paper observed that Sih has pointed out the inapplicability of J-integral to composite systems such as fiber reinforced concrete (55). However, no reason has been indicated in his paper for this conclusion.

Chamberlain while reviewing the J-integral model has highlighted some questionable practices regarding its application (56). He feels that the J-integral which is essentially based on deformation plasticity is being inappropriately used in conjunction with incremental plasticity based on Prandtl-Reuss equations. The large yielding observed during experimental investigations does not offer a realistic base for assessing the small scale yielding occuring at crack tips. To overcome this difficulty, he recommends Fe and Jext integrals, which can be applied to incompatible strain fields unlike the J-integral.

R-CURVE ANALYSIS :

Velazco examined the applicability of various fracture mechanics approaches like the critical stress intensity factor, CMOD, compliance technique and R-curve analysis (54). The R-curve was proposed to identify the resistance to fracture of materials during slow-stable incremental crack extension. Fig. 6 shows a typical relation between total crack length and the applied stress intensity factor. The applied stress intensity factor at the occurence of unstable crack growth is a function of the initial notch depth and not an independent material property as confirmed in fig. 6. The value K_c was plotted against actual measured crack extension to obtain a unique relationship regardless of the initial notch depth. Therefore, they recommended R-curve analysis for modelling fiber reinforced concrete.

Visalvanich and Naaman used several methods for determining fracture properties of asbestos cement, mortar and steel fiber reinforced mortar (SFRM) (57). The compliance energy and quasi static energy technique underestimated the true fracture energy of mortar and SFRM. However, an energy measuring approach taking into account the permanent deformation on unloading gave realistic results. K_c for SFRM was independent of the initial crack length but did not reach a plateau for the 610 mm long specimens tested.

Wecharatana and Shah observed that the strain energy release rates calculated from elastic, secant or reloading compliances did not accountably represent the fracture behavior of portland cement mortar (58). The straight line models and the actual curve models shown in fig. 7 show the overestimated and unestimated energy release rates. The fourth method is the modified method which includes both elastic and inelastic strain energy absorbed during crack extention. In addition to the reloading compliance the rate of change of permanent deformation with crack growth was necessary to formulate this modified analysis.

Visalvanich and Naaman confirmed their earlier observation (57) in a review of the fracture methods in cement composites (59). Fig. 8 shows the comparison of the fracture energy obtained by all the four methods reported in (57). As test specimen with different parameters showed the same trend, they could not confirm as to the best fracture energy that could be used for fiber reinforced concrete.

It is well known that in cementitious composites a micro-cracking

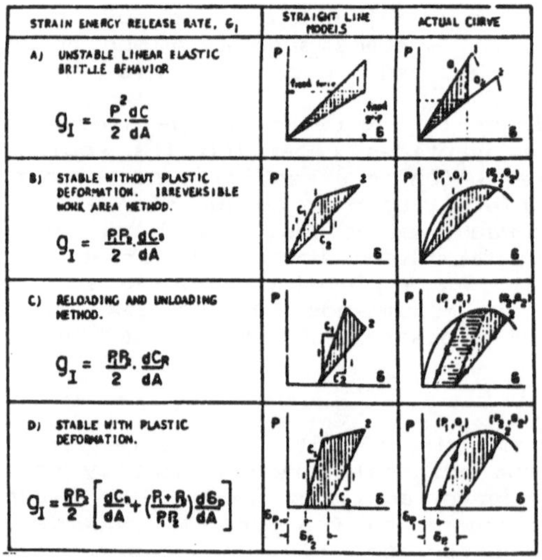

Fig 7 : Schematic models for different fracture behaviour and corresp -onding strain energy release rates - Ref. 58.

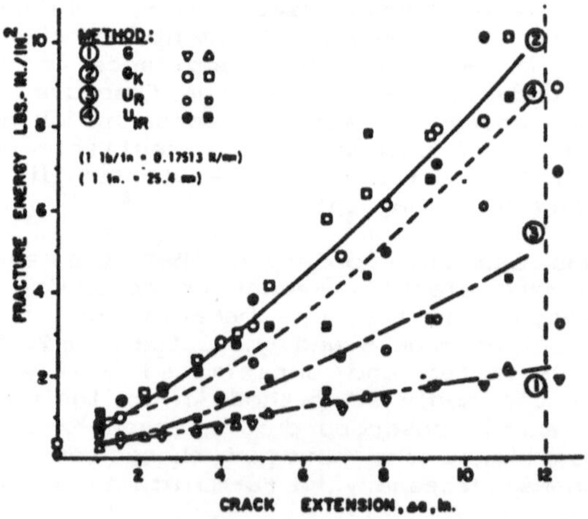

Fig 8 : Fracture energy of fiber reinforced mortar computed from four methods - Ref. 59.

zone or process zone of the order of several centimeters develop at the front of an advancing crack requiring a large size specimen. But as the specimen size was not sufficient no critical values could be found for steel fiber reinforced mortar. Also, the presence of fibers in fiber reinforced mortar precluded the possibility of deriving a crack opening angle as in the case of mortar.

Wecharatana and Shah extended their work (58) to concrete and fiber reinforced concrete (60). Irwin and Kies defined the strain energy release rate for an elastic material with an unstable crack propagation as follows (61).

$$G_I = \frac{P^2}{2t_n} \frac{dc}{da} \quad \ldots \quad 6$$

where P = load, t_n = thickness in the plane of crack, c = the compliance of the specimen i.e. (deflection/ load), a = crack length and dc/da = rate of change of compliance.

Shah et al modified this equation of (6), for a stable crack growth where the crack propagation occurs with a rising load-deflection curve. The modified model is shown in fig. 7D and is called the secant compliance Cs. The modified equation is as follows:

$$G_I = \frac{P_1 P_2}{2t_n} \frac{dC_s}{da} \quad \ldots \quad 7$$

in which P1 and P2 correspond to two neighbouring loads. As shown in Fig. 7B this modified model takes into account a continuous curve instead of the straight line model.

To account for the permanent or inelastic deformation an unloading reloading technique was used as shown in Fig. 7C . The strain energy release rate due to this reloading and unloading compliance C was defined as

$$G_I = \frac{P_1 P_2}{2t_n} \frac{dC_R}{da} \quad \ldots \quad 8$$

This definition does not account for the area (shown in horizontal lines in Fig. 7C) which is the inelastic energy during crack growth. The strain energy was further defined as:

$$G_m = \frac{P_1 P_2}{2t_n} \frac{dC_R}{da} + \frac{P_1 + P_2}{P_1 P_2} \frac{d\delta P}{da} \quad \ldots\ldots \quad 9$$

in which δP = the permanent deformation associated with loads P1 and P2. dC_R/da $\delta P/da$ = the changes of reloading compliance and permanent deformations with respect to crack extension and Gm the modified strain energy release rate. This modified strain energy release rate is shown in fig. 7D . The authors used double torsion and double cantilever specimens and found that

1) Modified R-curves for mortar and concrete were independent of the specimen size.

2) For fiber reinforced mortar, the strain energy release rate increased with the increase in crack opening. R curves for double cantilever specimen were higher than those for double torsion specimen.

3) In the case of mortar the strain energy release rate values were similar for cantilever and double torsion specimen for the same amount of crack opening.

4) The values of R-curves increased with the increase in volume of

aggregates and fibers.

In the earlier works of Naaman et al (59) a critical value could not be defined for fiber reinforced mortar because of the formation of pseudoplastic and process zone requiring a large specimen size. Fig. 9 shows the fracture zone and stress distribution for a fiber reinforced composite.

From several investigations it is now well known that if the steady state fracture energy of SFR composites are required then the specimen should be large enough to allow the formation of the process zone or the pseudo-plastic zone. Therefore, Visalvanich and Naaman proposed a comprehensive model to take into account the process of 1) crack initiation 2) slow stable crack growth and 3) crack extension prior to rapid propagation and failure (62). From the energy balance concept; the apparent critical fracture energy and steady state energy were defined as follows

$$Ga = \int_0^{\delta t} \sigma(\delta)\, d\delta \qquad \ldots \quad 10$$

$$Gc = \int_0^{\bar{\delta}} \sigma(\delta)\, d\delta \qquad \ldots \quad 11$$

Using the load vs displacement curve for FRM they defined the post cracking stress (maximum) in terms of the fiber reinforcing index $V_f\, l / \phi$ as

$$\sigma = \alpha\, \tau\, V_f \frac{l}{\phi} \qquad \ldots \quad 12$$

where τ is the interfacial bond strength and α the efficiency factor of discontinuous fibers. To relate the stress to the displacement they derived an analytical relation as follows,

$$\frac{\sigma}{\alpha \tau V_f \frac{l}{\phi}} = \left[0.1\left(\frac{2\delta}{l}\right)+1\right]\left[\left(\frac{2\delta}{l}\right)-1\right]^2 \qquad \ldots \quad 13$$

δt is defined as the initial crack tip boundary at a_0 and is related as „$\delta t = \beta c$" and $\bar{\delta}$ the maximum displacement at pseudoplastic zone = BR". For FRC, it is reported $\bar{\delta}$ can be taken equal to half the fiber length. From $\sigma - \delta$, relation pertaining to steel fiber reinforced mortar the apparent critical fracture energy has been derived and shown as

$$Ga = \alpha \tau V_f \frac{l}{\phi} \beta_c^n \left[1 - 0.9\left(\frac{\beta_c^n}{l}\right) + 1.067\left(\frac{\beta_c^n}{l}\right)^2 + 0.2\left(\frac{\beta_c^n}{l}\right)^3\right] \leq Gc \qquad \ldots \quad 14$$

For evaluation of Gc; $\bar{\delta}$ was replaced by $l/2$ and shown as

$$Ga = 0.171\, \alpha\, \tau\, V_f\, l^2 / \phi \qquad \ldots \quad 15$$

The model so recommended was checked with experimental data obtained using the offset fracture energy method of computation for a double cantilever beam.

Wecharatana and Shah used a crack model similar to the one shown in fig. 9 to predict non-linear fracture process zone in concrete (63). This model is essentially an extension of their previous work (60) with a correction for the observed crack extention. A model has been developed to calculate the length of the process zone which is essentially an iteration procedure and is as follows (fig 10)

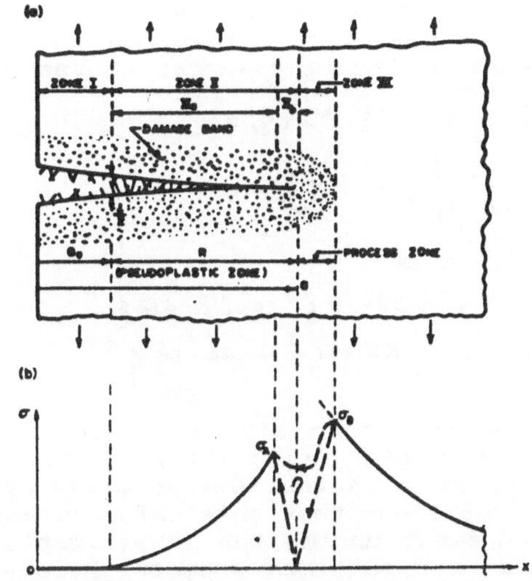

Fig 9 : Crack model for fiber reinforced cementitious composites a) fracture zone & b) possible stress distribution - Ref. 59.

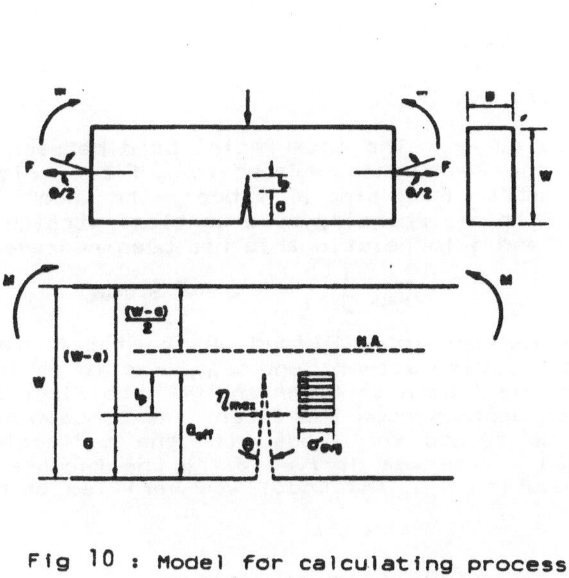

Fig 10 : Model for calculating process zone length for notched beam specimens - Ref. 63.

The change in rotation of the beam from the effective crack, a_{eff} has been given as

$$\theta = \Delta \gamma_{FM} F + \Delta \gamma_M M \qquad \ldots \quad 16$$

In which θ = the relative rotation of the crack front; F = axial force = $\sigma_{avg} \ell_p \beta$; M = resultant moment due to the external load and $\Delta \gamma_{FM}$ & γ_M = increments of the compliances caused by the presence of a crack are given by

$$\Delta \gamma_{FM} = 2 \left(\frac{1-\nu^2}{BE} \right) \frac{6}{W} \int_0^{\xi} \xi \gamma_F(\xi) \gamma_M(\xi) \, d\xi \qquad \ldots \quad 17$$

$$\Delta \gamma_M = 2 \frac{(1-\nu^2)}{BE} \left(\frac{6}{W} \right)^2 \int_0^{\xi} \xi \gamma^2 M(\xi) \, d\xi \qquad \ldots \quad 18$$

In which $\xi = a_{eff} / W$ and

$$\gamma_F(\xi) = 1.99 - 0.41\xi + 18.7\xi^2 - 38.48\xi^3 + 53.85\xi^4 \qquad \ldots \quad 19$$

$$\gamma_M(\xi) = 1.99 - 2.47\xi + 12.97\xi^2 - 23.17\xi^3 + 24.80\xi^4 \qquad \ldots \quad 20$$

For a given bending moment, beam size and notch length, the length of the fracture zone is calculated using iteration such that the value of θ, calculated using equation 16 is the same as given by h_{max}/ℓ_p. A plot of this ℓ_p with w-a, has shown that this relation was independent of the type of loading arrangement during the investigation. Thus the R-curve which was dependent on the specimen geometry because of the large size of the process zone was made independent by modifying the R-curves.

Shah and Jenq developed a fracture mechanics model to predict the resistance to crack propagation of steel fiber reinforced cement based composites (64). They divided the entire region into four stages as :(Fig.11) .

1) linear range
2) non-linear crack growth
3) steady state
4) completely cracked matrix.

Model for load-slip curve : The interfacial bond between the fibers and matrix determines the resistance offered by a fiber. Fig. 12 shows the load-slip relationship for single fibre with hook at the ends. Approximating the initial debonding by a vertical straight line upto the maximum load, the load slip relationship has been represented as

$$\frac{\sigma(w)}{\sigma_{max}} = \left[1 - \frac{w}{w_{max}} \right]^m, \quad 0 \leq w \leq w_{max} \qquad \ldots \quad 21$$

where σ_{max} is the maximum load divided by the total cracked area, w = slip or crack opening displacement and w_{max} = slip at zero load which is assumed as half the length of fiber and $\sigma(w)$ the fiber pullout stress. 'm' is a constant which depends upon the fiber. The concept of global energy balance was used to relate the loads with the deflection. The total strain energy release rate was derived using the terms shown earlier in fig. 7D and in equation 9 . The model was verified using experimental investigation.

LIMITATIONS :

In the case of metals with high yielding, as the zone with plane stress conditions widen; the crack growth resistance increases with the advance of the crack. Therefore, it may be possible to study the crack growth resistance (F_c, K_c, J_c etc.) with the change in crack length of a specimen. This phenomenon does not occur in concrete and hence may not be applicable to FRC also.

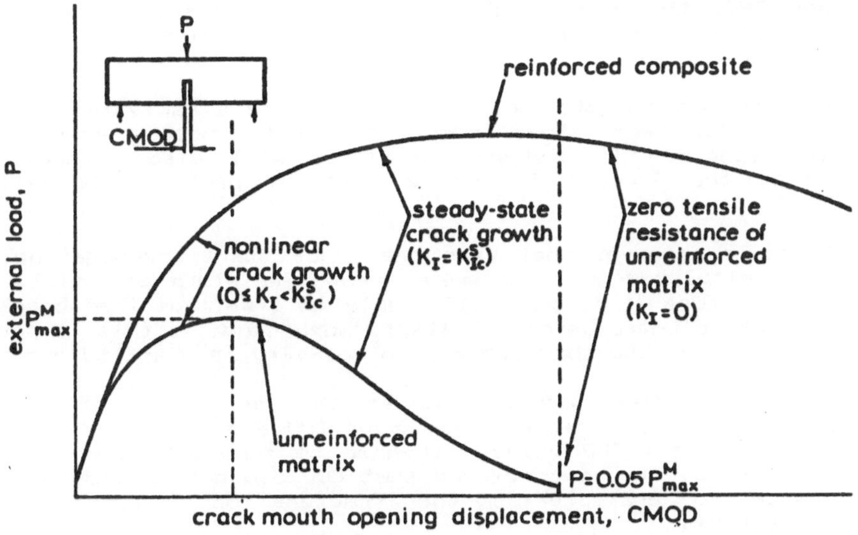

Fig 11 : Fracture resistance mechanisms of fiber reinforced concrete - Ref. 64.

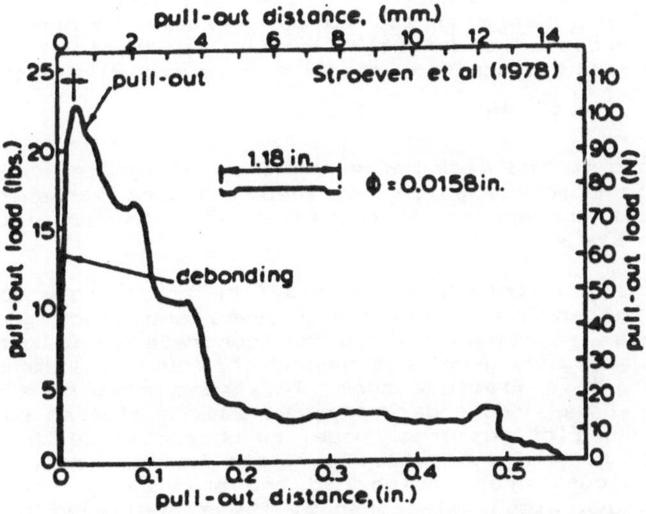

Fig 12 : Pull-out Vs load -slip relationship of single fiber with hook at the ends - Ref. 64.

The R-curve analysis is based on the observation of a crack growth. If it is not possible to observe a well defined crack tip, then the basis for an R-curve analysis may not be present. In the case of concrete and FRC; though a crack may be visible the crack may still continue to transfer stress. Hence the R-curve analysis is not very relevant to concrete as in metals (32).

Tests on specimen with a constant size and a variable notch may give R-curves which seem rather independent of the notch depth. If the size of the specimen is changed, R-curves will also change. This indicates that the R-curve is not a material property as believed earlier.

Hillerborg observes that when the crack starts propagating, the crack opening displacement is a measure of the width of the blunted crack (32). Therefore it is meaningful only if a well defined crack tip exists which gets blunted before it starts advancing. As this is not the case for concrete the COD approach may not be very relevant to concrete.

Moreover, studies carried out at the University of Illinois compared a variety of parameters like Kc,Jc,COD and R curves (54). They found that,in general, COD values depended to some extent on specimen geometry. Therefore, it was concluded that COD did not provide a useful critical value to characterize the fracture of fiber reinforced cementious materials.

FICTITIOUS CRACK MODEL :

In 1976, Hillerborg et al extended the concept of fracture mechanics to finite element analysis (65). He used a tied crack model similar to the Dugdale model for fracture to express energy absorption. This is shown in fig. 13A &13B. The crack was assumed to propagate when the stresses at the crack tip reached the tensile strength. When the crack opened, the stresses were assumed to gradually decrease with increasing crack width. At the crack where $w < wl$, the "crack" in reality corresponded to a micro-cracked zone with some remaining ligaments for stress transfer. As there was a stress to be overcome in opening the crack, energy was absorbed. The amount of energy so absorbed per unit crack area in widening the crack from zero to or beyond Wl was shown as :

$$\int_0^{W_l} \sigma \, dW \qquad \ldots \quad 22$$

and corresponded to the area between the curve and the co-ordinate axis in Fig. 14. This method which combined fracture mechanics and finite element analysis appears to be the first step in applying the same to FRC at a later stage.

In 1978, Hillerborg proposed his fictitious crack model which was then applied to bending of uncracked beams and crack growth (66). He assumed a linear relation of $\sigma - \varepsilon$ for concrete valid upto a limiting strain. A fracture zone developed beyond this and additional deformation was confined to this fracture zone. This gave rise to σ- crack width relationship and the area under σ-w curve was considered equal to G_c. He then extended the fictious crack model to characterize the relationships in the fracture zone of concrete. The whole fracture zone was assumed to act as a "fictious" crack which had the ability to transfer stresses according to crack width relationship. It was believed that this model could adequately explain the behaviour of 1) crack formation 2) crack growth & 3) rupture of concrete.

In 1980, Hillerborg applied his fictitious crack model to FRC (67). While the stress displacement curve was introduced to describe the crack process in the fracture zone , (σ-ε) relations were used for the

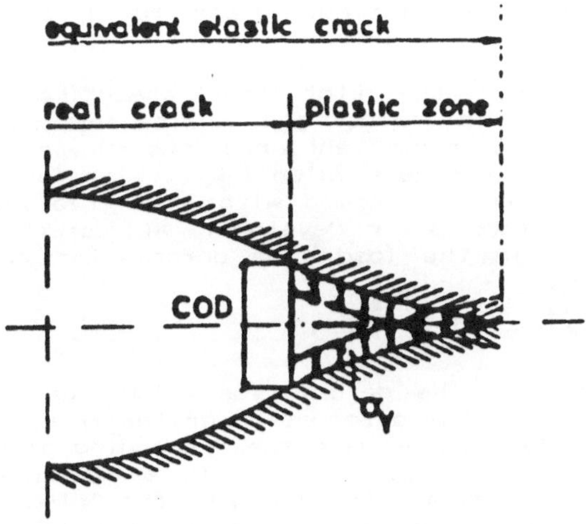

Fig 13A : Dugdale model for plasticity - Ref. 65.

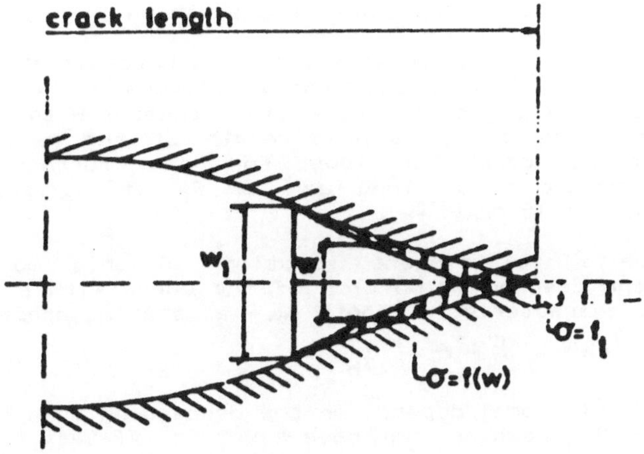

Fig 13B : Hillerborg model - Ref. 65.

rest of the material. As this model was eventually based on numerical methods it needed large computer storage and time. Different types of fracture and crack stability could be analysed using this technique. Peterson used Hillerborg's fictious crack model by modelling the fracture zone near a crack tip (68).

The basic idea of the fictitious crack model proposed by Hillerborg can best be explained using a tension test shown in fig. 15. For a homogeneous specimen of constant area, the curves A and B coincide till they reach the tensile stress value (f_t). With further increase in load, a fracture zone will be formed with a limited width within the specimen. As the fracture zone develops, microcracks occur with a corresponding decrease in the load. The deformation within the gauge length B is described as

$$\Delta l = \varepsilon l_0 \quad \ldots \quad 23$$

and is shown in fig. 15. The deformation within gauge length A is, $\Delta l = \varepsilon l_0 + w$, in which w is the deformation of the fracture zone and can be computed as the difference between the descending branches of curves A and B. The width of the fracture zone is assumed as zero. As the fracture zone in reality has a certain width, the "tied crack" which is introduced as a simplified description is actually no real crack but only a "fictious crack". Hence the name to this model.

The energy absorbed in the fictitious crack (fig.16) has been shown as

$$A \int_0^{w_c} \sigma \, dw = A G_F \quad \ldots \quad 24$$

where A = projected cross sectional area, w_c = w-value for σ, = 0 ; and G_F = observed energy per unit crack area as shown in fig. 17.

The stress distribution in front of a notch is shown in fig. 18. The relation between $\sigma + w$ within the fictitious crack is given by the $\sigma + w$ curve. The stresses at the crack tip increases with the increase in deformation reaching the tensile strength of the material. At this point, fictitious crack develops for any further increase in deformation. Therefore, as long as the fictitious crack grows the stresses at the tip is only f_t .

For the finite element solution of this model Hillerborg approximated the descending portion of the curve with a straight line. He defined the characteristic length l_{ch} - a material property as follows

$$l_{ch} = E G_F \,/\, f_t^2 \quad \ldots \quad 25$$

As accuracy of this model depends on the descending branch of the curve, the tensile testing machine must have a very stiff test rig.

LIMITATIONS :

At the fictitious crack tip a three dimensional state of stress is present which may actually influence the tensile strength. In the case of concrete, the difference between plane stress and plane strain is insignificant. Therefore, the same $\sigma - \varepsilon$ and $\sigma - w$ curves have been used in the case of crack growth as in the case of a tension test. But as Hillerborg himself admits, this point needs confirmation by further experimental investigation. If the compressive stress acting normal to the tensile stress is quite predominant then the compressive stress will influence the tensile strength and the $\sigma - \varepsilon$ and σ -w curves.

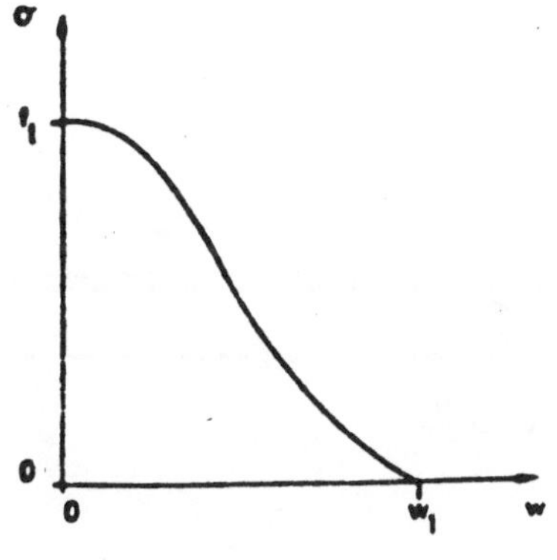

Fig 14 : Assumed variation of stress σ with crack width w , general case
- Ref. 65.

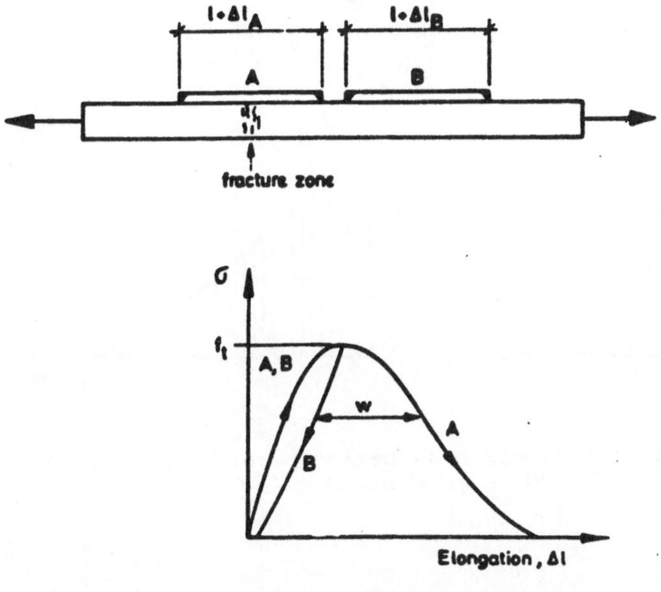

Fig 15 : σ - ε diagram and σ - w diagram
- Ref. 32.

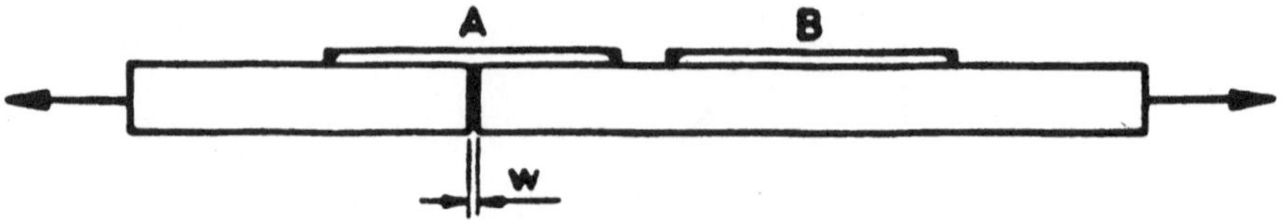

Fig 16 : Simplified description of the fracture zone as a " Fictitious Crack " with width w - Ref. 32.

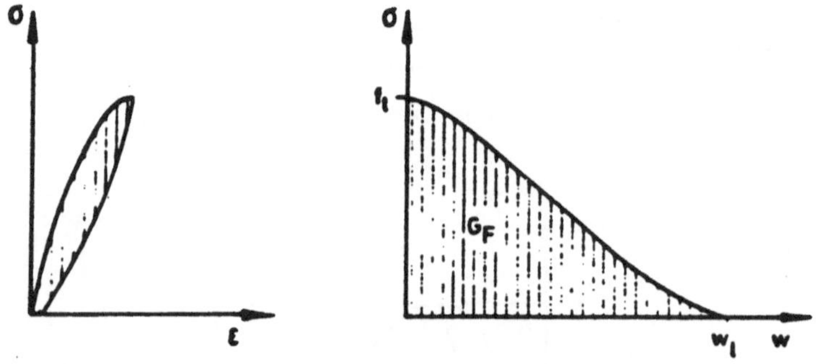

Fig 17: Energy dissipation related to the $\sigma - \varepsilon$ and $\sigma - w$ diagram - Ref. 32.

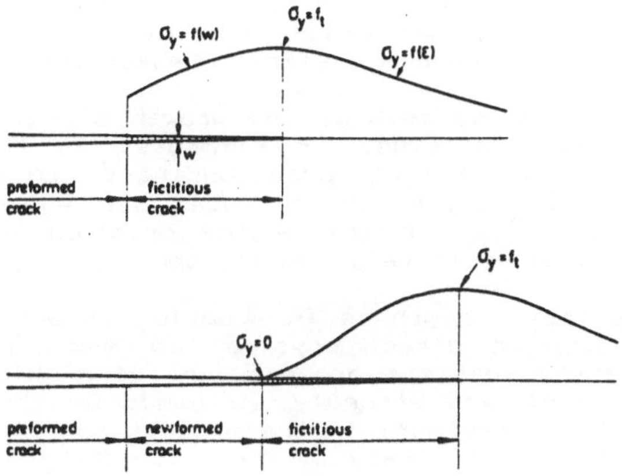

Fig 18 : Stress distribution in front of a crack tip before and after growth of the real crack - Ref. 32.

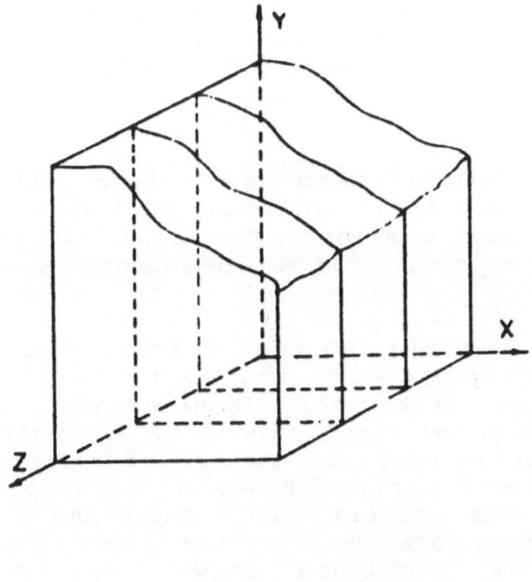

Fig 19 : Region of continuum divided into laminae - Ref. 76.

The fictitious crack model is very efficient in analysing the formation and the growth of fracture zones and cracks irrespective of whether the fracture starts from a crack, a notch, an irregularity or a plain surface. Such an efficient model lacks an analytical solution and all calculations are performed by the numerical method, using the finite element approach. This sometimes places limitation on computer time and memory especially when the specimen size increases considerably.

In the finite element method, the growth of fictitious and real cracks coincides with the sides of elements. As the elements are separated by distance 'w'; forces corresponding to σ from σ-w curve are introduced across the crack. It is stated that while it is possible to analyze cracks that do not follow the original element mesh, it appears that this procedure is very complicated.

The characteristic length as in equation 25 was shown as a pure material property with no direct physical correspondence. Victor and Erwin recently observed that the process zone size which is a function of characteristic length was strongly influencd by the material model (the stress-separation behaviour) in the process zone (69). Therefore, the characteristic length is likely to vary with the shape of σ-w curve.

Though the fictitious crack model predicts the failure load it apparently fails to explain the failure mechanisms involved. It has been pointed out by many that this may be an important requirement from a designer's point of view.

ENDOCHRONIC THEORY :

The Endochronic Theory proposed by Valanis's was applied to marine fiber reinforced concrete by Reddy and Gopal (70). They calculated the Endochronic constants by curve fitting the stress-strain curve determined by the experimental investigations on steel fiber reinforced concrete. Using the endochronic constants and fiber effect the flexural behaviour of beams were predicted.

MICROMECHANICAL MODEL:

The numerical analysis technique relates the local stresses in the matrix, the fiber and their interface to the overal macrocomposite behaviour. This approach dealing with local stresses identifies the crack growth in the matrix, interfacial debonding and then the final pullout of the fibre.

Taya and Mura investigated the effect of fiber and crack on the overall stiffness and strength of the composite (71). Takao, Chou and Taya investigated the effect of longitudinal Young's modulus of composites with misoriented short fibers (72). Their theoretical model acounted for the fiber interaction at various orientation angles. Argon and Shack analysed the interface transfer of stresses in a composite material (19). The shear stress was transferred along the interface length within a short distance from the fiber end. All these three micromechanical models mentioned above use Eshelby's equivalent inclusion concept. Gopalaratnam and Shah observes that this concept may not be well suited to problems where interfacial cracks and fiber pull out play a dominant role in the composite materials (73). As a result, the conclusion of Takao, Chou and Taya (72) may not be applicable to fiber reinforced concrete.

Mandel et al studied the stress conditions near a crack tip in a fiber reinforced isotropic material (74). They used a two dimensional parametric element for their analysis. The results were compared with experimental investigations using specimens made out of methyl methacrylate matrix reinforced with aligned fibers. They demonstrated that the roles of higher modulus fibers at the crack tip were to inhibit crack propagation and increase the fracture toughness. They also observed that the shear stresses in the matrix adjacent to the fibers and the bond stresses at the interface were larger for a shear mode loading than for an opening mode loading. Therefore, they recommended a mixed mode loading for studies of effective fracture toughness (75).

In plain strain and plane stress problems the strain energy associated with one displacement component is small compared to the total strain energy. For such problems, Pack and Mandel showed that the given domain may be divided into laminae parallel to an X-Y plane as shown in fig. 19.(76).

Tolf investigated the stress field in short-fiber composites with different mechanical properties of the constituents and different geometrical configuration (77). From this calculation; conclusions were drawn about the mechanical behaviour like critical fiber length and fracture toughness.

Fig. 20 represents the material modelled by a regular two-dimensional pattern of fibers embedded in a matrix. The unit cell consists of one fiber in the middle surrounded by matrix. The dimensions of the cell and fiber is in fig. 20. For static plane stress problems, the displacement is given by Navier's equations.

$$(1+k)\frac{\partial^2 u}{\partial x^2} + \frac{\partial^2 u}{\partial y^2} + k\frac{\partial^2 v}{\partial x \partial y} = 0 \qquad \ldots 26$$

$$\frac{\partial^2 v}{\partial x^2} + (1+k)\frac{\partial^2 v}{\partial y^2} + k\frac{\partial^2 u}{\partial x \partial y} = 0 \qquad \ldots 27$$

In this expression the displacement in X and Y directions are represented as U(x,y) and V (x,y). For the co-ordinate system oriented asin fig. 20 the following boundary conditions were imposed. On a line parallel to the Y-axis it was shown

$$\bar{u} = \bar{\varepsilon} x \quad (\bar{\varepsilon} \text{ is a constant}) \qquad \ldots 28$$

$$\sigma'_{xy} = 0 \qquad \ldots 29$$

and on a line parallel to X-axis

$$\sigma'_{xy} = \sigma'_{yy} = 0 \qquad \ldots 30$$

A representative domain was selected to simplify the boundary condition. The elasticity equations representing this domain were solved using the finite difference method. Two kinds of calculations were performed. The first calculation was done to find the stress field dependence on the elastic properties of the constituents. The second was performed to find the dependence on the geometrical configuration. The author concluded that the critical fiber length increased with the overlap of fibers at a particular volume fraction and decreased with the volume fraction at constant overlap. Using these results, the mechanical properties of the composites were predicted qualitatively.

Paramasivam et al studied the linear and non-linear behaviour of fiber reinforced cement composites under uniaxial loading using a three dimensional finite element model (78). Experiments have shown that predictions of the yield and ultimate stresses of the composite material were in reasonable agreement with the test results.

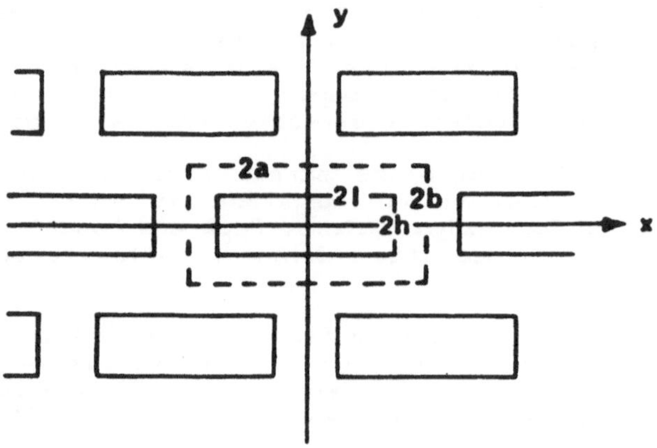

Fig 20 : Model of the fiber reinforced composite - Ref. 77.

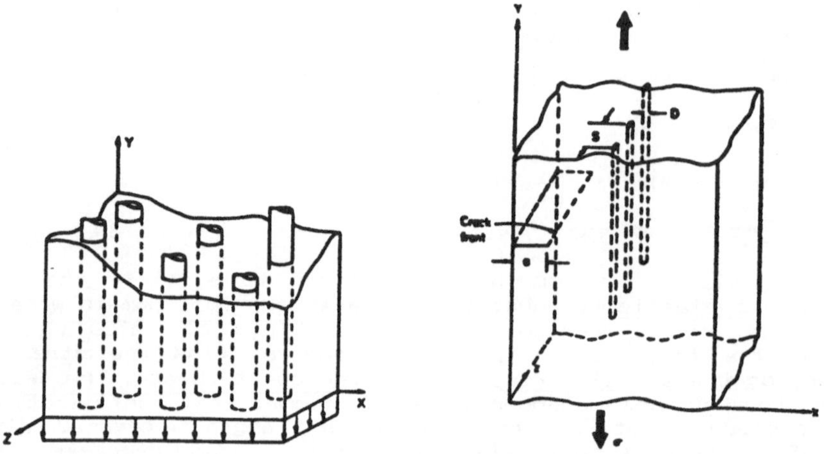

Fig 21: Model of the fiber reinforced specimen - Ref. 79.

Pack and Mandel used the 2-D multiplane method of analysis (76) to investigate the micromechanical failure mechanisms and determine an effective fracture toughness for reinforced materials (79). The failure of fiber reinforced materials is generally preceded by local failures such as yield in matrix of fibers, crack extension in the material and fiber delamination. Therefore, a sequential finite element analysis was performed to model this progressive failure mechanism (fig. 21). As the fibers were equally spaced in the Z-direction, a typical portion of the thickness of the specimen included only one half of the fiber and one half of the matrix material. While, one plane contained matrix material the other contained both matrix material and fiber. The specimen was also symmetric about the crack axis and hence only one half of the specimen was modelled. Both matrix and fiber materials were modelled using an eight noded isoparametric quadratic quadrilateral elements. A zero thickness element to model the interface between the matrix material, fiber and a connector element between dual planes to model the displacement in X and Y directions were also introduced. The failure criteria used for characterising crack growth was based on LEFM. The tensile specimen analysed showed failure in three stages :

1) Initiation of rapid crack growth.
2) Crack arrest at the row of fibers
3) Slow crack growth past the fibers, followed by unstable rapid crack growth across the specimen.

This model is capable of showing the type of progressive failure as normally observed in FRC. If the fiber pull-out is also incorporated beyond the stage of unstable rapid crack growth the model will prove to be effective.

In most experimental studies on the fiber matrix bond behavior, a uniform shear stress distribution is assumed and the interface strength is calculated as a material property. Gopalaratnam and Shah used two parameters namely the shear strength of the interface and the frictional resistance of the debonded portion in their interfacial model (73). The stable debonding process prior to pull-out of the fiber together with the matrix softening behavior were shown as the reasons for non-linearity in fiber composites. This indicates, interface strength will depend on the length and diameter of the fiber and hence it is not a material property as shown by other investigators.

LIMITATIONS :

Micromechanical models are useful in determining the causes of failure of a composite material. But there is certain amount of uncertainity in the mathematical representation of the cracking process. While some cracks are assumed to be discrete many others are modeled as smeared cracks.

Discrete crack modelling does not limit the level at which a material may be considered. If large computational facilities are available it is even possible to model the detailed particle matrix. However in structural design the fracture is considered only in terms of the overall material properties which in turn are functions of the composition. The smeared crack model gives an idea of the constitutive relationship and approximates the overall state of fracture. However, it can be observed that this model does not conform to the basic notion of fracture mechanics with regard to the propagation of discrete cracks (56).

Several investigations in the past had assumed pure mode I type of crack extension instead of either mode II or mode III. In mode II type of failure, the energy increments were computed from the strain energy release rates. This approach does not predict the actual direction of crack propagation. Sih using the distribution of the strain energy density around the crack tip predicted the direction of crack propagation. Sih states that the two fundamental hypotheses of crack extention are :

i) The crack will spread in the direction of the maximum potentialenergy density and
ii) The critical intensity, S_{CRIT} , of this potential field governs the crack propagation.

It is reported that the counterpart of this method has been proposed by Hussain et al (56). Both these methods offer a crack propagation theory which is much nearer to reality than any other theory.

The accuracy of the numerical solution vastly depends on the size and the number of elements about the crack tip. A group of eight triangles about the crack tip for the pure mode I case has proven to be satisfactory (56). However, the results show notable improvement if a group of twelve equal triangles are used instead of eight equal triangles. If the number of triangles used exceeds twelve then the radial solution field begins to deteriorate due to the excessive aspect ratio of the individual elements.

Finite element programs provide output data for node and integration point locations. These are used to determine the stress intensity factors etc. Though this can lead to a valid solution, enough care must be exercised at the time of interpretation of these results. While the pattern and density of the integration points do not affect the solution, the output sample at different patterns of output points may yield different values.

CONCLUSIONS :

Several constitutive relations and models have been discussed pointing out the various limitations in them. The application of a particular model depends on the type of failure mechanism which the designer chooses and the type of solution required to the problem. As all of the models are based on certain assumptions it is important to recognise these assumptions and their limitations before using the model.

Of all the available models, the concept of Fictitious Crack Model appears to be simpler in its approach with very little experimental data required for its use. Moreover, the model simulates the inelastic behavior of concrete and fibre reinforced concrete with a few limitations as already discussed. Also, the numerical analysis technique lends considerable support in the application of this model with the possibility of refining the existing solution and improving the model to explain the overall behavior of fibre reinforced concrete. However, for the design applications like determination of flexural strength, fatigue strength, cracking, tensile strength, impact strength and toughness, other models may be more appropriate. Many of the currently proposed models could be modified and refined for this purpose.

ACKNOWLEDGEMENTS :

This report is part of an extensive research on the " Constitutive Relationships and Performance Characteristics Of Concrete Fibre Composites ",.with the financial assistance of the National Science Foundation of USA, under Grant No. MSM - 8612082.

REFERENCES:

1. Ramakrishnan. V., Terje, Bradshaug ., Coyle, W.V., and Ernest, K., Schrader., "A Comparative Evaluation of Concrete Reinforced With Straight Steel Fibers and Fibers with Deformed Ends Glued Together into Bundels", ACI Journal Proceedings, V.77, No.3, May-June 1980, pp 135-143.

2. Ramakrishnan, V., Coyle, W.V., Kulandaisamy, V., and Ernest K. Schrader., "Performance Characteristics of Fiber Reinforced Concretes with Low Fiber Contents", ACI Journal, Proceedings V.78, No.5, Sept.-Oct., 1981, pp 384-394.

3. Balaguru, P., and Ramakrishnan, V., " Mechanical Properties Of Superplasticised Fiber Reinforced Concrete Developed For Bridge Decks and Highway Pavements ", American Concrete Institute, SP-93, 1986, pp. 563-584.

4. Swamy, R.N., "Linear Elastic Fracture Mechanics Parameters of Concrete", Fracture Mechanics of Concrete, edited by F.H.Wittmann, Elsevier Science Publishers, 1983, pp 411-461.

5. Kaplan, M.F., "Crack Propagation and the Fracture of Concrete", Journal of the American Concrete Institute, 58, 1961, 591-610.

6. Blakey, F.A. and Beresford., "Discussion of the Paper Crack Propagation and the Fracture of Concrete by M.F.Kaplan", Journal of the American Concrete Institute, 59, 1962, 919-923.

7. Glucklich, J., "Discussion of the Paper Crack Propagation and the Fracture of Concrete, by M.F.Kaplan", Journal of the American Concrete Institute, 59, 1962, pp 919-923.

8. Irwin, G.R., "Discussion of the Paper Crack Propagation and the Fracture of Concrete by M.F.Kaplan", Journal of the American Concrete Institute, 59, 1962, pp 929.

9. Romualdi, J.P., and Batson, G.B.,"Mechanics of Crack Arrest in Concrete", Journal of the Engineering Mechanics Division, ASCE, 89, 1963, 147-168.

10. Romualdi, J.P., and Batson, G.B.,"Behaviour of Reinforced Concrete Beams with Closely Spaced Reinforcement", Journal of the American Concrete Institute, 60(1963), pp 775-790.

11. Romualdi, J.P., and Mandel, J.A., "Tensile Strength of Concrete Affected by Uniformly Distributed and Closely Spaced Short Lengths of Wire Reinforcement", Journal of the American Concrete Institute, 61, 1964, pp 657-670.

12. Romualdi, J.P.,"The Static Cracking Stress and Fatigue Strength of Concrete Reinforced with Short Pieces of Thin Steel Wire, in A.E.Brooks and K.Newman eds", The Structure of Concrete, Proceedings of an International Conference, London, 1965, pp 190-201.

13. Romualdi, J.P., Ramey, M., and Sanday, S.C., "Prevention and Control of Cracking by use of Short Random Fibers", Mechanism and Control of Cracking in Concrete, SP-20, American Concrete Institute, Detroit, 1968, pp 179-203.

14. Shah, S.P., "Micromechanics of Concrete and Fiber Reinforced

Concrete, in M.Te'emi ed.", Structure, Solid Mechanics and Engineering Design, Proceedings of the Southampton 1969 Civil Engineering Materials Conference, Wiley-Interscience, 1971, pp 367-376.

15. Brown,E.T. and Hudson,J.A.,"Discussion on Griffith Fracture Criterion and Concrete, by Shah,S.P.", Journal of Engineering Mechanics Division,ASCE,98, 1972, 1310-1312.

16. Parimi, S.R., and Rao, J.J.S., "Effectiveness of Random Fibers in Fiber Reinforced Concrete, in Mechanical Behaviour of Materials," Kyoto 1971, The Society of Material Science, Japan, 1972, vol.V, pp 176-186.

17. Harris, B., Varlow, J., and Ellis, C.D., "The Fracture Behaviour of Fiber Reinforced Concrete", Cement and Concrete Research, vol.2, 1972, pp 447-461.

18. Synder, M.J., and Lankard, D.R., "Factors Affecting Cracking Strength of Steel Fibrous Concrete", ACI Journal, Proc V.69, No.2, Feb.1972.

19. Argon, A.S., and Shack, W.J., "Theories of Fiber Cement and Fiber Concrete, in Fiber Reinforced Cement and Concrete", Rilem Symposium 1975, The Construction Press Ltd, Lancaster, England, 1975, pp 39-53.

20. Spurrier, J., and Luxmoore, A.R., "The Critical Constrained Crack Length in Fiber Reinforced Cementious Materials", Fiber Science and Technology, vol.9, Applied Science Publishers Ltd, England, 1976, pp 225-236.

21. Mangat, P.S., "Tensile Strength Of Steel Fiber Reinforced Concrete", Cement and Concrete Research, Vol 6, 1976, pp. 245-252.

22. Gustaffson, P.J., and Brottmekaniska Studier, "Lattebetong ock Fiber-Armerad Betong", Rapport TVBM-5001, Dept. of Building Materials, The Lund Institute of Technology, 1977, pp 131.

23. Bunsell, A.R., "The Nature of Crack Growth in Composite Materials, in G.C. Sih and V.P. Tamuzus eds", Fracture of Composite Materials, Proceedings of the First USA - USSR Symposium, Riga, 1978, Sijthoft and Noordhoff, The Netherlands, 1979, pp 349-359.

24. Djabarov, N., Wehrstedt, A., Weiss, H.J., and Yamboliev, K., "Investigations on the Fracture Behaviour of Steel Fiber Reinforced Concrete", Proceedings of the Second National Conference on Mechanics and Technology of Composite Materials, Sofia, Bulgaria, 1979, pp 593-596.

25. Javan, L., and Dury, B.L., "Fracture Toughness of Fiber Reinforced Concrete", Concrete, 13(1979), pp 31-33.

26. Stroeven, P., "Crack Developement in Concrete as Influenced by the Addition of Short Steel Fibers", Seminar on Fracture Mechanics of Fiber Reinforced Cement Based Composites, Delft, 1979, pp C1-45.

27. Bazant, Z.P., "Material Behaviour Under Various Types of Loading in S.P.Shah (ed), High Strength Concrete", Proceedings of a Workshop held at the University of Illinois at Chicago Circle, December, 1979, pp 79-92.

28. Bazant, Z.P., and Cedolin, "Fracture Mechanics of Reinforced Concrete, in W.F.Chen and E.C.Ting (eds.) Fracture in Concrete",

Proceedings of an ASCE session, Florida, 1980, ASCE, New York, 1980, pp 28-35.

29. Swamy, R.N., "Influence of Slow Crack Growth on the Fracture Resistance of Fiber Cement Composites", International Journal of Cement Composites, 2 (1980), 43-53.

30. Saouma, V.E., Ingraffea, A.R., and Catalano, D.M., "Fracture Toughness of Concrete: KIc Revisited", Journal of Engineering Mechanics Division, Proceedings of the ASCE, vol.108, No.EM6, Dec. 1982, pp 1152-1166.

31. Newman, J.C., "An Improved Method of Collocation for the Stress Analysis of Cracked Plates with Various Shaped Boundaries", TN D-6376, NASA Technical Note, Aug. 1971.

32. Hillerborg, A., "Analysis of One Single Crack", Fracture Mechanics of Concrete, Edited by F.H. Wittmann, Elsevier Science Publishers, 1983.

33. Johnston, C.D., " Toughness of Steel Fiber Reinforced Concrete ", Steel Fiber Concrete, US-Sweden Joint Seminar, Stockhlom, June 3-5, 1985, pp. 333-360.

34. ASTM Designation: C1018-85, Standard Test Method for Flexural Toughness and First-Crack Strength of Fiber Reinforced Concrete (Using Beam With Third-Point Loading), Annual Book of ASTM Standard, Vol. 04-02, 1985.

35. Holiday, L., " Composite Materials ", Elsevier, Amsterdam, 1966.

36. Cox, H.L., " The Elasticity and Strength of Paper and other Fibrous Materials ", British Journal Of Applied Physics, 3, 1952, pp.72-79.

37. Krenchel, H., " Fiber Reinforcement ", Akademish Forlag, Copenhagen, 1964.

38. Allen, H.G., " Glass-Fibre Reinforced Cement, Strength and Stiffness ", CIRIA Report 55, Sept. 1975.

39. Shah, S.P., and Rangan, B.V., "Fiber Reinforced Concrete Properties", ACI Journal, Proc vol.68, No.2, Feb.1971, pp 126-135.

40. Naaman, S.R., Argon, A.S., and Moavenzadeh, "A Fracture Model for Fiber Reinforced Cementitious Materials", Cement and Concrete Research,vol 3, 1973 , pp 397-411.

41. Swamy, R.N., Mangat, P.S., and Rao, C.V.S.K., "The Mechanics of Fiber Reinforcement of Cement Matrices", Fiber Reinforced Concrete, SP-44, ACI, Detroit, 1974, pp 1-28

42. Swamy, R.N., and Mangat, P.S., " A Theory for the Flexural Strength of Steel Fibre Reinforced Concrete ", Cement and Concrete Research, Vol 4, 1974, pp. 313-325.

43. Soroushian, P., Cha-Don Lee., and Ziad Bayasi., " Fiber Reinforced Concrete : Theoretical Concepts and Structural Design ", Seminar Proceedings , Fiber Reinforced Concrete : Design and Applications, eds Parviz Soroushian and Ziad Bayasi, Michigan State University, East Lansing, Michigan, February 1987, pp.8.1

44. Aveston, J., Mercer, R.A., and Silwood, J.M., "The Mechanism of Fiber Reinforcement of Cement and Concrete", National Physical

Laboratory Report (Great Brtain), No.S2, No.90/11/98 Part I, Jan.1975, Part II, DMA 228, Feb.1976.

45. Soroushian, P., and Ziad Bayasi., " Prediction of the Tensile Strength of Fiber Reinforced Concrete : A Critique of the Composite Material Concept ", ACI Convention on Fiber Reinforced Concrete, Baltimore, November 1986.

46. Kelly,A.., "Interface Effects and the Work of Fracture of a Fibrous Composite", Proc. Royal Society of London, A319, 1970, pp 95-116.

47. Johnson, C.D., and Coleman, R.A., "Strength and Deformation of Steel Fiber Reinforced Mortar in Uniaxial Tension", Fiber Reinforced Concrete, SP-44, ACI, Detroit, 1974, pp 177-207,

48. Naaman, A.E., and Shah, S.P., "Pull-out Mechanism in Steel Fiber Reinforced Concrete", Journal of the Structural Division, ASCE, Aug.1976, pp 1537-1549.

49. Pinchin, D.J., and Tabor, D., "Interfacial Phenomena in Steel Fiber Reinforced Cement II: Pull-out Behaviour of Steel Wires", Cement and Concrete Research, vol.8, 1978, pp 139-150.

50. Morton, J., "The Work of Fracture of Random Fiber Reinforced Cement", Materiaux et Constructions, 12, 1979, pp 393-396.

51. Mindess, S., Lawrence, F.V., and Kesler, C.E.,"The J-Integral as a Fracture Criterion for Fibre Reinforced Concrete ", Cement and Concrete Research, 7, 1977, pp 731-742.

52. Brandt, A.M., "Crack Propagation Energy in Steel Fiber Reinforced Concrete", International Journal of Cement Composites, 2(1980), 35-42.

53. Carrato, J.L., "Experimental Evaluation of the J-Integral", M.S. Thesis, Dept. of Civil Engineering, University of Illinois at Urbana-Champaign, 1980.

54. Velazco, G., Visalvanich, K., and Shah, S.P., "Fracture Behaviour and Analysis of Fiber Reinforced Concrete Beams", Cement and Concrete Research, Vol.10, 1980, pp 41-51.

55. Mindess, S., "The Fracture of Fiber Reinforced and Polymer Impregnated Concretes: A Review", Fracture Mechanics of Concrete, Edited by F.H.Wittmann, Elsevier Science Publishers, Amsterdam, 1983, pp 481-501.

56. Chamberlain, D.A., and Boswell, L.F., "Numerical Methods in Fracture Mechanics", Fracture Mechanics of Concrete, Edited by F.H.Whittmann, Elsevier Science Publishers, Amsterdam, 1983, pp 503-533.

57. Visalvanich,K.,and Naaman, A.E., "Evaluation of Fracture Techniques in Cementitious Composites, in W.F.Chen and E.C.Ting (eds)", Fracture in Concrete, Proceeding of an ASCE Session, Florida, 1980, American Society of Civil Engineers, New York, 1980, pp 65-81.

58. Wecharatna, M., and Shah, S.P., "Double Torsion Tests for Studying Slow Crack Growth of Portland Cement Mortar", Cement and Concrete Research, vol.10, pp 833-844, 1980.

59. Visalvanich, K., and Naaman, A ., "Fracture Methods in Cement Composites", Journal of Engineering Mechanics Division, Proceedings of ASCE, Vol. 107, EM 6, Dec. 1981.

60. Wecharatna, M, and Shah, S.P., "Slow Crack Growth in Cement Composites", Journal of the Structural Division, ASCE , vol. 108, No.ST6, June 1982, pp 1400-1413.

61. Irwin, G.R., and Kies, J.A., "A Critical Energy Rate Analysis of Fracture Strength", Welding Research Supplement, 1954, pp 193-198.

62. Visalvanich, K., and Naaman, A.E., "Fracture Model for Fiber Reinforced Concrete", ACI Journal, March-April, 1983, pp 128-138.

63. Wecharatna, M., and Shah, S.P., "Predictions of Non-Linear Fracture Process Zone in Concrete", Journal of Engineering Mechanics, Vol. 109, No.5, October 1983.

64. Shah, S.P., and Jenq, Y.S. "Fracture Resistance of Steel Fiber Reinforced Concrete", Steel Fiber Concrete, US-Sweden Joint Seminar, Stockholm, 3-5 June, 1985.

65. Hillerborg, A., Modeer, M., and Petersson, P.E., "Analysis of Crack Formation and Crack Growth In Concrete by means of Fracture Mechanics and Finite Elements", Cement and Concrete Research, Vol.6, 1976, pp 773-782.

66. Hillerborg, A., "A Model for Fracture Analysis", Report TVBM-3005, Division of Building Materials, Lund Institute of Technology, Lund, Sweden, 1978.

67. Hillerborg, A., "Analysis of Fracture by Means of the Fictitious Crack Model, Particularly for Fiber Reinforced Concrete", International Journal of Cement Composites, 2,1980, 177-184.

68. Petersson, P.E., "Fracture Mechanics Calculations and Tests for Fiber Reinforced Cementitious Materials", Advances in Cement Matrix Composites, Proceedings, Materials Research Society, Symposium L, Boston, Nov.1980, pp 95-106.

69. Victor, C., Li., and Erwin Liang., "Fracture Processes in Concrete and Fiber Reinforced Cementitious Composites", Journal of Engineering Mechanics, Vol. 112, No.6, June 1986, pp 566-586.

70. Reddy, D.V., and Gopal, K.R., "Endochronic Constitutive Modeling of Marine Fiber Reinforced Concrete ", Computational Modelling of Reinforced Concrete Structures, Edited By Ernest Hinton and Roger Owen, Pineridge Press, 1986, pp 154-186.

71. Taya, M., and Mura, T., "On Stiffness and Strength of an Aligned Short-Fiber Reinforced Composite Containing Fiber-End Cracks Under Uniaxial Applied Stress", Journal of Applied Mechanics, Vol.48, June 1981, pp 361-367.

72. Takao, Y., Chou, T.W., and Taya, M., "Effective Longitudinal Young's Modulus of Misoriented Short Fiber Composites", Journal of Applied Mechanics, Vol. 49., Sept. 1982, pp 536-540.

73. Gopalaratnam, V.S., and Shah, S.P., "Failure Mechanisms and Fracture of Fiber Reinforced Concrete", Proceedings, Fiber Concrete Symposium, ACI Convention, Baltimore, Nov. 1986.

74. Mandel, J.A., and Pack, S.C., "Crack Growth in Fiber Reinforced Materials", Journal of the Engineering Mechanics Division, ASCE, vol.108, No.3 & 4, 1982, Page 509.

75. Mandel, J.A., Pack, S.C., and Tarazi, S., "Micromechanical Studies of Crack Growth in Fiber Reinforced Materials", Engineering Fracture Mechanics, vol.16, No.5, 1982, pp 741-754.

76. Pack, S.C., and Mandel, J.A., "2-D Multiplane Finite Element Technique for Solving a Class of 3-D problems", International Journal for Numerical Methods in Civil Engineering, Vol. 19, 1983, 113-124.

77. Tolf, G., "Mechanical Behaviour of Short-Fiber Composite", Fiber Science and Technology, 1983, pp 91-109

78. Paramasivam, P., Curiskis, J.I., and Valliappan, S., " Micromechanics Analysis of Fiber Reinforced Cement Composites", Fiber Science and Technology, 1984, pp 99-120.

79. Pack, S.C., and Mandel, J.A., "Micromechanical Multiplane Finite Element Modelling of Crack Growth in Fiber Reinforced Materials", Engineering Fracture Mechanics, Vol. 20, 1984, pp 335-349.

Proceedings of the International Symposium on Fibre Reinforced Concrete, December 16-19, 1987, Madras, India

BEHAVIOUR OF FIBRE REINFORCED CONCRETE BEAMS WITH EQUAL TENSION AND COMPRESSION REINFORCEMENT

V.S. Parameswaran T.S. Krishnamoorthy
K. Balasubramanian N. Lakshmanan

Scientists
Structural Engineering Research Centre
Madras, INDIA

SYNOPSIS

The behaviour of steel fibre reinforced concrete beams having equal compression and tension steel adequately designed to avoid shear failure has been under investigation at the Structural Engineering Research Centre, Madras, in order to find their application in the design of blast-resistant structures. The paper presents the experimental data obtained from tests carried out on five SFRC beams. Analysis of the experimental data reveals that SFRC beams have much better load dispersion characteristics as compared to normal reinforced concrete beams. Empirical expressions for the static rigidity of beams when subjected to various levels of bending moment have been derived based on the experimental data.

Ductility is one of the major factors to be reckoned with in a blast loading environment. SFRC beams have been observed to fail by the rupture of tensile steel thereby ensuring very large curvature at failure. However, the improved bond properties obtainable by the use of small diameter main steel bars in steel fibre reinforced concrete beams may reduce the effective length of plastic hinge at the failure load. This factor is to be borne in mind while detailing of reinforcement in SFRC beams. The commonly observed mode of buckling of compression steel with higher percentages of main steel reinforcement in reinforced and laced reinforced concrete beams is absent in SFRC beams. The results of the investigation are encouraging and have convincingly proved that SFRC beams provided with equal amount of tension and compression steel can be used in the design of blast-resistant structural elements.

INTRODUCTION

Introduction of steel fibres is generally found to improve several properties of reinforced concrete structural elements, particularly the initial cracking strength, ductility, energy-absorption and resistance to impact [1]. Structural Engineering Research Centre, Madras, has been in the forefront in developing new materials, products and construction techniques using reinforced and prestressed concrete and concrete composites for use in various structural applications. Recently, an experimental programme was undertaken by the Centre to study the suitability of steel fibre reinforced concrete for use in blast-resistant design of structures.

It is well known that the ductility ratio, defined by the ratio of the failure deformation to the yield deformation, of a structural element like beam, slab etc., controls the resistance requirements of the element under impulsive type of loading. Assuming that the duration of the pulse is very short in comparison to the time period of the structural element under consideration, it can be shown that the resistance requirement can be obtained using the relation

$$R = i\,\omega / \sqrt{2\mu - 1} \qquad \dots (1)$$

where i = impulse,

ω = natural frequency of the structural element,

μ = ductility ratio,

and R = resistance requirement.

Earlier investigations on reinforced and laced reinforced concrete structural elements[2] have shown that beam elements provided with equal compression and tension steel reinforcement exhibit vastly superior ductile characteristics at failure. In these cases, the failure was triggered by the rupture of the main tension steel and the strain in steel at failure generally exceeded 10% to 15%. One of the major drawbacks in using higher percentages of steel levels however was the large-scale spalling of concrete in the compression zone observed during the failure stages which resulted in the buckling of the compression bars. It was, therefore, found necessary to introduce closely-spaced transverse reinforcement either in the form of stirrups or lacing to prevent the buckling. By using steel fibres in doubly-reinforced beams, it may be possible to confine the concrete in the compression zone, thereby preventing its spalling at failure stages.

Al-Taan and Al Feel [3] have made a comprehensive review of the state of knowledge on the behaviour of SFRC beams. The role of steel fibre reinforcement in flexural elements has been well analysed by them. One of the major conclusions derived by them is that the inclusion of fibre reinforcement in conventionally reinforced concrete beams imparts a modest increase in ultimate strength, considerable increase in energy-absorption capacity, and far more ductility. These improvements increase with the fibre content and aspect ratio and decrease with main steel reinforcement ratio. The presence of the fibre reinforcement preserved the unity and structural integrity of the member upto and even after failure.

PRESENT INVESTIGATIONS

The test programme reported in this paper is comprehensive and is devised in a manner as to provide full information on the structural behaviour of SFRC beams provided with equal main reinforcement at top and bottom. The important characteristics that were studied include the moment-curvature relationships for the test beams, the static and dynamic flexure rigidities at various levels of preloading, the load-deformation behaviour, the ductility ratio (defined by the ratio of failure deformation to yield deformation), the failure rotation achieved at the plastic hinge and the dispersion phenomenon observed under the action of a single concentrated load on the test beams. The main variables studied are the percentage of main reinforcement and volume percentage of steel fibre reinforcement. A constant aspect ratio of 100 was maintained in all the test beams. The nominal cross-sectional dimensions of the test beams were 200 mm x 100 mm. The effective span of the test beams was 2060 mm. The grade of concrete used was M 20. Other relevant details of the test beams are given in Table 1. Fig.1 shows the schematic diagram of

TABLE 1 DETAILS OF TEST BEAMS

Serial Number	Designation	Main Reinforcement (at top and bottom)		Volume Percentage of Fibres	Transverse Steel
1	B1	2 nos.	10 #	0.50	2 legged 8#at 150 mm centres
2	B2	2 nos.	10 #	0.75	2 legged 8#at 150 mm centres
3	B3	2 nos.	10 #	1.00	2 legged 8#at 150 mm centres
4	B4	2 nos.	12 #	1.00	2 legged 8#at 150 mm centres
5	B5	2 nos.	16 #	1.00	2 legged 8#at 125 mm centres
6	B6	2 nos.	10 #	0.00	2 legged 8#at 150 mm centres

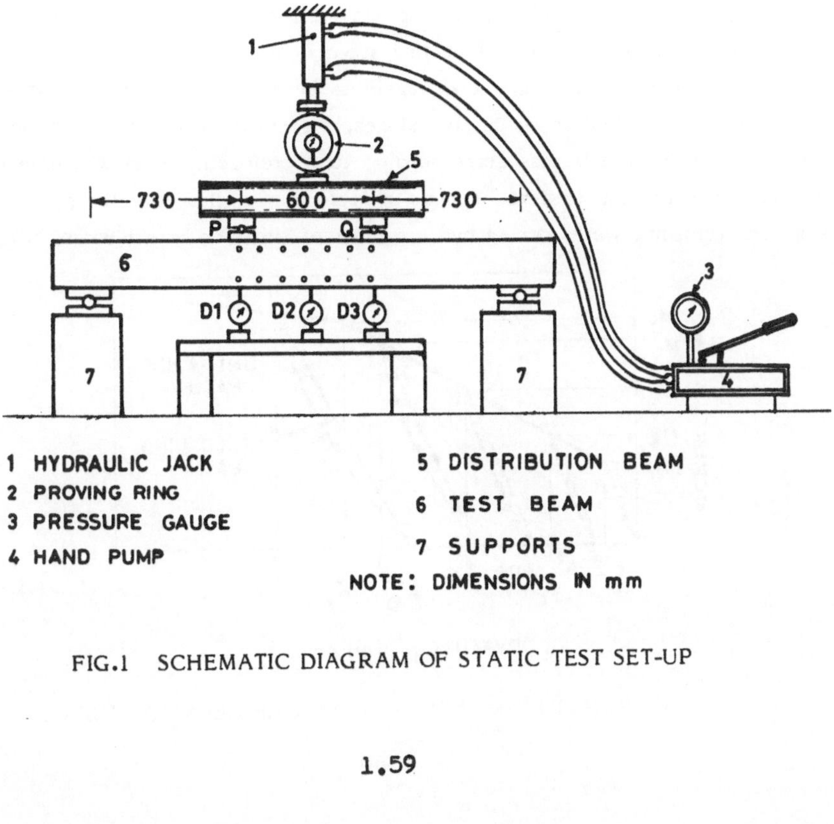

FIG.1 SCHEMATIC DIAGRAM OF STATIC TEST SET-UP

the test set up used for carrying out static tests on the beams. The instrumentation essentially consisted of a hydraulic jack of 30t capacity actuated by a high pressure hand pump fitted with a Bourdon-type pressure gauge. A proving ring of 10t capacity was used to measure the applied loading accurately. Dial gauges having a least count of 0.01 mm and mounted on magnetic bases were used to record the deformations at various stages of loading. Demountable mechanical strain gauge of the Pfender type having a nominal gauge length of 100 mm was used to measure the strains in concrete. The least count of the mechanical strain gauge used was 10^{-5} mm/mm. A crack-measuring microscope having a least count of 0.02 mm was used to record the widths of typical cracks at the level of the tension steel. The test programme also envisaged the determination of the static and dynamic rigidities of SFRC beams at various pre-loading levels.

The general programme of testing was as follows. After the beam was fully set on the test span, the dynamic test was conducted first to determine the dynamic response of the uncracked test beam. The details of dynamic test and the discussion on the behaviour of test beams is given elsewhere [4]. After conducting the dynamic test, the dynamic test set-up was dismantled, and the static test set-up was assembled. The loading on the beam was incremented in stages. The strains and deformations at predetermined locations were measured at each loading stage. The initial cracking load of the SFRC beams tested was observed to be between 1.0 to 1.6 tonnes. The test was continued beyond the cracking stage and the loading was released at 2.0t, which represents the first pre-loading level. The permanent set was recorded by means of the dial gauges. The static test set-up was then dismantled. The dynamic test set-up was reassembled, and the test to determine the dynamic response was conducted at this stage. After finishing the dynamic test, the dynamic test set-up was disengaged once again as before, and the static testing restarted. As the loading was applied in successive stages, the deformations of the beam was found to increase. The static stiffness corresponding to a preloading level was determined from the deformation behaviour observed during the reloading cycle. Similar cycles of loading, releasing, and reloading were carried out a number of times on a test beam. Fig.2 schema-

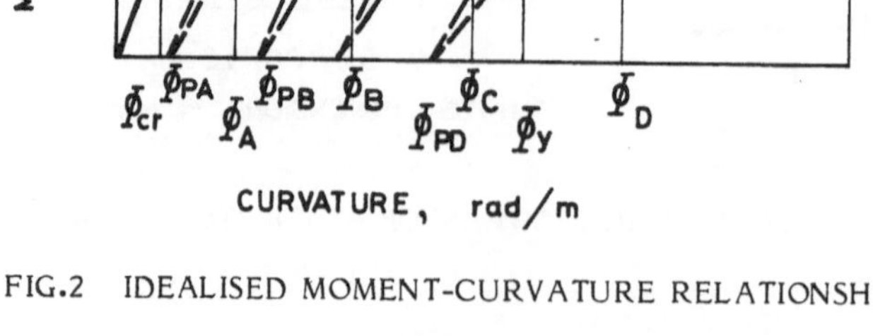

FIG.2 IDEALISED MOMENT-CURVATURE RELATIONSHIP

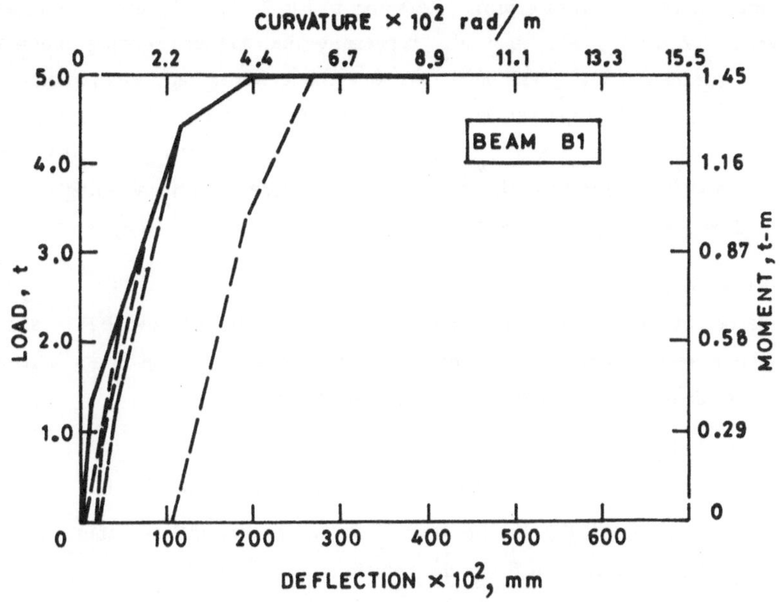

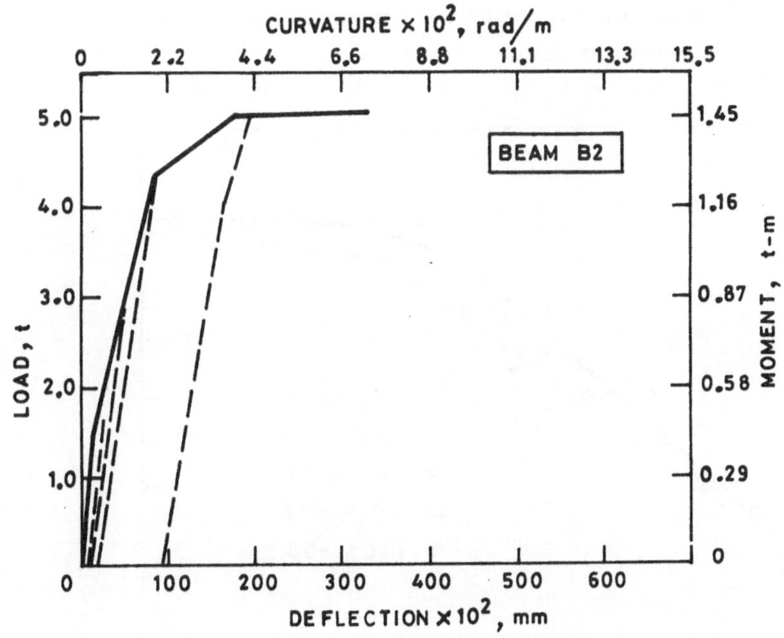

FIG.3 LOAD-DEFORMATION BEHAVIOUR OF SFRC BEAMS

tically shows a typical moment-curvature plot. Refering to Fig.2, A, B, C, and D represent the pre-loading levels. M_A, M_B, M_C, and M_D represent the applied bending moments at the centre of the test beam. M_{cr} and M_y represent the cracking and yield moments, respectively. Φ_A, Φ_B, Φ_C, and Φ_D represent the total curvatures corresponding to the pre-loading stages A, B, C, and D. Φ_{cr} and Φ_y represent the cracking and yield curvatures while Φ_{PA}, Φ_{PB}, Φ_{PC}, and Φ_{PD} represent the permanent curvature-set on unloading corresponding to the pre-loading levels A, B, C, and D.

DISCUSSION OF TEST RESULTS

Typical load-deformation behaviour of two SFRC test beams B1 and B2 is shown in Fig.3. Load in the present context means the total load applied on the beam and the deflection represents the differential deflection, 'δ' represented by

$$\delta = \delta_2 - \left(\frac{\delta_1 + \delta_3}{2}\right) \qquad \text{... (2)}$$

where δ_1, δ_2, δ_3, are the deflections recorded in dial 1, 2, and 3 shown in Fig. 1.

Refering to Fig.1, the region PQ in the test beam represents the constant moment zone. The curvature 'Φ' can be evaluated using the relation

$$\Phi = [\, 8\delta /(PQ)^2 \,] \qquad \text{... (3)}$$

Fig.4 and 5 show the variation of total deflection with load for test beams B1 to B5. The failure load for test beams B1, B2, and B3 was between 5.0 and 5.25 tonnes. Assuming that the loads are concentrated, the failure moments lie in the range of 1.83 to 1.92 tm. The peak force that can be resisted by the 2 nos. of 10 mm diameter HSD bars is only

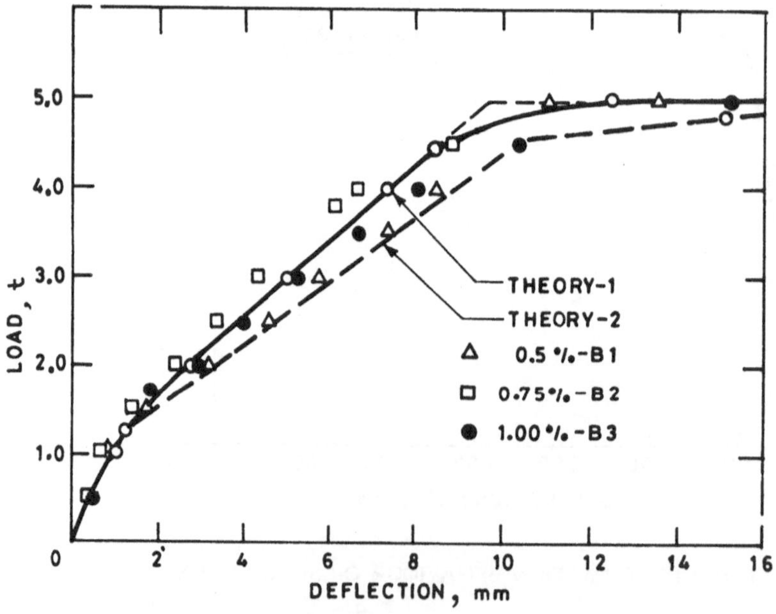

FIG.4 LOAD Vs TOTAL DEFLECTION DIAGRAM FOR SFRC BEAMS

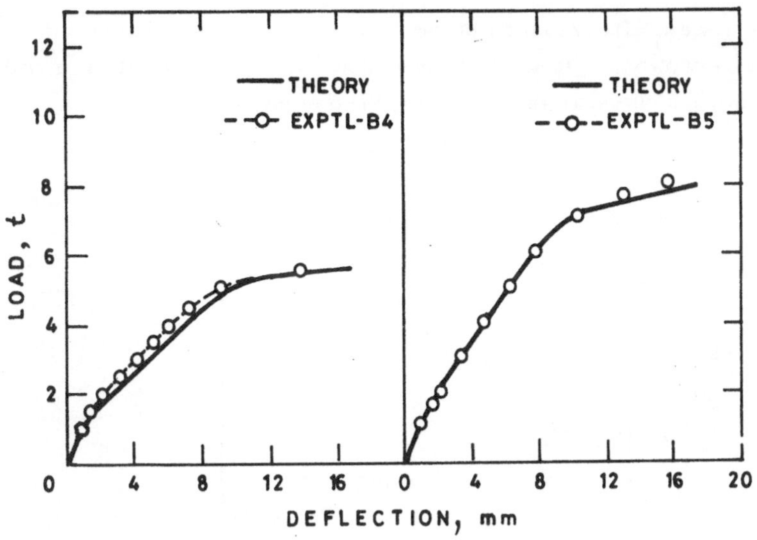

FIG.5 LOAD Vs TOTAL DEFLECTION DIAGRAM FOR SFRC BEAMS

10.4 t, and this leads to a moment capacity of only 1.45 tm. This indicates that an increase in the capacity of the beam to the tune of about 30% is possible by including a small percentage of fibres in the beams. This does not seem practical. Hence, a rational interpretation for the increase in the moment capacity of the SFRC beams could be that, eventhough the load was applied on the beam as two point-concentrated loads, the loads were actually getting distributed. Assuming that the peak moment capacity of the beams is 1.45 tm, the effective width of dispersion works out to 1.80 m. This indicates that the inclusion of steel fibres leads to a very good dispersion of concentrated loads. Normally, it can be assumed that the loads are dispersed at an inclination of one horizontal to one vertical upto the level of tension steel. However, in the case of SFRC beams having a volume fraction of fibres of the order of 1%, the load dispersion (distribution) is around three horizontal to one vertical. Assuming that the above dispersion is possible, the theoretical load-deflection variation is plotted and indicated as 'theory-1' in Fig. 4 and 5 in which the experimental load-deflection variation obtained for test beams B3, B4, and B5 have also been shown. It may be seen that the agreement between the theory and experiment is good. Assuming that the loads are concentrated, the theoretical load-deflection variation is shown by theoretical line-2 plotted in Fig.4. All the deflections obtained in the tests on SFRC beams are found to be lower than the values represented by this line which again goes to prove that the reduction in deflection under load in SFRC beams is due to the ability of the fibres in causing dispersion of any concentrated load applied on them, thereby reducing the magnitude of the applied moment in the constant moment zone.

As explained earlier, there were a number of stages of loading and unloading during the course of a test on a SFRC beam. The loads were successively released from prefixed, pre-loading stages. The variation of the permanent set in the beams with the pre-loading curvature is shown in Fig.6. It is seen that the permanent set is around $0.23\bar{\Phi}$ where $\bar{\Phi}$ is the maximum curvature reached during the pre-loading stage.

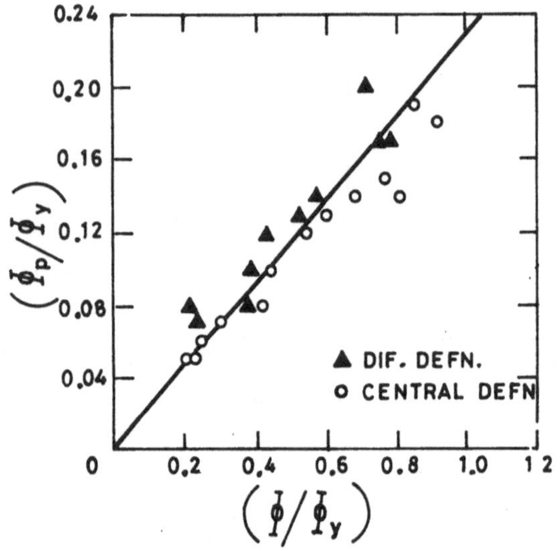

FIG.6 PERMANENT SET ON SFRC BEAMS

The static stiffness can be assumed as the reloading stiffness. Defining 'C' as the ratio of the static stiffness at any particular stage to the pre-cracking stiffness of the SFRC beam, Fig.7 and 8 show the variation of C with M/M_u. The theoretical value of

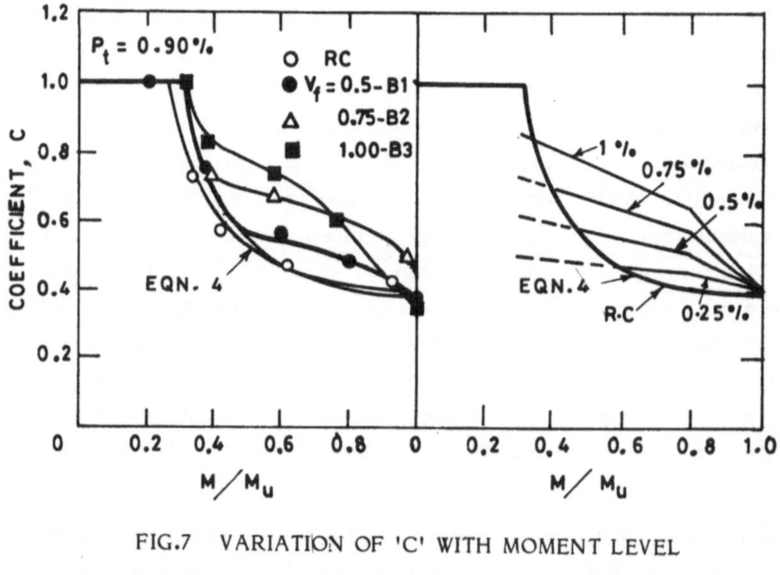

FIG.7 VARIATION OF 'C' WITH MOMENT LEVEL

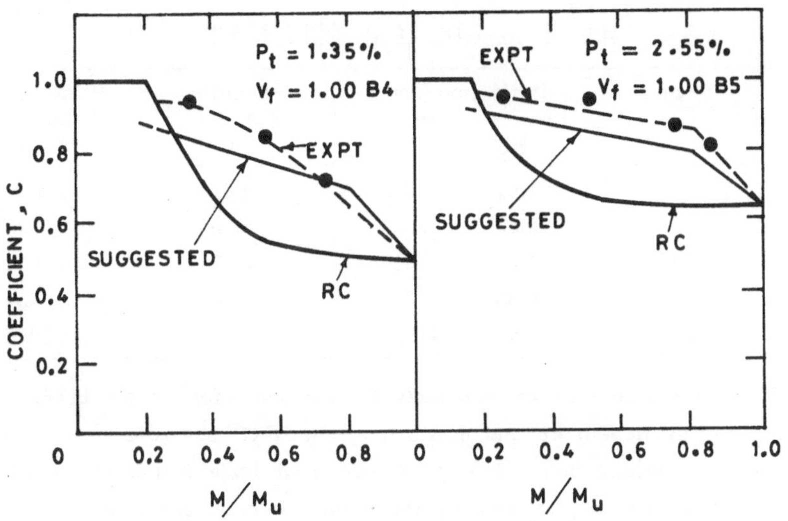

FIG.8 VARIATION OF 'C' WITH MOMENT LEVEL

C for reinforced concrete beams is given by the relation

$$C = \left[\left(\frac{M_{cr}}{M} \right)^3 + \left\{ 1 - \left(\frac{M_{cr}}{M} \right)^3 \right\} \left(I_{cr}/ I_g \right) \right] \quad \ldots \quad 4a$$

When $M = M_u$, the static stiffness 'C_u' is given by

$$C_u = \left[\left(\frac{M_{cr}}{M_u} \right)^3 + \left\{ 1 - \left(\frac{M_{cr}}{M_u} \right)^3 \right\} \left(I_{cr} / I_g \right) \right] \quad \ldots \quad 4b$$

where C is a constant, M_{cr} is the cracking moment, M is the moment at any particular stage, M_u is the moment at ultimate stage, I_{cr} is the cracked moment of inertia of beam section, and I_g is the gross moment of inertia of beam section. It can be easily seen from Fig.7 and 8 that the introduction of steel fibres improves the static stiffness. This improvement is more for SFRC beams having higher volume percentage of steel fibres. Assuming that C_u is given by the ratio (I_{cr}/I_g), the following equations can be empirically derived for $0 < V_f \leq 1.0$ from the experimental variation of C vs M/M_u plotted in Fig.7 and 8.

$$\left. \begin{aligned} C &= C_u (1 - 0.75 V_f) + 0.75 V_f \leq 1.00, && \text{at } M = M_{cr} \\ &= 0.42 V_f + C_u (1 - 0.42 V_f) && \text{at } M/M_u = 0.80 \\ &= C_u && \text{at } M/M_u = 1.00 \end{aligned} \right] \quad \ldots \quad (5)$$

The variation of C with M/M_u as derived using Eq.5 is also shown in Fig.7 and 8.

Analysis of test data on strain measurements is not being presented in this paper. The SFRC beams tested withstood full loading even beyond the stage of crushing of concrete. Unlike reinforced concrete and laced reinforced concrete beams, large-scale spalling of compression concrete was not observed in SFRC beams. Table 2 gives the yield deflection, failure deflection, failure rotation at the plastic hinge, and the ductility ratio of all the tested SFRC beams. The rotation of SFRC beams at the plastic hinge was found to vary between 6.65° to 12.5°. It is seen that the plastic hinge rotation increases with the increasing diameter of the main steel bars. As all the failures were caused by the rupture of tensile

TABLE 2 DUCTILITY OF SFRC BEAMS

Serial Number	Beam	Deflection (cm)		Ductility Ratio	Plastic Hinge Rotation (Degrees)
		δ_y	δ_u	μ	
1	B1	1.04	6.01	5.87	6.76°
2	B2	0.87	6.0	6.90	6.65°
3	B3	0.85	7.3	8.59	8.09°
4	B4	0.85	7.0	8.24	7.76°
5	B5	0.90	11.3	12.56	12.52°

steel, the failure curvature may approximately be the same for all the beams. Hence, it is apparent that the length of the plastic hinge in SFRC beams is influenced by the diameter of the main tensile bars. It can be seen from Table 2 that the ductility ratio increases in proportion to the diameter of the main reinforcement used which indicates that increased bond is the critical factor that reduces the length of the plastic hinge. It can also be concluded that, for a given percentage of steel, use of smaller diameter bars will lead to lower ductility. The ductility is comparatively higher in beam B5 in which 16 mm bars were used. As the concrete in the compression zone did not disintegrate, the chances of buckling of the compression bars, which was an observed mode of failure in reinforced and laced reinforced concrete beams during earlier tests carried out at Structural Engineering Research Centre, Madras, does not appear feasible in SFRC beams. Also, the increased shear resistance of SFRC beams can also be advantageously used in design.

CONCLUSIONS

The suitability of SFRC beams reinforced with equal amounts of tension and compression steel for use in blast-resistant design of structures is being studied as part of a long range research programme. Based on the investigations conducted on five SFRC beams reported in this paper, the following major conclusions can be drawn.

1. SFRC beams give far higher load dispersion for concentrated loads. The dispersion can be taken as three horizontal to one vertical upto the level of tensile steel for the volume fraction of fibres reported in the investigation.

2. The static stiffness on the reloading path can be obtained using the relation

$$(EI)_{eff} = C\,(EI)$$

where C is as defined in Eq.(5), and EI is the uncracked stiffness.

3. For the range of main percentage reinforcement studied (upto 2.5%), the failure of SFRC beams is by rupture of tensile steel.

4. The failure ductility and rotation at the plastic hinge of SFRC beams have been found to be significantly different from normal reinforced concrete beams. The ductility can be improved by using larger diameter bars.

5. The commonly observed failure by buckling of compression steel at higher percentages of reinforcement in RC and LRC beams is not observed in SFRC beams which seems to be a decisive advantage.

6. At the present state of knowledge, SFRC seems to be ideally suited for blast-resistant design of structures. However, care must be taken to increase the length of the plastic hinge as much as possible by proper detailing. Further studies are necessary in this direction before concrete recommendations can be made.

ACKNOWLEDGEMENT

This paper is being published with the permission of Director, Structural Engineering Research Centre, Madras.

REFERENCES

1. Swamy, R.N. and Stavrides, H., Flexural behaviour of fibre concrete with conventional steel reinforcement, RILEM Symposium, 1975, pp 187-196.

2. Parameswaran, V.S., Lakshmanan, N., Srinivasulu, P., Krishnamoorthy, T.S., Balasubramanian, K., Thandavamoorthy, T.S. and Arumugam, M., 'Application of laced reinforced concrete construction techniques to blast resistant structures', Report No. RCC-SR-86-1, October 1986.

3. Al Taan, S.A., and Al-Feel, J.R., Role of steel fibre reinforcement in flexure failure, RILEM Symposium - FRC 86, July 1986.

4. Srinivasulu, P., Lakshmanan, N., Muthumani, K., and Sarma, B.S., Dynamic behaviour of fibre reinforced concrete beam elements, paper presented at the International Symposium on Fibre Reinforced Concrete, Madras, December 16-19, 1987.

Proceedings of the International Symposium on Fibre Reinforced Concrete, December 16-19, 1987, Madras, India

ULTIMATE STRENGTH OF STEEL FIBRE REINFORCED CONCRETE DEEP BEAMS

Indradatta I Pandya and Shreeram K Damle
Department of Applied Mechanics, Faculty of Technology and Engineering, M.S. University of Baroda, Baroda, India.

This paper describes the experimental investigations carried out on fifteen deep beams consisting of plain and fibre reinforced concrete. The beams contained conventional tensile steel reinforcement and web reinforcement. The first cracking load, crack patterns, ultimate load and modes of failure are reported. On the basis of these investigations, empirical formulae are proposed for estimating the first cracking load and ultimate shear strength of beams. The addition of steel fibres in reinforced concrete resulted in a significant increase in the ultimate shearing capacity of beams.

INTRODUCTION

A reinforced concrete beam can be defined as a deep beam whose span to depth ratio is smaller in comparison to the span to depth ratio of beams encountered in normal beam design. The Indian Standard Code [1] specifies that the beams with span to depth ratios less than 2.5 for continuous spans or less than 2.0 for simple spans shall be designed as deep beams taking into account the nonlinear distribution of stress, lateral buckling and other pertinent effects. The stress distribution in deep beams is different from that given by the simple bending theory of beams because of the presence of vertical stresses due to support reactions, normal stresses due to applied loads and shearing deformations.

The behaviour of reinforced concrete deep beams has been studied both analytically and experimentally by numerous investigators during the last few decades [2, 3, 4, 5]. The investigations covered aspects such as crack patterns and modes of failure, diagonal cracking load, ultimate load, etc. Recent experimental investigations have shown that the shear strength of deep beams could be significantly inreased by the addition of steel fibres [6, 7, 8]. The main objective of this paper is to study experimentally the behaviour of steel fibre reinforced concrete deep beams with conventional te nsile and web reinforcement and based on the test results to propose empirical formulae for diagonal cracking load and ultimate shear load.

EXPERIMENTAL INVESTIGATION

Test Specimens : Fifteen steel fibre reinforced concrete deep beams, simply supported on a span of 600 mm, were tested under two point loading. The width of the web in all the beams was kept (75 mm) constant. The beams were divided into five series. Each series consisted of three beams with span to depth ratios of 1, 1.2 and 1.5 respectively. The first series of beams (P series) was of plain cement concrete, while the second series (F series) was concrete with steel fibres. The last three series of beams comprised of conventional longitudinal tensile reinforcement (FS series), horizontal web reinforcement (FSH series) and horizontal and vertical web reinforcements (FSHV series) respectively.

The main longitudinal tensile reinforcement consisted of one 16 mm nominal diameter round mild steel bar. The web reinforcement was of 10 mm nominal diameter round mild steel bars. All the joints of reinforcement were provided with spot welding before positioning into the mould. In order to prevent premature bond and anchorage failure at support region, the main bar of each beam was anchored to steel plates at the ends with nut and washer arrangement. The support and loading points of the deep beams were adequately reinforced to prevent failure due to local crushing of concrete. Details of test specimens, the arrangement of reinforcement and support details are shown in Figure-1.

The steel fibres used in this study were having a mean diamter of 1.2 mm and an aspect ratio of 80. Steel fibres of one percent by volume were used in all the beams. The cement used was ordinary portland cement. Ordinary river sand and crushed basalt were used as fine and coarse aggregates respectively. A concrete mix of 1:2:4 by weight with a water/cement ratio of 0.55 was used for all the beams. Three 150 mm cubes and three 150 mm x 300 mm cylinders were made from each mix.

Testing : All the beams were tested on a specially provided bending bench of the WPM Compression Strength Testing Machine for building materials under two point top loading applied at the third span points until they no longer sustained any further loading. Midspan deflection was recorded with the help of dial gauge at an interval of 10 kN load throughout the testing. Crack progressions were traced in pencil and the end points were marked to correspond with the loading stage at which they appeared or progressed further until the complete failure of the beam was observed.

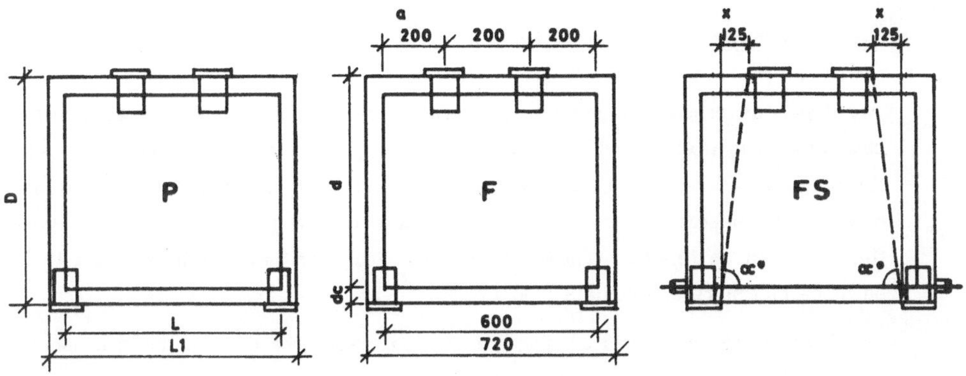

SECTIONAL FRONT VIEW OF ALL 5 SERIES

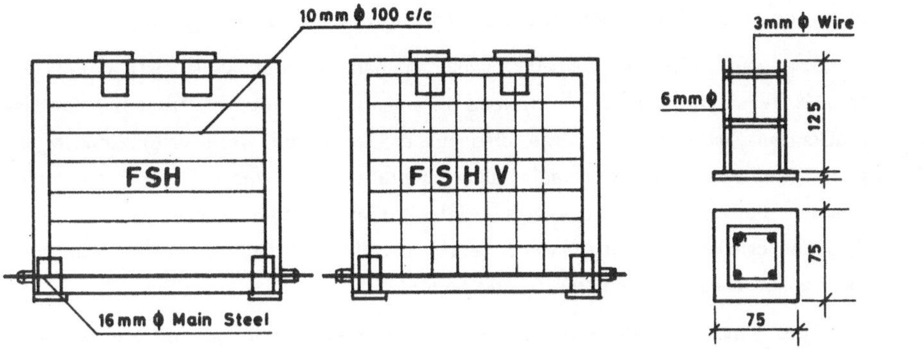

ALL DIMENSION ARE IN MM

OTHER PROPERTIES OF BEAMS

L	D	d	dc	B	a	X	X/D	L/D
600	600	560	40	75	200	125	0.208	1.0
600	500	460	40	75	200	125	0.250	1.2
600	400	360	40	75	200	125	0.312	1.5

Figure 1.

TYPICAL DETAILS OF SIMPLY SUPPORTED DEEP BEAMS

TEST RESULTS

Crack Patterns and Modes of Failure : The crack patterns of all the beams tested in the present investigation are shown in Figure 2. In all the beams two types of cracks namely flexural cracks and diagonal tension cracks were observed. Flexural cracks developed in the region of maximum moment. All the flexural cracks were observed in the lower half depth of the beams. In most of the beams (except P series) the width of the flexural cracks was less than 0.1 mm at failure. These cracks were found to have little effect either on mode of failure or ultimate load.

There were two types of diagonal tension cracks observed in the test specimens. The first type of diagonal tension cracks originated from the inner edge of the reaction bearing plate to the outer edge of the loading plate. In beams of F series without web reinforcement, the appearance of this type of crack was gradual. These cracks were either immediate cause of failure of the beam or they brought the beam to its eventual collapse.

The other type of diagonal cracks opened up first at a distance of D/2 to D/4 from the soffit. With load inrements, the rate of progress of these cracks was as gradual as the diagonal cracks which were originated in the vicinity of inner edge of loading bearing plate. Any web reinforcement which crossed the diagonal cracks was helpful to arrest their propogations. These diagonal cracks so formed were nearly parallel to each other with a "strut like" appearance between the loading points.

The complete failure of the beams was observed in one of the following ways :

1. All plain concrete beams collapsed by flexure with a flexural crack near the midspan. This type of failure was observed in beams of P series.
2. The diagonal tension failure, observed in all the beams of F and FS series was indicated by splitting the beam in a direction of a line nearly joining the inner edge of the reation bearing plate at the support to the outer edge of the loading plate.
3. The shear compression failure was indicated by crushing of the "strut like" portion of conrete between two adjacent parallel diagonal cracks accompanied by splitting of concrete along the plane of the diagonal cracks and followed by heavy crushing and bursting in the web. This type of failure was qbserved in beams of FSH and FSHV series.

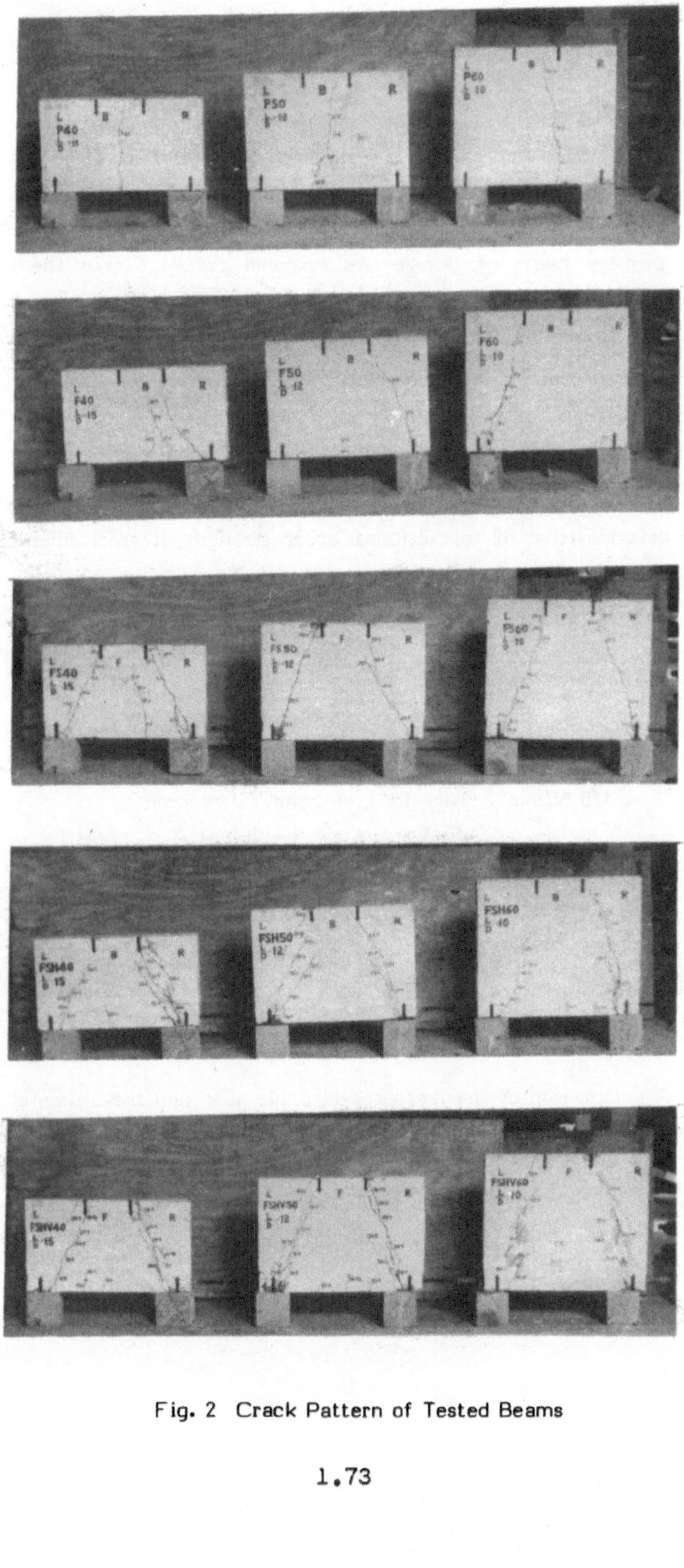

Fig. 2 Crack Pattern of Tested Beams

Diagonal Cracking Load and Ultimate Load : For P and F series of beams, the loads which cause the formation of the first flexural crak and/or the first diagonal crack is considered to be the ultimate load. However, for deep beams, after the development of diagonal cracks, the beam can support considerable additional load before failure depending upon the effectiveness of the main longitudinal tensile reinforcement and web reinforcement which could arrest the growth of diagonal cracks. It was observed from the test results that the primary cause of failure was diagonal cracks. Thus the ultimate (shear strength) load of the beam is mainly dependent on the diagonal cracking load and the subsequent reserve of strength beyond this load due to the presence of longitudinal and web reinforcement.

PROPOSED EMPIRICAL FORMULAE

On the basis of the test results of the deep beams in this study and the formula presented by P.J. Robins [9], the following formula is proposed for the determination of the ultimate shear strength of steel fibre reinforced concrete deep beams :

$$Vu1 = C_1 f_t bD + C_2 \frac{X}{D} f_t bD + C_3 \sum A \frac{Y}{D} \sin^2\alpha \quad \ldots\ldots (1)$$

where f_t is the concrete split cylinder tensile strength.
The coefficients C_1, C_2 and C_3 are determined from the test results by using the method of least squares. They were found to be $C_1 = 1.84$, $C_2 = -1.26$ and $C_3 = 320\ N/mm^2$. Therefore, equation (1) becomes

$$Vu1 = 0.6\left(1 + 2.2 \frac{X}{D}\right) f_t bD + 320 \sum A \frac{Y}{D} \sin^2\alpha \quad \ldots\ldots (2)$$

$$= \frac{Wu1}{2} \text{ for two point loading.}$$

In equation (2), the term $\sum A \frac{Y}{D} \sin^2\alpha$ should be summed up for all web bars including the main longitudinal tensile reinforcement.

A comparison of the measured ultimate load and the ultimate load computed in accordance with equation (2) is given in Table 1. It can be observed that the suggested equation predicts ultimate load for both the diagonal tension failure and shear compression failure satisfactorily.

The P series beams which failed in flexure were analysed by using the flexure formula given in reference [10].

TABLE 1

MEASURED AND COMPUTED DIAGONAL CRACKING LOADS AND ULTIMATE LOADS

Beam Mark	Diagonal Cracking Loads			Ultimate Loads					Type of Failure
	Measured Wc kN	Computed Wc1 kN (Eq. 4)	$\frac{Wc1}{Wc}$	Measured Wu kN	Computed Wu1 kN (Eq. 2)	Computed Wu2 kN (Eq. 3)	$\frac{Wu1}{Wu}$	$\frac{Wu2}{Wu}$	
P 40 *	72.5	65.05	0:907	-	-	-	-	-	Flexure
P 50 *	120.0	101.6	0.850	-	-	-	-	-	Flexure
P 60 *	145.0	146.3	1.008	-	-	-	-	-	Flexure
F 40	130.0	130.10	1.0	130.0	130.10	151.0	1.0	1.16	Direct Tension
F 50	175.0	149.20	0.852	175.0	149.20	202.0	0.852	1.15	Direct Tension
F 60	170.0	168.50	0.991	170.0	168.50	252.5	0.991	1.32	Direct Tension
FS 40	170.0	145.8	0.857	230.0	251.3	245.9	1.092	1.069	Direct Tension
FS 50	200.0	167.4	0.837	280.0	282.4	314.0	1.008	1.12	Direct Tension
FS 60	220.0	289.0	1.31	305.0	328.2	371.2	1.076	1.21	Direct Tension
FSH 40	190.0	231.4	1.218	390.0	391.4	394.0	1.005	1.01	Shear Compression
FSH 50	250.0	265.1	1.060	517.5	455.3	515.9	0.879	0.996	Shear Compression
FSH 60	260.0	300.0	1.153	580.0	518.7	633.7	0.894	1.092	Shear Compression
FSHV 40	190.0	171.5	0.902	307.5	331.5	324.3	1.078	1.054	Shear Compression
FSHV 50	180.0	196.9	1.093	367.5	387.1	422.7	1.053	1.14	Shear Compression
FSHV 60	200.0	222.28	1.111	427.5	440.8	517.2	1.031	1.20	Shear Compression

* Remarks : Beams failed by flexure. Failure load calculated by the flexure formula given in Reference [10]. For P series beams ultimate load is the same as the first cracking load.

Another formula, as proposed by S.N. Patel [10] and given in equation (3) below, gives the ultimate shear capacity of concrete deep beams provided the value of f_t is known. This formula was developed on the basis of splitting of an elliptical section whose major axis lies on a line joining the load point and support. The formula is simple and straight forward. There is no necessity of assuming values of various constants as is usually done in various other known formulae. The ultimate shear capacity is given by

$$Vu2 = \frac{1.5\ f_t\ bd}{\sqrt{1 + 0.75\ (a/d)^2}} + \frac{f_y}{2\sqrt{1 + (a/d)^2}} \sum A\ (y/d + 0.4)\ \sin\alpha \quad \text{.... (3)}$$

$$= \frac{Wu2}{2} \text{ for two point loading.}$$

where f_y is the yield stress of reinforcement.

The ultimate load for all the beams (except P series) is also computed by using equation (3) and recorded in Table 1. The values so obtained show good agreement with the observed ultimate values.

Web reinforcement has a considerable effect on ultimate stength of deep beams; but its effectiveness in resisting shear can be considered only after the formation of the first diagonal crack. Based on this fact, the above equation (2) can be used for assessing the first diagonal cracking load of deep beams assuming $f_y = 0$ as follows :

$$Vc1 = 0.6\ (1 + 2.2\ \frac{X}{D})\ f_t\ bD \quad \text{..... (4)}$$

The diagonal cracking loads for all the beams based on equation (4) are calculated as shown in Table 1. The values so obtained show good agreement with observed diagonal cracking loads.

DISCUSSION AND CONCLUSIONS

The study of crack patterns of beams shows that the diagonal tensile cracks were always developed in shear span. Initially, a few flexural cracks were also developed in the pure moment zone. For a span/depth ratio of 1.5, the plain concrete beam failed at a load of 72.5 kN while the failure load of fibre reinforced concrete beam was 130.0 kN showing an increase of 80 percent. The corresponding failure loads for span/depth ratio of 1.2 were 120.0 kN and 175.0 kN respectively and hence an increase of 45.8 percent. Failure loads for span/depth ratio of 1.0 were 145.0 kN and 170.0 kN respectively, the increase being 32 percent. Thus, addition of steel fibres in concrete deep beams resulted in significant increase of ultimate load capacity of the beams.

A study of load deflection curves shows linear variation in the initial stages of loading which later become nonlinear. The larger area under the load deflection curves, especially for fibre reinforced concrete deep beams, indicates a more ductile type behaviour.

Inclusion of steel fibres in deep beams increases the ultimate flexural and/or shear capacity of beams. The type of failure noticed was more of ductile type, whilst it was sudden in case of plain concrete beams. Horizontal web reinforcement has little effect on the diagonal cracking load; but after the formation of diagonal cracks, it was effective in increasing the ultimate shear capacity of the beams. Equation (2) proposed in the present investigation can be suitably applied to estimate the ultimate shear strength of fibre reinforced concrete deep beams for two point loading having span/depth ratios in the range of 1 to 1.5 and X/D ratios in the range of 0.2 to 0.3. The first diagonal cracking load can be computed by equation (4). Equation (3) is quite general and can predict satisfactorily the ultimate strength of fibre reinforced concrete deep beams. Calculations of diagonal cracking load and ultimate load, on the basis of formulae proposed, give values which are about ± 15 percent of the observed values.

ACKNOWLEDGEMENT

This study was carried out in the Department of Applied Mechanics, Faculty of Technology & Engineering, M.S. University of Baroda, Baroda, India. The authors thank the authorities in providing necessary help and laboratory facilities.

REFERENCES

1. Indian Standard Code of Practice for Plain and Reinforced Concrete, IS:456 -1978. Indian Standard Institution, New Delhi, 1978.

2. Rawdon De Paiva, H.A. and Siess, C.P. Strength and behaviour of deep beams in shear, Journal of the Structural Division, ST 5, American Society of Civil Engineers, October 1965, Vol 91, pp 19-41.

3. Ramakrishnan, V. and Ananthanarayana, Y. Ultimate strength of deep beams in shear, Journal of the American Concrete Institute, February 1968, Vol 65, pp 87-98.

4. Kong, F.K., Robins, P.J., Singh, A. and Sharp, G.R. Shear analysis and design of reinforced concrete deep beams, The Structural Engineer, October 1972, Vol 50, No 10, pp 405-409.

5. Joint ASCE-ACI Task Committee on Shear and Diagonal Tension. The shear strength of reinforced concrete members, Journal of the Structural Division, ST 6, American Society of Civil Engineers, June 1973, Vol 99, pp 1091-1187.

6. Batson, G., Jenkins, E. and Spatney, R. Steel fibres as shear reinforcement in beams, Journal of the American Conrete Institute, October 1972, Vol 69, No 10, pp 640-644.

7. Roberts, T.M. and Ho, N.L. Shear failure of deep fibre reinforced concrete beams, The International Journal of Cement Composites and Light Weight Concrete, August 1982, Vol 4, No 3, pp 145-152.

8. Shanmugam, N.E. and Swaddiwudhipong, S. The ultimate load behaviour of fibre reinforced concrete deep beams, The Indian Concrete Journal, August 1984, Vol 58, No 8, pp 207-211.

9. Robins, P.J. Reinforced concrete deep beams studied experimentally and by the finite element method. Ph.D. Thesis, University of Nottingham, October 1971.

10. Patel, S.N. Behaviour of reinforced concrete deep beams in shear and flexure. Ph.D. Thesis, M.S. University of Baroda, December 1976.

Proceedings of the International Symposium on Fibre Reinforced Concrete, December 16–19, 1987, Madras, India

FLEXURAL BEHAVIOUR OF REINFORCED STEEL FIBRE CONCRETE BEAMS

P. Paramasivam T.Y. Lim S.L. Lee
Department of Civil Engineering
National University of Singapore
Singapore

SYNOPSIS: Idealised stress-strain curves are given for steel fibre concrete (SFC) and reinforcing steel and simplified analytical expressions are derived for the moment-curvature ($M-\phi$) and load-deflection ($P-\delta$) behaviour of simply supported reinforced steel fibre concrete (RSFC) beams in flexure. Analytically predicted $M-\phi$ and $P-\delta$ curves for test beams were found to agree well with the experimental data. The RSFC beams showed higher flexural strength and curvature ductility at ultimate load when compared to similarly reinforced plain concrete beams. The approach is a useful tool in the analytical study and design of RSFC members.

INTRODUCTION

Various methods had been proposed to calculate the ultimate flexural strength of Reinforced Steel-Fibre-Concrete (RSFC) beams [1,2,3]. Basically, the methods applied the usual sectional analysis, using fully plasticised stress blocks and forces. To account for fibre resistance in tension, an idealised tensile stress block was added. The magnitude of this stress block, although expressed in differing formulae, was usually based on the pullout bond-stress, aspect ratio and volume fraction of the fibres. Fibre contribution to strength and ductility in compression were neglected.

Investigations [1,2] reported in the literature have consistently showed that RSFC beams have better deformational characteristics than similarly reinforced plain concrete beams. There are however no convenient analytical model whereby this may be evaluated. Such a model is important in the design of such members under serviceability requirements.

In this paper, explicit expressions are given for the idealised tensile and compressive stress-strain behaviour of Steel-Fibre Concrete (SFC), taking into account fibre contribution in both. Together with a standard idealised stress-strain curve for the reinforcing steel, a simplified analytical moment-curvature ($M-\phi$) relationship for beams with rectangular cross section is proposed. Analytically predicted $M-\phi$ and load-deflection ($P-\delta$) curves are compared with experimental data from simply supported test beams and good agreement is found. An example problem is also presented.

STRESS-STRAIN RELATIONSHIPS

Stress-Strain Curve of SFC in Compression

The stress-strain curve of SFC in compression can be idealised as in Fig. 1. The composite compressive modulus of elasticity, E_{cc} is given by [4,5]

$$E_{cc} = E_{mc}V_m + n_\ell \eta_o E_f V_f \tag{1}$$

where E_{mc} is the modulus of elasticity of the concrete matrix in compression, V_m the matrix volume fraction, η_ℓ the length efficiency factor, η_o the orientation factor, E_f the modulus of elasticity of the fibres and V_f the fibre volume fraction.

η_ℓ accounts for the varying fibre stress at the end portions of the fibres and is given by [6,7]

$$\eta_\ell = \begin{cases} 0.5 & \text{for } \ell_f < \ell_c \\ 1 - \dfrac{\ell_c}{2\ell_f} & \text{for } \ell_f > \ell_c \end{cases} \tag{2}$$

where ℓ_f is the fibre length and ℓ_c is the critical length required to develop the ultimate fibre stress when a uniform ultimate bond stress, τ_u between the fibre and matrix is assumed [4,6,7].

The orientation factor, η_o is a function of the depth h and width b of the composite section and the fibre length ℓ_f [4,6] and is expressed as

$$\eta_o = \frac{\int_o^{\bar{\rho}} \int_o^{\bar{\theta}} \cos^4\theta \cos^4\rho d\theta d\rho}{\int_o^{\bar{\rho}} \int_o^{\bar{\theta}} d\theta d\rho} \tag{3}$$

where $\bar{\theta} = \sin^{-1}(h/\ell_f) \leqslant \pi/2$; $\bar{\rho} = \sin^{-1}(b/\ell_f) \leqslant \pi/2$.

The failure stress in compression for plain concrete is taken as 0.85 σ_{cu}, where σ_{cu} is the cylinder strength [8]. For SFC the failure stress is 0.9 σ_{cu}. This accounts for the greater ductility and slight increases in strength due to the addition of fibres [9,10]. As suggested by Swamy [11], the failure strain for SFC in compression is modified to be 0.004 in contrast to the 0.003 [8] assumed for plain concrete.

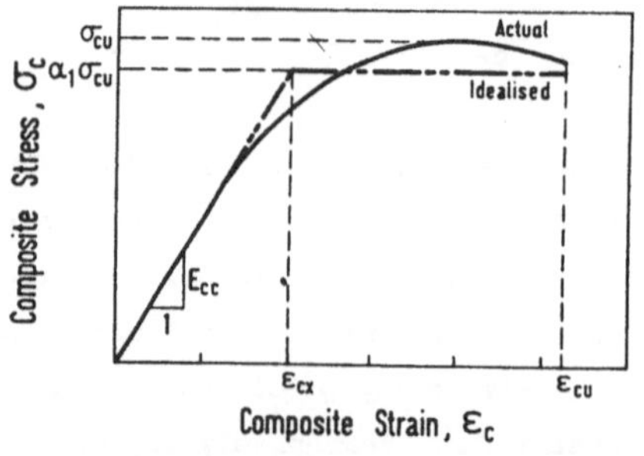

Fig. 1 Stress-Strain Curve (Compression)

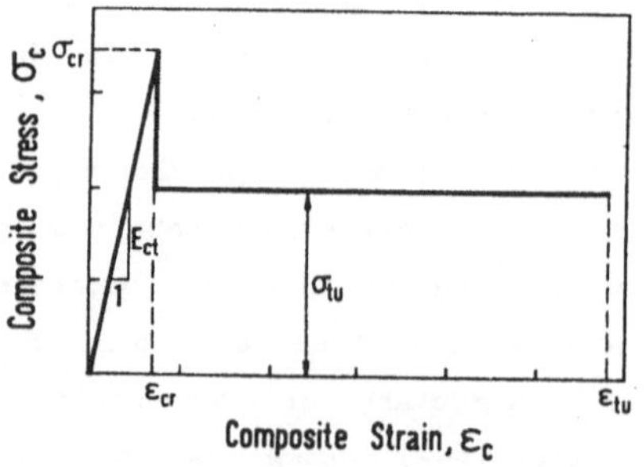

Fig. 2 Stress-strain Curve (Tension)

Stress-Strain Curve of SFC in Tension

The idealised tensile stress-strain curve of SFC is shown in Fig. 2 [6]. The tensile modulus of elasticity, E_{ct} can be defined as [4,6]

$$E_{ct} = E_{mt}V_m + \eta_\ell \eta_o E_f V_f \tag{4}$$

where E_{mt} in this case is the modulus of elasticity of the concrete matrix in tension. The cracking strain, ε_{cr} is empirically related to the cracking strain of the matrix, ε_{mp} and the strain at proportionality limit, ε_{fp} of the fibres [7]

$$\varepsilon_{cr} = \eta_\ell \eta_o' V_f(\varepsilon_{fp} - \varepsilon_{mp}) + \varepsilon_{mp} \tag{5}$$

Eq. (5) indicates that the cracking strain equals that of the matrix when the fibre volume fraction is zero. As V_f increases, the cracking strain increases proportionately. In the flexure mode, ε_{mp} is taken as the strain of concrete matrix at initial flexural cracking.

The strength after cracking, σ_{tu} is governed by the ultimate pullout bond stress, τ_u of the fibre as [6]

$$\sigma_{tu} = \eta_\ell \eta_o' V_f \ell_f \frac{\tau_u}{2r} \tag{6}$$

where r is the ratio of the fibre cross-sectional area to its perimeter.

In Eqs. (5) and (6), the orientation factor is η_o' because at cracking, the fibres realign across the cracks. η_o' is given by

$$\eta_o' = \frac{\int_o^{\bar{\rho}} \int_o^{\bar{\rho}} \cos\theta \, \cos\rho \, d\theta d\rho}{\int_o^{\bar{\rho}} \int_o^{\bar{\theta}} d\theta d\rho} \tag{7}$$

which is the same as that proposed by Romualdi and Mandel [12].

The tensile load carried by SFC after cracking degenerates from a maximum at initiation of fibre pullout to half the value when the associated crack width reaches approximately $\frac{1}{8}\ell_f$ [6,13]. In considering the post-cracking load as constant and sustained at the maximum level (Fig. 2), a limit need to be placed on the maximum crack width allowed. A reasonable limit is for a crack width of $\frac{1}{16}\ell_f$, i.e. when the load sustained had dropped to approximately $\frac{3}{4}$ of the maximum. In view of this, ε_{tu} in Fig. 2 can be set equal to $\ell_f/(16\ell^*)$ where ℓ^* is the average crack spacing. Given the characteristic close crack spacing found on RSFC beams in flexure [1,2], it is unlikely that tensile strain of such magnitude could be reached prior to the beam attaining failure strain of ε_{cu} in compression.

Stress-Strain Curve of Reinforcing Steel

The idealised stress-strain curve for reinforcing steel given by the ACI code [8] is utilized (Fig. 3).

SIMPLIFIED MOMENT-CURVATURE RELATIONSHIP OF RSFC BEAMS

In the following analysis, only under-reinforced beams are considered. It is also assumed that the concrete (SFC or plain) is effective till the level of the main reinforcement only and there is no secondary reinforcement in compression.

The development of the sectional strain and stress profiles of a beam with nominal depth h and width b at various stages of loading is shown in Fig. 4. ϕ_{cr} is the cracking curvature when the extreme tensile strain reaches ε_{cr}. ϕ_{cr1} is the curvature when the extreme compressive strain reaches yield value, ε_{cx}. ϕ_{cr2} is the curvature when the strain in the reinforcing steel just reaches the yield value ε_{sy}. ϕ_u is the curvature at ultimate failure. By taking equilibrium in the section at each stage, the position of the neutral axis can be determined and the complete analytical M-ϕ relationship derived [14]. The analytical M-ϕ curves obtained in this way for actual test beams takes the form of Fig. 5(a). Note that the curve has two distinct segments, one rather steep portion prior to and another flat portion after the yielding of the main steel reinforcement at (ϕ_{cr2}, M_{cr2}).

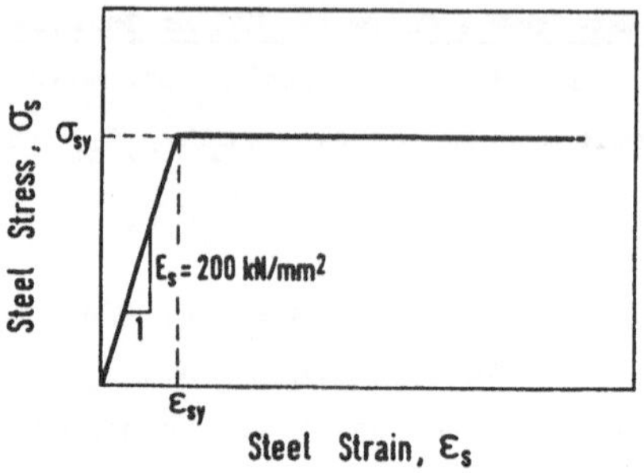

Fig. 3 Stress-Strain Curve (Steel)

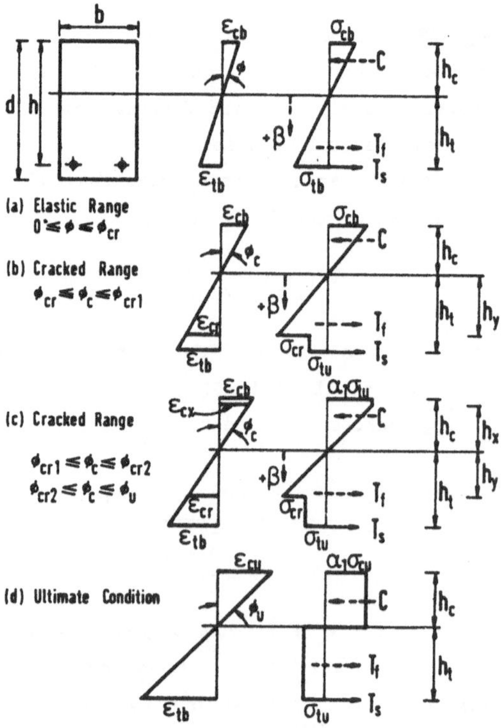

(α_1= 0.85 for plain concrete, 0.90 for SFC)

Fig. 4 Stress-Strain Profiles

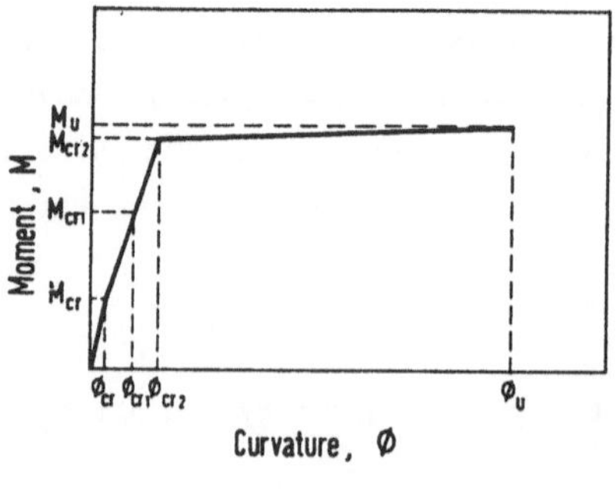

Fig. 5(a) Analytical M-ϕ Relationship

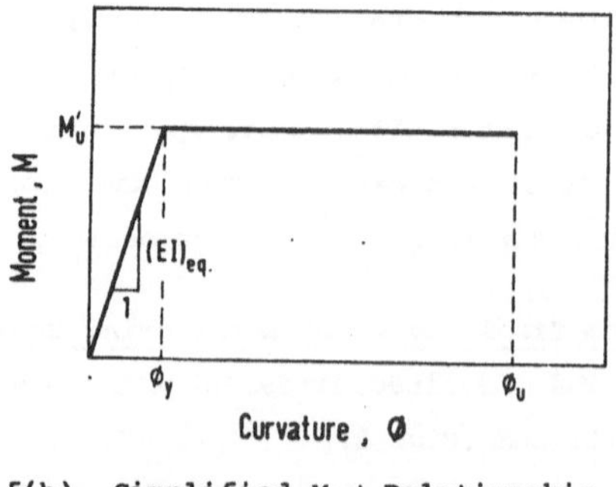

Fig. 5(b) Simplified M-ϕ Relationship

Simplified M-ϕ Relationship

For ease of analysis, the analytical M-ϕ relationship can be simplified as an elastic-perfectly plastic curve shown in Fig. 5(b). The slope of the elastic part in the simplified curve is defined by the equivalent flexural rigidity $(EI)_{eq}$ where

$$(EI)_{eq} = \frac{M_{cr2}}{\phi_{cr2}} \tag{8}$$

To find the quantities ϕ_{cr2} and M_{cr2}, first let the neutral axis be located by the parameter μ where

$$\mu = \frac{h_t}{h_c} \tag{9}$$

h_t and h_c being the depth of the tensile and compressive zone respectively. When the curvature is ϕ_{cr2}, by considering the equilibrium of forces as in Fig. 4(c), the value of μ is given by the expression

$$\mu = \frac{a_1}{a_2} \tag{10}$$

where $a_1 = 2\alpha_1 \sigma_{cu} \varepsilon_{sy} - \frac{2A_s E_s}{bh} \varepsilon_{sy}^2$

$$a_2 = 2\sigma_{tu}\varepsilon_{sy} + \frac{2A_s E_s}{bh} \varepsilon_{sy}^2 + \frac{(\alpha_1 \sigma_{cu})^2}{E_{cc}} - 2\sigma_{tu}\varepsilon_{cr} + E_{ct}\varepsilon_{cr}^2$$

ϕ_{cr2} is related to μ by

$$\phi_{cr2} = \varepsilon_{sy} \frac{(1+\mu)}{h\mu} \tag{11}$$

The moment M_{cr2} in this case is given by

$$M_{cr2} = b[\int_{-h_c}^{o} \phi_{cr2}E_{cc}\beta^2 d\beta - \int_{-h_c}^{-h_x} (\phi_{cr2}E_{cc}\beta^2 - \alpha_1 \sigma_{cu}\beta) d\beta + \int_{o}^{h_t} \phi_{cr2}E_{ct}\beta^2 d\beta$$
$$- \int_{h_y}^{h_t} (\phi_{cr2}E_{ct}\beta^2 - \sigma_{tu}\beta) d\beta] + A_s E_s \varepsilon_{sy} h_t \tag{12}$$

where β is the distance away from the neutral axis taken as positive downwards, A_s, E_s, ε_{sy} is the area, Young's modulus and yield strain of the reinforcement respectively, h_y is the depth of the tensile zone remaining elastic and h_x the depth of the compression zone remaining elastic. h_y and h_x can be related to h_t and h_c respectively by

$$\frac{h_y}{h_t} = \lambda = \frac{\varepsilon_{cr}(1+\mu)}{\phi_{cr2} h\mu} \tag{13}$$

$$\frac{h_x}{h_c} = \nu = \frac{\varepsilon_{cx}(1+\mu)}{\phi_{cr2} h} \tag{14}$$

where ε_{cx}, the compressive yield strain is equal to $\frac{\alpha_1 \sigma_{cu}}{E_{cc}}$

In view of Eqs. (9), (13), (14), Eq. (12) after integration becomes

$$M_{cr2} = \frac{b\phi_{cr2}h^3}{3(1+\mu)^3}[E_{cc}\nu^3 + E_{ct}\lambda^3\mu^3]$$

$$+ \frac{bh^2}{2(1+\mu)^2}[\alpha_1\sigma_{cu}(1-\nu^2) + \sigma_{tu}(1-\lambda^2)\mu^2 + \frac{2A_sE_s\phi_{cr2}\mu^2}{b}] \quad (15)$$

The value of M_{cr2} amd M_u for a rectangular section do not usually vary much and the plastic ultimate moment M_u can be taken as

$$M_u' = \frac{1}{2}(M_{cr2} + M_u) \quad (16)$$

which gives the yield curvature ϕ_y in the idealised curve as

$$\phi_y = \frac{M_u'}{(EI)_{eq}} \quad (17)$$

With little loss of accuracy, the ultimate condition can be represented by the stress blocks and strains as shown in Fig. 4(d), hence

$$M_u = \sigma_{tu}\frac{bh\ h_t}{2} + \frac{A_s\sigma_{sy}}{2}(h+h_t) \quad (18)$$

$$\phi_u = \frac{\varepsilon_{cu}}{(h-h_t)} \quad (19)$$

where σ_{sy} is the yield stress of the reinforcement and h_t from considering the equilibrium of the section is

$$h_t = \frac{\alpha_1\sigma_{cu}h - \frac{A_s\sigma_{sy}}{b}}{\alpha_1\sigma_{cu} + \sigma_{tu}} \quad (20)$$

A measure of the curvature ductility of the section is given by ζ, where

$$\zeta = \frac{\phi_u}{\phi_y} \quad (21)$$

LOAD-DEFLECTION RELATIONSHIP OF SIMPLY SUPPORTED RSFC BEAMS

Consider a simply supported beam of length L with two load points at equal distance αL from the ends. The curvature diagrams corresponding to the constant moment span attaining the curvature and moment (ϕ_y, M_u') and (ϕ_u, M_u') are given by Figs. 6(a) and (b). Midspan deflection can be determined by taking moment, about one support, of the area under each curvature diagram between the support and midspan.

The midspan deflection δ_y corresponding to Fig. 6(a) is given by

$$\delta_y = \phi_y L^2 [\frac{1}{8} - \frac{\alpha^2}{6}] \qquad (22)$$

The midspan deflection at ultimate condition, δ_u is

$$\delta_u = \phi_y L^2 [\frac{\alpha^2}{3} + \zeta(\frac{1}{8} - \frac{\alpha^2}{2})] \qquad (23)$$

In both instances, the load P_u is given by

$$P_u = \frac{2M_u'}{\alpha L} \qquad (24)$$

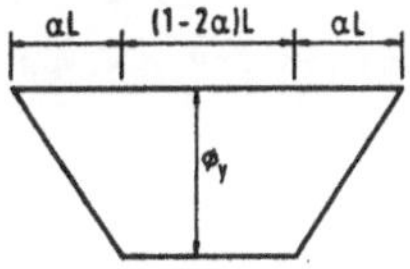

Fig. 6(a) Curvature Diagram (ϕ_y, M_u')

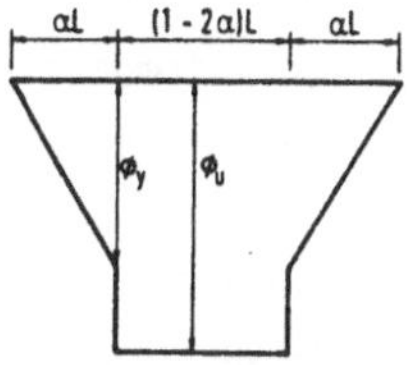

Fig. 6(b) Curvature Diagram (ϕ_u, M_u')

EXPERIMENTAL INVESTIGATION

To verify the present model, analytical predictions were made and compared to data from simply supported test beams having the cross sectional dimensions shown in Figs. 7(a) and 7(b). The beams were designated 2/0.0, 2/0.5, 2/1.0, 4/0.0, 4/0.5 and 4/1.0. The first number in the designation refers to the number of 16 mm diameter main reinforcement bars and the second to the volume fraction of fibres incorporated. All beams were 2.5 m in length and were tested over a simply supported span of 2.2 m with a shear span of 0.86 m. Beams with 4 bars have the bars placed in pairs so that the space between them does not obstruct movement of fibres when the mix is placed and consolidated.

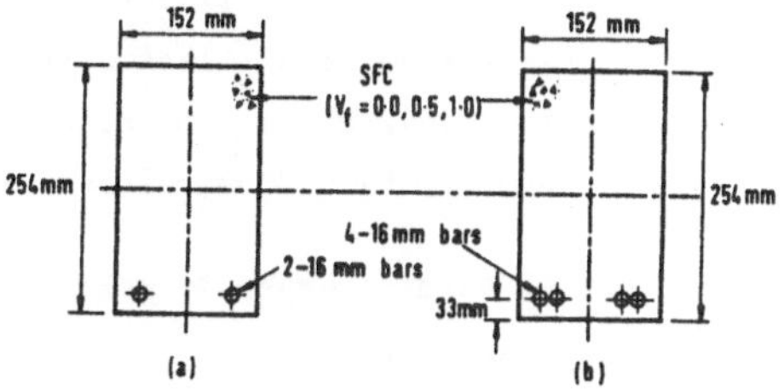

Fig. 7 Cross-Section of Beams

Mix Proportions and Preparation of Test Specimens

The mix proportions by weight of 1:1.5:2.5:0.5 cement:sand:coarse aggregate:water was used throughout. Ordinary Type I portland cement and washed natural sand falling within the Zone 1 grading requirements of BS 882 were used. The aggregate was crushed granite of 10 mm nominal size. An accelerator was added at a dosage of 1 litre per 22 kg of cement to obtain high early strength. Steel fibres were commercially available hook-ended fibres of length 30 mm and diameter 0.5 mm.

Cement, sand and 10 mm size aggregates were first mixed dry before the mix water including accelerator was introduced. For SFC mixes, the fibres which came in glued bundles were then slowly added and thoroughly mixed to ensure proper dispersion. Beams were cast in pairs in steel moulds, one with 2 and the other with 4 main reinforcement bars [Figs. 7(a) and 7(b)]. Control specimens for testing in compression and direct tension were also cast simultaneously with each pair of beams. All specimens were demoulded within 24 hrs and cured under damp hessian cloth for 4 days. They were then allowed to surface dry and prepared for testing on the 7th day.

Material Properties

Physical test properties of materials used are listed in Table 1. The mix and fibre parameters were obtained from an earlier small scale study reported elsewhere [5,6] where the same mix proportions and type of fibre was used.

Table 1: Material Properties from References 5 and 6

Steel Fibre Properties			
Length,	ℓ_f	mm	30.0
Diameter,	d	mm	0.50
Modulus of Elasticity,	E_f	kN/mm^2	200
Strain at Proportionality Limit,	ε_{fp}	$\mu\varepsilon$	5650
Ave. Ultimate Bond Stress,	τ_u	N/mm^2	6.90
Concrete Matrix Properties (7 days)			
Mix Proportions 1:1.5:2.5:0.5 Cement:Sand:Coarse aggregate:Water			
Ave. Modulus of Elasticity (Ten.),	E_{mt}	kN/mm^2	25.3
Ave. Modulus of Elasticity (Compr.),	E_{mc}	kN/mm^2	28.5
Ave. Ten. Strain at Cracking (flexural),	ε_{mp}	$\mu\varepsilon$	170
Ave. Cylinder Compressive Strength,	σ_{cu}	N/mm^2	34.0
Steel Reinforcement Properties - (High Strength Deformed Bars)			
Diameter,	d_s	mm	16.0
Modulus of Elasticity,	E_s	kN/mm^2	200
Yield Stress,	σ_{sy}	N/mm^2	450

Beams Tests

Deflections at midspan and the load points were measured by means of displacement transducers (accurate to 0.002 mm) placed beneath the beam. In addition, support movements were detected by a transducer placed directly above each support. Measured midspan deflections were corrected by substracting measured values from the average support movements recorded. Curvature in the constant moment span was measured in terms of strains using a pair of detachable yokes. Displacements between the yokes at the top and bottom of the beam were measured by transducers and were converted into strains by dividing over the guage length bridged by the yokes. Readings were converted into curvature values by assuming a linear strain variation from the top to bottom. These readings were checked with manual measurements using demac gauge and pins at mispan. The load was applied through a 500 kN MTS actuator at a stroke rate of 0.8 mm/min.

RESULTS AND DISCUSSION

Figs. 8(a) and 8(b) compare the experimental and predicted M-ϕ curves for the 2 groups of beams. Table 2 gives relevant values for constructing the predicted curves. It is seen that the simplified model gives satisfactory prediction of the M-ϕ behaviour.

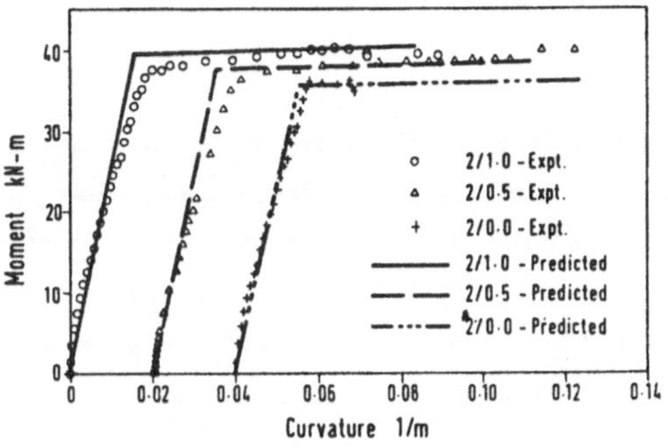

Fig. 8(a) Beams with 2 Bars (M-ϕ)

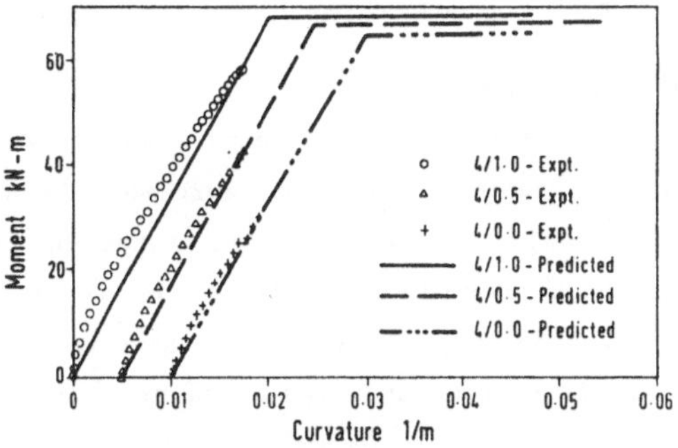

Fig. 8(b) Beams with 4 Bars (M-ϕ)

Table 2 Predicted Quantities for Constructing Simplified M-ϕ and P-δ Curves

Beam	Eq (6) σ_{tu} N/mm^2	Eq (10) μ	Eq (11) ϕ_{cr2} x10^{-3} 1/m	Eq (15) M_{cr2} kN.m	Eq (16) M_u kN.m	Eq (17) ϕ_y x10^{-3} 1/m	Eq (19) ϕ_u	Eq (21) ζ	Eq (22) δ_y mm	Eq (23) δ_u mm	Eq (24) P_u kN
2/1.0	1.68	1.75	15.85	39.5	40.5	16.05	83.0	5.17	7.73	23.46	93.0
2/0.5	0.84	1.85	15.52	37.7	38.6	15.71	92.0	5.86	7.57	25.51	88.7
2/0.0	0.00	1.96	15.24	35.6	36.3	15.39	73.0	4.74	7.41	20.94	83.6
4/1.0	1.68	1.01	20.09	67.9	68.5	20.18	47.2	2.34	9.72	16.08	158.6
4/0.5	0.84	1.05	19.68	66.7	67.2	19.75	49.3	2.50	9.51	16.48	155.7
4/0.0	0.00	1.05	19.70	64.8	65.1	19.74	36.7	1.86	9.51	13.50	151.0

Figure 7(a) Configuration

All beams with two reinforcing bars failed in flexure. The slight deviations between predicted and experimental curves just before the yielding of reinforcing steel can be attributed the assumed linear idealisation of material beahviour. The curvature ductility predicted also agrees well with experimental data. Beam 2/0.0 apparently failed prematurely prior to reaching the full extent of predicted curvature ductility whereas RSFC beams showed curvature ductility comparable to prediction. In terms of energy absorption this means that RSFC beams failing in flexure can absorb significantly more energy than plain concrete ones. This makes the use of RSFC members an attractive alternative for structures in earthquake zones or for blast protection. The ultimate moment predicted are also close to the test values.

Figure 7(b) Configuration

All beams with four reinforcing bars failed in shear, nonetheless the M-ϕ behaviour before failure was also well predicted. Note that beam 4/1.0 failed at the highest load followed by 4/0.5 and 4/0.0 in that order. Since the beams were tested over the same shear span, the results indicate that the fibres act as effective shear reinforcements. In this case the shear capacity was doubled by the addition of 1.0% volume fraction of fibres. Fibres could possibly replace stirrups and the compression steel needed to anchor them [15].

Load-deflection Curves

Figs. 9(a) and 9(b) compare predicted load-midspan deflection with experimental results. Again satisfactory agreement is noted. Where there are deviations, they can be attributed to difficulties inherent in deflection measurement as well as the simplifying assumptions made in the theoretical derivations. Additionally, for beams that fail in shear a different mechanism is operating, hence the differences between the predicted and experimental values near the failure load.

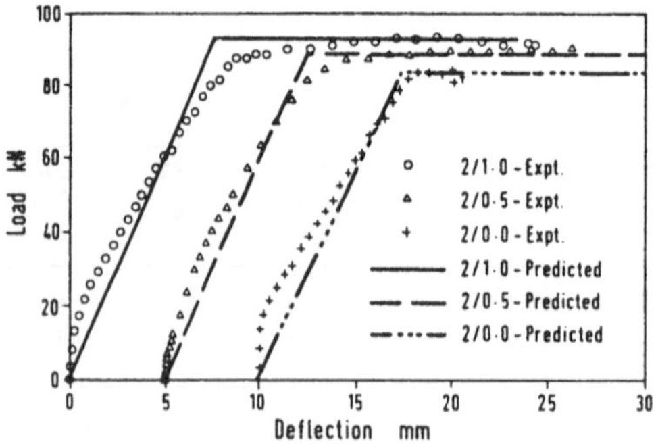

Fig. 9(a) Beams with 2 Bars (P-δ)

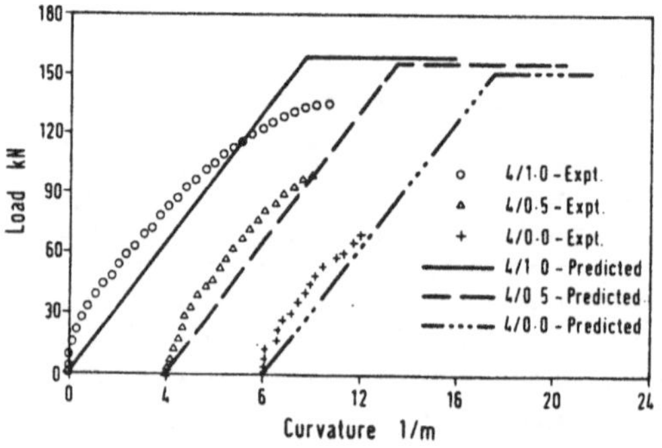

Fig. 9(b) Beams with 4 Bars (P-δ)

EXAMPLE PROBLEM

A RSFC beam having the cross-section shown in Fig. 10 is made from a SFC mix with properties given below. Obtain the simplified M-ϕ curve for the section and calculate the curvature ductility.

Matrix Properties

E_{mt} = 25 kN/mm^2

E_{mc} = 28 kN/mm^2

σ_{cu} = 35 N/mm^2

ε_{mp} = 200 $\mu\varepsilon$ (flexure)

Reinforcement Properties

E_s = 200 kN/m^2

f_s = 440 N/mm^2

A_s = 1470 mm^2

Fibre Properties

ℓ_f = 30 mm

d_f = 0.4 mm

σ_{fu} = 1200 N/mm^2

ε_{fp} = 6000 $\mu\varepsilon$

E_f = 200 kN/mm^2

τ_u = 5 N/mm^2

V_f = 1.5%

Solution:

(a) Critical length $\ell_c = 0.5 \times 0.4 \times 1200/5 = 48$ mm $> \ell_f$

Eq. (2), $\eta_\ell = 0.5$

(b) Orientation factors

Eq. (3), $\eta_o = 0.14$; Eq. (7), $\eta_o' = 0.41$

(c) In compression

Eq. (1), $E_{cc} = 28 \times 0.985 + 0.5 \times 0.14 \times 200 \times 0.015$

$= 27.8$ kN/m^2

(d) In tension

Eq. (4), $E_{ct} = 25 \times 0.985 + 0.5 \times 0.14 \times 200 \times 0.015$

$= 24.8$ kN/m^2

Eq. (5), $\varepsilon_{cr} = 0.5 \times 0.41 \times 0.015 \times (6000 - 200) + 200$

$= 218 \ \mu\varepsilon$

Eq. (6), $\sigma_{tu} = 0.5 \times 0.41 \times 0.015 \times 30 \times \dfrac{5}{0.2}$

$= 2.3$ N/mm^2

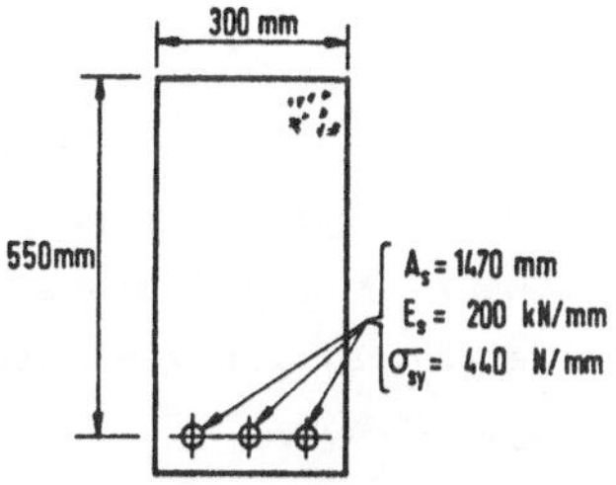

Fig. 10 Beam Section

(e) To obtain $(EI)_{eq}$, ϕ_y, ϕ_u and M_u'

(i) Eq. (10), $$\mu = \frac{2\text{x}0.9\text{x}35\text{x}2.2\text{x}10^{-3} - 0.0172}{2\text{x}2.3\text{x}2.2\text{x}10^{-3} + 0.0172 + \dfrac{(0.9\text{x}35)^2}{27.8\text{x}10^3} - 2\text{x}2.3\text{x}218\text{x}10^{-6} + 1.179\text{x}10^{-3}}$$

$= 1.92$

(ii) Eq. (11), $\phi_{cr2} = 6.08 \times 10^{-6}$ 1/mm; Eq. (13), $\lambda = 0.099$; Eq. (14), $\nu = 0.989$

(iii) Eq. (15), $M_{cr2} = 4067[27.8\text{x}10^3\text{x}(0.989)^3 + 24.8\text{x}10^3\text{x}(0.099)^3(1.919)^3]$

$+ 5.33\text{x}10^6[0.9\text{x}35(1-0.989^2) + 2.3(1-0.099^2)(1.919)^2 + 43.88]$

$= 392.3$ kN.m

(iv) Eq. (8), $(EI)_{eq} = 6.45 \times 10^{13}$ Nmm^2

(v) Eq. (20), $h_t = 448.8$ mm

(vi) Eq. (18), $M_\mu = 408.2$ kN.m

(vii) Eq. (16), $M_u = 400.3$ kN.m

(viii) Eq. (17), $\phi_y = 6.21 \times 10^{-6}$ 1/mm

(ix) Eq. (19), $\phi_u = 39.5 \times 10^{-6}$ 1/mm

(f) Curvature Ductility

Eq. (21), $\zeta = 6.36$

Fig. 11 shows the simplified M-ϕ curve obtained.

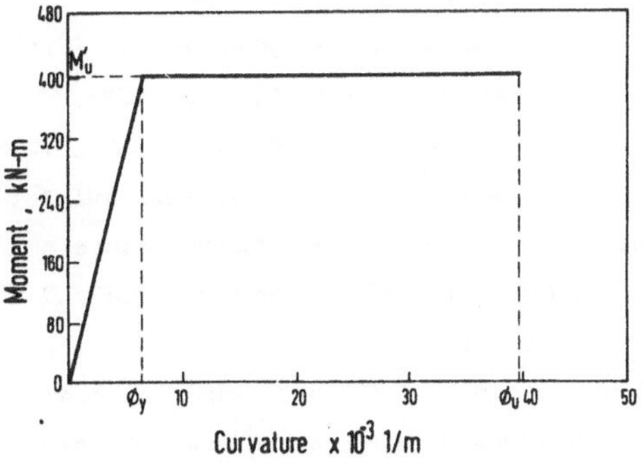

Fig. 11 Simplified M-ϕ Curve

CONCLUSION

Analytical expressions were derived to predict the deformational behaviour and strength of RSFC beams. The analytical predicted M-ϕ and P-δ curves for simply supported beams were found to agree well with test data. It was shown that the inclusion of fibres lead to higher strength and curvature ductility in flexural failure of RSFC beams when compared to similarly reinforced plain concrete beams.

The model presents a simple analytical framework to evaluate the actual fibre contribution and is useful to check for ductile failure, ensuring moment distribution is possible and to obtain the curvature ductility for design.

The fibres also act as effective shear reinforcement and in many cases may eliminate the need for stirrups reinforcement and compression steel [15].

REFERENCES

1. Henager, C.H. and Doherty, T.J. Analysis of reinforced fibrous concrete beams, *Proceedings ASCE*, Vol. 102, ST1, Jan. 1976, pp. 177-188.
2. Swamy, R.N. and Al-Ta'an, S.A. Deformation and ultimate strength in flexure of reinforced concrete beams made with steel fiber concrete, *ACI Journal*, Sept. 1981, pp. 395-405.
3. Jindal, R.L. Shear and moment capacities of steel fiber reinforced concrete beams, International Symposium on Fiber Reinforced Concrete, *ACI SP-81*, 1984, pp. 1-16.
4. Pakotiprapha, B., Pama, R.P. and Lee, S.L. Mechanical properties of cement mortar with randomly oriented short steel wires, *Mag of Conc Res* , Vol. 26, No. 86, Mar. 1974, pp. 3-15.
5. Lim, T.Y., Paramasivam, P. and Lee, S.L. Bending behaviour of steel fibre concrete beams, *ACI Structural Journal*, (in press).
6. Lim, T.Y., Paramasivam, P. and Lee, S.L. An analytical model for tensile behaviour of steel fibre concrete, *ACI Materials Journal* (in press).
7. Nathan, G.K., Paramasivam, P. and Lee, S.L. Tensile behaviour of fiber reinforced cement paste, *Journal of Ferrocement*, Vol. 7, No. 2, Oct. 1977, pp. 59-79.
8. ACI Committee 318 *Building Code Requirements for Reinforced Concrete (ACI 318-71)*, ACI, Detroit, 1971, 78 pp.
9. Shah, S.P. and Rangan, B.V. Effects of reinforcements on ductility of concrete, Proceedings, ASCE, Vol. 96, ST6, Jun 1970, pp. 1167-1184.
10. Fanella, D.A. and Naaman, A.E. Stress strain properties of fiber reinforced mortar in compression, *ACI Journal*, Jul 1985, pp. 475-483.
11. Swamy, R.N. Steel fibre concrete for bridge deck and building floor applications, *US-Sweden Joint Seminar (NSF-STU) on Steel Fiber Concrete*, Swedish Cement and Concrete.
12. Romauldi, J.P. and Mandel, J.A. Tensile strength of concrete as affected by uniformly distributed and closely spaced short lengths of wire reinforcement, *ACI Journal*, Vol. 61, No. 6, Jun 1964, pp. 657-671.
13. Argon, A.S. and Shack, W.J. Theories of fibre cement and fibre concrete, *RILEM Symposium 1975 on Fibre Reinforced Cement and Concrete*, The Construction Press, 1975, pp. 39-53.
14. Lim, T.Y., Paramasivam, P. and Lee, S.L. Behaviour of reinforced steel fibre concrete beams in flexure, *Journal of the ASCE Structural Div* , (in press).
15. Lim, T.Y., Paramasivam, P. and Lee, S.L. Shear and moment capacity of reinforced steel fibre concrete beams, accepted for publication in *Mag of Concrete Research*

Proceedings of the International Symposium on Fibre Reinforced Concrete, December 16-19, 1987, Madras, India

STEEL FIBRE REINFORCED CONCRETE COLUMNS WITH AND WITHOUT CONVENTIONAL BAR REINFORCEMENT UNDER UNIAXIAL BENDING

D.L.N. Rao
ANDHRA UNIVERSITY
VISAKHAPATNAM, INDIA

S.V. Rao
URBAN DEVELOPMENT AUTHORITY
VISAKHAPATNAM
INDIA

R. Ramachandra Rao
URBAN DEVELOPMENT AUTHORITY
VISAKHAPATNAM, INDIA

SYNOPSIS

Tests are reported on the influence of fibre reinforcement on the ultimate strength in combined axial compression and uniaxial bending of concrete columns, with and without bar reinforcement. The test specimens comprised of short columns 1200 mm in length and 100 x 150 mm in cross section. The principal variable used was eccentricity of applied compression; the eccentricity relative to the central axis of the columns was varied from 0 to 150 mm. Tests indicate only small improvement in ultimate strength of the columns with bar reinforcement as a result of fibre addition. However in the case of columns without bar reinforcement ultimate strength increase has been significant. Substantial improvement has been observed in reduction of transverse deformations and tensile strains and crack control. An ultimate strength theory similar to the one proposed by Swamy and Al - Ta'an [1] for beams, is presented and shows very good agreement with the test data.

INTRODUCTION

It has been established [1-9] that of all the fibres currently in use to reinforce cement matrices, steel fibres are the only fibres that can be used for long term load carrying properties. Addition of steel fibres have been reported to give significant improvement in flexural strength, impact strength and fracture toughness of concrete with smaller improvements in compressive strength. Tests on fibre reinforced concrete elements, to date have been conducted predominantly to study unit actions - axial compression, tension, flexure or torsion. The influence of combined actions that are normally present in actual structural components received little attention. Further only limited data are available on the reinforcing effect of steel fibres when they are used in combination with bar reinforcement in structural members failing in combined axial compression and bending. Niyogi and Chawla [10] have reported tests on fibre reinforced concrete columns without conventional reinforcement under uniaxial bending. These tests have shown that, the relative increase in strength for fibre reinforced column over the corresponding plain concrete column depends upon eccentricity of load, being higher the greater the eccentricity. The increase was as high as 100%. Swamy and Sa'ad [1] have presented an analytical method to predict the ultimate flexural strength of steel fibre concrete beams with bar reinforcement, based on their flexural tests. The beams were made with 20 mm maximum size of aggregates.

The present paper reports the results of a test programme to study the influence of steel fibres on the strength in uniaxial bending of reinforced concrete columns containing steel bar reinforcement. Based on the theory of Swamy and Sa'ad [1] equations are developed to predict the strength of such columns.

EXPERIMENTAL PROGRAMME

Details of Test Columns

The experimental work consisted of tests on 24 columns of which 12 were plain concrete and 12 were conventionally reinforced. All the columns were of the same size, 100 x 150 x 1200 mm. The main variable in the tests was eccentricity of column load. The bar reinforcement was 6.85 mm in diameter with an average yield strength of 540 N/mm^2. Ties were of 6 mm diameter at 100 mm centres; yield strength being 260 N/mm^2 The Volume fraction of fibre used was 1.3%. Fibres were used over the whole length of the columns. The details of the test specimens are shown in Fig. 1.

Details of Materials

Both plain and fibre concrete mixes used in the specimens were identical except for the presence of fibre in the latter. The matrix proportions by weight for cement, sand and gravel were respectively 1:2:3 with a water cement ratio of 0.55 by weight. Steel binding wire of 1.0 mm diameter with a tensile strength of 548 N/mm^2 was cut into pieces of 50 mm by shear cutting machine to obtain fibre for use. The aspect ratio of the fibres was 50. Table - I lists the details of the test specimens.

Test Details

The columns were cast in wooden moulds and compacted on a table vibrator. They were cured along with control specimens under uncon-

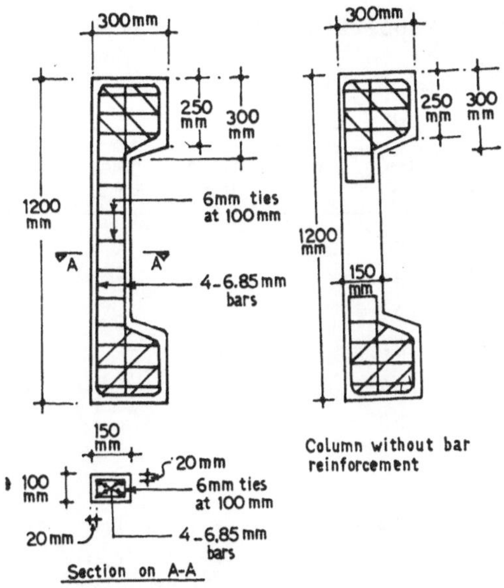

Fig. 1 Details of test specimens

trolled laboratory environment. The specimens were divided into four groups; each consisting 6 specimens. In each group the first specimen was tested under axial force and the sixth specimen in pure bending. Other four columns were subjected to eccentric loading. The eccentricity of the applied load with respect to central axis of column was varied from 0 to 150 mm. First and second groups did not contain conventional bar reinforcement; the difference between them was that while the first group didnot have steel fibres, the latter group had steel fibres. Third and fourth groups' specimens were provided with bar reinforcement, and only the fourth group was provided with steel fibres. Pure bending test was conducted with the specimen in horizontal position, under two point loading, creating a pure moment zone of 37 cm at centre of span. The load for the columns was applied through ball and socket bearings from both ends of the columns. Taken were measurements of transverse deflection, concrete strain and crack width at various sections of the specimen throughout the loading range. Loading arrangement and typical locations of the various measurements for columns and beams are shown in Fig. 2.

ULTIMATE STRENGTH

A large amount of data was obtained from the tests. Only strength aspects are presentd in this paper. The principal test results are shown in Table - II.

Table - I : Properties of Test Specimens

Column No.	Concrete strength Compression (N/mm^2)	Concrete strength Split cylinder (N/mm^2)	Age at test (days)	Longitudinal steel No	Longitudinal steel Bar dia. (mm)	Fibre Diameter (mm)	Fibre Aspect ratio	Fibre Volume (%)	Load eccentricity e/D
P001	22.90	1.80	36	-	-	-	-	-	0.000
P002	22.90	1.75	42	-	-	-	-	-	0.133
P003	23.10	1.75	42	-	-	-	-	-	0.200
P004	23.60	1.75	36	-	-	-	-	-	0.350
P005	21.90	1.75	42	-	-	-	-	-	0.500
P006	23.10	1.50	48	-	-	-	-	-	PB
PBH1	23.90	3.30	42	-	-	1.0	50	1.3	0.000
PBH2	22.00	2.90	48	-	-	1.0	50	1.3	0.133
PBH3	23.00	3.30	48	-	-	1.0	50	1.3	0.200
PBH4	23.20	2.90	48	-	-	1.0	50	1.3	0.350
PBH5	21.90	3.30	45	-	-	1.0	50	1.3	0.500
PBH6	22.90	2.90	46		-	1.0	50	1.3	PB
R001	23.60	1.80	35	4	6.85	-	-	-	0.000
R002	23.50	1.75	36	4	6.85	-	-	-	0.200
R003	22.00	1.75	36	4	6.85	-	-	-	0.400
R004	21.90	1.75	36	4	6.85	-	-	-	0.500
R005	21.90	1.75	32	4	6.85	-	-	-	1.000
R006	23.10	1.50	51	4	6.85	-	-	-	PB
RBH1	23.90	3.30	38	4	6.85	1.0	50	1.3	0.000
RBH2	22.00	2.90	74	4	6.85	1.0	50	1.3	0.200
RBH3	22.05	3.30	39	4	6.85	1.0	50	1.3	0.400
RBH4	22.00	2.90	33	4	6.85	1.0	50	1.3	0.500
RBH5	21.90	3.30	38	4	6.85	1.0	50	1.3	1.000
RBH6	21.90	2.90	81	4	6.85	1.0	50	1.3	PB

Note : All specimens have a cross section of 100 x 150 mm. Column ties 6 mm at 100 mm c/c . PB = Pure Bending

Equations are derived to predict the ultimate strength of fibre reinforced concrete columns provided with or without conventional bar reinforcement under compression plus bending. The method has been developed by Swamy and Al-Ta'an to determine the ultimate strength in flexure of reinforced concrete beams made with steel fibre concrete. In their analysis the contribution of steel fibres in the tensile zone was accounted for, to the moment of resistance of the beam. Here

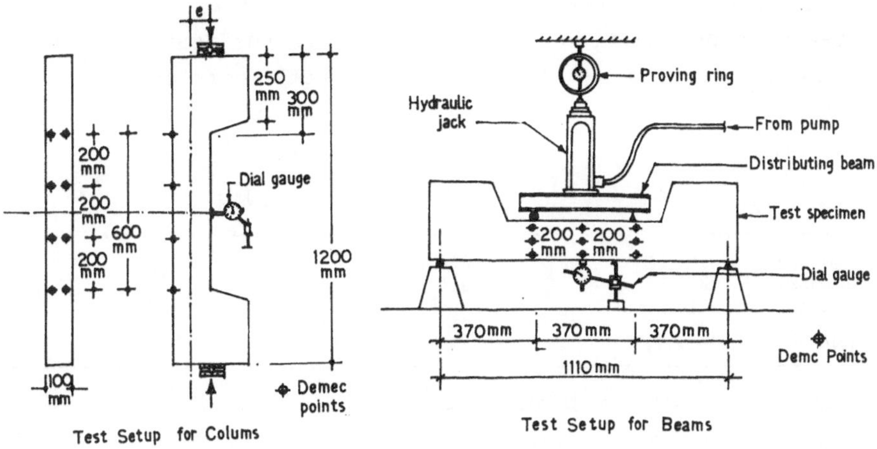

Fig. 2 Test setup

these equations are modified for columns under combined axial load and bending.

The cracking strength of two phases composite is given by law of mixtures.

$$\sigma_{cr} = \sigma_{mu} V_m + \sigma_f V_f \tag{1}$$

where σ_{cr} = cracking stress, V_f = volume fraction of fibres, σ_{mu} = cracking stress of the matrix and σ_f = stress in fibres at matrix cracking.

The ultimate strength of the composite is given by

$$\sigma_{cu} = 2\,\tau\,(l_f/d_f)\,V_f \tag{2}$$

where, τ is interfacial bond stress and σ_{cu} is ultimate strength of the composite . As most favourable ideal conditions are not realised relative to orientation of fibres, interfacial bond stress and short length of fibres, correction factors are to be applied to Eq (2). These correction factors include orientation factor, η_0, length correction factor, η_1 and bond efficiency factor, η_b. Eq (2) becomes

$$\sigma_{cu} = \eta_0\;\eta_1\;\eta_b\;2\tau\,(l_f/d_f)\,V_f \tag{3}$$

The factor derived by Romualdi and Mandel [11] assuming equal probabilities to the angles in both vertical and horizontal planes works out to 0.405 (= 0.41). This value is used in the present study. A length correction factor, η_1 proposed by Cox [12] is used in this study.

$$\eta_1 = \frac{1 - \tan h\,(\beta\, l_f/2)}{\beta\, l_f/2} \tag{4}$$

where, l_f is fibre length and

$$\beta = \left| \frac{2\pi\, G_m}{E_f\, A_f\, \text{In}\,(s/r_f)} \right|^{1/2} \tag{5}$$

where, G_m is the shear modulus of matrix, E_f is the fibre modulus, A_f is the area of cross section of the fibre, s is the mean centroid spacing, r_f is the radius of fibre. The length correction factor is a function of critical length which depends on the characteristics of the matrix fibre.

Table - II : Principal Test Results

Column No.	Concrete strength Compression (N/mm^2)	Split cylinder (N/mm^2)	$\frac{e}{D}$	Volume fraction (%)	P_c k.N	P_u k.N	M_u N.m
P001	22.90	1.80	0.000	-	-	258.0	-
P002	22.90	1.75	0.133	-	-	190.0	3790
P003	23.10	1.75	0.200	-	-	160.0	4800
P004	23.60	1.75	0.350	-	-	80.0	4200
P005	21.90	1.75	0.500	-	-	39.0	2925
P006	23.10	1.50	PB	-	-	-	900
PBH1	23.90	3.30	0.000	1.30	-	300.0	-
PBH2	22.00	2.90	0.133	1.30	210.00	225.0	4480
PBH3	23.00	3.30	0.200	1.30	178.00	198.0	5940
PBH4	23.20	2.90	0.350	1.30	95.00	110.0	5775
PBH5	21.90	3.30	0.500	1.30	50.00	60.0	4500
PBH6	22.90	2.90	PB	1.30	-	-	1516
R001	23.60	1.80	0.000	-	-	370.0	-
R002	23.50	1.75	0.200	-	221.70	240.0	7200
R003	22.00	1.75	0.400	-	116.00	138.0	8280
R004	21.90	1.75	0.500	-	90.50	107.0	8025
R005	21.90	1.75	1.000	-	26.70	43.5	6525
R006	23.10	1.50	PB	-	-	-	5058
RBH1	23.90	3.30	0.000	1.30	-	390.0	-
RBH2	22.00	2.90	0.200	1.30	238.70	248.0	7440
RBH3	22.05	3.30	0.400	1.30	120.00	160.0	9600
RBH4	22.00	2.90	0.500	1.30	94.00	125.0	9375
RBH5	21.90	3.30	1.000	1.30	42.00	52.2	7830
RBH6	21.90	2.90	PB	1.30	-	-	6353

P_c = cracking load; P_u = ultimate load; M_u = ultimate moment; PB = pure bending.

Fibre spacing formulae were proposed by several investigators, and the spacing equation suggested by Swamy et al [13] is used here.

$$S = 25\ (d_f / \rho\ l_f)^{0.5} \tag{6}$$

where s is the fibre spacing in mm and ρ is volume percentage of fibre in the matrix.

The surface characteristics of the fibre determines the bond of fibre with matrix. The bond efficiency factors were based on pull out tests. But this factor is to be refined on the basis of flexure tests incorporating not only pull out tests but also flexure tests. In the present investigation a unit value is used. The choice of the bond efficiency factor is not critical; a 20% change of this factor would result only in about 4% change in the ultimate strength of the column.

Method of Analysis

The method of analysis is similar to that of a column subjected to axial force and bending moment. The analysis is based on the ACI Code 318-1971 [14]. The strain distribution, the actual stress distribution and the idealised stress distribution in tension and compression are shown in Fig. 3. The maximum concrete compressive strain at ultimate is assumed equal to 0.003.

Moment equilibrium about tensile steel is

$$P_u\ [e + (d/2) - d'] = 0.85 \times f'_c \times b \times a\ (d - a/2) + f'_s\ A'_s\ (d-d') - \sigma_{cu}\ b(D-c)\ (\frac{D-c}{2} - d') \tag{7}$$

Force equilibrium equation

$$P_u \quad P_u = 0.85 \times f'_c \times b \times a + f'_s A'_s - f_s A_s - \sigma_{cu}\ b(D-c) \tag{8}$$

where P_u is external load at ultimate; D is overall depth of column; d is effective depth of column; d' is cover to the longitudinal reinforcement on tension and compression sides; e is eccentricity of axial load to the centre of column section; f'_c is cylinder strength of concrete; σ_{cu} is ultimate strength of composite; ε_c is compressive strain at ultimate in concrete; ε_s is strain in tensile reinforcement at ultimate; ε'_s is strain in compression reinforcement at ultimate; A_s is area of steel on tension side; A'_s is area of steel in compression side and C_s, C_c, T_f and T_s are forces at ultimate in compression steel, concrete in compression, fibre concrete in tension, and tensile steel,. c is depth of neutral axis.

To start the solution, the depth of stress block, a is assumed in Eq (7) and the axial load P_u is computed as a first approximation. The latest value of P_u is substituted in the force equilibrium equation and solved again for depth of stress block, a. The calculations are repeated till latest value of the depth of stress block, a from

Eq (8) is close to the assumed value of depth of stress block in Eq (7).

The theoretical ultimate strengths are shown in Table - III.

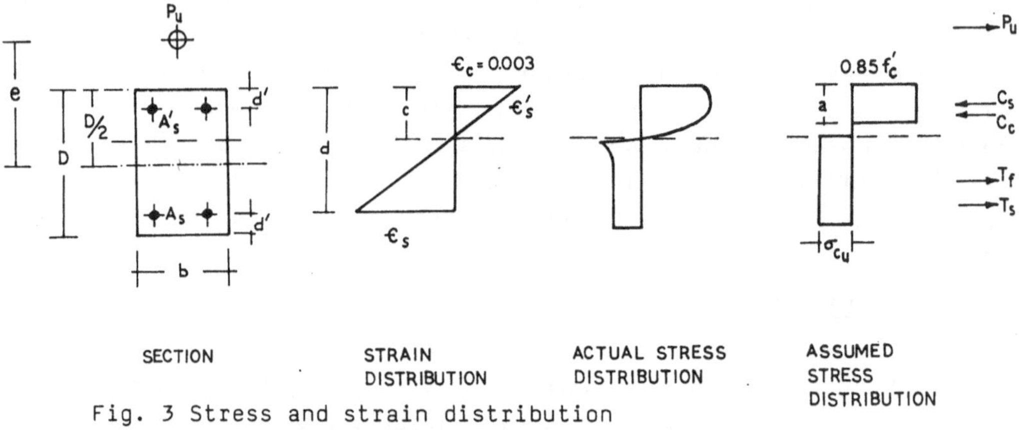

Fig. 3 Stress and strain distribution

Table - III Comparison of Test and Theoretical Results

Column No.	Volume fraction %	Eccen-tricity e/D	Axial load at ultimate Test kN	Theory kN	Test/Theory	Ultimate moments Test Nm	Theory Nm	Test/Theory
PBH1	1.3	0.000	300	244	1.23	-	-	-
PBH2	1.3	0.133	225	167	1.35	4480	3330	1.35
PBH3	1.3	0.200	198	146	1.36	5940	4380	1.36
PBH4	1.3	0.350	110	86.5	1.27	5775	4541	1.27
PBH5	1.3	0.500	60	44.6	1.35	4500	3345	1.35
PBH6	1.3	PB	-	-	-	1516	1087	1.39
RBH1	1.3	0.000	390	323.5	1.21	-	-	-
RBH2	1.3	0.200	248	190.6	1.30	7440	5718	1.30
RBH3	1.3	0.400	160	120.9	1.32	9600	7254	1.32
RBH4	1.3	0.500	125	100.7	1.24	9375	7553	1.24
RBH5	1.3	1.000	52.2	48.9	1.07	7830	7335	1.07
RBH6	1.3	PB	-	-	-	6353	6001	1.06

Note: PB = Pure Bending

Columns without bar reinforcement: $\frac{\text{Test Strength}}{\text{Theorl.Strength}}$ = 1.33

Columns with bar reinforcement: $\frac{\text{Test Strength}}{\text{Theorl. Strength}}$ = 1.20

DISCUSSION OF RESULTS

For columns without bar reinforcement and fibre, at cracking, the failure occurred. For fibre columns cracking and failure loads could be distinguished. The difference between the cracking and ultimate loads depended upon the eccentricity of applied load, the difference being greater for increased eccentricity. The ratio of cracking load to ultimate load varied from about 0.93 for small eccentricity column to about 0.8 for the column having largest eccentricity. For the columns with bar reinforcement cracking loads were 61 to 90% of ultimate loads when fibre was there and 75 to 96% when fibre was not there.

The ultimate compressive load for columns decreased with increased eccentricity of applied load; this was true whether the column contained fibres or bar reinforcement. For equal eccentricities of applied loads, the ultimate strength of column with fibre was always higher than that of its counterpart without fibre. Fibre addition resulted in very large increase in strength in columns without bar reinforcement in comparison with columns having bar reinforcement. The strength increase varied from 16% for axially loaded column to 68% for the specimen under pure moment, for columns without bar reinforcement. In the case of columns with bar reinforcement, the increase was only 8% for axially loaded column and 35% for the specimen under pure moment.

The theory predicted accurately the test strengths of the columns. The ratios of test to theoretical strengths varied from 1.23 to 1.39 for the columns without bar reinforcement; the theory underpredicted the strength by 33%. These ratios for columns with bar reinforcement varied from 1.07 to 1.32 and the average was 1.20.

CONCLUSIONS

The following conclusions are drawn from the results of this investigation.

1. Addition of steel fibres into plain concrete columns improved the ductility, cracking strength and ultimate strength substantially. The greater the eccentricity of applied load, the higher was the relative increase in strength.

2. Fibre inclusion in columns with bar reinforcement improved cracking strength and ultimate strength in a limited measure, though improvement in deformation characteristics was substantial.

3. The theory predicted accurately and conservatively the strength of fibre reinforced concrete columns under axial load and bending, with or without bar reinforcement. The theory underpredicted the strength of columns without bar reinforcement by 33% and the strength of columns with bar reinforcement by 20 %.

ACKNOWLEDGEMENTS

The investigation reported in this paper was carried out in the Civil Engineering Department of Andhra University, Waltair. The authors wish to thank the Heads of the Department Prof.T. Shivaji Rao, Prof.R.P. Ranganatha Rao and Prof.C. Veeraiah for the helpful way in which they accommodated the research work in their respective periods.

REFERENCES

1. Swamy, R.N. and Sa'ad, A. Al-Ta'an; *ACI Journal,* Sept. Oct. 1961, pp.395-405.
2. Romualdi, J.P. and Batson, G.B., *Journal Engg. Mech. Div. Proc. Am. Soc. Civ. Engrs.* June 1963, pp.147-168.
3. Romualdi, J.P. and Batson, G.B., *Journal Am. Conc. Inst.* June 1963, pp.775-790.
4. Romualdi, J.P. The static cracking and fatigue strength of concrete reinforced with short pieces of thin steel wire, International Conference on the Structural Concrete, London, 1965, pp. 1-12.
5. Shah, S.P. and Rangan, B.V., *Journal Struct. Div. Proc. Am. Soc. Engrs.* June 1970, pp. 1167-1184.
6. Shah, S.P. and Rangan, B.V., *Journal Am. Concrete Inst.* February 1971, pp. 126-135.
7. Harries, B. et al.,*Cem. Conc. Res.,* July 1972, pp.447-461.
8. Swamy, R.N. and Mangat, P.S., *Cem. Conc. Res.,* May 1974, pp. 451-465.
9. Swamy, R.N. and Mangat, P.S., *Cem. Conc. Res.,* January 1975, pp. 37-53.
10. Niyogi, S.K. and Chawla, A.P., *Institution of Civil Engineers,* March 1982, pp. 199-206.
11. Romualdi, J.P. and Mandel, J.A., *ACI Journal.* Proc. V. 61, June 1964, pp. 657-671.
12. Cox, H.L., *British Journal of Applied Physics,* V. 3, Mar. 1952, pp.72-79.
13. Swamy, R.N. Mangat, P.S. and Rao, C.V.K. The mechanics of fibre re-inforcement of cement matrices, Fibre Reinforced Concrete, SP-44, American Concrete Institute, Detroit, 1974, pp.1-28.
14. ACI Committee 318 'Building Code Requirements for Reinforced Concrete' (ACI 318 - 71), American Concrete Institute, Detroit, 1971, 78 pp.

Proceedings of the International Symposium on Fibre Reinforced Concrete, December 16–19, 1987, Madras, India

INFLUENCE OF STEEL FIBRES ON THE SHEAR TRANSFER IN CRACKS

Van de Loock L.
Department of Construction, Katholieke Universiteit Leuven, Belgium

SYNOPSIS

In order to investigate the transmission of forces over small cracks in fibre reinforced concrete tests were carried out on push-off specimens. These specimens were precracked in the shear plane and were reinforced by external bars normal to the crack plane. Shear stress in the crack was created by an external load. The variables studied were the restraint stiffness and the fibre content. Volume percentages Vf=0, Vf=0.005 or Vf=0.01 were used. The stresses normal and parallel to the crack and the displacements normal and parallel to the crack were measured.

The results of the experiments are given and the influence of the parameters mentioned is discussed. Provisional relations between the stresses in the crack and the displacements are proposed. The steel fibres have a clear influence on the shear force. The normal force and the crack opening path are less dependent on the fibre content.

INTRODUCTION

It is clear that steel fibres can have a beneficial influence on the capacity of reinforced concrete structures to sustain so-called shear forces. The research carried out in this field is too comprehensive to be treated here in detail. We mention the work of Muhidin and Regan [1] and Uomoto, Weeraratne, Furukoshi and Fujino [2] on the use of steel fibres in the shear span of reinforced beams, the studies of Walraven, Pat and Markov [3] and Swamy and Ali [4] on the punching shear in slabs with steel fibres and the experiments of Fattuhi [5] on corbels. In all these articles it is demonstrated that steel fibres can increase the admissible shear force and can replace the shear reinforcement or part of it.

The explanation of the shear strength of fibre reinforced concrete is mostly based on the post-cracking tensile strength of the material. This strength is defined as the stress remaining in the material after cracking. It can be measured in tensile tests or derived from the results of bending tests. Usually it will be expressed as a function of the fibre content. Truss models are proposed to calculate the influence of the fibres in the shear span of beams. The resulting force attributed to the steel fibres is supposed to be orthogonal to the crack face [2] or vertical [1]. In analysing the results of the punching resistance of steel fibre reinforced lightweight concrete the contribution of the fibres is calculated as the product of the effective area of the punching cone and the post-cracking strength [3]. A fibre term is used in the shear-friction model to calculate the contribution of the steel fibres to the shear capacity of corbels [5].

More elaborate models are being developed at this moment to analyse the forces in a crack in reinforced concrete under the combined action of normal and shear loads. In these models the forces in the reinforcement including dowel forces, the forces in the uncracked concrete and the forces transmitted over the cracks by mechanisms such as aggregate interlock are taken into account. It may be supposed that steel fibres will also have a contribution. Longitudinal forces are created in the fibres. As stated by Uomoto et. al. [2] steel fibres can possibly increase the effects of aggregate interlock by reducing the crack width. Swamy and Bahia [6] demonstrated steel fibres can improve the dowel action of reinforcing bars.

This article describes push-off tests to study the transmission of forces over small cracks by steel fibres under the combined action of normal and shear forces. The tests differ from the shear tests on fibre reinforced concrete made by Barr, Hasso and Liu [7] in the existence of a force normal to the crack faces during the tests.

OUTLINE OF EXPERIMENTS ON PUSH-OFF SPECIMENS

Concrete Mix

Gravel concrete with a maximum particle diameter of 25 mm was used. The grading of the aggregates and the composition of the mix were adjusted by trial mixes to ensure a good workability for the fibre concrete. No plasticizer was used. The cement was ordinary Portland cement (300 kg/m^3). The same concrete mix with a compressive strenght of 35 N/mm^2 measured on cubes of 200 mm side was used for all the specimens. The steel fibres had hooked ends, a diameter of 0.8 mm and a length of 60 mm. They were glued together in plates of approximately 30 fibres. The mixing was continued until complete separation of the fibres and a good distribution over the mix were observed.

Test Specimens

The geometry and the dimensions of the test specimens are given in Fig. 1 . A considerable reinforcement was provided to prevent failure of the specimen in other parts than the shear plane. There was no reinforcement across the shear plane except for the steel fibres. The blocks were cast in steel moulds. Two days after casting the specimens were demoulded and stored in a climate conditioned room with a temperature of 21 °C and a relative humidity of 95 %.

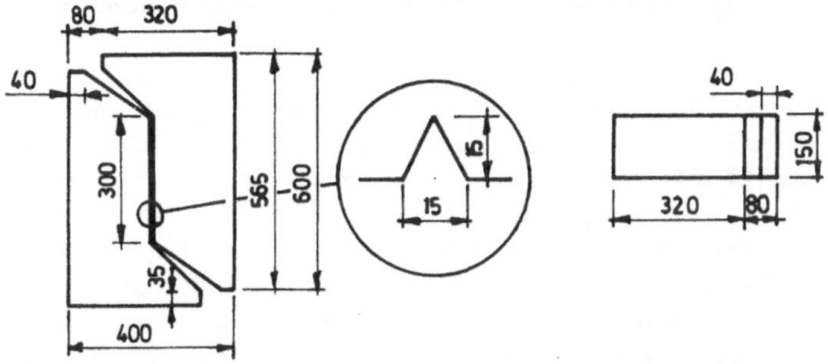

Fig. 1 Geometry and Dimensions of Test Specimens

Precracking

In order to crack the specimen along the shear plane it was placed on a frame with the faces in horizontal position (Fig. 2). A line load was applied through a steel rod in the groove of the top face to create tensile stresses in the bottom face. The elongation perpendicular to the crack was continuously recorded. After cracking the specimen was turned with the cracked face upside and the other side was cracked. During these operations external reinforcement prevented separation of the two halves of the block. Small crack widths could be obtained. The resulting crack width was, however, not always the same at both faces. If the difference was substantial or if the crack width was more than 0.02 mm, the specimen was rejected. The crack width before the actual test was between 0.015 and 0.02 mm.

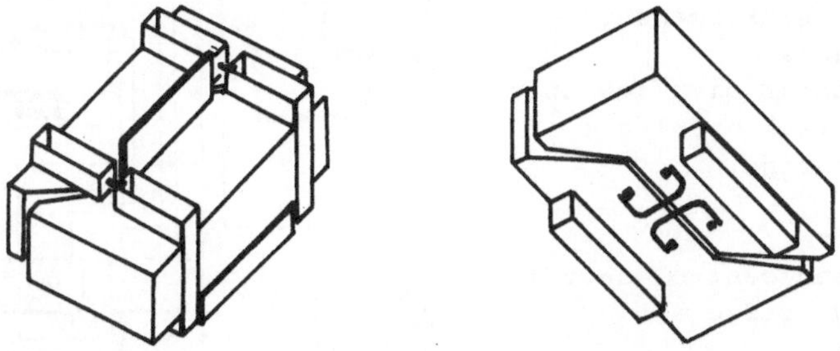

Fig. 2 Precracking of the Test Specimen, View of the External Reinforcement (Left) and the Measuring System (Right)

Actual Test

Push-off tests were carried out on the precracked specimens. The blocks were loaded parallel to the shear plane by a hydraulic jack. Parasite forces were prevented by a ball hinge at the top and a roller bearing at the bottom of the specimen. Normal forces were caused by external reinforcement consisting of four bars perpendicular to the crack fastened in stiff plates at the side faces of the specimen. The normal force changed during the test. The relation between the normal force and the displacements was dependent on the diameter of the bars and the stiffness of the whole frame and could not be changed during the test. The external restraint stiffness was a variable in the test program. The test set-up described here is similar to the set-up used previously by Walraven [8] to study aggregate interlock. During the tests the displacements normal to the crack and parallel to the crack were measured at both faces by displacement transducers. The stresses in the four bars were measured by strain gauges and were recorded together with the vertical force. An outline of the test set-up is given in Fig. 3 . The shear displacement ratio was kept constant at 0.02 mm/min up to a shear displacement of 0.1 mm and at 0.1 mm/min afterwards. The tests were stopped at a shear displacement of 2 to 3 mm depending on the external restraint stiffness.

1. hydraulic jack
2. hinge
3. external reinforcement
4. strain gauges
5. LVDT (shear displacements)
6. LVDT (crack opening)
7. roller bearing
8. oil pump
9. servo valve
10. Losenhausen control device
11. measuring bridge
12. amplifier
13. amplifier
14. XY-recorder
15. data collector

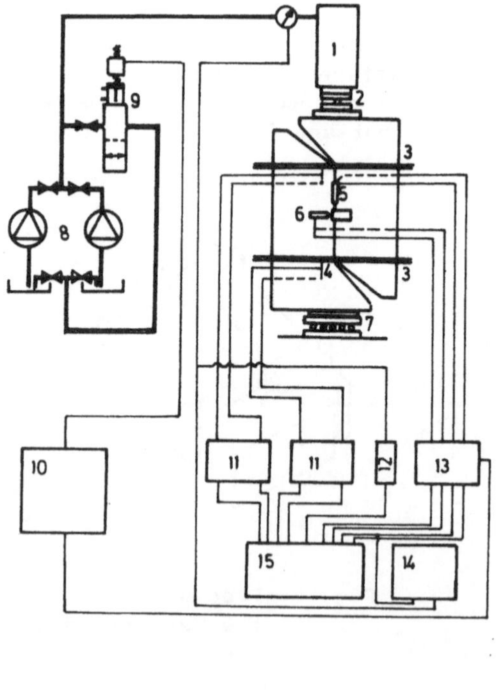

Fig. 3 Test Set-up

Variables

Studies on aggregate interlock have shown that the concrete strength and the restraint stiffness influence the behaviour of cracked concrete under shear loading. The concrete strength and the volume fraction of fibres have an important influence on the strength of the fibre reinforced concrete. As the size of this test program had to be limited it was decided to investigate only two of the parameters mentioned above : the volume fraction of steel fibres and the restraint stiffness. The concrete quality, the type of aggregates and the initial crack width were kept constant during the whole test program.

To be able to compare the fibre reinforced concrete with the normal concrete it was necessary to carry out tests on specimens without steel fibres. Two fibre contents were used : a volume fraction Vf=0.005 and Vf=0.01.

TEST RESULTS

The shear displacement and the crack opening were taken as the mean value of the values measured at both faces. The normal stress is the sum of the forces in the bars divided by the net area of the shear plane. The shear stress is the vertical force divided by the area of the shear plane. If the recorded crack opening path was considerably different at both faces of the specimen or if the restraining forces in the bars differed too much, the specimen was rejected because it was not sure that in this case taking the mean value could not produce errors in the interpretation of the results. No secondary cracks in the central zone of the specimen were seen. Altogether 15 specimens were considered in good order. This is approximately half of the blocks cast.

Fig. 4 and 5 each give the results of the shear tests for three specimens with the same volume fraction of steel fibres. The shear stress and the normal stress are presented as a function of the shear displacement and the crack opening. The relation between the normal and the parallel displacements (crack opening path) is also represented. As could be expected, the shear strength and the normal stress for the same crack opening or shear displacement increase with increasing restraint stiffness. In general the crack opening path was steeper and the influence of the fibres was smaller as the normal force was increased.

Fig. 6 shows the results for three specimens with nearly the same restraint stiffness but different fibre fractions. The steel fibres have a noticeable influence on the shear stress of the concrete. The normal stress and the crack opening path are quite unaffected by the steel fibres. The same conclusions hold true when another restraint stiffness is chosen, although the influence of the fibres is smaller when the restraint stiffness is increased.

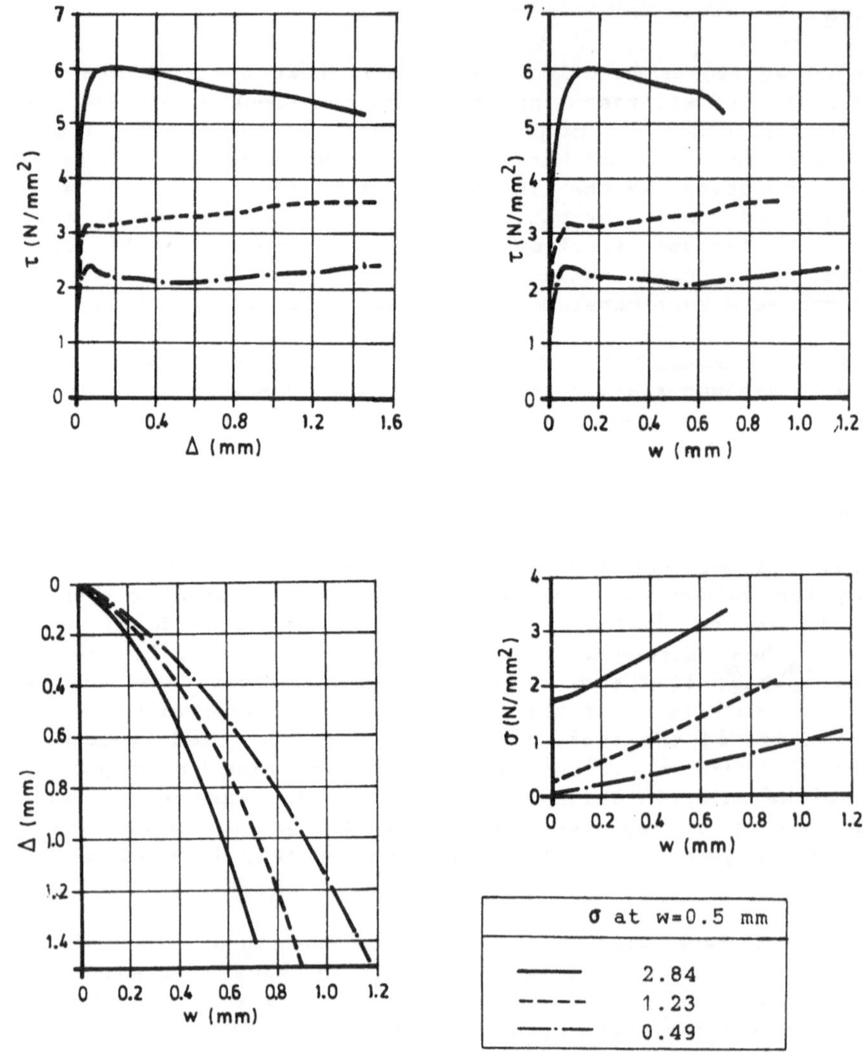

Fig. 4 Test Results for 3 Specimens Without Steel Fibres, Shear Stress τ as Function of Shear Displacement Δ Shear Stress as Function of Crack Opening w , Crack Opening Path, Normal Stress σ as Function of Crack Opening

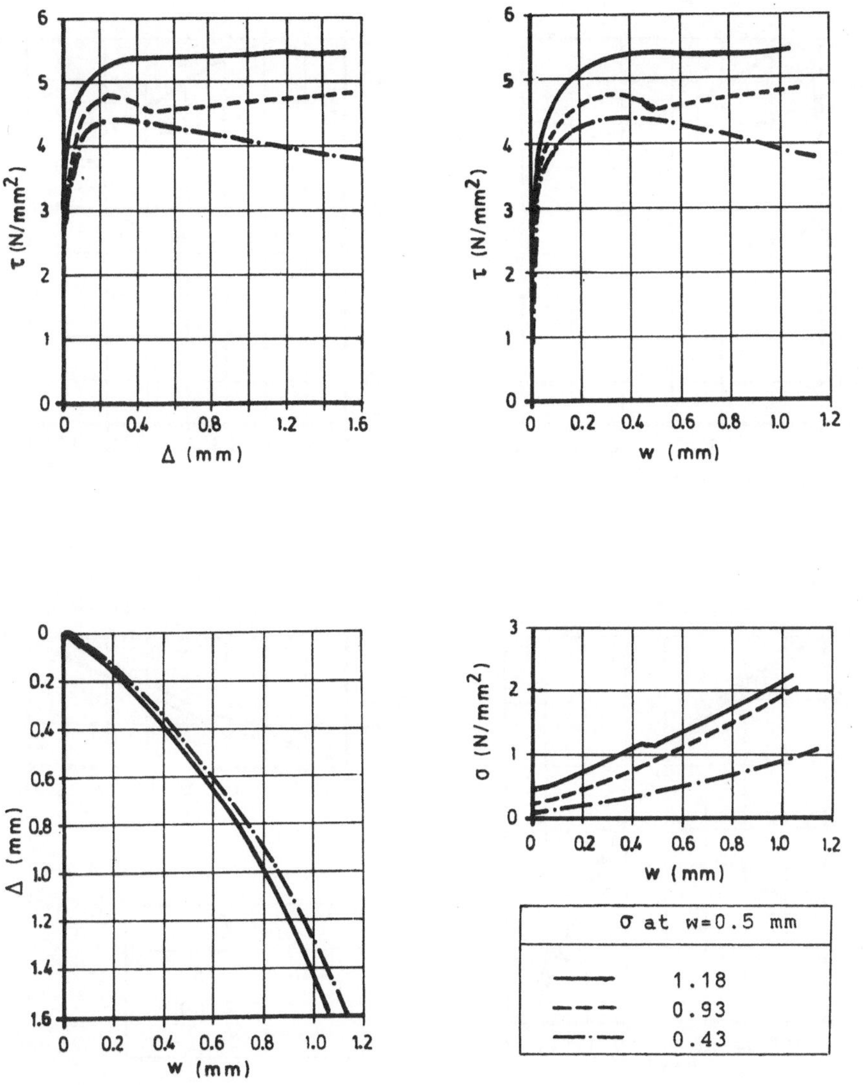

Fig. 5 Test Results for 3 Specimens with Vf=0.01, Shear Stress τ as Function of Shear Displacement Δ, Shear Stress as Function of Crack Opening w, Crack Opening Path, Normal Stress σ as Function of Crack Opening

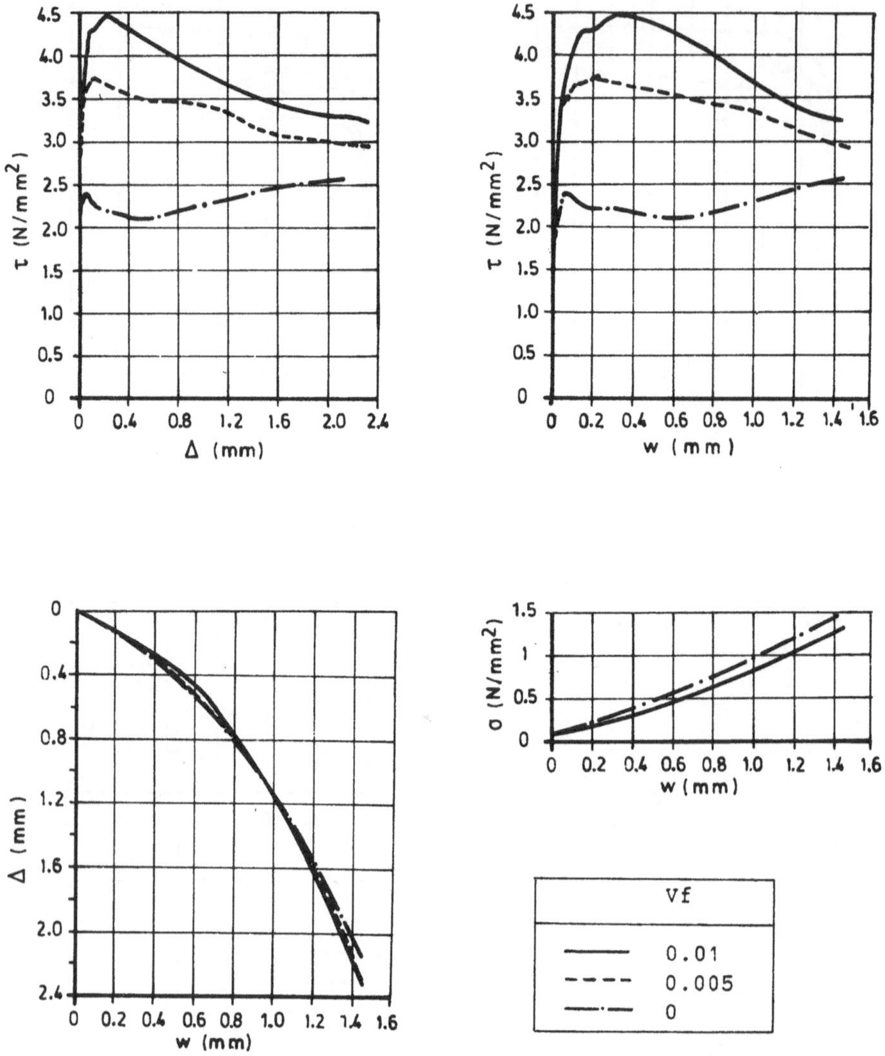

Fig. 6 Test Results for 3 Specimens with Different Fibre Contents, Shear Stress τ as Function of Shear Displacement Δ , Shear Stress as Function of Crack Opening w , Crack Opening Path, Normal Stress σ as Function of Crack Opening

The experimental results are summarized in Fig. 7, 8 and 9 for the different fibre contents. The normal stress and the shear stress are represented as a function of the shear displacement and the crack opening. The test results are compared with lineair equations. A regression analysis for $0.1 \leq w \leq 1$ yielded the following formulas for the concrete without fibres :

$$\tau = -1.21 + (3.83 \times w^{-1.26} - 1) \times \Delta \qquad \tau \geq 0 \qquad (N/mm^2 , mm)$$
$$\sigma = -1.82 + (3.27 \times w^{-1.03} - 1) \times \Delta \qquad \sigma \geq 0 \qquad (N/mm^2 , mm)$$

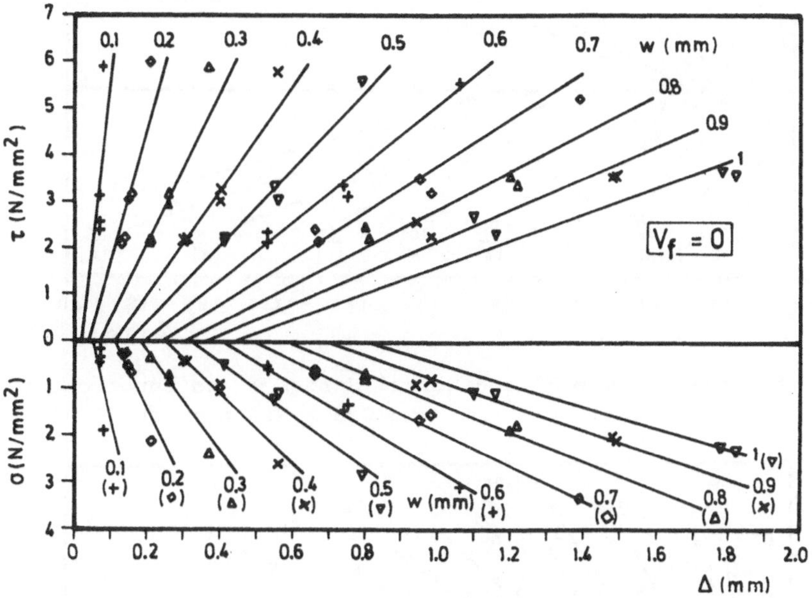

Fig. 7 Shear Stress τ and Normal Stress σ as Function of Shear Displacement Δ and Crack Width w for Vf=0

As the influence of the fibres on the normal stress is small, the same relation between the normal stress and the displacements can be used for the steel fibre reinforced concrete. A relation between the shear stress and the displacements for the steel fibre reinforced concrete ($0.005 \leq Vf \leq 0.01$) is proposed :

$$\tau = -1.21 + ([4 + 128 \times Vf] \times w^{-1.18} - 1) \times \Delta \qquad \tau \geq 0 \qquad (N/mm^2, mm)$$

This formula mostly underrates the shear stress in comparison with the experimental values. All these formulas only hold for a matrix strength of 35 N/mm^2 , an initial crack width of approximately 0.02 mm, the aggregates used in this program and the range of crack openings and shear displacements tested. It may be assumed that they will undergo

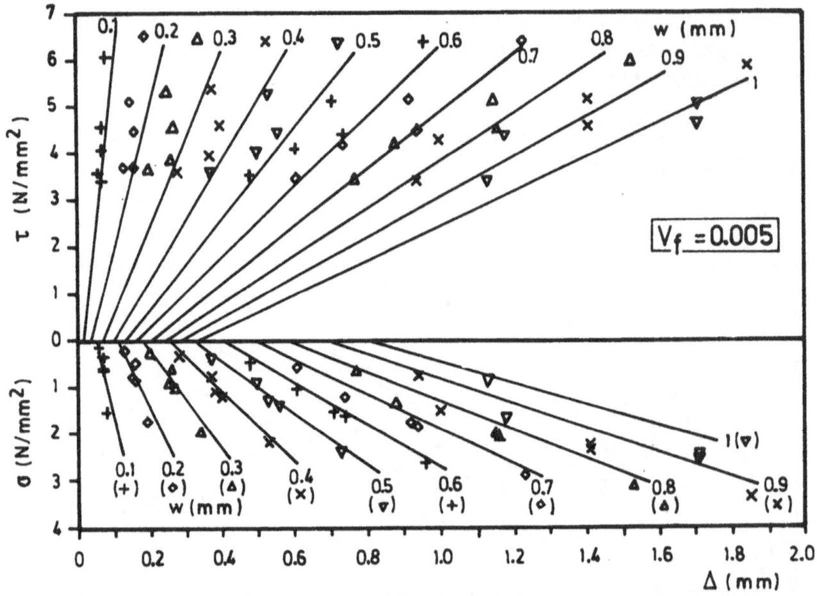

Fig. 8 Shear Stress τ and Normal Stress σ as Function of Shear Displacement Δ and Crack Width w for Vf=0.005

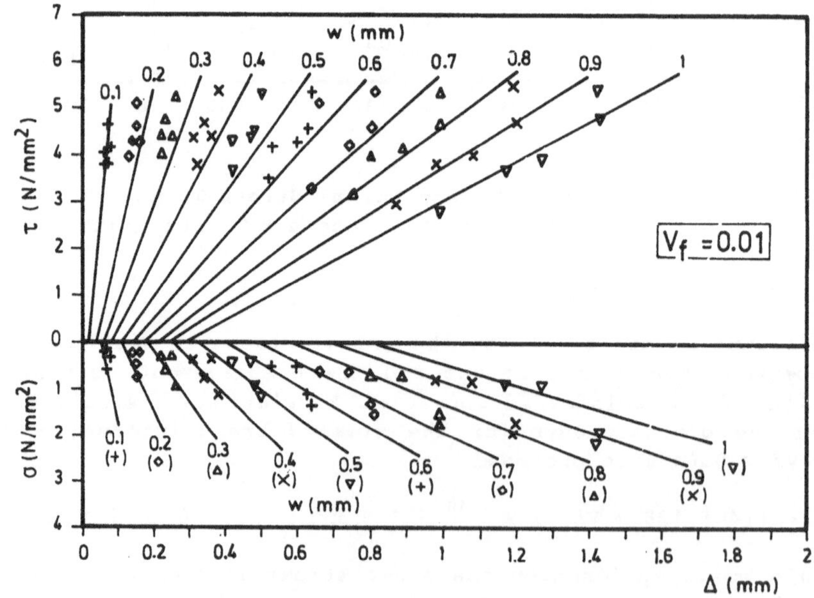

Fig. 9 Shear Stress τ and Normal Stress σ as Function of Shear Displacement Δ and Crack Width w for Vf=0.01

alteriations if one of these parameters is changed and if more test results become available.

CONCLUSIONS

The following conclusions may be derived from the experimental work performed :

1. Steel fibres increase the shear strength of push-off specimens. The influence of the fibres decreases with increasing normal restraint stiffness. The influence on the normal stress and the crack opening path is small for the fibre contents under investigation.
2. A first estimation of the stresses in a crack as a function of the displacements can be made with the formulas presented.
3. More research is needed to come to a constitutive model for the cracked region of fibre reinforced concrete. The influence of other parameters such as the concrete quality and the initial crack width should be investigated.

As the results of the tests have been encouraging up to now, the program will be continued. The main purpose is to get simple relations between stresses and displacements as functions of parameters such as the fibre content. These relations may be used in a model for the shear region of reinforced concrete with steel fibres as secondary reinforcement.

ACKNOWLEDGEMENTS

The research has been carried out at the Reyntjens Laboratory of the Department of Construction of the Katholieke Universiteit Leuven (Belgium) as part of a specialization study of the author under the promotion of prof. dr. ir. F. Mortelmans. The study was financially supported by the Institute for Scientific Research in Industry and Agriculture (IWONL).

REFERENCES

1. Muhidin, N.A. and Regan, P.E., Chopped steel fibres as shear reinforcement in concrete beams, Fibre reinforced Materials, Institution of Civil Engineers, London 1977, pp 163-168.
2. Uomoto, T., Weeraratne, R.K., Furukoshi, H. and Fujino, H., Shear strength of reinforced concrete beams with fibre reinforcement, Third International Symposium on Developments in Fibre Reinforced Cement and Concrete, Sheffield, 13-17 July 1986, Vol II, paper 8.7.
3. Walraven, J., Pat, T. and Markov, I., The punching shear resistance of fibre reinforced concrete slabs, Third International Symposium on Developments in Fibre Reinforced Cement and Concrete, Sheffield, 13-17 July 1986, Vol II, paper 8.9.
4. Swamy, R.N. and Ali, S.A.R., Punching shear behaviour of reinforced slab-column connections made with steel fibre concrete, ACI-Journal, Proceedings, Vol 79, No 5, Sept-Oct 1982, pp 392-406.

5. Fattuhi, N.I., Corbels with shear reinforcement in the form of stirrups and fibres, Third International Symposium on Developments in Fibre Reinforced Cement and Concrete, Sheffield, 13-17 July 1986, Vol II, paper 8.8.
6. Swamy, R.N. and Bahia, M., Influence of fibre reinforcement on the dowel resistance to shear, ACI-Journal, Proceedings, Vol 76, No 2, February 1979, pp 327-355.
7. Barr, B., Hasso, E.B.D. and Liu, K., Shear strength of FRC materials, Composites, Vol 16, No 4, October 1985, pp 326-334.
8. Walraven, J.C., Aggregate Interlock, An experimental and analithical analysis, Doctor's Thesis, Delft University of Technology, Oct 1980.

Proceedings of the International Symposium on Fibre Reinforced Concrete, December 16–19, 1987, Madras, India

STEEL FIBERS AS SHEAR REINFORCEMENT IN REINFORCED CONCRETE T-BEAMS

G.B. BATSON, CLARKSON UNIVERSITY, POTSDAM, NEW YORK, USA
C. ALGUIRE, SARGENT LUNDY, CHICAGO, ILLINOIS, USA

SYNOPSIS

The effectiveness of hooked end steel fibers and/or vertical stirrups as shear reinforcement was evaluated by testing eleven reinforced concrete T-beams simply supported and loaded at mid span. Only the volume percentage of the steel fibers and the spacing of the stirrups were varied. The test data for the T-beams indicated that steel fibers were most effective when combined with widely spaced vertical stirrups, three times the normal design spacing, and provided greater shear strength than could be obtained by either vertical stirrups or steel fibers separately. Test results are compared with published equations proposed to predict the shear capacity and the moment capacity of steel fiber reinforced concrete beams. There is a short discussion of the applicability of the lower bound theorem of limit state analysis for predicting the shear capacity of the beams that fail in shear.

INTRODUCTION

Investigations of the effectiveness of steel fibers to enhance the strength and ductility of plain and reinforced concrete have been conducted for nearly twenty-five years. These investigations have covered a wide range of strength properties; such as flexure, tensile, compressive, shear, and torsion, impact resistance, ductility and flexural fatigue [1]. The enhancement of the strength properties by steel fibers compared to plain concrete ranges from an order of magnitude increase for impact resistance and ductility to a negligible increase in the compressive strength. Steel fibers also improves the corrosion resistance of concrete to salt water and deicing agents, retards or controls the cracking due to shrinkage of the concrete, increases the resistance of concrete and refractory cements to thermal gradients, and reduces the spacing and the width of cracks in concrete.

Steel fibers in sufficient quantity can increase the shear capacity to the extent that beams will fail in flexure [2]. Also, there is the potential for savings in the construction costs of fabrication and placing of stirrups in the forms by the reduction in the number or possibly the elimination of vertical stirrups for shear reinforcement in beams. However, flexural reinforcing steel will still be required in beams to resist the bending stresses from applied loads.

Testing of reinforced concrete beams in flexure has shown that steel fibers are effective as shear reinforcement either by themselves or in combination with vertical stirrups in simply supported beams and two span continuous beams [3]. The minimum volume percentage of steel fibers required for shear reinforcing of a concrete beam that fails in flexure depends on the shape and size of the fibers; the necessary volume percentage of deformed fibers being less than for straight fibers. Nearly all test data published on shear capacity indicates that flexural failures develop when the fiber volume percentage exceeds some minimum value. However, a limited number of beams tests with large cross sections, prototype size, develop increasing shear capacity with increasing fiber content, but not enough to prevent a shear failure [4].

Only test results for beams with a solid rectangular cross section had been published for assessing the effectiveness of steel fibers as shear reinforcement. Therefore a testing program was initiated using T-beams with steel fibers and/or vertical stirrups for the shear reinforcement to resist the more severe shear stress distribution that exists in T-beams than beams with a solid rectangular section. The volume percentage of steel fibers and the spacing of the stirrups were varied, but the span, cross section, area of flexural steel and the loading condition were held constant. With the neutral axis in the flange, the T-beam acted like a solid rectangular beam in flexure but the shear was resisted by the smaller cross section of the stem and the flange. (It should be noted that these tests were ongoing at the time Swamy's extensive research on the effectiveness of steel fibers for rectangular beams and T-beams was published [5].)

TESTING PROGRAM

The testing program consisted of eleven T-beams varying only the volume percentage of steel fibers and the stirrup spacing as shown in Table 1. The dimensions of the T-beams are shown in Figure 1.

TABLE 1 T-BEAM SPECIMENS

BEAM NO.	FIBER VOLUME %	STIRRUP SPACING mm (in)	f'_c MPa (ksi)
1	0.2	---	58.5 (8.49)
2	0.3	---	47.1 (6.83)
3	0.5	---	50.0 (7.25)
4	0.2	368 (14.5)	50.3 (7.29)
5	0.3	368 (14.5)	46.8 (6.79)
6	0.5	368 (14.5)	49.4 (7.16)
7	0.0	---	56.5 (8.19)
8	0.0	368 (14.5)	54.4 (7.89)
9	0.0	127 (5.0)	59.8 (8.67)
10	1.0	368 (14.5)	45.9 (6.95)
11	1.5	368 (14.5)	42.0 (6.10)
		Average	51.2 (7.42)

Specimen 9 was designed according to the requirements of ACI 318-83 [6]. The flexural reinforcing steel had an average yield stress of 496 MPa (72 ksi) and the stirrups had an average yield stress of 434 MPa (63 ksi). The steel fibers were 30 mm (1.18 in) long with hooked ends and 0.50 mm (0.02 in) in diameter. The concrete mix is given in Table 2.

TABLE 2 CONCRETE MIX

9.5 mm (3/8 inch) peastone	823.4 kg/m^3	(51.4 $\#/ft^3$)
sand	825.0 "	(51.5 ")
cement	408.5 "	(25.5 ")
water	245.1 "	(15.3 ")
admixture (water reducer)	4.3 L/m^3	(122.3 ml/ft^3)

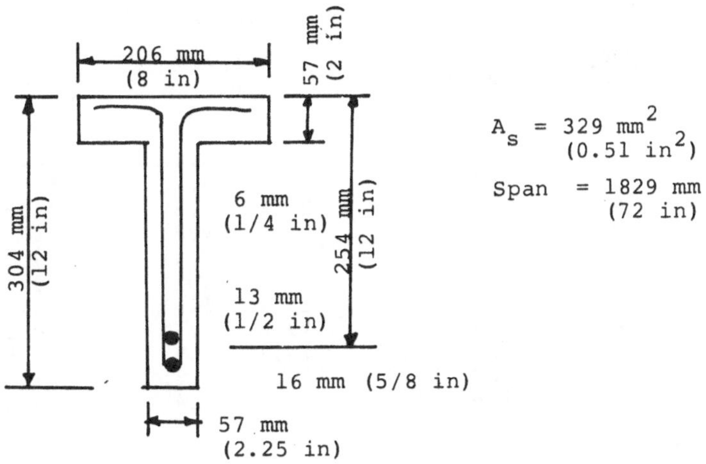

Figure 1. T-Beam Section

T-beams were cast in wood forms in three lifts and externally vibrated to consolidate the fibrous concrete. One standard cylinder was made of each beam. Specimens remained in the forms for two days under plastic sheets and then removed from the forms and stored in a moisture curing room for at least 28 days before testing. Two days prior to testing, a beam was removed from the moisture room to dry in the laboratory before a coat of common white wash was applied to aid in locating cracks. Dial gages measured the deflection at the center line and third points on the bottom of the flange.

A T-beam was supported at the ends of its span on steel bearing plates resting on simple knife edge supports. A universal hydraulic testing machine applied the load at mid span by a round rod and a bearing plate resting on the top flange. Load increments of 0.89 kN (200 pounds) were used up to a load or 40 kN (9000 pounds) and there after 2.22 kN (500 pound) increments until failure. The number, location and extension of the cracks were marked on the white wash coating at the end of each load increment. Photographs of the crack patterns were taken after the completion of each beam test.

TEST RESULTS

The test data for the T-beams are given in Table 3 from which some inferences can be drawn. First, there is much less range in the values of the first crack load than for the ultimate load. It appears that beams with steel fibers combined with stirrups have a lower first crack load than for beams with steel fibers only, while the beam without stirrups or fibers had the highest first crack load. The ultimate load is greater for the beams with steel fibers and widely spaced stirrups than for beams with only steel fibers or stirrups. Secondly, the data indicates that a minimum volume percentage of about 0.5 is required to change the mode of failure from shear to flexure which is consistent with most of the published research [7].

Table 4A and 4B provide data on the shear capacity of the T-beams evaluated by ACI 318-83 [6] and equations published by Sharma [8]. Several researchers have published equations to predict the shear capacity of steel fiber reinforced concrete beams, the data available from this project was convenient to use with Sharma's equations [9,10].

TABLE 3 TEST RESULTS

BEAM #	FIBER VOL	STIRRUP SPACING mm	(in)	1ST CRACK LOAD kN	(K)	ULTIMATE LOAD kN	(K)	MODE OF FAILURE
1	0.2	--	--	28.5	(6.4)	55.6	(12.5)	S
2	0.3	--	--	29.1	(5.2)	55.6	(12.5)	S
3	0.5	--	--	21.4	(4.8)	54.3	(12.2)	S
4	0.2	368	(14.5)	21.4	(4.8)	74.3	(16.7)	S
5	0.3	368	(14.5)	22.2	(5.0)	79.2	(17.8)	S
6	0.5	368	(14.5)	18.7	(4.2)	103.6	(23.3)	F
7	--	--	--	26.7	(6.0)	42.3	(9.5)	S
8	--	368	(14.5)	19.6	(4.4)	60.0	(13.5)	S
9	--	127	(5.0)	14.2	(3.2)	92.5	(22.3)	F
10	1.0	368	(14.5)	25.8	(5.8)	99.2	(22.3)	F
11	1.5	368	(14.5)	19.6	(4.4)	103.6	(23.3)	F

S Shear Failure
F Flexure Failure

TABLE 4A SHEAR CAPACITY

BEAM (1)	MODE OF FAILURE (2)	ULTIMATE LOAD ACI (3) kN	(K)	TEST (4) kN	(K)	COL (4) / COL (3) (5)
1	S	79.2	(17.8)	55.6	(12.5)	0.70
2	S	78.7	(17.7)	55.6	(12.5)	0.71
3	S	78.7	(17.7)	54.3	(12.2)	0.69
4	S	78.7	(17.7)	74.3	(16.7)	0.94
5	S	78.7	(17.7)	79.2	(17.8)	1.01
6	F	78.7	(17.7)	103.6	(23.3)	1.32
7	S	79.2	(17.8)	42.3	(9.5)	0.53
8	S	79.2	(17.8)	60.0	(13.5)	0.76
9	F	79.2	(17.8)	92.5	(20.8)	1.17
10	F	78.7	(17.7)	99.2	(22.3)	1.26
11	F	78.3	(17.6)	103.6	(23.3)	1.32

TABLE 4B SHEAR CAPACITY

BEAM (1)	MODE OF FAILURE (2)	SHEAR CAPACITY SHARMA (3) kN	(K)	TEST (4) kN	(K)	COL (4) / COL (3) (5)
1	S	42.7	(9.6)	27.8	(6.25)	0.65
2	S	38.3	(8.6)	27.8	(6.25)	0.73
3	S	39.6	(8.9)	27.1	(6.1)	0.68
4	S	58.7	(13.2)	37.4	(8.4)	0.64
5	S	57.4	(12.9)	39.6	(8.9)	0.69
6	F	58.7	(13.2)	51.8	(11.7)	0.88
7	S	18.2	(4.1)	20.9	(4.7)	1.15
8	S	37.4	(8.4)	30.3	(6.8)	0.81
9	F	49.4	(11.1)	46.3	(10.4)	0.94
10	F	57.8	(13.0)	49.4	(11.1)	0.85
11	F	55.6	(12.5)	51.8	(11.7)	0.93

Sharma assumes that the ultimate shear stress is the sum of contribution to shear by the stirrups and fiber concrete.

$$v_{nf} = v_{cf} + v_s \qquad \text{Eq 1}$$

where

$$v_s = A_s f_y d/s \qquad \text{Eq 1a}$$

and

$$v_{cf} = kf' \, (d/a)^{0.25} \qquad \text{Eq 1b}$$

and k equals 0.67. (Note that there is a typographical error in Sharma's paper; the exponent 0.25 applies to the ratio (d/a).) Sharma suggests that the tensile splitting strength be obtained from compressive strength by

$$f_t' = 9.5(f_c')^{0.5} \qquad \text{Eq 1c}$$

where the units of f_c' and f_t' are psi and f_t' must be converted to N/mm for Eq. 1b.

As would be expected ACI procedures underestimate the ultimate load for beams with stirrups and a steel fiber content greater than about 0.5 volume percentage. Likewise ACI procedures overestimates the ultimate load for the T-beams that fail in shear because ACI designs provide ductile moment failures, not brittle shear failures. Sharma's equations appear to be good for predicting shear capacity of T-beams that fail in flexure, but not very good for the T-beams that fail in shear, particularly for steel fiber contents less than 0.5 volume percent.

Table 5 provides data on the moment capacity of the T-beams. ACI design procedure underestimates the ultimate moment capacity of the T-beams with sufficient steel fiber content and stirrups to fail in flexure, and overestimates the moment capacity of beams that fail in shear. Among the various methods proposed for computing the moment capacity of steel fiber reinforced concrete beams, the equations developed by Henager [11] is used for the T-beams. Henager's equations are good for moment failures, but they overestimate the moment capacity for the T-beams that fail in shear.

TABLE 5 MOMENT CAPACITY

BEAM # (1)	MODE OF FAILURE (2)	M_u ACI (3)		M_u TEST (4)		M_u HENAGER (5)		COL (5) / COL (4) (6)
		kN-m	(k-in)	kN-m	(k-in)	kN-m	(k-in)	
1	S	36.3	(321)	25.4	(225)	41.4	(366)	1.63
2	S	36.1	(319)	25.4	(225)	41.6	(368)	1.63
3	S	36.1	(319)	24.9	(220)	42.7	(378)	1.72
4	S	36.1	(319)	35.2	(301)	41.1	(363)	1.21
5	S	36.0	(318)	36.2	(320)	41.6	(368)	1.15
6	F	36.1	(319)	47.4	(419)	42.7	(378)	0.90
7	S	36.2	(320)	19.3	(171)	40.3	(356)	2.08
8	S	36.2	(320)	27.5	(243)	40.3	(356)	1.47
9	F	36.3	(321)	40.7	(360)	40.3	(356)	0.99
10	F	36.0	(318)	45.4	(401)	45.2	(400)	1.00
11	F	35.8	(317)	47.4	(419)	47.1	(417)	0.99

The shear and moment capacities of the T-beams reinforced with steel fibers and/or stirrups are consistent with other published research [5]. This data indicates that a minimum volume percentage of about 0.5 combined with widely spaced stirrups provides sufficient shear capacity for the T-beams to fail in flexure. Similar results have been obtained for solid rectangular simply supported beams and two span continuous beams [2,12].

Representative load deflection curves are shown in Figures 2, 3, 4 and 5 for the range of steel fiber contents and the wide stirrup spacings used for the shear reinforcement. The deflection plotted is the difference between the average of the third points and the mid span deflection. The curves show increasing stiffness with increasing fiber content.

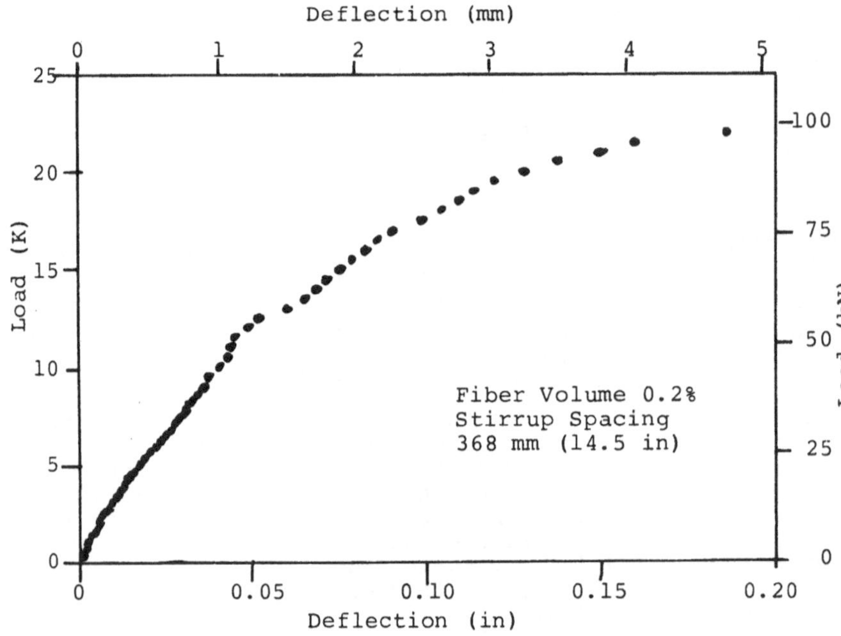

Figure 2.- Load Deflection

SHEAR CAPACITY OF FIBER CONCRETE AND PLASTICITY THEOREMS

The traditional shear theory is that the external shear of a diagonally cracked beam is resisted by several internal mechanisms; shear strength of the uncracked concrete in compression, aggregate interlock, dowel action of the flexural steel, stirrups and arch action for deep beams. Steel fibers enhance to some degree each of these internal mechanisms, but it is difficult to quantify the contribution of fibers to each mechanism [5]. The truss analog is commonly used to explain how vertical or inclined stirrups provide shear reinforcement [13]. Kani proposed that a series of internal arches form that are supported by the stirrups and the flexural steel [14]. More recent developments are based on compression field theory concepts from plasticity theory and truss analog. The plasticity concepts are well established for plain and reinforced concrete [15]. The 1984 Canadian Concrete Code has adopted the compression field concepts of plasticity as a rational method for the shear design of reinforced concrete beams [16].

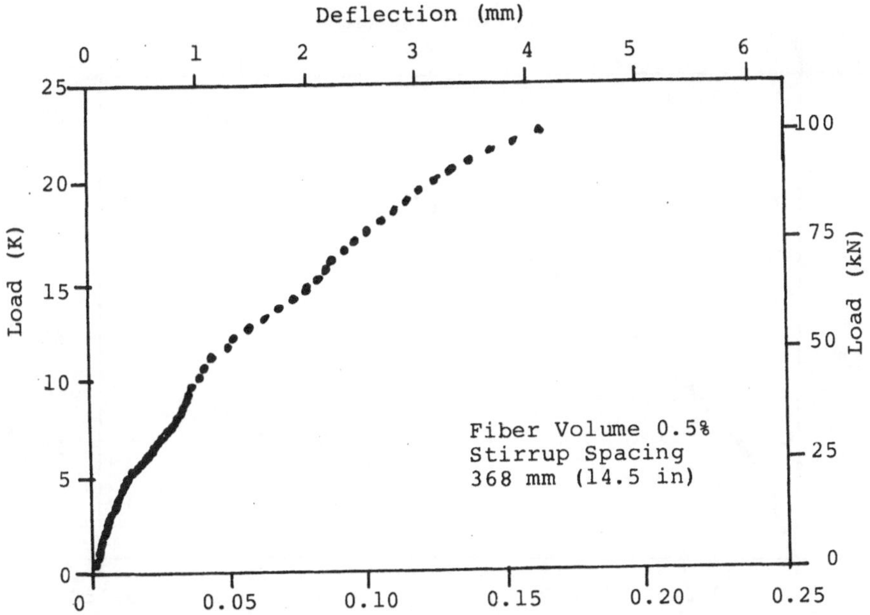

Figure 3. Load Deflection

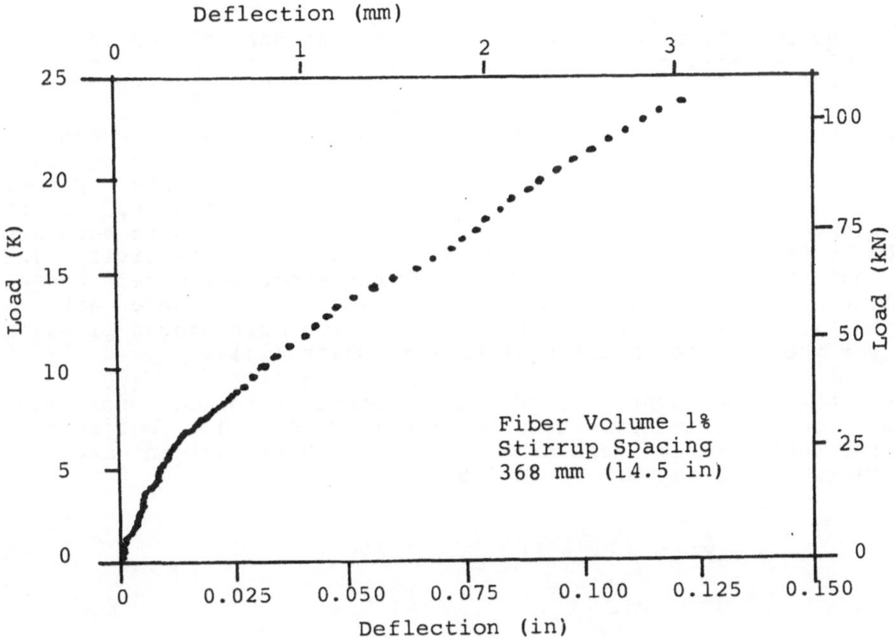

Figure 4. Load Deflection

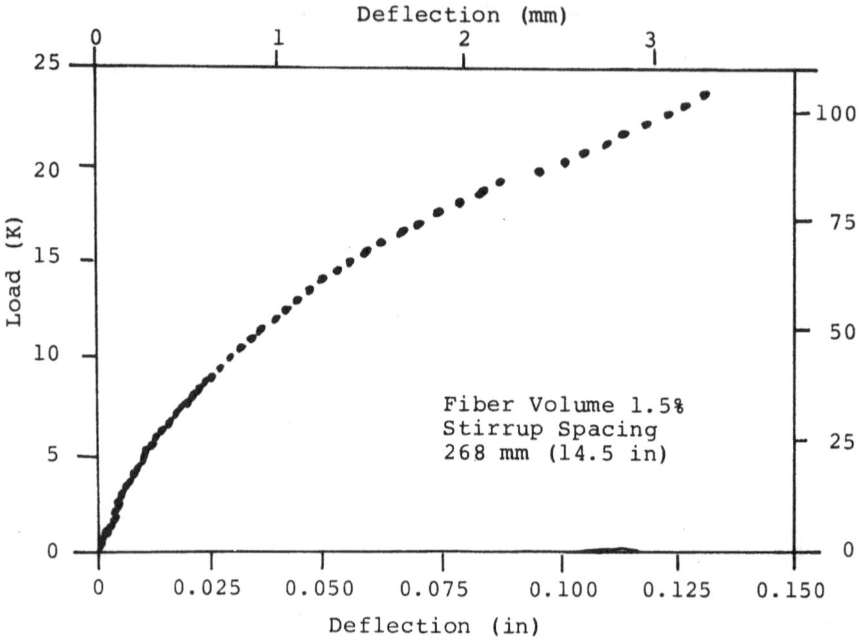

Figure 5. Load Deflection

The incorporation of steel fibers modifies many of the mechanical poperties of concrete, a brittle material with little tensile strength, into a relatively ductile material with much improved fracture toughness. The ductility is a significant aspect for many of the potential applications of fiber reinforced concrete in seismic and blast resistent structures. Thus it would appear that the plasticity theory for concrete might apply to steel fiber reinforced concrete. Research published on the torsional loading of steel fiber reinforced concrete beams indicates that the ultimate torsional strength is quite accurately predicted by Nadai's classical "sand heap" model of plasticity [17]. Noting that the equations used to predict the shear and moment capacity of the T-beams that failed in shear in this project were not very satisfactory, it was decided to consider the lower bound plasticity theorem for the concrete beams that fail in shear [15].

The lower and upper bound limit state solutions for simply supported beams with vertical or inclined stirrups subjected to symmetrical concentrated loads or a single concentrated load are given by M.P. Nielsen [15] and W.F. Chen [18].

$$\frac{\tau}{f'_c} = \begin{cases} \sqrt{\psi(1-\psi)} & \psi \le 1/2 \\ \frac{1}{2} & \psi > 1/2 \end{cases} \qquad \text{Eq 2}$$

where

$$\psi = \frac{\rho f_y}{f'_c} \qquad \text{Eq 2a}$$

is called the degree of shear reinforcement. Equation 2 for the lower bound is a circle with coordinates τ/f'_c and ψ shown in Figure 6.

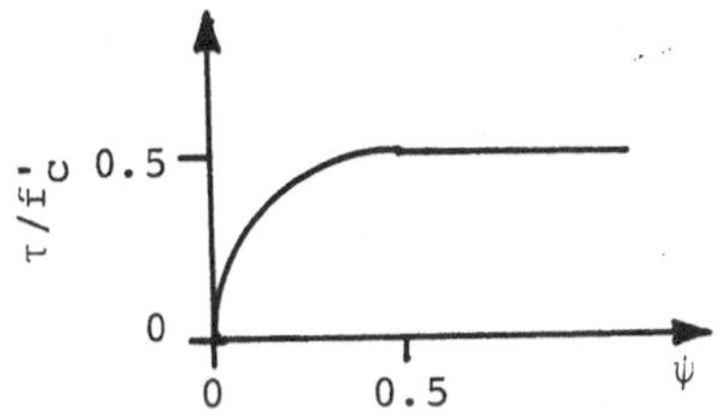

Figure 6. Lower Bound Solution

To compare the test data with the lower bound theorem requires the introduction of an effective strength factor, $0 < \nu < 1.0$ []. It is well known that for a good correlation between test data and plasticity solutions, it is necessary to use a modified or reduced compressive strength that is less than the compressive strength obtained from the standard compression cylinder test [18]. The compressive strength of the concrete in the web or stem of the beam must be $\nu f'$. The lower bound theorem in terms of the effective strength factor, ν, is

$$\frac{\tau}{f'_c} = \begin{cases} \sqrt{\psi(\nu-\psi)} & \psi \leq \frac{\nu}{2} \\ \frac{1}{2} & \psi > \frac{\nu}{2} \end{cases} \qquad \text{Eq 3}$$

where

$$\tau = \frac{Q}{bh} \qquad \text{Eq 3a}$$

where Q is the shear force acting on the cross section. Several researchers have proposed expressions for the effective strength factor. Nielsen [15] proposed

$$\nu = 0.8 - \frac{f'_c}{200} \text{ (MPa)} \qquad \text{Eq 4}$$

Based on the average value of the compressive strength of the concrete of 51.5 MPa (7.4 ksi), $\nu = 0.54$. To what extent the value of ν is effected by the fibers in the concrete is not known. Perhaps only slightly because the fibers do not increase the compressive strength very much. Research needs to be done on determining the appropriate value of ν for fiber concrete. The value of ψ in the lower bound theorem should change significantly because it depends directly on the percentage of shear reinforcement, ρ

$$\rho = \frac{A_s}{eb} \qquad \text{Eq 5}$$

where A_s = area of stirrups crossing the plane defined by eb
e = stirrup spacing
b = width of stem

To account for the steel fibers in the term ψ, the volume of fibers between stirrups was converted into an equivalent area of vertical steel stirrups and used in plotting the data for the beams that failed in shear in Figure 7. The data plotted in Figure 7 falls below the lower bound which should be lowest load for shear failure. It appears that improved values of the two parameters, and , are required before plasticity theory of concrete can be applied to steel fiber reinforced concrete.

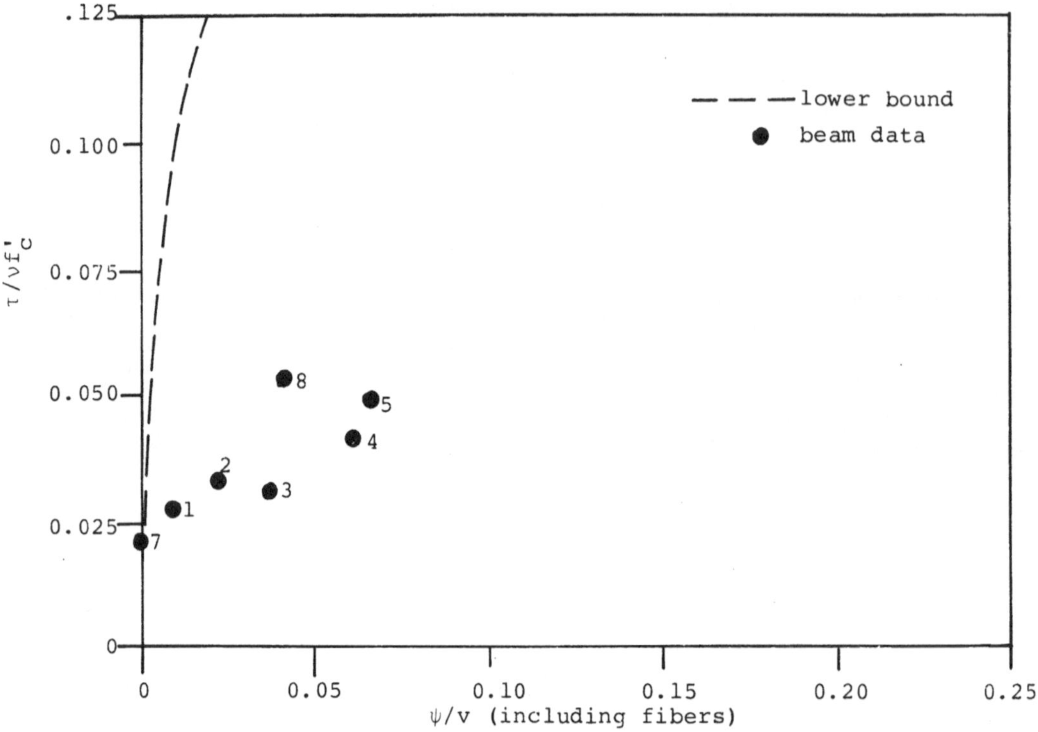

Figure 7. Lower Bound Section for T-Beams that Failed in Shear

CONCLUSIONS

For the size and loading of the T-beams investigated, a combination of steel fibers and widely spaced stirrups were more effective as shear reinforcement than either steel fibers or stirrups separately. It appears that a minimum volume percentage of fibers is about 0.5 to achieve flexure failures. These results are consistent with published results for solid rectangular beams.

Equations proposed to predict the shear and moment capacity were not satisfactory for T-beams that failed in shear, but were satisfactory for flexure failures.

The data for the T-beams that failed in shear did not satisfy the lower bound limit state theorem of plasticity for concrete beams with vertical stirrups. Research is needed to develop appropriate values for the two parameters, the effective strength factor and the degree of shear reinforcement.

REFERENCES

[1] ACI Committee 544, State-of-the-Art Report on Fiber Reinforced Concrete (ACI 544.1R-82) American Concrete Institute, Detroit, 1982, 16 pp.

[2] Batson, G., Jenkins, E., Spatney, R., "Steel Fibers as Shear Reinforcement in Beams," ACI Journal Proceedings, Vol. 69, No. 10, Oct. 1972, pp. 640-644.

[3] Batson, G., "Use of Steel Fibers for Shear Reinforcement and Ducitlity," US-Sweden Joint Seminar (NSF-STU), S. Shah and A. Skarendahl, editors, Swedish Cement and Concrete Research Institute, Stockholm, 1985.

[4] Williamson, G.R., Knab, L.I., "Full Scale Fibre Concrete Beam Tests," Fibre Reinforced Cement Concrete RILEM Symposium 1975, Editor A. Neville, The Construction Press Ltd., England.

[5] Swamy, R.N., Bahia, H.M., "The Effectiveness of Steel Fibers as Shear Reinforcement," Concrete Internaitonal, Vol. 7, No. 3, March 1985, pp. 35-40.

[6] American Concrete Institute, "Building Code Requirements for Reinforced Concrete ACI 318-83," American Concrete Institute, Detroit, Michgian, 1983.

[7] Mansur, M.A., Ong, K.D.G., Paramasivam, P., "Shear Strength of Fibrous Concrete Beams Without Stirrups," Journal of Structural Engineering, ASCE, Vol. 112, No. 9, September 1986, pp. 2066-2079.

[8] Sharma, A.K., "Shear Strength of Steel Fiber Reinforced Concrete Beams," J. of American Concrete Institute Proceedings, Vol. 83, No. 4, July-August 1986, pp. 624-628.

[9] Narayanan, R., Darwish, I.Y.S., "Use of Steel Fibers as Shear REinforcement," ACI Structural Journal, Vol. 84, No. 3, May-June 1987, pp. 216-227.

[10] Jindal, R.L., "Shear and Moment Capacities of Steel Fiber Reinforced Concrete Beams," Fiber Reinforced Concrete International Symposium, ACI Publication Sp-81, American Concrete Institute, Detroit, Michigan, 1984, pp. 1-16.

[11] Henager, C.H., Doherty, T.J., "Analysis of Reinforced Concrete Fibrous Concrete Beams," Proceedings, ASCE, V 102, ST-1, Jan. 1976, pp. 177-188.

[12] Bollana, R.D., "Steel Fibers as Shear Reinforcement in Two Span Continuous Reinforced Concrete Beams," Unpublished MS Thesis, Clarkson University, Potsdam, NY, May 1980.

[13] Wang, C.K., Salmon, _Reinforced Concrete Design_, 4th Edition, Harper & Row Publishers, New York, 1985.

[14] Kani, G.N.J., "A Rational Theory for The Funciton of Web Reinforcement," J. of American Concrete Institute Proceedings, Vol. 66, No. 3, Mar 1969, pp. 185-197.

[15] Nielsen, M.P., _Limit Analysis and Concrete Plasticity_, Prentice Hall, New Jersey, 1984.

[16] Collins, M.P., Mitchell, D., "A Rational Approach to Shear Design - The 1984 Canadian Code Provisions," J. of American Concrete Institute Proceedings, Vol. 83, No. 6, Nov.-Dec. 1986, pp. 925-933.

[17] Mansur, M.A, Paramasivam, P., "Steel Fiber Reinforced Concrete Beams in Pure Torsion," The Int. Journal of Cement Composites and Lightweight Concrete, Vol. 4, No. 1, Feb. 1982, p. 39.

[18] Chen, W.F., _Plasticity in Reinforced Concrete_, McGraw Hill, New York, 1982.

Proceedings of the International Symposium on Fibre Reinforced Concrete, December 16–19, 1987, Madras, India

FIBRE REINFORCED CONCRETE BEAMS SUBJECTED TO SHEAR FORCE

D.S.R. MURTY
Andhra University
Waltair, INDIA

T. VENKATACHARYULU
Regional Engineering College
Warangal, INDIA

SYNOPSIS

This investigation was designed to provide a comprehensive experimental and analytical evaluation of steel fibres as shear reinforcement. To obtain this information, a total of eleven rectangular beams were tested to failure. Variables that were studied included shear span/depth ratio, aspect ratio of fibres and volume percentage of fibres. The inclusion of steel fibres in the concrete provided effective reinforcement against shear failure. The shear resistance increased with increasing aspect ratio of fibres and volume fraction of fibres. The method incorporated in the Draft European Code [1] for concrete structures was adopted to predict shear strength of fibre reinforced sections. In this mehod, the post cracking tensile resistance of the web was estimated according to Muhidin [2]. Comparisons were presented for theoretical and test results at ultimate and the agreement was generally found to be good.

INTRODUCTION

Shear failure of a reinforced concrete beam loaded in flexure, occurs by the tensile failure of concrete in shear span. The use of steel fibres as web reinforcement in reinforced concrete beams appears attractive due to enhancement of first crack tensile strength and ultimate tensile strength, resulting in increase in shear strength and possible prevention of shear failures. Steel fibres have some advantage over vertical stirrups or bentup flexural reinforcement in resisting shear force. Conventional stirrup reinforcement requires a high labour input for bending and fixing in position. In thin webs, placement of stirrups becomes completely impractical. In long line prestressed concrete work provision of stirrups is difficult. Investigations [3-7] directed specifically to beams reinforced with steel fibres have been very few, but they have confirmed the effectiveness of fibres as shear reinforcement.

EXPERIMENTAL PROGRAMME

Test Specimens

The experimental programme involved tests on eleven reinforced concrete rectangular beams of same cross-section under two point loading. The beams were divided into two series based on shear span/depth ratio. Among the series either volume fraction or aspect ratio of fibres were varied. The beams were designed to fail in shear with flexural reinforcement and fail to moment capacity when vertical stirrups were provided. Out of the eleven beams, two beams were without any web reinforcement, two had conventional stirrups and the remaining seven were provided with fibres in the test zone. The non-test zone was reinforced against shear failure by providing web reinforcement in the form of vertical stirrups in all the beams. The nominal details of beam cross-sections, and the arrangement of reinforcement are shown in Fig. 1. The details of the test specimens are presented in Table-I.

Bar Reinforcement

Two numbers of 12 mm high yield strength deformed bars were used as flexural reinforcement. The test zone was on shear span with moment gradient and contained either no shear reinforcement, stirrups or steel fibres. 6 mm diameter mild steel bars were used for stirrups. The yield stress of flexural reinforcement and stirrup reinforcement was 510 and 260 N/mm^2 respectively.

Steel Fibres

The fibres were produced by chopping binding wire having diameter 0.54 mm. Same type of fibre was used throughout the programme. The fibres were circular in

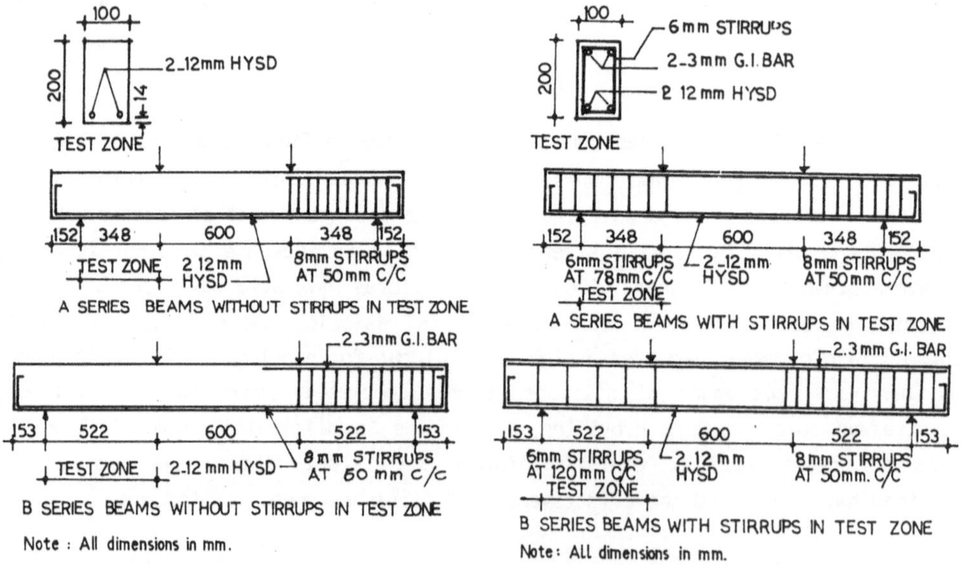

Fig. 1 : Reinforcement Details

TABLE - 1 : DETAILS OF TEST SPECIMENS *

Beam No.	Shear span/ depth	Fibre Details: Diameter (mm)	Length (mm)	Aspect ratio	Fibre content by volume (%)	Cube strength (N/mm^2)	Split cylinder strength (N/mm^2)	Age at test (days)	Shear reinforcement in test zone
A21	2	-	-	-	0.0	22.5	1.87	18	-
A22	2	0.54	27	50	0.5	28.7	2.65	25	Fibres
A23	2	0.54	54	100	0.5	32.2	2.88	25	Fibres
A24	2	0.54	27	50	1.0	29.0	2.67	23	Fibres
A25	2	0.54	54	100	1.0	32.6	2.90	23	Fibres
A26	2	-	-	-	0.0	27.5	1.93	25	Stirrups
B31	3	-	-	-	0.0	26.8	1.90	23	-
B32	3	0.54	27	50	1.0	32.1	2.87	23	Fibres
B33	3	0.54	54	100	1.0	32.3	2.88	25	Fibres
B34	3	0.54	27	50	1.5	32.8	2.92	23	Fibres
B35	3	-	-	-	0.0	25.6	1.91	23	Stirrups

* Longitudinal flexural reinforcement in all beams = 2 Nos. of 12mm diameter.

cross-section. The length of fibres was either 27 mm or 54 mm resulting in aspect ratios, L/D equal to 50 and 100 where L is length of fibre and D is diameter of fibre. The fibre percentages by volume were 0.5, 1 and 1.5.

Fabrication of Specimens

Beams were cast in steel moulds. The mixing was carried out on a horizontal platform manually. The procedure consisted of charging on the platform sand, gravel and cement. The fibre was added in dry condition and mixed for three minutes. Fibres were fed in small quantities by using a hand sieve in order to minimise balling and interlocking. All the water was then added and the mixing time continued for three more minutes. The casting schedule for each mix comprised one beam, three 150 mm cubes and three 150 x 300 mm cylinders. The specimens were compacted uniformly by a needle vibrator. All the beams were cast in the same direction in which they were tested subsequently. One day after casting, the beams and control specimens were demoulded and cured in water. The age at test, varied from 23 to 25 days except one beam which was tested at 18 days.

Instrumentation of Specimens

The test specimens were equipped to measure vertical deflections, rotations at the support and concrete surface strains in the test zone. Dial gauges were used to measure vertical deflections at the load points and mid point of the beam. For measurement of rotation, two dial gauges were mounted, one at the support and the other at a distance of 100 mm from the support. The two dial gauges recorded the vertical deflection of the top face of the beam. The difference between the two gauges divided by the distance between them yielded rotation. Concrete surface strains were obtained by demec gauge readings. The full length of the test zone, which was in shear, was instrumented with demec gauges on one side face of the beam.

Test set-up and Procedure

A general view of test set-up and typical locations of various measurements are shown in Fig.2. All the beams were simply supported and tested under two point loading. The same procedure was used for testing all the beams. The transverse load was applied and measured by the universal testing machine. Each beam was tested to failure by applying the transverse load in a series of increment. The entire test for a beam took about 2 hours.

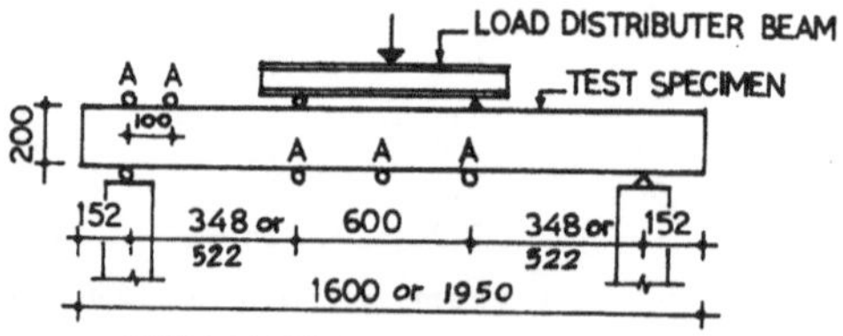

BEAM LOADING ARRANGEMENT

Note : All dimensions in mm.

Fig. 2 : Test Setup

TEST RESULTS AND DISCUSSION

The principal test results are furnished in Table-II. A large amount of data was obtained from the tests. Only those relevant to the paper and representative of the strength of the beams are presented here.

Initial cracks in all the beams originated either in uniform moment zone or moment gradient zone. Initial cracks were flexure shear cracks relative to shear failure; the shear failure of the members did not follow immediately inclined cracking, both in beams without any web reinforcement and in the fibre reinforced beams. The inclined cracks stabilised and subsequently more shear force was applied before the member failed. The margin between the cracking and the ultimate loads for shear failures varied from 38% to 109%; 109% related to one percent fibre beams and 38% to plain concrete beams. The margin between cracking and ultimate strengths increased as the aspect ratio of fibres increased, while fibre concentration was same. As the fibre content increased, the cracking strength increased, whether shear cracking or flexure cracking. Initial flexural cracking occured in the beams provided with stirrup reinforcement, leading to flexural failure. There was considerable margin between the first crack and ultimate load, the margin varied from 147 to 178%.

Ultimate Strength

Flexural failure occured in beams with conventional stirrup reinforcement. Flexural failure instead of shear failure occured in beams that developed sufficient shear strength due to incorporation of steel fibres. In other words, a beam that should have failed in shear, failed in flexure. Flexural failures of fibre beams were not sudden. The margin between cracking and ultimate loads for flexural failure varied from 147 to 178%. In such cases, increase in volume fraction of fibres resulted in increase of ultimate strength. In A series, A25 with one percent fibre reached nearly the strength of the beam with stirrups; the mode of failure changed from shear to flexure. Similarly in B series, B34 with one and a half percent fibre, reached strength 26% more than that of the corresponding stirrup beam, B35; and the mode of failure changed to flexure. As the fibre content increased from 0 to 1% the strength increase was 165% in A series and for the fibre content increase from 0 to 1.5% the strength increase was 102% in B series. Increase in fibre content showed increased neutral axis depth at failure. Influence of aspect ratio, relative to strength can be observed comparing beams A22 and A23, A24 and A25 and B32 and B33. For higher aspect ratio the increase was from 15 to 25%. The influence of fibre parameters in increasing the shear strength in order of significance was as follows. (a) Fibre volume, (b) Fibre aspect ratio.

ULTIMATE STRENGTH ANALYSIS

The shear strength of the beams without any web reinforcement and the beams with stirrup reinforcement is computed using the IS 456-1978[8]. The ultimate strength theory as incorporated in Draft European Code [1] and modified by Muhidin [2] to suit fibre reinforced members is applied to the specimens of the present investigation.

Tensile failures of the type of fibre concrete used here are primary failures of the fibre matrix bond, generally accompanied by some disruption of the matrix. The post cracking tensile resistance of fibre reinforced webs was derived by Muhidin [2]

$$f_{ct} = 0.125\ PL/D\ (1+0.57\ L/b_w)\sqrt{f_{ck}} \qquad (1)$$

where f_{ct} = post cracking tensile resistance of web, P = volume fraction of fibre, b_w = breadth of web, f_{ck} = cube strength of concrete.

In deriving this formula, the effective bond resistance was assumed to be proportional to the square root of the cube strength as in most codes. For narrow sections, two dimensional orientation for the fibres was assumed. Truss system is a reasonable model of behaviour of a beam under bending and shear, once cracking is well developed. The tensile resistance of the web lattice is limited by the strength of the fibre at the cracks. With such small diameter reinforcement, it is reasonable to assume that the direction of tension is that of the movement at the cracks and thus tends to be vertical at failure. Expression was derived [7] from the test data, used together with the equilibrium conditions for the tension chord.

$$\text{Cot}\ \theta = 1 + 0.25\ \sqrt{f_{ck}}/f_{ct} \qquad (2)$$

where θ = angle of web compression.

In the truss model, ultimate conditions of a fully cracked web are described fully in terms of f_{ct} and Cot θ. The secondary effects have to be considered, namely (a) contribution of compression zone to shear resistance (b) any residual flexural stiffness of the web portion. Though among shorter beams the compression zone supports considerable transverse shear force at failure, this large resistance is unreliable. In lightly reinforced beams web portions are not fully cracked even at failure and could be assumed to have had appreciable rigidity.

The situation described above is similar to that arising in beams with stirrups. In beams with stirrups there is a significant addition to the pure truss resistance, when the web reinforcement is light and only a negligible effect when the web is heavily reinforced. This behaviour is represented in the Draft CEB Code [1] by an addition V_c to the truss resistance, V_w.

$$V_u = V_c + V_w \tag{3}$$

where $V_w = 0.09\ f_{ct} b_w d \mathrm{Cot}\,\theta$

$V_c = 0.2 \sqrt{f_{ck}}\ b_w d$ when $V_w = 0$

$V_c = 0$ when $V_w \geqslant 0.6 \sqrt{f_{ck}}\ b_w d$

where V_u = ultimate shear strength, V_w = truss resistance, V_c = additional resistance to truss resistance, d = effective depth of beam.

V_c varies linearly between these values. The shear strength of the beams without any web reinforcement and the beams with stirrup reinforcement is computed using the IS 456-1978 [8]. The shear strength of all the fibre beams is predicted by Eq.(3).

TABLE - II : PRINCIPAL TEST AND THEORETICAL RESULTS

Beam Designation	Shear span/ depth	Fibre Details: Aspect ratio	Fibre Details: Fibre content by volume(%)	Fibre Details: Cube strength (N/mm^2)	Test Cracking Shear(N): Flexure zone	Test Cracking Shear(N): Shear zone	Shear at ultimate (N): Test	Shear at ultimate (N): Theory	V_u test / V_u theory
A21	2	-	0	22.5	-	13836	19174	11832	1.62
A22	2	50	0.5	28.7	17173	20509	30518	34627	0.88
A23	2	100	0.5	32.2	17200	17840	36078	39340	0.92
A24	2	50	1.0	29.0	20509	22733	40527	36858	1.10
A25	2	100	1.0	32.6	20509	27182	50758	43782	1.16
A26	2	-	0.0	27.5	20509	22733	51648	41220	1.25
B31	3	-	0.0	26.8	-	16145	22261	11832	1.88
B32	3	50	1.0	32.1	-	18367	29487	38550	0.77
B33	3	100	1.0	32.3	13253	14142	36830	44342	0.83
B34	3	50	1.5	32.8	17257	22372	45059	41355	1.09
B35	3	-	0.0	25.6	14809	15032	35718	30199	1.18

Strength claculation as per IS code = 1.48

V_u test/V_u theory for Fibre beams failing in shear = 0.92

Fibre beams failing in flexure = 1.03

Strength prediction by the IS code for the beams without shear reinforcement and the beams with stirrup reinforcement was conservative. V_u test/V_u theory varied from 1.18 to 1.88. The shear strength predicted by Eq.(3) is applicable to the fibre beams failing in shear. Beams A22, A23 and A24 in A series and beam B32 in B series were fibre beams and failed in shear. For these specimens the ratio of test to theoretical shears varied from 0.77 to 1.1 and the average is 0.92. The remaining fibre beams changed their mode of failure from shear to flexure due to addition of fibres. For these beams as flexure would be more than the shear force recorded corresponding to flexural failure. In such a situation, the theoretical prediction by Eq.(3) would be conservative, in the present case, shear strength is under predicted by three percent.

CONCLUSIONS

The following conclusions are drawn, based on the findings of the tests reported here,

1. Steel fibres appear suitable as shear reinforcement for beams.
2. The replacement of vertical stirrups by round steel fibres provided effective reinforcement against shear failure.
3. Shear strength prediction by the IS code for beams without any shear reinforcement and stirrup reinforcement is conservative.
4. Shear strength prediction for beams with fibre as shear reinforcement is predicted by the equation proposed by Muhidin [2]. Though it is adequate, it appears to over estimate, the influence of concrete strength.

ACKNOWLEDGEMENTS

The authors gratefully acknowledge the facilities offered by the Department of Civil Engineering, Regional Engineering College, Warangal for carrying out the research. The authors wish to express their thanks to Prof. K. Koteswara Rao, Principal, R.E.C., Warangal, for his help through out the course of research.

REFERENCES

1. CEB, Systeme International de Reglementation Technique Unifee des Structures (2 3m 3 Draft) Vol.II, Code de Reference Pour les Structures en Beton, Bulletin d' Information No.III, Paris, 1975.

2. Muhidin, N.A., 'Design Criteria for Fibre Reinforced Concrete', Ph.D. thesis, Polytechnic of Central London, 1977.

3. Ramey, R.M.,. The Flexural Behaviour of Fibre Reinforced Concrete Beams, Ph.D. thesis, Carnegic Mellon University, 1967.

4. Batson, G.B., Jenkins, E. and Spatney, R., Steel Fibres as Shear reinforcement in beams, Journal of the American Concrete Institute, Vol.69, Oct. 1972, PP 640-644.

5. Williamson, G.R. and Knab, L.I. Full scale fibre concrete beam tests, Rilem Symposium on Fibre Reinforced Cement and Concrete, London, September 1975.

6. Lafraugh, R.W. and Monstafa, S.E., Experimental investigation of the use of steel fibres for shear reinforcement, Concrete Technology Associates, Tacoma, Washington, January 1975, PP 53.

7. Muhidin, N.A. and Regan, P.E. Chopped steel fibres as shear reinforcement in concrete beams, Proceedings of the Conference on Fibre Reinforced Materials, Design and Engineering Applications (London, March 1977), Institution of Civil Engineers, London 1977, PP 149-163.

8. Indian Standards Institution, Indian Standard Code of Practice for Plain and Reinforced Concrete, IS 456-1978, New Delhi.

Proceedings of the International Symposium on Fibre Reinforced Concrete, December 16–19, 1987, Madras, India

BEHAVIOUR OF FIBER REINFORCED CONCRETE BEAMS IN SHEAR

S.K. KAUSHIK, Professor
V.K. GUPTA, Reader
Civil Engineering Department
University of Roorkee, Roorkee, India.

N.K. TARAFDAR, Professor
Civil Engineering Department
Bihar Engineering College,
Patna, India.

ABSTRACT

The increase in shear strength of concrete due to inclusion of fibres in the concrete is well recognised. This paper reports an experimental investigation on twenty rectangular simply supported fibre reinforced concrete beams with conventional flexural reinforcement and designed to fail in shear. The beams had identical nominal dimensions and were subjected to a single concentrated load placed in the shear span of the beam.

Theoretical expressions have also been derived to predict the strength in shear and the shear moment capacity. The expressions yield results which agree well with the experimental values in most of the cases. The results indicate that a strength ratio of 1.67 can be achieved with the addition of 1.5% volume fraction of fibres with an aspect ratio of 100.

INTRODUCTION

The use of fibres to enhance strength and deformation properties of the cement matrix is now well established. Plain concrete possesses a low tensile strength, little resistance to crack propagation as well as limited ductility. The development of micro-cracks occurs during the manufacture of concrete as a result of inherent volumetric and micro-structural changes. Hence, even before the application of external loads a discontinuous and heterogenous system exists. Under loads initial micro-cracks start developing at about 30-40 percent of the ultimate load in compression and subsequent loading results in uncontrolled and unstable cracks. The meagre resistance offered to tensile crack propagation results in low fracture toughness, resistance to impact and explosive loads.

In the two-phase fibrous composite materials the fibres inhibit cracks thus imparting improved properties of strength and stiffness. The fibres inhibit crack formation by exerting pinching forces at the crack-tip i.e., these fibres act as crack-arrestors. In recent years, it has been well established that the addition of small, closely spaced and uniformly dispersed fibres in concrete would substantially improve its static as well as dynamic properties. Recent laboratory investigations have shown that steel fibres can be used in place of stirrups for shear reinforcement of frames, beams and flat slabs [1,2,3]. Results of full-scale tests on beams hold promise in this respect [4,5]. Further the ductility of steel fibred concrete would permit the development of plastic hinges under overload conditions [1,2,4,5,6].

With the more prevalent use of high-strength steels in recent years and the increasing trend towards ultimate load and limit theory designs, control of cracking has become imortant. Flexural micro-cracking (not visible to the naked eye) can start at stress-levels lower than 633 kg/cm^2 (9000 psi) in the reinforcement. As steel corrodes in an adverse environment, the control of cracking plays an important role in the control of corrosion by limiting the possibilities for entry of the moisture and salts which together with oxygen can accelerate corrosion. Minimization of corrosion helps to maintain the reserve load-factor or safety factor used in the design.

SCOPE AND OBJECTIVES

This paper is an attempt to study the ultimate strength of fibre reinforced concrete beams vis-a-vis shear failure. The scope of the study was limited to observations on the gain in strength compared to ordinary R.C.C. beams without fibres, deflections, curvatures, rotations and the crack pattern.

ANALYSIS

Ultimate Shear Strength

The expressions in this section are based on the effective fibre spacing concept of fibre reinforced concrete and the shear mechanism of reinforced concrete beams alongwith the experimental evidence collected by the author.

An F.R.C. beam can be visualised as developing resistance against shear through two components :-

a) The shear-resistance of concrete, V_c
b) The resistance offered by the fibres, say, through interfacial bond-stress, V_s

Thus $V_u = V_c + V_s$

The shear resistance of concrete can be estimated by any of the several formula available in literature e.g. formula due to Zsutty [7,8] and others. The second part of the resistance is built up through fibres crossing a major diagonal crack or any such similar cracks due to shear. Using the Stepanova effective fibre-spacing equation [9], the total force developed through the fibres lying in one vertical plane, can be computed.

TEST PROGRAM

The test program consisted of testing of 20 beams with identical dimensions and reinforcement. The beams were classified as A,B,C,D,E,F,G,H,I and J. There were two beams in each series. The average results of two such beams have been reported.

Ordinary portland cement was used for all the beams. The coarse aggregate consisted of local crushed stone aggregate of 10 mm maximum size. The fine aggregate consisted of sand with a fineness modulus of 1.33. Tap water was used for both mixing and curing of concrete. The fibres were obtained by cutting 26SWG black annealed mild steel wires of suitable length. The longitudinal steel in the beams consisted of 12 mm diameter mild steel deformed bars. The properties of reinforcing steel and the fibres are given in Table 1.

The specimens were cast in the laboratory at a temperature of 30°C. The specimens were demoulded after 24 hours and cured in water at 27°C. The test set up for simply supported beams is shown in Fig. 1. The load was applied in suitable increments and the strains, reactions and deflections were recorded at each stage of loading. The propagation of cracks at each loading stage were also noted.

TEST DATA

The measurements and observations were taken for all the specimens. However, only typical results are presented. Figure 2 shows the location of demac gauge points for strains in concrete and steel studs for strain in reinforcing steel. The details at beams specimen are shown in Table 2. The crack pattern for beams A_1, B_1....., J_1 are shown in Fig. 3. Figure 4 shows typical load deflection curves under the load point and at mid span.

Figure 5a shows a typical strain distribution across the depth of the beam at the load section for various stages of loading. The curvature distribution along the beam length is also shown in Fig. 5b. The effect of volume fraction of fibres on the ultimate nominal shear stress is shown in Fig. 6.

The observed loads at first crack and ultimate stages of all beams are presented in Table 3. Ultimate moments and ultimate deflections under load have been tabulated in Tables 4 and 5. The ultimate curvatures under load are given in Table 6.

Discussion of Results

Figure 5 shows that the linearity of the strain distribution is fairly maintained upto first crack stage. The formation of shear cracks disturbs the linearity of the distribution and near failure the disturbance is predominant in the bottom part of the beam. The maximum concrete strain in fibre reinforced concrete can be taken as 0.007 as compared to 0.0035 for concrete without fibre.

Figure 7 shows the ultimate strain in fibre concrete ϵ_{cu} plotted against $L_f p_f / d_f$, the

TABLE 1 : PHYSICAL PROPERTIES OF STEEL USED

Bar diameter Steel type	Yield / Proof Stress N/mm²	Yield / Proof Strain	Ultimate Stress N/mm²	Young's modulus of Elasticity N/mm²	Ultimate bond stress between fibre and concrete
0.46mm black annealed steel wire (26 SWG)	312.9		410.4	2.11×10^5	1.52+
12mm tor steel deformed bars	670.0	0.002	718.5	2.35×10^5	

\+ Using simple pull out test

TABLE 2 : DETAILS OF BEAM SPECIMEN FOR SHEAR

Beam designation+	Volume fraction	Aspect ratio	Avg Comp. strength (N/mm²)	Remarks
A	0	-	20.99	Fibres mixed in entire concrete volume
B	1.5	100	27.77	
C	1.0	100	24.08	
D	0.5	100	21.88	
E	1.5	80	22.47	
F	1.0	80	22.95	
G	0.5	80	22.13	
H	1.5	60	22.58	
I	1.0	60	21.72	
J	0.5	60	21.22	

\+ Beam size ; 100 x 150 x 1220

Mix proportion 1 : 2.07 : 2.4 (w/c = 0.62)

Each series has two specimen such as A_1, A_2, B_1, B_2 etc.

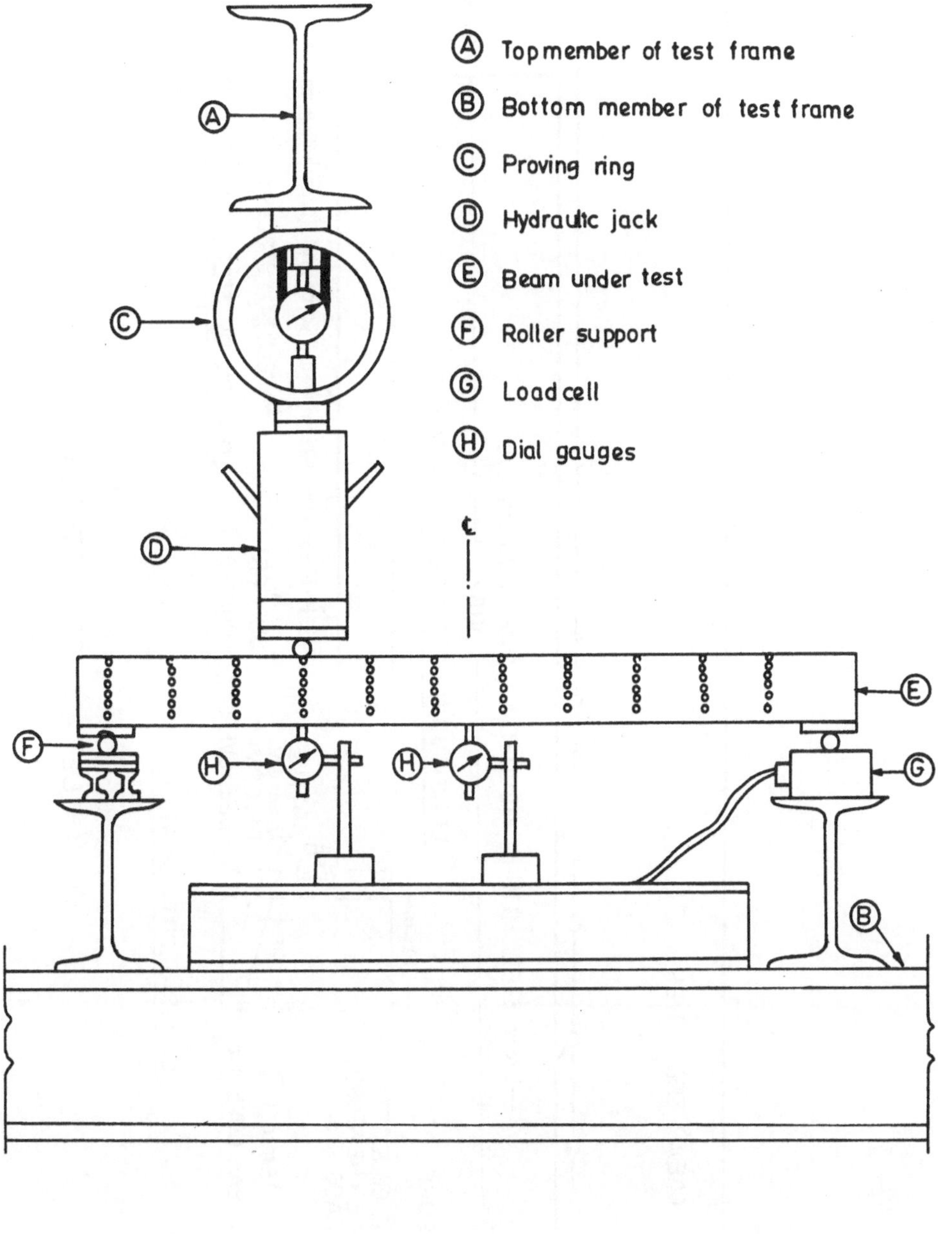

FIG. 1 — TEST SET-UP OF BEAM

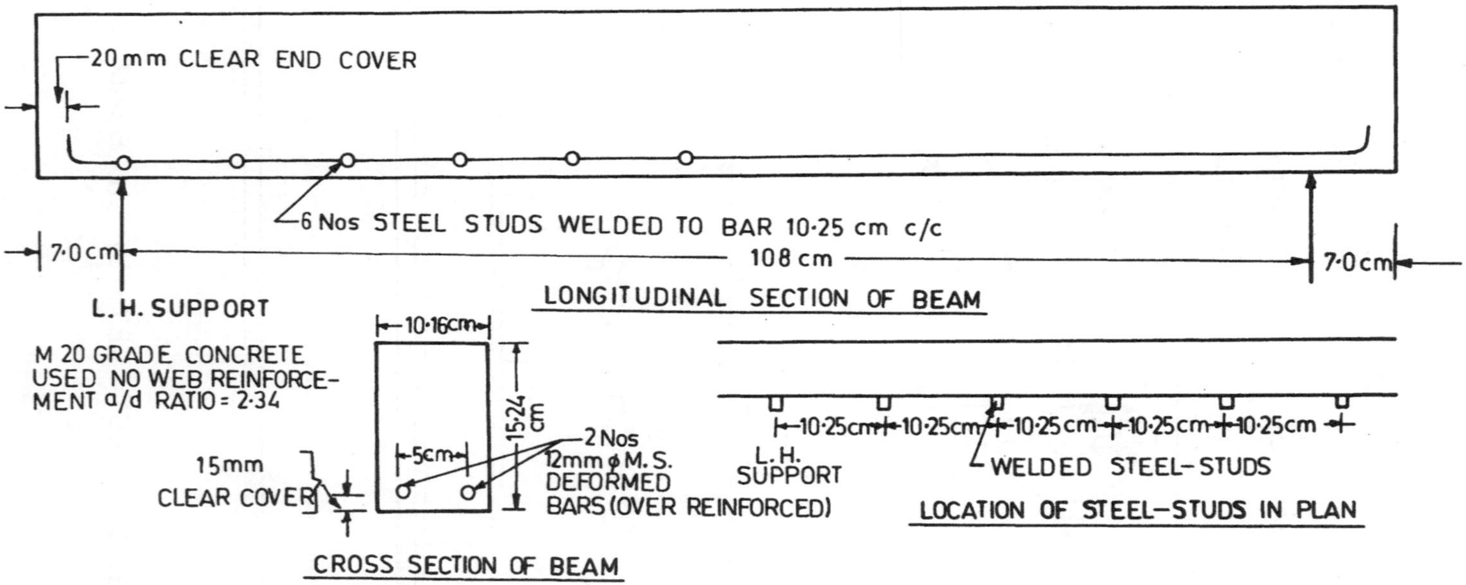

FIG. 2 — DETAILS OF TEST-BEAM

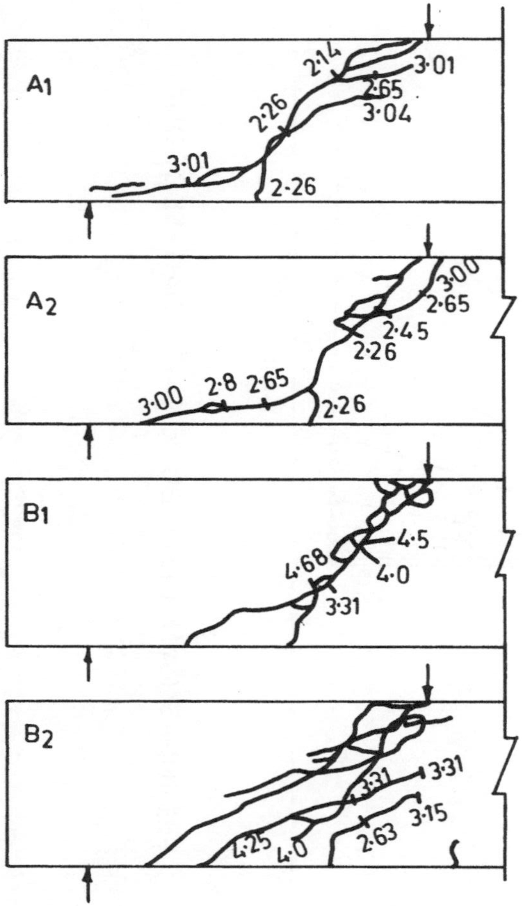

FIG. 3—CRACK–PATTERN OF BEAMS
(Loads shown in ton)

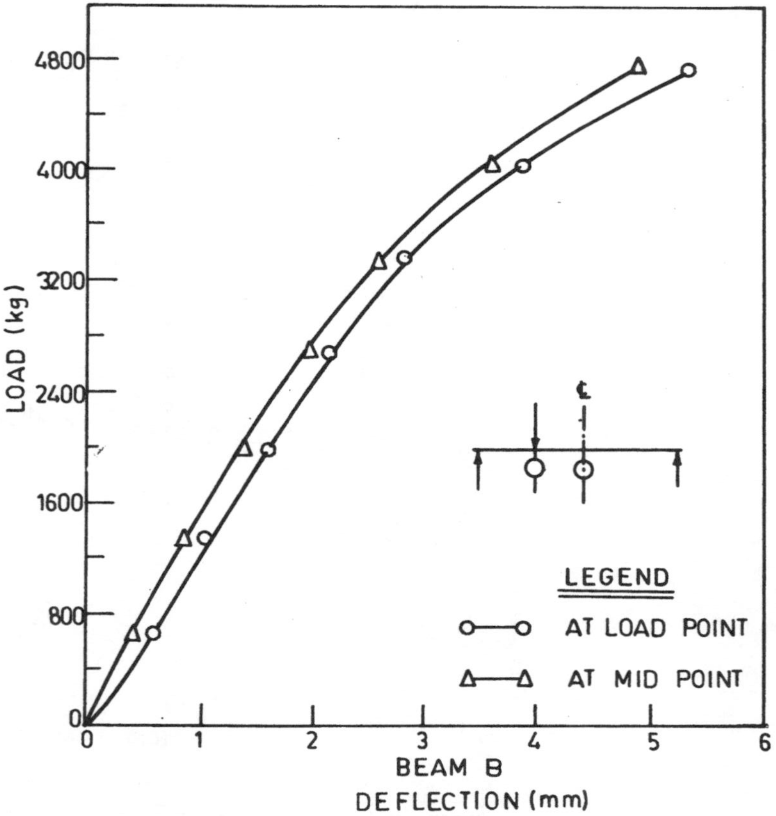

FIG. 4 —LOAD – DEFLECTION CURVE

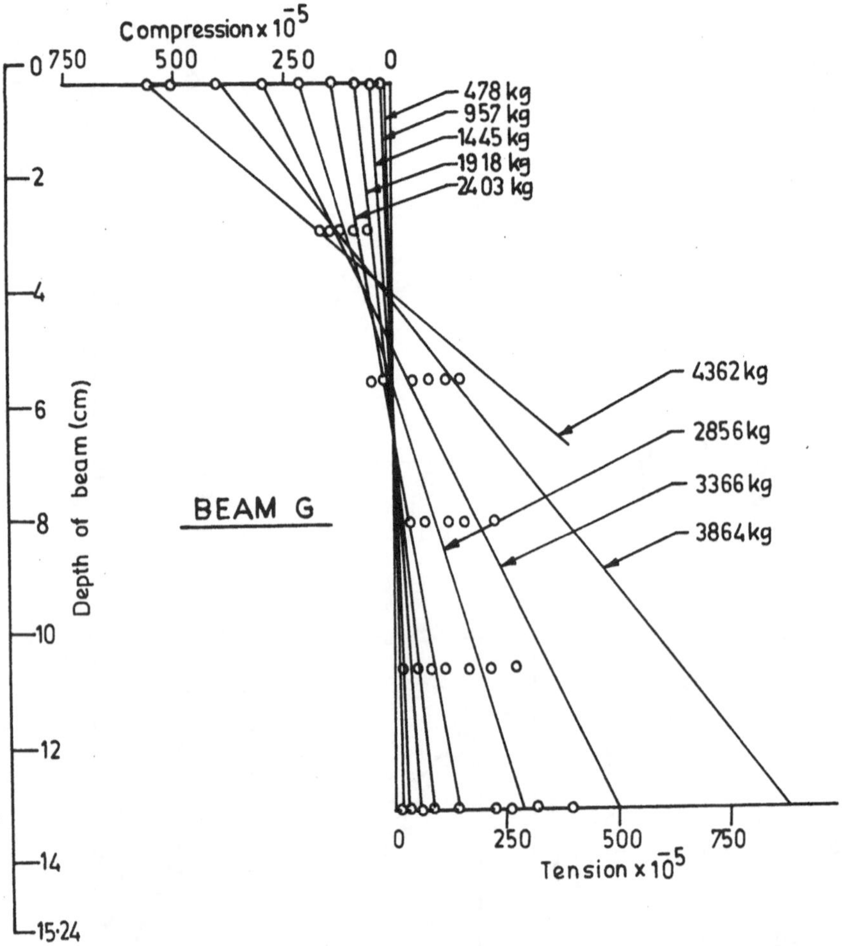

FIG.5(a)—STRAIN DISTRIBUTION AT LOAD-SECTION

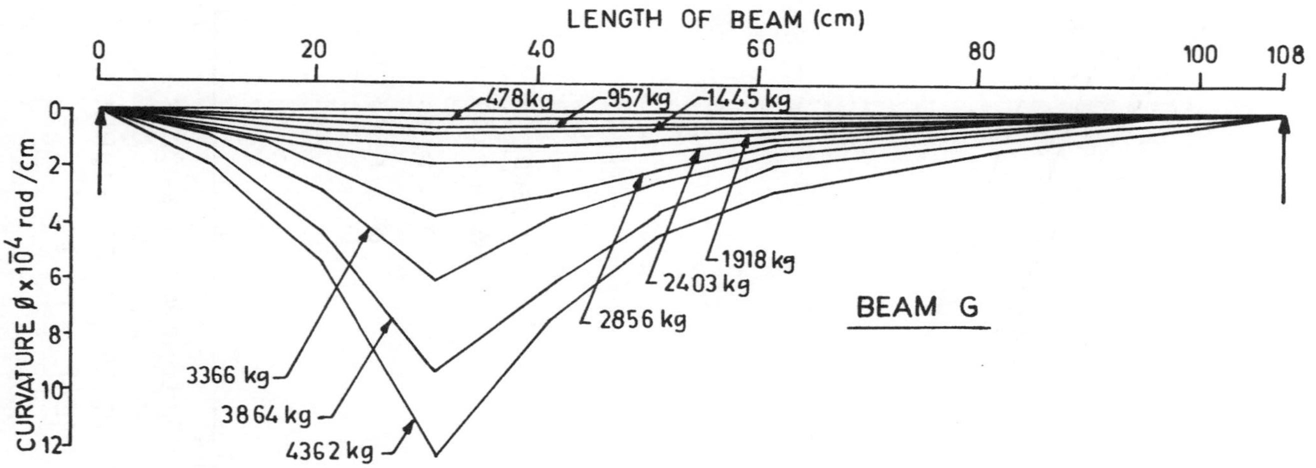

FIG. 5 (b) — CURVATURE DISTRIBUTION ALONG BEAM G

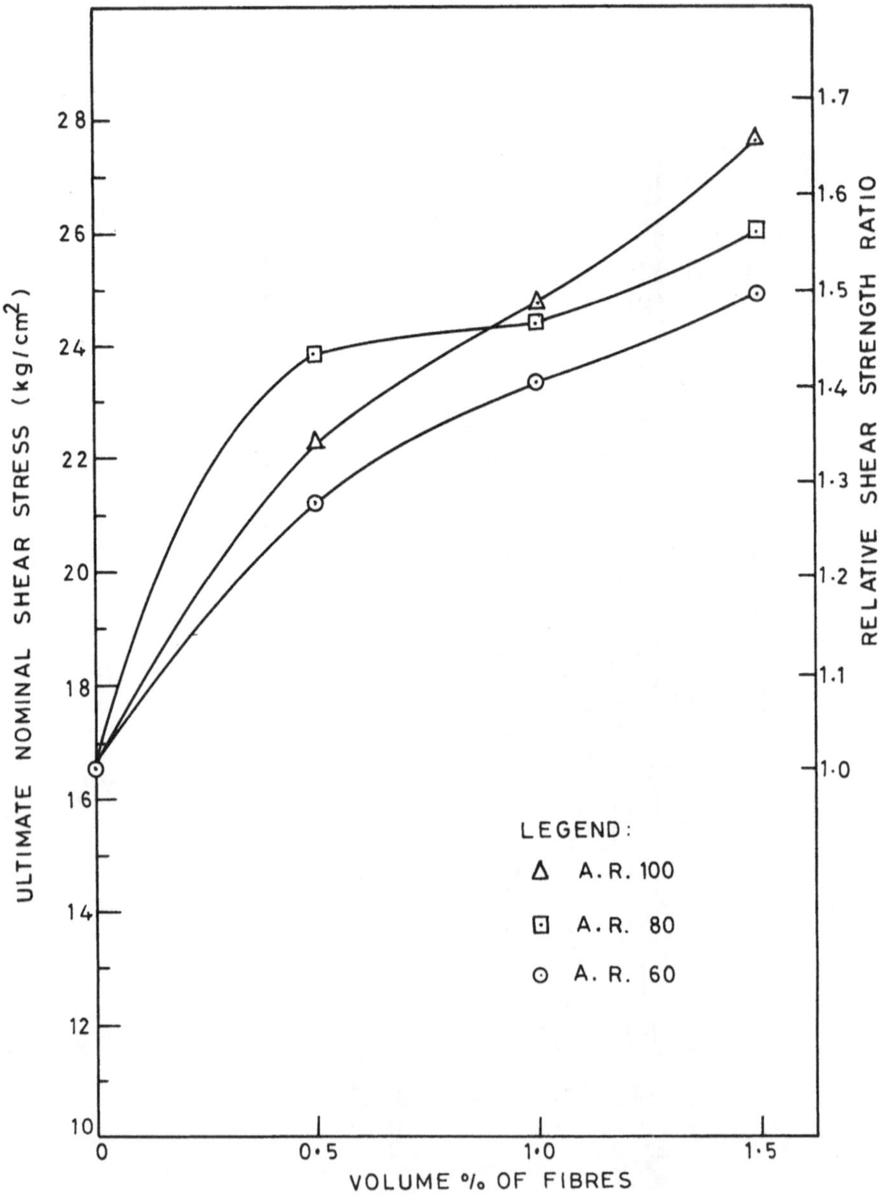

FIG.6 – EFFECT OF VOLUME % OF FIBRES ON ULTIMATE NOMINAL SHEAR STRESS

TABLE 3 : LOADS AT FIRST CRACK AND FAILURE

Beam Designation	Observed load at First Crack (kN)	Observed load at failure (kN)	Observed Maximum S.F. at ultimate (kN)	Shear Strength Factor	Relative $P_f L_f/d_f$	Remarks
A	22.95	30.88	22.09	1.0	-	
B	33.67	51.62	36.92	1.672	1.5	
C	25.79	46.12	32.99	1.494	1.0	
D	26.77	41.64	29.78	1.349	0.5	All beams
E	28.72	48.61	34.77	1.574	1.2	failed in
F	22.95	45.72	32.70	1.480	0.8	shear
G	22.95	44.60	31.90	1.444	0.4	
H	26.77	46.54	33.29	1.507	0.9	
I	22.95	43.62	31.20	1.413	0.6	
J	22.95	39.74	28.43	1.287	0.3	

TABLE 4 : ULTIMATE MOMENTS

Beam Designation	Ultimate Moment (kN -m)		Ratio = $\frac{\text{Theory}}{\text{Test}}$
	Theory	Test	
A	7.25	6.90	1.05
B	11.24	11.54	0.97
C	11.07	10.31	1.07
D	9.51	9.30	1.02
E	10.78	10.86	0.99
F	10.40	10.22	1.02
G	9.37	9.97	0.94
H	10.41	10.41	1.00
I	9.62	9.75	0.99
J	8.83	8.88	0.99

TABLE 5 : ULTIMATE DEFLECTIONS

Beam Designation	Ultimate, Deflection δ_u (mm) Theory	Test	Author's formula	Ratio = $\frac{\text{Theory}}{\text{Test}}$	Ratio = $\frac{\text{Author's formula}}{\text{Test}}$
A	2.93	3.22	-	0.91	-
B	5.55	5.35	5.95	1.04	1.11
C	5.48	5.72	4.54	0.96	0.80
D	3.20	3.42	3.29	0.94	0.96
E	3.45	3.80	4.53	0.91	1.19
F	4.60	5.94	4.54	0.78	0.77
G	5.98	5.40	5.12	1.11	0.95
H	3.48	4.56	4.54	0.76	0.99
I	4.90	4.50	4.53	1.09	1.01
J	3.80	4.20	3.29	0.91	0.99

TABLE 6 : ULTIMATE CURVATURES

Beam Designation	Ultimate Curvature+ ($x10^{-4}$ rad/cm) Theory	Test	Ratio = $\frac{\text{Theory}}{\text{Test}}$	ϕ_u++ calculated (10^{-4} rad/cm)	$\frac{\phi_{u\,Calc.}}{\phi_{u\,test}}$	θ_u Test Radx 10^{-2}
A	4.30	3.20	1.34	3.07	0.96	7.38
B	7.39	11.80	0.63	10.44	0.89	18.45
C	7.50	11.60	0.65	10.12	0.87	18.75
D	7.44	6.20	1.20	6.93	1.12	10.25
E	7.41	6.00	1.23	9.70	1.61	10.25
F	7.38	8.80	0.84	9.62	1.09	13.12
G	7.43	12.40	0.60	8.12	0.66	19.99
H	7.40	10.20	0.73	9.33	0.92	13.73
I	7.45	11.20	0.67	9.81	0.88	16.81
J	7.49	8.20	0.91	7.20	0.88	11.78

\+ For fibre reonforced concrete $\epsilon_{cu} = 0.007$

and for plain concrete $\epsilon_{cu} = 0.0035$

++ Results based ϵ_{cu} values obtained from Fig. 7.

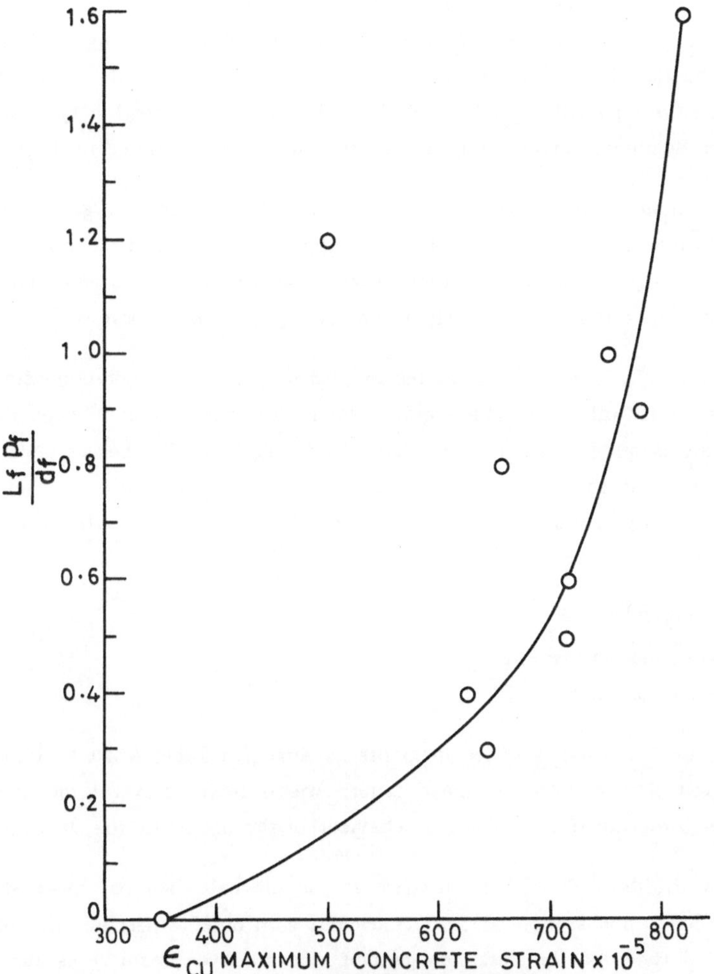

FIG. 7 – $\frac{L_f P_f}{d_f}$ VERSUS ϵ_c (EXTRAPOLATED)

maximum value of ϵ_{cu} can be obtained from this figure for computing the ultimate theoretical curvature. It was observed that after the formation of first crack, the strain in steel was maximum at the point where the crack crossed the steel level. The test data also clearly shows that for any particular load stage the FRC beams have a remarkably high value of curvature as compared to beams without fibre.

Except for beams C_2, D_1, H_1 and H_2 in all the other beams the crack started as a fine flexural crack which subsequently joined an inclined crack. In beams A_1, B_1, D_2, E_1, E_2, F_1, F_2, G_1, G_2 and I_2 flexural cracks were absent. These flexural cracks were only one or two in number in each beam. The inclination of the major diagonal crack was on an average 45° to the horizontal. Beams I_1, I_2, J_1, J_2, G_1 and G_2 had a shear compression type of failure while beam C_2 had a predominent diagonal tension failure. All the beams failed in shear.

A typical moment curvature relationship for beam B is shown in Fig. 8a. The plot shows that the contribution of fibres is practically negligible upto the first crack load whereas the contribution is significant beyond that stage. The figure also shows that the moment curvature relation can be represented fairly accurately by a bilinear curve.

Figure 8b shows a typical load deflection curve plotted in non-dimensional form for beam B. The results for other beams are quite similar. The agreement between the theoretical and observed value is quite close. For the same load stage all FRC beams underwent smaller deflections as compared to beams without fibres. This shows the stiffening effect of fibre inclusion. Based on test results, deflection under the load can be satisfactorily estimated by the equation.

$$Y = 1680\,(\delta)^{0.584}$$

where Y = Concentrated load, kg

δ = Deflection, mm

Theoretical and observed ultimate moments as shown in Table 4 are in close agreement. It is also observed that in case of shear failure the ultimate flexural moment capacity is not achieved due to the modification of the behaviour of the beams by the shear cracks.

All beams with fibres showed an increase in the ultimate nominal shear stress. Beam B with 1.5 percent fibres and aspect ratio 100 showed a gain of 67.2 percent over beams A without fibres. Beam J recorded the minimum gain of strength over beam A as 28.7%. The study shows that for more than 50 percent gain in shear strength an aspect ratio of 100 would yield the best result. The corresponding volume fraction of the fibres should be 1.2%.

CONCLUSIONS

Based on the studies conducted on beams under shear, the following conclusions are drawn:

1. The inclusion of 1.5% fibres by volume and with an aspect ratio of 100, increases the ultimate nominal shear stress by 67.2%.

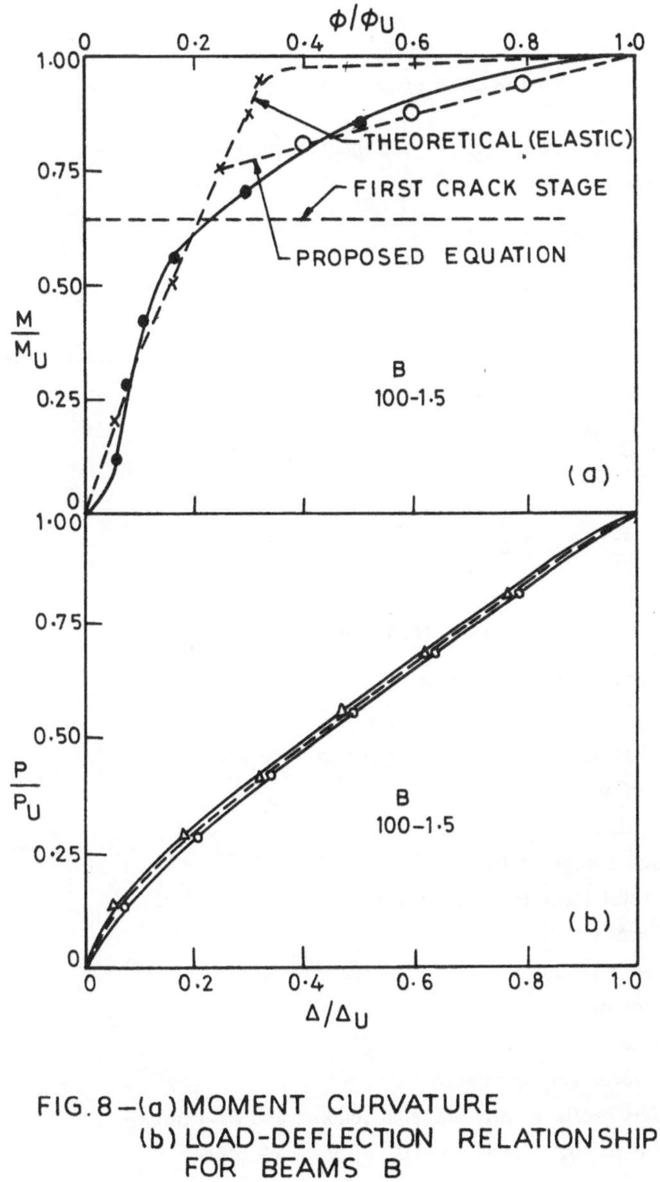

FIG. 8—(a) MOMENT CURVATURE
(b) LOAD-DEFLECTION RELATIONSHIP
FOR BEAMS B

2. The average maximum strain in fibre reinforced concrete beams were of the order of 0.007 as compared to 0.0035 for plain reinforced concrete beams.

3. The gain in shear strength of fibre reinforced concrete beams over plain concrete beams can be estimated by the following equation.

$$Y = 0.24\,U\,b\,d\sqrt{p_f\,l_f}$$

The expression gives an average variation of about 8% compared to the test results.

4. The deflection under the load for the beams subjected to shear span loading can be obtained from the equation

$$W = 1680\,(\delta)^{0.584}$$

The expression yields result with an average variation of ±10%, as compared to test results.

5. The moment curvature relationship for beams subjected to shear can be expressed as a bilinear curve by the following expressions:

(a) $$\frac{M}{M_u} = 3\frac{\phi}{\phi_u} \quad \text{for } \frac{M}{M_u} < 0.75$$

(b) $$\frac{M}{M_u} = 0.65 + 0.4\frac{\phi}{\phi_u} \quad \text{for } \frac{M}{M_u} > 0.75$$

NOTATIONS

δ	=	deflection under the load point
L	=	length of crack
U	=	ultimate interfacial bond stress between fibre and matrix
d_f	=	diameter of fibre
l_f	=	length of fibre
p_f	=	volume percentage of fibre
α	=	angle of total force F_{wl} with the neutral axis
b	=	width of beam
D	=	overall depth of the beam
d	=	effective depth of beam
L_s	=	length of beam
A_s	=	area of tension reinforcement
K, K_u	=	neutral axis factor at any stage of loading and at ultimate
F_{wl}	=	total force through fibres normal to the crack plane
ϕ_u, ϕ	=	ultimate curvature and curvature at any stage of loading
M_u, M	=	ultimate moment and moment in general
EI	=	flexural rigidity
ϵ_{cu}	=	ultimate compressive strain in extreme concrete fibre
ϵ_s	=	strain in tension reinforcement

R_1 = reaction at support

a = shear span

E_s = young's modulus of elasticity for steel

j = coefficient of lever arm.

REFERENCES

1. Tarafdar, N.K., 'Ultimate Strength Investigations of R.C.C. Beams with Fibre Reinforcement' M.E. Thesis, Civil Engg.Deptt., Univ. of Roorkee, Roorkee, 1978.

2. Kukreja, C.B., 'Structural Properties of Fibre Reinforced Concrete Beams' Ph.D. Thesis, Civil Engg.Deptt., Univ. of Roorkee, Roorkee, 1981.

3. Batson, G. et.al., Steel Fibre as Shear Reinforcement in Beams, ACI Journal, Proc. Vol.69, No.10, Oct. 1972, pp. 640-644.

4. Ramouldi, J.P. et.al., Effects of Impulsive Loads on Fibre Reinforced Concrete Beams, Contract No. OCD-PS-64-21, Carnegie Institute of technology, Pittsburgh, for office of Civil Defence, Secretary of Army, Deptt. of Defence, Oct. 1965, pp. 77.

5. Williamson, G.R. at.al., Full Scale Fibre Concrete Beam Tests, RILEM Symposium, 1975 on Fibre Reinforced Cement and Concrete, pp. 209-214, The Construction Press Ltd., Lancaster, U.K., 1975.

6. Untraner, R.E. et.al., Continuous Reinforced Concrete Beams using Concrete Reinforced with Short Lengths of Steel Wire, Presented at the ACI Fall Convention, Cleveland, Ohio, 1965.

7. Zsutty, T.C., 'Beam Strength Prediction by Analysis of Existing Data' ACI Journal, Proc. Vol.65, No.11, Nov. 1968, pp. 943-951.

8. Zsutty, T.C., 'Shear Strength Prediction for Separate Categories of Simple Beam Tests' ACI Journal, Proc. Vol.68, No.2, Feb. 1971, pp. 138-143.

9. RILEM Symposium, 1975 on Fibre reinforced Cement and Concrete. The Construction Press Ltd., Lancaster, U.K., 1975.

Proceedings of the International Symposium on Fibre Reinforced Concrete, December 16–19, 1987, Madras, India

A PHYSICAL SIMULATION OF THE FIBER REINFORCED CONCRETE BEHAVIOR UNDER COMPRESSION

Parviz Soroushian, Ph.D, Assistant Professor
Cha-Don Lee, Research Assistant

Department of Civil and Environmental Engineering
Michigan State University
East Lansing, Michigan 48824

SYNOPSIS

Reinforcement by randomly distributed fibers can improve, among other properties, the compressive constitutive behavior of concrete. This effect of fiber reinforcement is comparable to that of confinement by the conventional transverse reinforcement. Analytical constitutive models of fiber reinforced and confined concretes were used to derive the confinement levels (represented by the transverse steel ratio and spacing) capable of producing compressive constitutive performances comparable to those of concretes reinforced with certain fiber levels (represented by the volume fraction times the aspect ratio of a certain fiber type). A comparative cost analysis was also performed on the fiber reinforcement and conventional confinement techniques, considering only the effects of these techniques on the compressive constitutive behavior of concrete.

INTRODUCTION

Short, randomly distributed fibers, when added to concrete, confine the material and delay the crack propagation, thus producing noticeable increases in the strain at the peak stress, ductility and energy absorption capacity as described by descending portion of the stress-strain relationship (Fig. 1) [1,2,3,4]. Fiber reinforcement does not significantly modify the pre-peak ascending branch and the ultimate strength of concrete in compression.

The fiber type, volume fraction and aspect ratio play important roles in determining the compressive ductility and energy absorption capacity of fiber reinforced concrete. The material behavior is generally enhanced as the volume fraction and aspect ratio of fibers increase up to limits after which the problems with fresh mix workability and fiber dispersability start to damage the hardened material properties. As the increases in both fiber volume fraction (V_f), and aspect ratio (ℓ/d) lead to improvements of the same nature in the

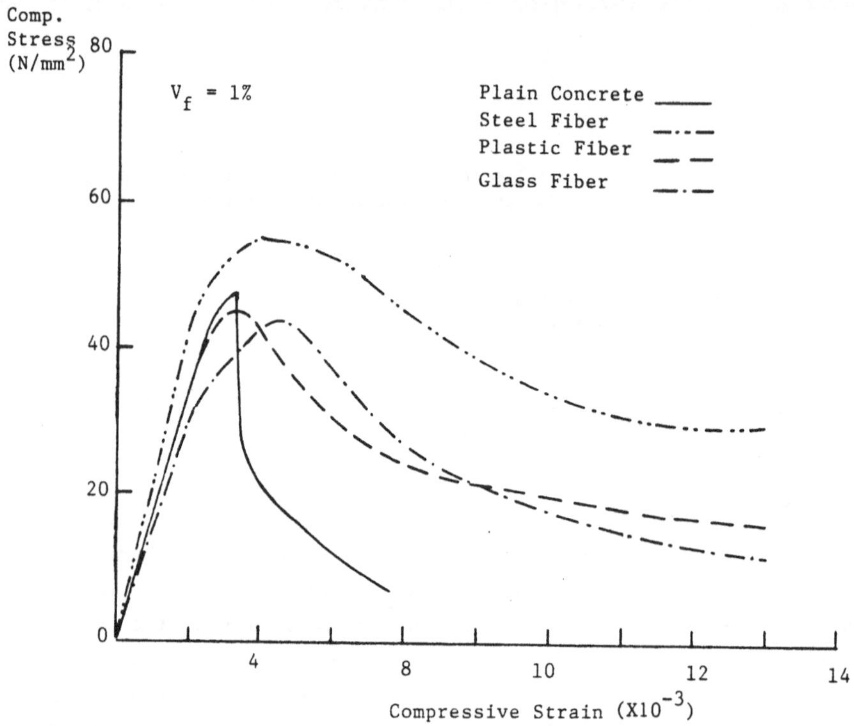

Figure 1. Typical Compressive Stress-Strain Curves for Concrete Reinforced with 1% Volume Fraction of Difference Fiber Types [3].

compressive behavior of the material, their combined effect has been generally analyzed using the fiber reinforcing index $V_f\ell/d$. In general, the higher the fiber reinforcing index, the higher the ductility and energy absorption capacity of fiber reinforced concrete. However, for high values of fiber reinforcing index, the problems with workability and fiber dispersability of fresh mix tend to deteriorate the compressive behavior of the hardened material.

The improvements in ductility and energy absorption capacity resulting from the increase in fiber reinforcing index are comparable to those improvements resulting from the increased confinement of non-fibrous concrete by transverse reinforcement. This is shown in Fig. 2, where the effects of increasing the fiber reinforcing index (Fig. 2a) are compared with the effects of the increase in confinement of non-fibrous concrete by transverse reinforcement (Fig. 2b). The degree of confinement by transverse steel is represented in this figure by the volumetric ratio of transverse reinforcement (ρ_s) and the ratio of the lateral dimension of the confined specimen (h') to the spacing of transverse steel (s).

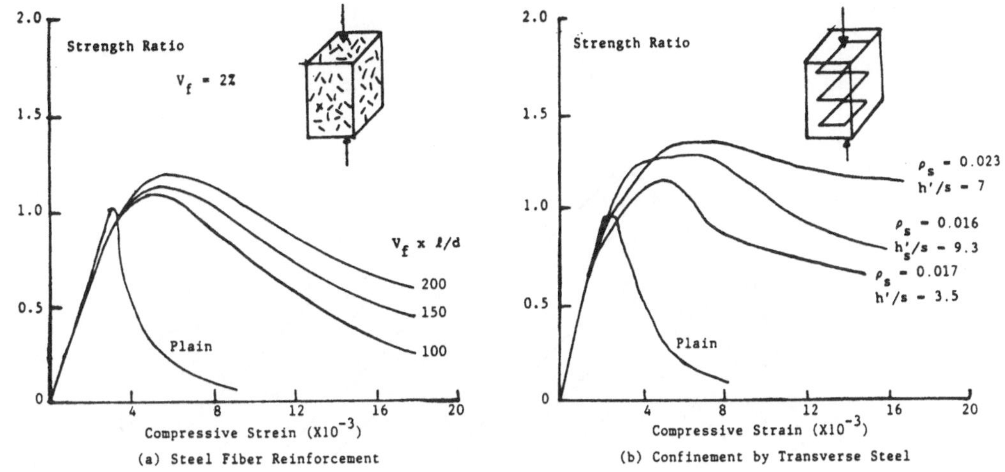

Figure 2. Effects of Fiber Reinforcement and Confinement by Transverse Steel on the Compressive Behavior of Concrete.

In this study, for different reinforcement indices of steel, monofilament polyprophylene and glass fibers, the equivalent volumetric ratios and spacings of transverse reinforcement which induce comparable improvements in the concrete compressive constitutive behavior were determined. The comparative cost analysis was performed in order to assess the economical advantages of substituting transverse steel with fiber reinforcement for achieving comparable compressive performance characteristics between the confined and fibrous concretes. The investigation reported herein was concerned only with the material behavior under monotonic compression loads.

FIBER REINFORCEMENT VS. CONFINEMENT

Since confinement by transverse steel produces improvements of the same nature as fiber reinforcement in the compressive behavior of concrete, for a certain reinforcement index of each fiber type there exists a confinement condition (transverse steel spacing and volumetric ratio) which results in comparable compressive stress-strain relationships of concrete. This section illustrates an approach for finding the equivalent transverse reinforcement condition for a given reinforcement index of a certain fiber type. As shown in Fig. 3, the compressive stress-strain relationship of concrete can be characterized by the peak stress (f'_p), the secant stiffness from origin to the peak (K), the total area (A_1+A_2) under stress-strain relationship up to a strain beyond the peak, and the ratio of this area to the area under stress-strain curve up to its peak ($(A_1+A_2)/A_1$), which is a measure of the material toughness. Equivalent stress-strain relationships were defined as those with minimum errors in the weighted sum-total errors of these characteristic values. In this study, the total area

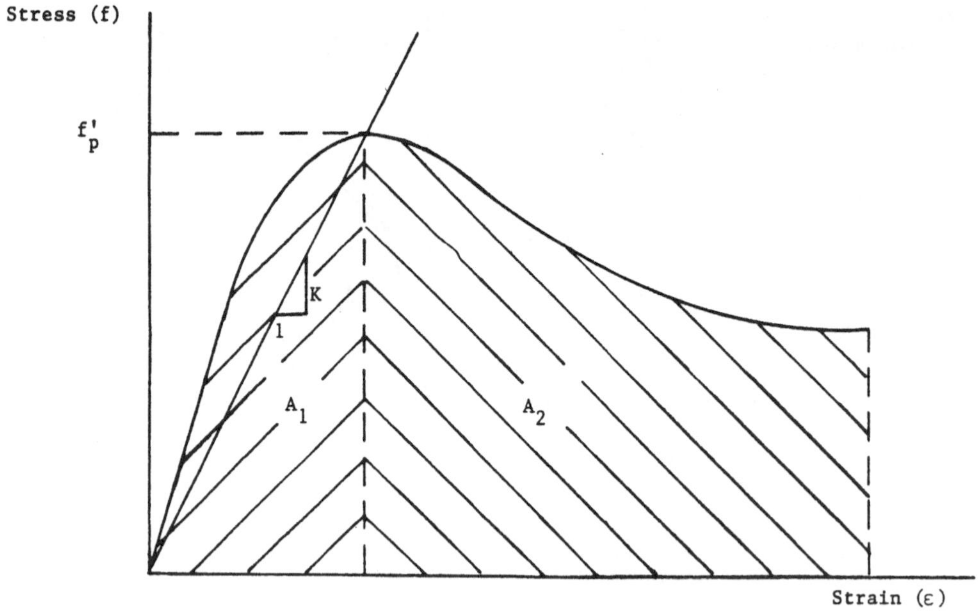

Figure 3. Characteristics of the Compressive Stress-Strain Relationship of Concrete.

under stress-strain curve and the toughness of the material were calculated for a maximum post peak strain of 0.02. The weights given to different characteristics of the stress-strain curves were selected to emphasize on those aspects of the constitutive relationship which are significantly enhanced by confinement and fiber reinforcement. These weights are: 0.5 for the area under the curve, 0.2 for the peak stress, 0.2 for the secant stiffness and 0.1 for the toughness index. The weighted sum-total of the normalized errors of the compressive stress-strain characteristics of confined concrete with respect to fibrous concrete was then calculated as follows:

Total Error = 0.5 x (error in the area) + 0.2 x (error in peak stress) + 0.2 x (error in secant stiffness) + 0.1 x (error in toughness)

All the errors in characteristic values of the confined concrete compressive behavior (when compared with those of fibrous concrete) were normalized with respect to the peak values of these errors. For each reinforcing index of a certain fiber type, the total error defined above was expressed as a function of two confinement variables: transverse steel spacing and volumetric ratio. Those confinement conditions which minimize the weighted error were chosen as the equivalent ones for a specified fiber reinforcement index.

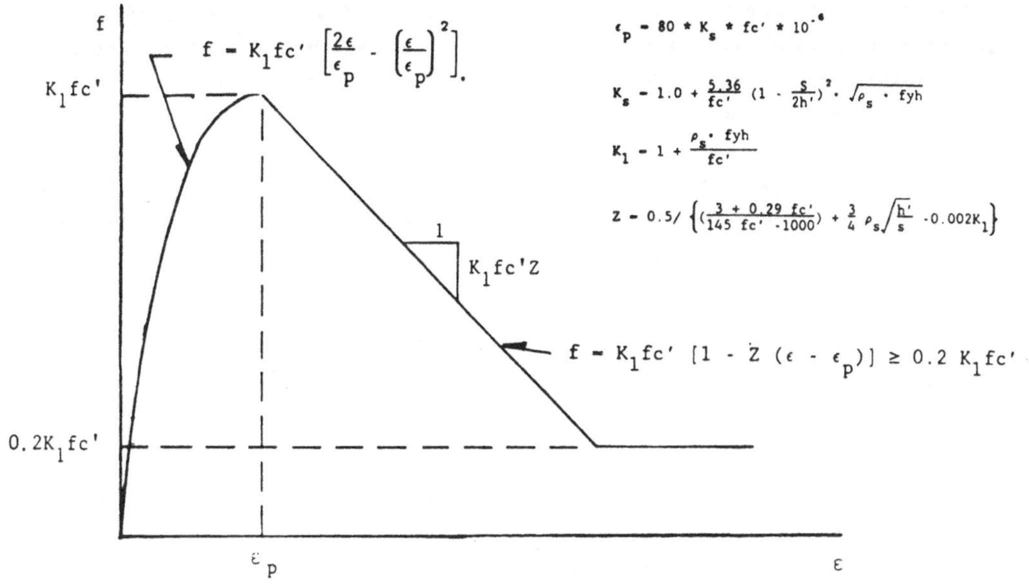

Figure 4. Confined Concrete Constitutive Model

The stress-strain relationships of both confined and fibrous concretes used in this study were produced using the empirical constitutive models presented in Ref. 3 for fibrous concrete and in Ref. 5 for confined concrete. These constitutive models are briefly described below.

The confined concrete constitutive model used in this study, which is a slightly modified version of the one given in Ref. 5, is shown in Fig. 4. This model consists of a curvilinear ascending branch and a linear descending branch followed by a flat tail. For this model, the degree of confinement is represented by the volumetric ratio (ρ_s) of the confining reinforcement (with respect to the confined concrete core) and the ratio of the confined core lateral dimension to the longitudinal spacing of the confining steel. The constitutive model for fiber reinforced concrete was directly adopted from Ref. 3. This model (Fig. 5) consists of two curvilinear ascending and descending branches, both of which are presented by the same equation, but different empirical coefficients are used in equation for different branches. The fiber reinforcement effects are considered in this model by making all its characteristic parameters functions of the fiber reinforcement index, $V_f \ell/d$, where V_f is the volume fraction and ℓ/d the aspect ratio of fibers. The effects of fiber type (steel, monofilament or twisted polyprophylene and glass) are considered in this model.

For specific volume fractions ($V_f \ell/d$ = 1 to 3) of different fiber types, Table 1 presents the equivalent confining (transverse) steel volume fraction and spacing which results in the minimum weighted error in the confined concrete constitutive behavior when compared with that of fiber reinforced concrete. The reasonable comparison between the compressive constitutive relationships of the fibrous and the equivalent confined concretes is obvious in Fig. 6. This figure presents typical comparisons for different types and volume fractions of fibers.

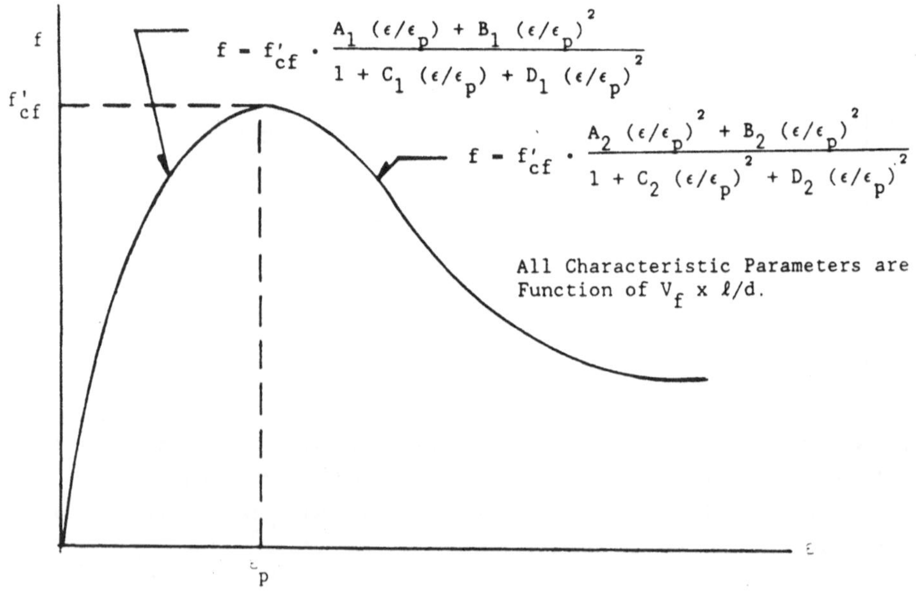

Figure 5. Fiber Reinforced Concrete Constitutive Model.

Table 1: Equivalent confinement conditions (transverse steel volume fraction and spacing) in a 40.6^{cm}x40.6^{cm} (16"x16") R/C column for different fiber volume fractions and types.

Fiber $V_f \ell/d$ (ℓ/d=100)	1.0		1.5		2.0		3.0	
Fiber Type	ρ_s	h'/s	ρ_s	h'/s	ρ_s	h'/s	ρ_s	h'/s
Steel	0.005	3.0	0.0075	3.0	0.01	3.5	0.0175	4.0
Twisted Polyprophylene	0.0025	0.5	0.0025	2.0	0.0025	5.0	0.005	3.5
Glass	0.005	0.5	0.005	1.0	0.005	1.5	0.005	3.0

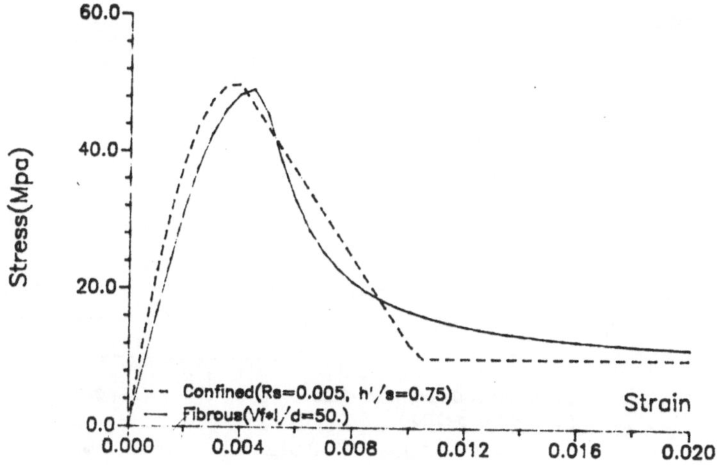

(a) Steel Fiber

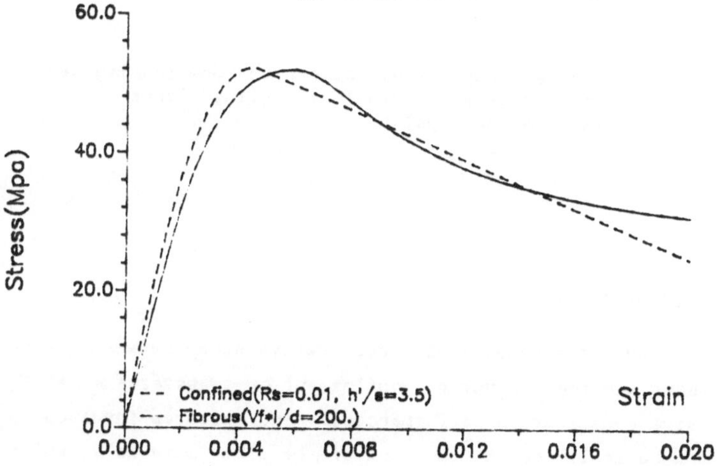

(b) Steel Fiber

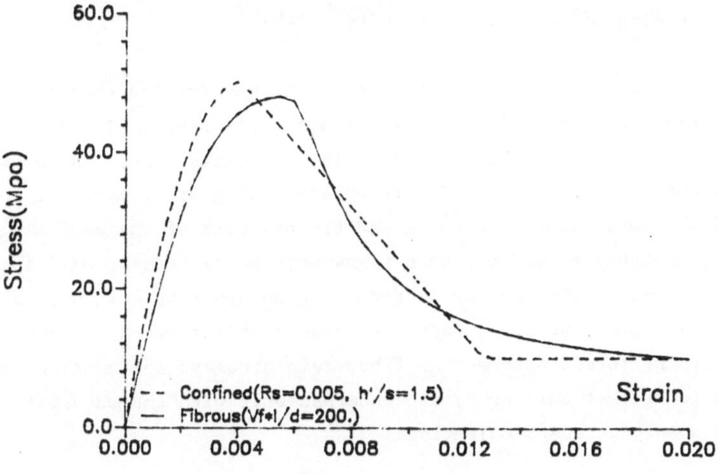

(c) Glass Fiber

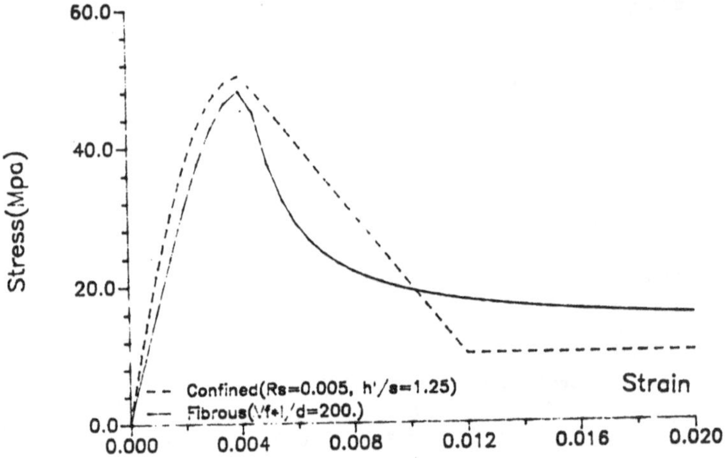

(d) Twisted Polyprophylene

Figure 6. Typical Comparisons between the Compressive Behavior of Fibrous and Equivalent Confined Concretes. (f'_c = 48.28 M pa)

COMPARATIVE COST ANALYSIS

This section presents the results of a comparative study between the costs of fiber reinforcement and the equivalent confinement by transverse steel (giving similar compressive constitutive performances). The expenses involved in fiber reinforcement are basically due to the cost of the fiber materials, while the confinement expenses are caused by the material cost as well as the spendings for bending and tieing of the transverse steel bars.

Based on a survey of some fiber manufacturers, it was decided that the fiber material costs are: $0.45/lb. for straight round steel fiber, $3.00/lb. for monofilament polyprophylene fiber, and $2.10/lb. for glass fiber. A survey of construction companies also indicated that the confining steel material costs $0.20/lb., and the construction (bending and tieing) cost of the confining reinforcement is roughly $0.20/lb. for a transverse steel spacing of 6 inches or more, which increases as the spacing is reduced (construction cost per lb. = $1.20/inches of spacing ≥ $0.20). Based on these cost estimates, a comparative cost analysis was performed between the fiber reinforcement and confinement techniques, noticing that this study is concerned only with the material behavior under monotonic compression loads.

Table 2: Comparison between the fiber reinforcement and the equivalent continement costs.

FIBER			CONFINEMENT		
Type	$V_f \ell/d$	Cost/ yd^3 of Conc.	ρ_s	h'/s	Cost/ yd^3 of Conc.
Steel	1.0	59.5	0.005	3.0	59.0
	1.5	89.0	0.0075	3.0	59.0
	3.0	178.6	0.0175	4.0	92.5
Twisted Polyprophylene	1.5	68.9	0.0025	2.0	32.63
	3.0	138.0	0.005	3.5	73.64
Glass	1.5	143.1	0.005	1.0	16.2
	3.0	286.1	0.005	3.0	59

Table 2 compares the costs of fiber reinforcement and the equivalent confinement (giving similar constitutive performances in compression) per cubic yard of concrete for different fiber types and fiber reinforcement indices ($V_f \ell/d$).

It may be concluded from Table 2 that confinement by transverse steel is generally more economical than fiber reinforcement in improving the compressive constitutive behavior of the material. This is more obvious for higher degrees of confinement (corresponding to higher fiber volume fractions), and for the twisted polyprophylene and especially the glass fibers. It should, however, be emphasized that the advantages of fiber reinforcement go far beyond improving the compressive constitutive behavior of concrete. Increasing the tensile strength and post-cracking tensile resistance, enhancing the shear resistance across the cracks, and improving the concrete durability and dimensional stability as well as its integrity and performance under inelastic cyclic loads are some other key advantages of fiber reinforcement, many of which can not be achieved effectively through confinement by transverse steel or other conventional reinforcement techniques. There are also situations, like in seismic-resistant beam-column connections, where the high confinement requirements result in a high congestion of the transverse steel. Here, fiber reinforcement might be a more economical technique, even for improving the compressive constitutive behavior, due to the extreme construction difficulties and costs involved in tieing transverse bars in a congested reinforcement cage. In short, the comparative cost analysis presented in this study is just the first step in comparing the effectiveness of fiber reinforcement versus the conventional reinforcement techniques. More comprehensive studies in this direction are needed in order to distinguish the specific applications through which major cost benefits can be achieved by fully or partially replacing the conventional reinforcement with randomly distributed fibers.

SUMMARY AND CONCLUSIONS

Compressive constitutive models of fiber reinforced and conventionally confined concretes were used to derive the confinement levels (represented by the volumetric ratios and spacings of the transverse steel) which result in compressive constitutive performances comparable to those achieved at certain fiber reinforcement levels (represented by the volume fraction of fibers times their aspect ratio, for steel, twisted polyprophylene and glass fibers). The material and construction costs corresponding to fiber reinforcement and the equivalent confinement were estimated and the results were used to compare the cost effectiveness of the fiber reinforcement and conventional confinement techniques in improving the compressive constitutive behavior of concrete.

Results indicated that, as far as we are strictly concerned with improving the compressive constitutive behavior under monotonic loads, confinement by transverse steel is generally more economical than fiber reinforcement. This is more obvious for the higher degrees of confinement (corresponding to higher fiber volume fractions), and for the twisted polyprophylene and especially the glass fibers.

This study was restricted to the fiber reinforcement effects on the constitutive behavior of concrete under monotonic compression. A comprehensive comparative economical study should account for the improvements in tensile behavior, durability and dimensional stability, and the integrity of concrete under monotonic and cyclic loads resulting from fiber reinforcement which might not be effectively achievable through confinement by transverse bars or other conventional reinforcement techniques. Such a comprehensive comparative economical study can help in distinguishing those concrete applications in which the partial or full substitution of the conventional reinforcement with fibers can lead to major cost savings.

ACKNOWLEDGEMENT

This investigation was performed in the Composite Materials and Structures Center of Michigan State University with financial support of the State of Michigan through the Research Excellence and Development (RE/ED) Fund. The computational facilities used in this study were provided by the Albert H. Case Center for Computer Aided Design in the Engineering College of Michigan State University. The authors are also thankful to Dr. Lawrence T. Drzal, the director of the Composite Center, for his continuing support of the fiber reinforced concrete research program in Michigan State University.

NOTATIONS

A_1 - area under stress-strain curve up to peak stress

A_2 - area under stress-strain curve from peak stress to final stress.

ϵ - strain

ϵ_p - strain at peak stress

f - stress

f'_{cf} - peak stress of fiber reinforced concrete

f_p' - peak stress

h' - concrete core width measured to outside of the transverse reinforcement

K - secant stiffness from origin to the peak stress

ℓ/d - aspect ratio

ρ_s - volumetric ratio of transverse reinforcement

s - center to center spacing of transverse reinforcement

V_f - fiber volume fraction

$V_f \ell/d$ - fiber reinforcing index

REFERENCES

1. Bayasi, Z., 'Mechanical Properties and Structural Applications of Steel Fiber Reinforced Concrete,' PH.D Thesis, Volume II, Michigan State University, 1987.

2. ACI Committee 544, 'State-of-the-Arc Report on Fiber Reinforced Concrete,' Report: ACI 544 IR-82, American Concrete Institute, Detroit, May, 1982, pp. 16.

3. Fanella, D., and Naaman, A., 'Stress-Strain Properties of Fiber Reinforced Mortar in Compression,' *ACI Journal*, V. 82, No. 4, July-August 1985, pp. 475-483.

4. Craig, R., McConnell, J., Germann, H., Dib, N., and Kashani, F., 'Behavior of Reinforced Fibrous Concrete Columns,' American Concrete Institute, sp-81: Fiber Reinforced Concrete, 1981, pp. 69-105.

5. Parviz Soroushian, Ki-Bong Choi, and Abdulaziz Alhamad, 'Dynamic Constitutive Behavior of Concrete,' *ACI Journal*, March-April 1986, pp. 251-259.

Proceedings of the International Symposium on Fibre Reinforced Concrete, December 16–19, 1987, Madras, India

TENSILE FAILURE MECHANISMS AND MECHANICAL PROPERTIES OF FIBER REINFORCED CONCRETE

Li, V.C.[1], S. Backer[2], Y. Wang[2], C. Leung[1] and K. Yamanobe[3]

[1] Department of Civil Engineering, MIT, Cambridge, MA 02139

[2] Department of Mechanical Engineering, MIT, Cambridge, MA 02139

[3] Visiting Department of Civil Engineering, MIT, Cambridge, MA 02139, from Shimizu Institute of Technology, Shimizu Construction Company, Ltd., Tokyo, JAPAN

SYNOPSIS

Tensile properties of Fiber Reinforced Concrete, particularly that of fracture toughness characterization, are investigated at the macroscopic (composite) level and at the fine structure (constituents) level. A broad based research program, including both theoretical and experimental components, with particular focus on synthetic fibers, is being carried out jointly between the Center of Advanced Construction Materials and the Fibers and Polymers Laboratory at MIT. This paper presents a brief overview of the research program and reports some interesting preliminary results obtained to date.

INTRODUCTION

It is known that the tension-softening curve controls the fracture behavior of a structure made from quasi-brittle material, such as concrete and certain short fiber reinforced composites [1,2,3,4,5]. In particular, Li and Liang [6] showed that for a quasi-brittle material with the same toughness, the shape of the tension-softening curve can control the apparent ductility of the structure. They also showed that the shape of the tension-softening curve determines the transition from brittle fracture behavior describable by linear elastic fracture mechanics to ductile behavior describable by simple yield strength concept. Thus it would be important to be able to measure this tension-softening property as well as to investigate how the micro/fine material structures govern the details of this property.

In addition, our recent experience with cement reinforced with synthetic fibers indicates that the ductility of the composite is greatly improved [7]. This can be seen qualitatively from the load-deformation curves of FRC compact tension specimens, shown in Figure 1. In order to quantify this ductility improvement, it has been suggested that their tension-softening curve should be determined. This curve can be converted back to fracture toughness [8,9].

Thus an objective of this research program is to devise a method to measure the tension-softening curve for quasi-brittle material in general, and for fiber reinforced composites in particular. The research program also aims at investigating at a basic level the contributions of fundamental mechanisms to composite tensile behavior and composite tension-softening and toughness. In this respect our approach is to carry out theoretical studies of plausible operating mechanisms, specifically the details of the fiber breakage and/or pull-

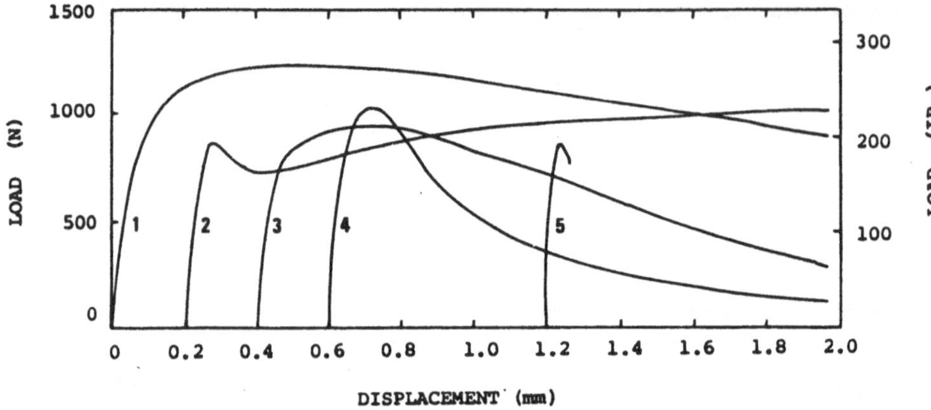

1 -- ARAMID FRC, MIX #K1 (V_f=1.85%, L_f=25.4 mm)
2 -- NYLON FRC, MIX #N1 (V_f=2.0%, L_f=38.1 mm)
3 -- ACRYLIC FRC, MIX #A5 (V_f=4.0%, L_f=6.4 mm)
4 -- ACRYLIC FRC, MIX #A6 (V_f=2.0%, L_f=12.7 mm)
5 -- PLAIN CONCRETE, MIX #C1 - #C10 (V_f=0)

Figure 1. Load vs. Displacement Curves for FRC CTS Tests

out processes when the composite is under nominally tensile loading. Various experimental tests at the fiber/matrix interface level are simultaneously carried out to provide essential data for the theoretical studies.

FRACTURE RESISTANCE MEASURING TECHNIQUE

The tension-softening σ–δ relation is defined as the functional relationship between the traction acting across a crack plane and the separation distance of the crack faces, in a uniaxial tension specimen loaded quasi-statically to failure. By definition, such a relation can be obtained from a direct tension test. However, due to the inherent instability of this test after the peak load is reached, modifications to stiffen up testing machines are required. Such modifications are usually too complicated to be widely adopted in normal testing laboratories.

In the current research program, the σ–δ relation for fiber reinforced concrete was obtained by an indirect technique first introduced by Li [5] and further developed by Li et al [9] for plain mortar. A thorough explanation of the theory can be found in [9]. A schematic of the experimental and data analysis procedure is shown in Figure 2. Two precracked specimens of slightly different crack lengths are loaded in the same way and for each specimen, simultaneous measurement of the applied load (P), the load point displacement (Δ) and the crack tip opening (δ) are made. From the area between the two P-Δ curves, we can obtain the J-integral [8], as a function of Δ. This, together with the δ–Δ curves, enables us to get J as a function of δ. Differentiating J with respect to δ and deleting the rising part of the resulting curve (to correct for inelastic energy absorption by materials outside the crack) will then leave us with the σ–δ curve.

Before actual application to synthetic fiber reinforced concrete, the validity of this experimental

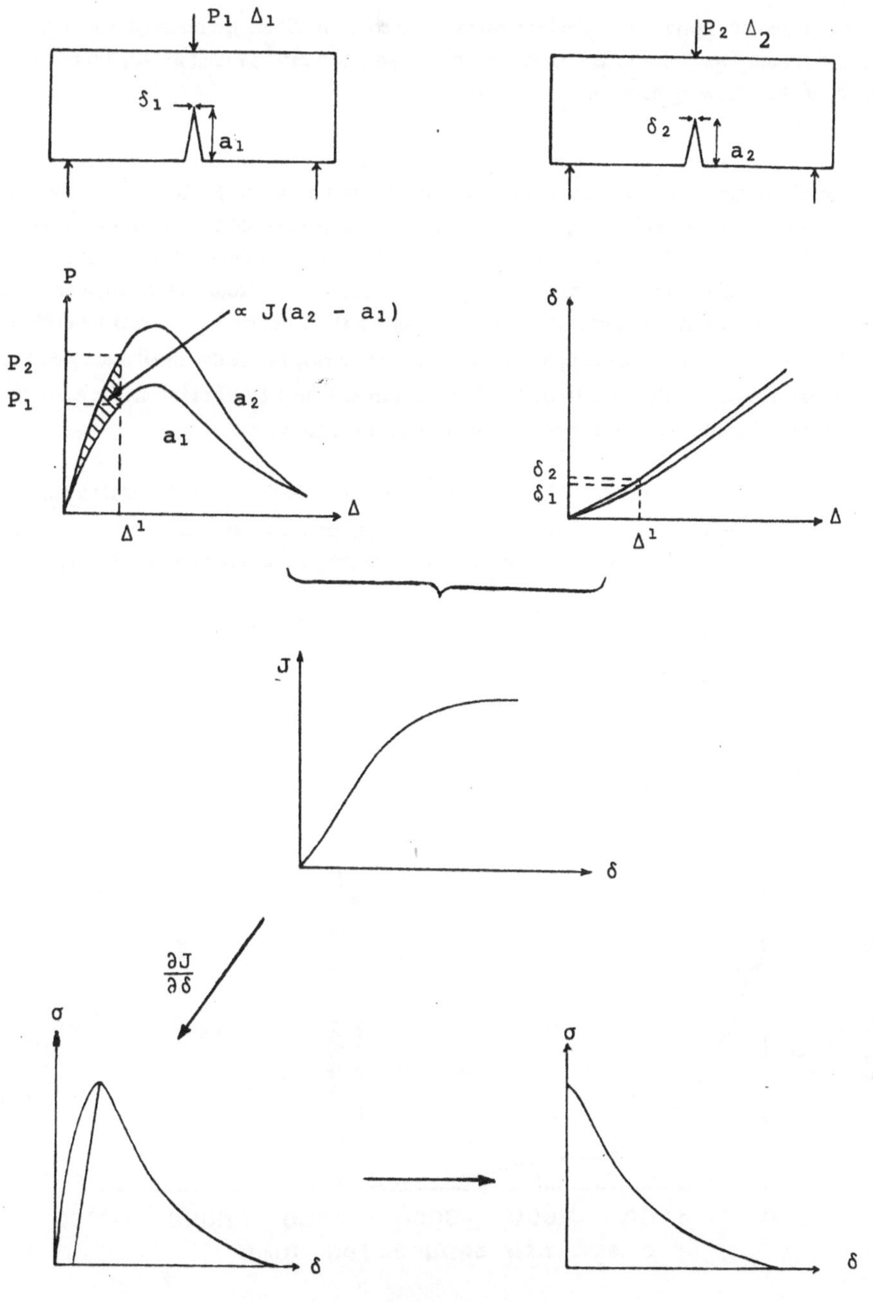

Figure 2. Deduction of the σ–δ Curve by the Indirect Testing Technique

technique was checked by applying it to steel fiber-reinforced concrete (SFRC), a material with reported σ–δ curve obtained from the direct tension test. Details of the investigation can be found in [10] but major findings are summarized in the following paragraph.

The σ–δ curve of a steel fiber reinforced mortar with 1 volume % of 9.525mm (3/8 in.) steel fiber was obtained. The result (Figure 3) was compared with direct tension test results on SFRC obtained by other researchers [11, 12] (Figures 4 and 5). Despite the difference in matrix material as well as fiber used in the various investigations, good qualitative agreement can be found between the three sets of results. They all exhibit a sharp drop from the peak followed by more gentle stress decrease. Note that the apparent 'flatness' in Figure 5 compared with the other figures is due to the expanded scale in Figure 5. The total area below the σ–δ curve is denoted by J_c and is equivalent to the critical energy release rate under small scale yielding condition. From our result, J_c was found to be 1.20 kJ/m^2, compared to 1.87 kJ/m^2 from the curve in Figure 4. Due to the use of different materials, a direct comparison is not possible.

From our experimental data on SFRC, we are confident that the indirect experimental technique for σ–δ curve determination is applicable to fiber reinforced concrete. Hence, we are ready to apply this technique to synthetic fiber reinforced concrete, which is one of the main object of the current research program.

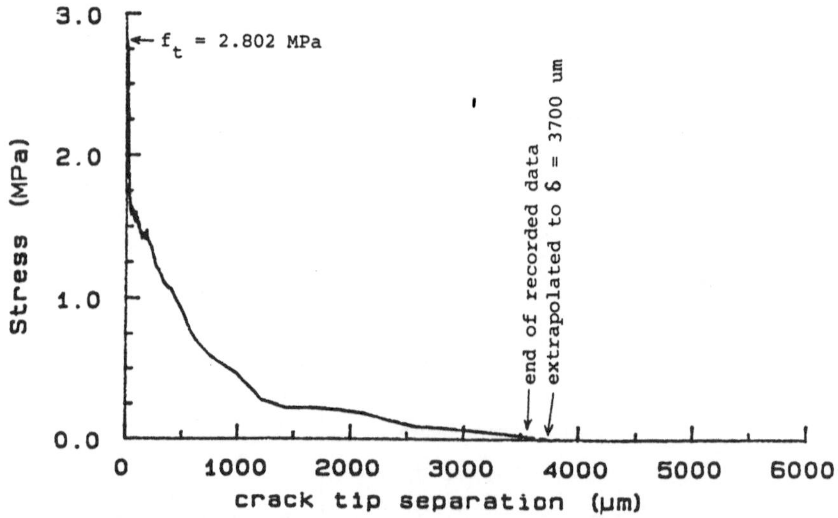

Figure 3. Deduced Tension-Softening Curve for SFRC

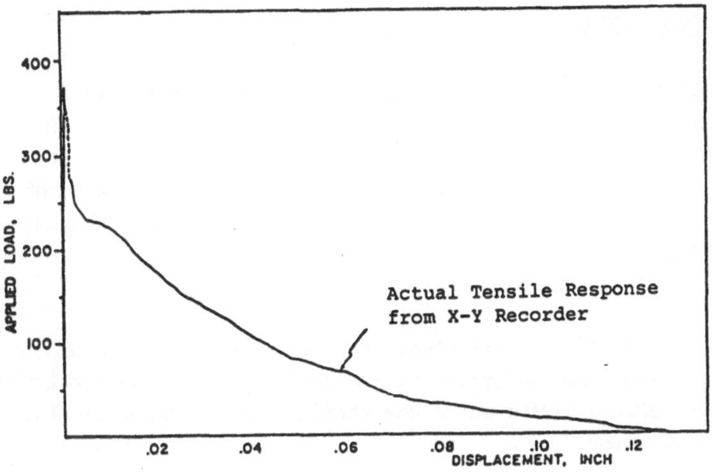

Figure 4. Tensile Test Curve for Steel Fiber Reinforced Mortar, V_f=1%, L_f=6.35 mm (0.25"), L_f/d_f=42 (From Visalvanich and Naaman [11])

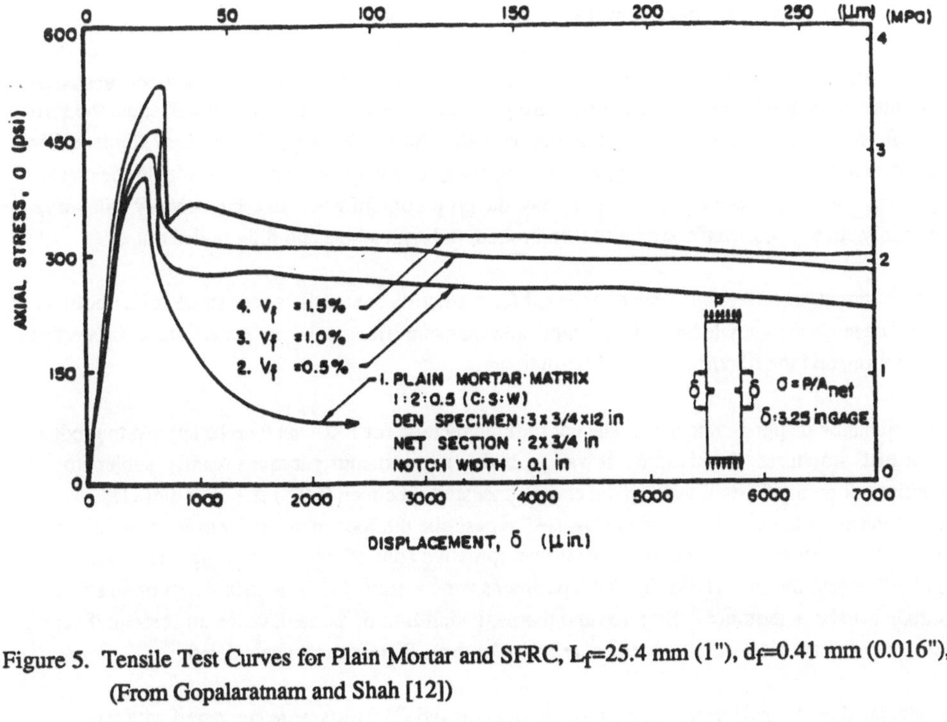

Figure 5. Tensile Test Curves for Plain Mortar and SFRC, L_f=25.4 mm (1"), d_f=0.41 mm (0.016"), (From Gopalaratnam and Shah [12])

CONSTITUTIVE MODELING OF FRC

It has been suggested [6] that the best way to describe the crack resistance of FRC, which improvement is the greatest contribution of the fibers, is by means of a tension-softening curve. In the previous section, an experimental technique to determine such tension-crack separation curve was discussed. By constitutive modeling of FRC in terms of the fiber and matrix parameters, the tensile behavior of FRC, before and after matrix first cracking, can also be described. The constitutive model of FRC is not only useful for optimal design of FRC materials and for structural analysis of FRC members, but also helpful for understanding the mechanisms of fiber reinforcement in FRC.

In this study, a preliminary statistical model has been developed to predict the tensile behavior of short fiber reinforced concrete, by considering an FRC specimen under uniaxial tensile test. The model, while promising, still requires further refinement. Here we will give a brief description of this model. The details of the model have been reported elsewhere [13, 14].

For strains in the FRC composite smaller than the matrix cracking strain, the Law of Mixture is assumed to apply. After matrix cracking, the deformation of the tensile specimen consists of two parts: the elastic deformation of the specimen, and the crack opening. Because concrete reinforced with randomly distributed fibers generally fails in a single fracture mode, it is assumed that only one crack is developed in the matrix during the tensile test.

On the assumption of randomly distributed fibers, the model establishes the probability density functions of embedded fiber lengths and of fiber orientation relative to the crack plane. These are used in subsequent composite stress computation. It is further assumed that the fiber/matrix bond shear stress is constant along the fiber (as if held by friction) so that a given fiber force will determine the magnitude of the stretched segment of an embedded fiber, as well as its elastic elongation.

As the tensile test proceeds, the stretched segment of each fiber will increase until the fiber breaks, or slips along its entire embedded length. At any time during the tensile test, the crack width will equal the sum of the elastic deformation of the fiber within the matrix on both sides of the crack, plus the fiber slippage from the matrix and the elastic extension of this slipped fiber segment, for each fiber currently bridging the crack. Clearly, as the crack opening in the tensile test proceeds, the proportion of fibers in each category will vary, including: partially stretched, partially stretched then broken, fully stretched, and fully pulled out.

A computational procedure has been established for predicting the tensile stress-strain behavior of an FRC specimen, based on the knowledge of fiber length and diameter, fiber and matrix mechanical properties, fiber volume fraction, and the fiber/matrix bond strength.

The computational approach to the tensile constitutive relation for FRC can then be applied to predict the behavior of FRC structures. In this study, it was applied to two test configurations readily subject to laboratory confirmation: the flexural test and the compact tension specimen (CTS) test. For these test configurations, the tensile constitutive relation was used to describe the specimen local tensile behavior, and linear elasticity was assumed in the specimen over the compressive zone of the specimen ligament. For computational efficiency, the strain fields in these specimens were assumed to be a linear function in space, but its magnitude is to be determined. By force and moment equilibrium, moment vs. beam center deflection for the flexural test, and the load vs. displacement relation for the CTS test can be calculated.

The calculated load-displacement curves for the flexural and CTS tests were compared with the experimental results. Descriptions of the fibers used in the experiment, the method of determining the fiber/matrix bond strength, and the preparation and test of the flexural and CTS specimens can be found in [7]. Figure 6 presents an example of the theoretical and the experimental moment-displacement curves for the flexural test of acrylic and nylon FRC. It was found that in both the flexural and the CTS tests, the shapes of the load-displacement curves vary significantly for FRC specimens using different types of synthetic fibers and with different fiber volume fractions, and the form of the computed curves corresponds closely to the experimental ones. However there still exists a difference in the load magnitude between the theory and the experiment. Clearly additional study is needed on the assumptions built into the constitutive model and on the assumptions of the strain distribution in the modeled structures. And these are some of the problems we currently focusing on.

To reexamine the assumption of constant fiber/matrix shear bond strength, experiments of monofilament pull-out from a cement matrix were conducted. It was found that for certain types of fibers (e.g. nylon and polypropylene), the fiber/matrix bond strength increases with the distance of fiber slippage, resulting from the fiber surface abrasion which can be seen under a Scanning Electron Microscope (Figure 7). The extra amount of energy absorption due to fiber abrasion has important implications in composite toughness or resistance to crack propagation. The increase in fiber/matrix bond strength also has great importance in reducing the difficulties in the composite fabrication process, for it permits the use of lower fiber aspect ratio in FRC. A theoretical analysis based on these observations has been conducted to describe the

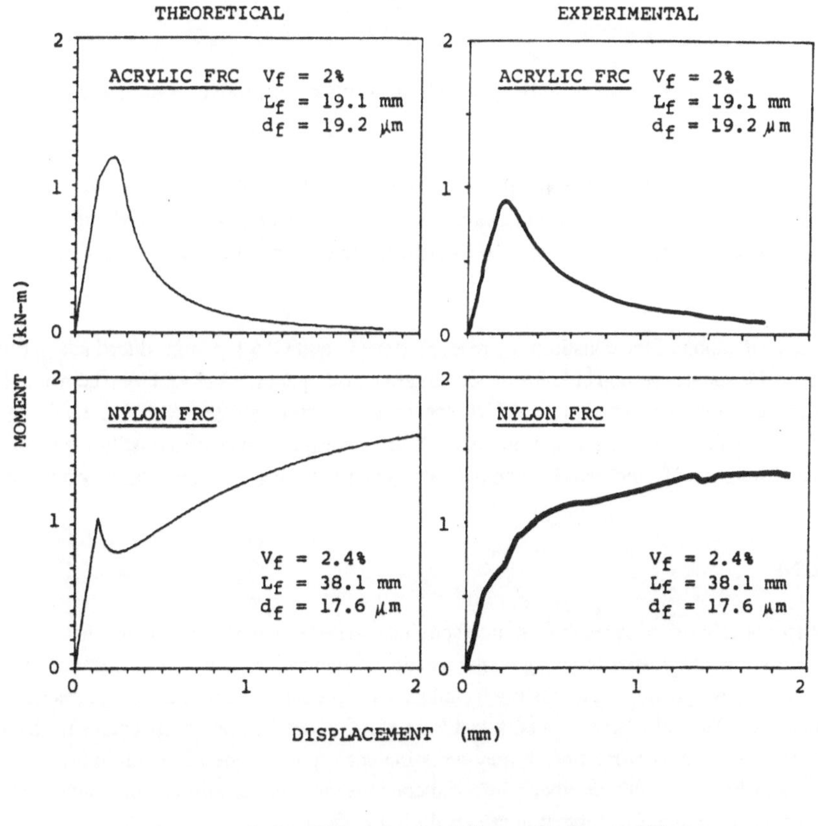

Figure 6. Theoretical and Experimental Curves for Flexural Tests

Figure 7. Surface Damage of Nylon Fiber due to Pull-Out from Cement Matrix

fiber pull-out behavior, and this result will be applied in the further constitutive modeling to improve the model.

The assumption of random fiber distribution was also found to differ from the actual situation where fibers in the FRC specimens often appeared in bundle form. Such effect is being considered in the theoretical analysis, and, at the same time, being reduced by improving the equipment as well as the specimen forming technique.

Indirect verification of the constitutive model by flexural and CTS tests introduced added difficulties in identifying the major sources responsible for the quantitative discrepancy. A closed loop controlled testing frame is being built for uniaxial tensile test of FRC specimens. Such tensile test would certainly provide more useful information about the FRC behavior, and provide direct verification of the constitutive model. These test results will also serve to further validate the indirect J-integral testing technique described in the previous section.

CONCLUSIONS

This paper briefly summarizes the research activities carried out at MIT on the mechanics of fiber reinforced cementitious composites focusing on the tensile mechanical properties and failure mechanisms. These investigations have yielded useful testing techniques that properly determine fracture resistance of quasi-brittle materials. They also provide a better understanding of detailed failure mechanisms that could be related to composite toughness properties. Future work includes applying the J-integral indirect testing technique to various types of synthetic fiber reinforced concrete, and further refinements in the constitutive model to include, for example, fiber abrasion effects during pull-outs.

Ultimately these materials with improved tensile properties may be used in members of structures which may be required to withstand seismic loads. In this regard, we have begun an investigation in the behavior of FRC beams subjected to shear loading. Experimental tests will be carried out for beams with different volume fractions of fiber reinforcement. This series of tests together with the tensile tests will form a data base from which models could be formulated in relating tensile properties of the FRC to shear properties of FRC structural components. We expect that such results could eventually be useful for aseismic design using FRC as a structural material.

ACKNOWLEDGEMENTS

The authors would like to acknowledge the support of the Shimizu Institute of Technology, Shimizu Construction Company, Ltd., the Systems Engineering for Large Structures Program and the Solids and Geomechanics Program of the National Science Foundation.

REFERENCES

1. Petersson, P.E., Crack Growth and Development of Fracture Zones in Plain Concrete and Similar Materials, Report TVBM-1006, Lund Institute of Technology, Division of Building Materials, 1981.

2. Hillerborg, A., In: *Fracture Mechanics of Concrete*, F.H. Wittman, Ed. Elsevier Science Publisher, B.V., Amsterdam 1983, p.223-249.

3. Reinhardt, H.W., HERON, Delft Univ. of Technology Vol.29, No. 2, 1984.

4. Ingraffea, A.R. and Gerstle, W.H., In: *Application of Fracture Mechanics to Cementitious Composites*, S.P.Shah, Martinus Nijhoff Publisher, Dordrech, 1985, pp247-285.

5. Li, V.C., In: *Application of Fracture Mechanics to Cementitious Composites*, S.P.Shah, Martinus Nijhoff Publisher, Dordrech, 1985, pp.431-449.

6. Li, V.C. and Liang, E., *ASCE J. of Engineering Mechanics*, Vol 112, No. 6, June, 1986, pp 566-586.

7. Wang, Y., Backer, S., and Li, V.C., An Experimental Study of Synthetic Fiber Reinforced Cementitious Composites,*J. of Materials Science*, in press, 1987.

8. Rice, J.R., In: *Fracture: An Advanced Treatise*, Vol.2, Academic Press, New York 1968, pp191-311.

9. Li, V.C., Chan, C.M. and Leung, C.K.Y., *Cement and Concrete Research*, Vol 17, 1987, pp.441-452.

10. Leung, C.K.Y. and Li, V.C., Determination of Fracture Toughness Parameter of Quasi-Brittle Materials with Laboratory-Size Specimens, International Conference on Fracture of Concrete and Rock, Houston, TX, USA, June 1987.

11. Visalvanich K. and Naaman, A., *ACI Journal*, Vol 80, 1983, pp 128- 138.

12. Gopalaratnam V.S. and Shah, S.P., *ASCE Journal of Engineering Mechanics*, Vol 113, No 5, 1987, pp 635-652.

13. Wang, Y., Mechanics of Fiber Reinforced Concrete, SM Theses, Department of Mechanical Engineering, MIT, Cambridge, MA, USA, 1985.

14. Wang, Y. and Backer, S., A Statistical Model of Fiber Reinforced Cementitious Composites, Paper presented at the Fiber Society Spring 1987 Technical Conference, Greensboro, N.C., USA, May 1987.

Proceedings of the International Symposium on Fibre Reinforced Concrete, December 16–19, 1987, Madras, India

THE FRACTURE TOUGHNESS OF FIBRE REINFORCED CONCRETE

B.B. Sabir, S.D. Summerhayes and M. Asili

Department of Civil Engineering and Building,
The Polytechnic of Wales,
Pontypridd,
Mid. Glamorgan,
CF37 1DL, UK.

SYNOPSIS

The paper demonstrates the usefulness of applying fracture mechanics in evaluating the enhanced toughness properties of fibre reinforced OPC/ microsilica concrete. Both numerical modelling, by finite elements, and laboratory testing on standard concrete cubes modified by the introduction of two notches on opposite faces are carried out. The test specimen results in unstable fracture and linear elastic fracture mechanics (LEFM) is used to evaluate the fracture toughness on the basis of the stress intensity factor. The post-cracking performance is assessed by a toughness index determined from the load-deflection relationship.

INTRODUCTION

Recent years have seen a growing interest in the use of microsilica as a partial replacement material for cement. The reasons for this are the marked improvements obtained in the durability and strength of the resulting concrete. Unfortunately these improvements have been accompanied by an increase in the concrete's brittleness [1] and measures have to be taken to reinforce the basic matrix with chopped fibres, for example. Under these conditions, the enhanced toughness and the post-cracking ductility imparted by the fibres to the concrete are evaluated on the basis of fracture mechanics theory.

Amongst the earliest papers on the application of fracture mechanics to concrete is that by Kaplan [2]. Many attempts have since been made to develop testing procedures for evaluating the fracture toughness of concrete materials. For example, recent work at the Lund Institute of Technology [3] has led to the RILEM recommendations for fracture energy evaluation for concrete materials from beam specimens notched at midspan. More recently Karihaloo[4] used compression splitting tests to evaluate the fracture energy of concrete. Both of these works[3,4] employ relatively large specimens and if fracture tests are to be accepted by industry an effort must be made to develop more compact specimens.

There are two problem areas in applying fracture mechanics theory to concrete. The first is the effect of slow crack growth which takes place during testing and the second is related to the size of specimen necessary for valid fracture toughness measurement. Swamy[5] discussed the problem of slow crack growth and highlighted the importance of its inclusion in fracture toughness testing. The second problem is the difficulty which can arise in the use of small specimens[6]. The general theory of fracture mechanics requires that the plastic zone is small in comparison with the crack length, the thickness of the specimen and the uncracked ligament length. The effect of specimen size has been reported by Mindess[7]. Similar size effects were reported by Barr and Sabir[8] using the compact compression test specimen. Karihaloo[4] states that using the compression splitting test, provided certain conditions are met, the fracture toughness is independent of specimen size.

The influence of microcracking ahead of the crack front was considered by Petersson[9] who developed a fracture model which takes account of the progressive microcracking occuring prior to the propogation of the pre-existing crack. Wecharatana et al[10] extended these ideas and in their model the crack was augmented by an additional length due to the zone of microcracking. Using LEFM they analysed the results of tests performed on double cantilever, double torsion and beam specimens. Bazant and Cedolin[11] found it necessary to couple the idea of equivalent crack length with the R-curve concept in order to obtain satisfactory fit from data obtained from three and four-point bend specimens.

The two main approaches for determining the toughness of a material are those based on the fracture energy required to create a crack extension of unit area, i.e. G_c and the critical stress intensity factor K_c. Nallathambi et al[12] used both approaches and their results were in some disagreement. Whereas they argued that the fracture energy approach is more appropriate, there are advantages, discussed by Swamy[13], in measuring the fracture toughness on the basis of the stress intensity factor.

In this paper the fracture toughness of fibre reinforced concrete is evaluated from tests conducted on pre-notched 100 mm cubes using the stress intensity factor approach. The post-cracking ductility imparted by the fibres is assessed by a toughness index determined from the load-deflection relationship. Fibrillated polypropylene fibres are used to reinforce the basic Portland cement concrete in which microsilica is employed as an admixture.

EXPERIMENTAL STUDY

Specimen Geometry

The test specimen used in the present study is that which was originally proposed by Barr et al [14]. The specimen is a 100 or 150mm cube modified by the introduction of two notches on opposite faces and loaded eccentrically as shown in figure 1. The attractiveness of this specimen lies in the fact that it is based on standard quality control samples requiring little additional preparation and that the test is simple to perform.

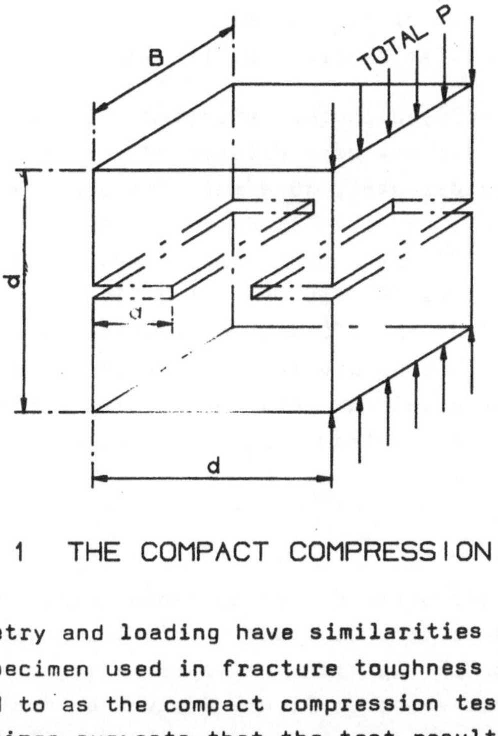

FIGURE 1 THE COMPACT COMPRESSION SPECIMEN

The specimen geometry and loading have similarities with those of the compact tension specimen used in fracture toughness testing and is therefore referred to as the compact compression test specimen. Experience with this specimen suggests that the test results in unstable crack propogation and slow crack growth is not considered to be a problem. Initial work employed cubes with symmetrical notches. Later Barr and Sabir[8]extended the work to include unsymmetrical notches to investigate the effect of the size of the uncracked ligament on the fracture toughness. The authors[8]also examined the influence of specimen size by conducting a series of tests on specimens of varying lengths. It was found that the fracture toughness was independent of notch depth and test specimen size.

The results reported in this paper were obtained from 100mm cubes with symmetrical notches. The notches were introduced by means of an adapted horizontal milling machine equipped with a masonry cutting disc which produced notches to within 0.5mm of the desired depth. The specimens were tested at normal room temperature using an Instron testing machine. Loading was under deflection control with a cross-head movement of 0.5 mm/min. The load was applied via two 6mm square steel bars placed along the edges of the specimen.

Mix Details

The materials used for the basic mix were as follows:

Cement	ordinary Portland cement
Fine aggregate	local dredged sand - zone 2
Coarse aggregate	10mm single size limestone

The aggregates were employed in the saturated surface-dry condition. The proportions for the basic mix were determined by the DoE method [15] for a 28 day strength of approximately 40 N/mm^2. The basic mix was modified by the addition of microsilica as an 8% direct replacement by weight of OPC. The water/cement (OPC plus microsilica) ratio was kept constant at 0.47 throughout the tests. The fibres, being 12mm long having nominal thickness and width of 30 and 200 microns respectively, were added at the final mixing stage and no problems were encountered in dispersing them into the mix. The specimens were cured under water for 21 days at 20^oC when the notches were introduced. The specimens were further cured until testing at 28 days.

FINITE ELEMENT SOLUTIONS

Solutions for the test specimens were obtained using special crack tip finite elements together with constant strain triangular (CST) elements. Full details of the crack tip elements appear elsewhere [16] and only a brief description is given here. Figure 2 gives a schematic representation of the various elements employed. The elements I and II exhibit strain singularities of $r^{-\frac{1}{2}}$ type and are suitable for employment near the tip of the crack where linear elastic conditions are assumed. In both elements the radial and circumferential displacements u and v are expressed in terms of simple polynomial functions in the angular coordinate θ . In addition the displacements for element I contain the trigonometric terms associated with the rigid body movements. When analysing a fractured specimen, a cluster of elements II, all having a common

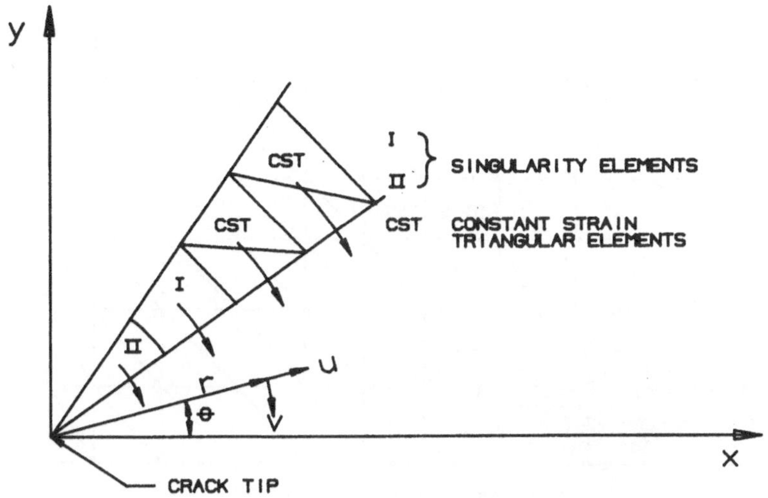

FIGURE 2 THE FINITE ELEMENTS USED IN THE ANALYSIS

point at the tip of the crack, are surrounded by an annulus of elements I as shown in figure 2. The remainder of the body is represented by CST elements. It was shown [16] that in analysing a compact tension fracture toughness specimen accurate estimates for the stress intensity factors are obtained from the displacements of the nodal line immediately surrounding the crack tip using a simple averaging process. It is pointed out, however, that using such idealisations, the results for the stresses are of inferior accuracy to that for displacements and that unless very fine meshes are employed, the CST elements are not capable of representing the steep gradients of the strains in the vicinity of the crack tip. In situations where the stress distribution is required it would be more economical to use the well known isoparametric elements.

Figure 3 gives the finite element mesh employed for the upper half of a 100mm cube with 25mm notches. The region surrounding the tip of the notch in the tension zone was represented by twelve triangular elements II and twelve trapezoidal elements I. The area covered by elements I and II extended to a radial distance of notch depth/70 from the tip of the notch. In this idealisation the solution employed a total of 438 degrees of freedom. Continuity conditions were imposed along the uncracked ligament line of the specimen. As deformation took place during testing, the load was concentrated at the inside edges of the loading bars and in the analysis, therefore, the load P was applied at 6mm from the edge of the specimen.

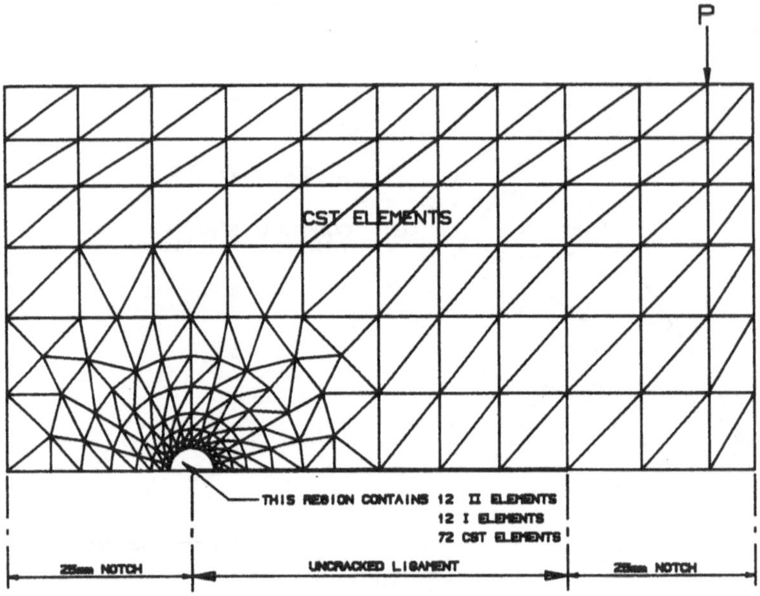

FIGURE 3 FINITE ELEMENT MESH FOR THE COMPACT COMPRESSION SPECIMEN

Solutions were obtained for a range of notch depths employing similar meshes in which the total number of degrees of freedom ranged between 438 and 446. This enabled an expression to be found for the stress intensity factor K as a function of notch depth/width, i.e. a/d, ratio. This is:

$$K = PY/Bd^{\frac{1}{2}}$$

where Y is a polynomial given by;

$$Y = 18.3\ R^{\frac{1}{2}} - 430\ R^{3/2} + 3445\ R^{5/2} - 11076\ R^{7/2} + 12967\ R^{9/2}$$

and R = a/d .

RESULTS AND DISCUSSION

The concrete mix used throughout the work reported here contained micro-silica as an 8% direct replacement by weight of OPC. In all four mixes were prepared, three of which contained varying amounts of polypropylene fibres. The fibres were added to the mix in proportions by weight of the total wet solids of 0.035, 0.05 and 0.065. From each mix eighteen cubes were cast, six were used to determine 7 and 28 day compressive strengths and the remaining twelve were used for the notched specimens. Notch depths varying between 25 and 40mm in increments of 5mm were introduced with three specimens being produced for each geometry.

Mix No.	Fibre content (%)	Compressive strength N/mm²		Comp. strength ratio
		7 days	28 days	
1	0	37.1	56.0	
2	0.035	35.2	51.7	0.92
3	0.050	34.4	51.8	0.93
4	0.065	32.5	49.1	0.88

Table 1 Variation of compressive strength with fibre content

Fibre content (%)	Notch depth (mm)	Failure load (kN)	V (%)	K_c $MN/m^{3/2}$
0	25	9.99	2.1	0.59
	30	5.80	3.7	0.53
	35	2.88	4.1	0.51
	40	1.35	4.1	0.56
0.035	25	9.39	2.8	0.55
	30	5.42	4.1	0.50
	35	2.80	1.9	0.50
	40	1.31	2.7	0.54
0.050	25	9.09	2.0	0.54
	30	5.24	3.1	0.48
	35	2.72	4.2	0.49
	40	1.15	1.7	0.47
0.065	25	9.57	0.9	0.56
	30	5.64	4.1	0.52
	35	2.83	5.7	0.51
	40	1.22	5.5	0.50

V = coefficient of variation

Table 2 Variation of fracture toughness with fibre content

Table 1 gives the compressive strengths for the four mixes. The result for the plain concrete mix, i.e. mix 1, reflects the extra strength due to the addition of microsilica. The compressive strength ratio was determined by dividing the average compressive strength of cubes containing fibres by the average strength of cubes taken before the fibres were added to the mix. It is seen that adding polypropylene fibres to the mix results in a decrease in the compressive strength.

The results for the fracture loads and fracture toughness are shown in table 2. The fracture toughness was evaluated from the expression given

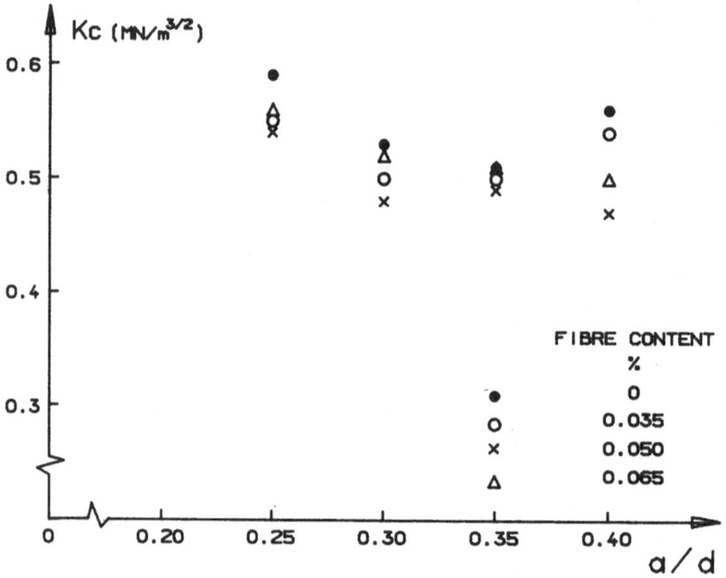

FIGURE 4 VARIATION OF FRACTURE TOUGHNESS WITH NOTCH DEPTH

as a function of notch depth. The results demonstrate that the addition of the fibres leads to similar reductions in the fracture toughness to those obtained for the compressive strengths. Such small reductions can be justified when considering the advantages to be gained from the crack inhibiting properties provided by the fibres for the concrete matrix.

Load-deflection graphs have not been included here but it is pointed out that, following an initial bedding of the specimen at the loading edge, all curves exhibited a simple linear relationship until first cracking was produced. Both the plain and the fibre reinforced specimens failed under a distinctive maximum load. Whereas in the case of plain concrete complete failure took place under this load, the fibre concrete specimens developed a residual strength after initial failure. This residual strength forms the basis for evaluating the post-cracking toughness of the material. A number of approaches have been proposed for evaluating the post-cracking toughness of fibre concrete [17-19]. The most convenient approach is that which employs the areas under the load-deflection curves. The toughness index definition proposed by Barr et al [20], which is shown in figure 5, is adopted in this paper. This definition is preferred because it has no restriction on specimen type and does not require the prescribed deflection necessary for the ACI toughness index [17]. The only requirement is that the specimen is loaded up to twice the deflection at first crack.

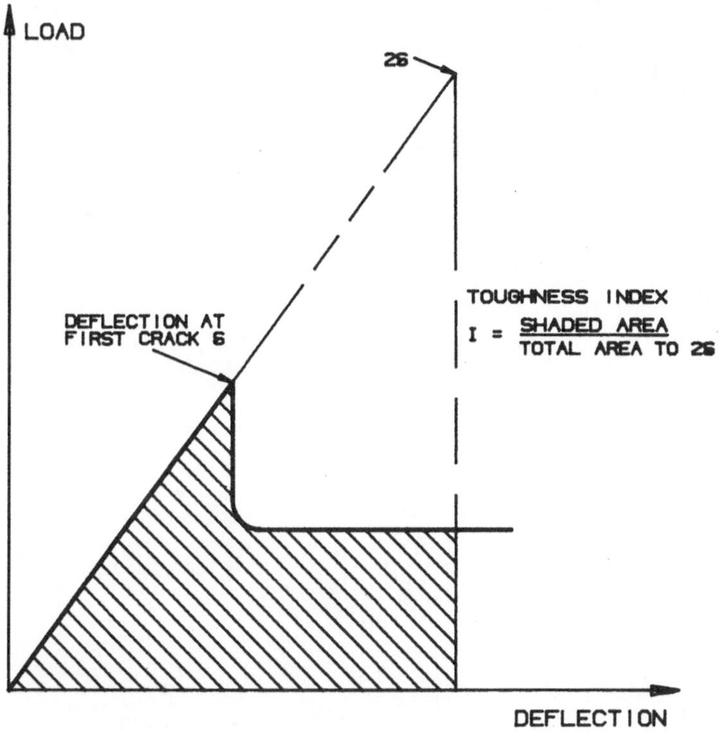

FIGURE 5 DEFINITION OF TOUGHNESS INDEX
Barr et al 19

The results for the toughness index for the 35 and 40mm notch depths are given in table 3 and graphically in figure 6. Results from other notch depths were not available at the time of writing. It is seen that even with the small percentage additions of the fibres used in the present study, considerable improvement in the post-cracking performance is obtained. Similar toughness index values were obtained in [20] using conventional polypropylene fibres but with concentrations of as high as 0.2% by weight.

Notch depth (mm)	Fibre content (%)	Toughness index	Average toughness index
35	0.035	0.313,0.305,0.301	0.306
	0.050	0.341,0.333,0.340	0.338
	0.065	0.371,0.344,0.381	0.365
40	0.035	0.344,0.340,0.372	0.352
	0.050	0.353,0.403,0.369	0.375
	0.065	0.434,0.404,0.397	0.412

Table 3 Variation of toughness index with fibre content

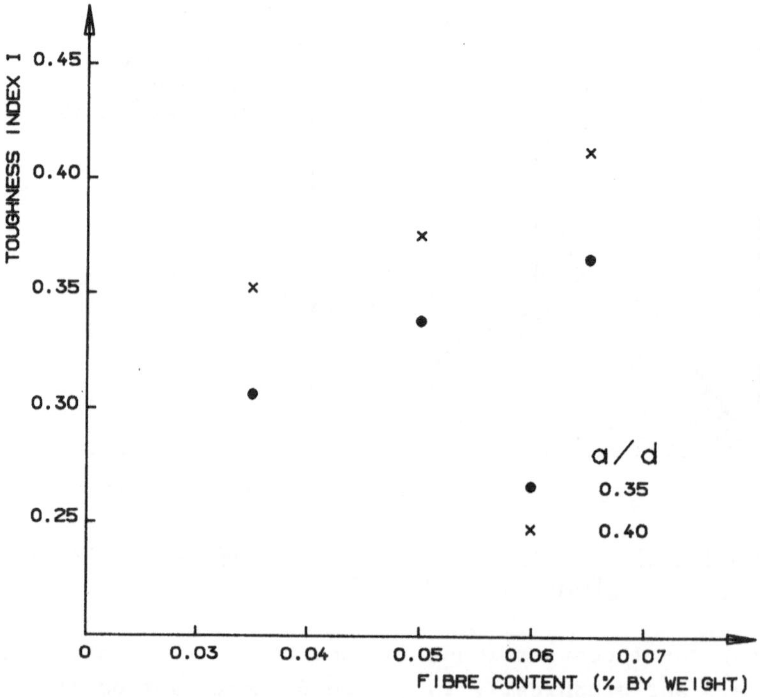

FIGURE 6 VARIATION OF TOUGHNESS INDEX WITH FIBRE CONTENT

CONCLUSIONS

1. This paper gives some results from an ongoing research programme investigating the performance of fibrillated polypropylene fibres in OPC/microsilica concrete.

2. The addition of the fibres leads to reductions in the fracture toughness of the concrete which are similar to those obtained in the compressive strengths.

3. The post-cracking toughness index for the fibre concrete employed in the present study varies between 0.3 and 0.4. The toughness index is zero for plain concrete and has a theoretical maximum value of unity for fibre concrete.

REFERENCES

1. Bache, H., Concretes in Engineering, International Congress on Utilisation of Materials Know-how in the Engineering Industry, Finland, 10-13 September,1985.
2. Kaplan, M.F., Crack propogation and the fracture of concrete, Proceedings, ACI Journal, Vol. 58, No. 5, Nov. 1961, pp 591-610.
3. Lund Institute of Technology, Division of Building Research, Proposed RILEM recommendation: Determination of the fracture energy of mortar and concrete by means of three-point bend tests on notched beams, Revised Proposals, Lund, June 1982.
4. Karihaloo, B.L., Fracture toughness of plain concrete from compression splitting tests, Int. Journal of Cement Composites and Lightweight Concrete, Vol. 8, No. 4, Nov. 1986, pp 251-259.
5. Swamy, R.N., Influence of slow crack growth on the fracture resistance of fibre composites, Int. Journal of Cement Composites, vol. 2, No. 1, February 1980, pp 43-53.
6. Walsh, P.F., Fracture of plain concrete, The Indian Concrete Journal , Vol. 46, No. 11, Nov. 1972, pp 469-472.
7. Mindess, S., The effect of specimen size on the fracture energy of concrete, Cement and Concrete Research, Vol. 14, No. 3, May 1984, pp 431-436.
8. Barr, B.I.G. and Sabir, B.B., Fracture toughness testing by means of the compact compression test specimen, Magazine of Concrete Research , Vol. 37, No. 131, June 1985, pp 88-94.
9. Petersson, P.E., Fracture energy of concrete: method of determination, Cement and Concrete Research, Vol. 10, 1980, pp 78-89.
10. Wecharatana, M. and Shah, S.P., Prediction of non-linear fracture process zone in concrete, Journal of Engineering Mechanics, ASCE,

Vol. 109, No. 5, October 1983, pp 1231-1246.

11. Bazant, Z.P. and Cedolin, L., Approximate linear analysis of concrete fracture by R-Curves, Journal of Structural Engineering, ASCE, vol. 110, No. 6, June 1984, pp 1336-1355.

12. Nallathambi, P., Karihaloo, B.L. and Heaton, B.S., Effect of specimen and crack sizes, water/cement ratio and coarse aggregate texture upon fracture toughness of concrete, Magazine of Concrete Research, Vol. 36, No. 129, December 1984, pp 227-236.

13. Swamy, R.N., Fracture mechanics applied to concrete, In: Developments in Concrete Technology, Vol. 1, Applied Science Publishers Ltd., London 1979, pp 221-281.

14. Barr, B.I.G. , Evans, W.T. and Dowers, R.C., Fracture toughness of polypropylene fibre reinforced concrete, Int. Journal of Cement Composites and Lightweight Concrete, Vol. 3, No. 2, May 1981, pp 115-122.

15. Teychenne, D.C., Franklin, R.E. and Erntroy, H.C., Design of Normal Concrete Mixes, H.M. Stationary Office, London, 1975.

16. Sabir, B.B., Stress intensity factors using simple strain-based finite elements, Proceedings, Int. Conference on Computational Mechanics, Tokyo, May 25-29 1986, pp V99-V105.

17. American Concrete Institute, Committee 544, Measurement of properties of fibre reinforced concrete, Journal, American Concrete Institute, Proceedings, Vol. 75, No. 7, July 1978, pp 283-289.

18. Henager, C.H., A toughness index of fibre concrete, Proceedings, RILEM Symposium on Testing and Test Methods of Fibre Cement Composites, The Construction Press, Lancaster, 1978, pp 79-86.

19. Henager, C.H., Steel fibrous concrete - A review of testing procedures, Proceedings Concrete International , Symposium on Fibrous Concrete, London, The Construction Press, Lancaster,1980, pp 16-28.

20. Barr, B.I.G., Liu, K. and Dowers, R.C., A toughness index to measure the energy absorption of fibre reinforced concrete, Int. Journal of Cement Composites and Lightweight Concrete, Vol. 4, No. 4, Nov. 1982, pp 221-227.

Proceedings of the International Symposium on Fibre Reinforced Concrete, December 16-19, 1987, Madras, India

STATISTICAL EVALUATION OF THE EFFECTS OF CEMENT CONTENT, WATER/CEMENT RATIO AND FIBRE CONTENT ON THE PHYSICAL-MECHANICAL PROPERTIES OF A STEEL FIBRE REINFORCED CONCRETE

G. Corradini, G. Scoccia, R. Volpe

Dipartimento di Chimica, Ingegneria Chimica e Materiali

Università dell'Aquila

Italy

SYNOPSIS

This paper discusses the statistical evaluation of the effects of cement content, water/cement ratio and fibre content on 28 day compressive strength, tensile strength, secant modulus of elasticity, first crack moment and failure bending moment of a steel fibre reinforced concrete.

The statistical approach was adopted both to design the tests and to interprete the results obtained. The present work has been performed on the basis of a factorial experiment programme.

Particular interest was devoted to the study of flexure in these concretes; with this end in view, the tenacity index and the ratio of the fibrous concrete strain work to the matrix strain work were determined.

Therefore, a model of the behaviour of steel fibre reinforced concrete in flexure is reported. Mathematical-statistical techniques were used for the construction of the models.

INTRODUCTION

There are numerous studies in steel fibre reinforced concretes, due to the growing interest in these materials.

The statistical approach is an interesting way to evaluate the physical and mechanical behaviour of a fibre reinforced concrete. Statistical criteria were adopted not only to interprete the results obtained from the experimental tests and to define mathematical models of the behaviour of these concretes, but also to plane the set of performed experiments [1].

In this paper we report statistical evaluation of the effects of cement content (C), water/cement ratio (W/C) and fibre content (F) on 28 day compressive strength, tensile strength, secant modulus of elasticity, first crack moment, failure bending moment, tenacity index [2] and ratio of the fibrous concrete strain work to the matrix strain work [2].

The behaviour in flexure of these concretes was particularly investigated and an attempt was made to set up a correlation between the tensile strength, the failure bending moment, the tenacity index, the (FRC strain work)/(matrix strain work) ratio and the factors characterizing the fibrous conglomerate: cement content, water/cement ratio and fibre content [3].

The work was performed on the basis of a factorial experiment with the principal aim to evaluate correctly the eventual interactions among the considered factors (C, W/C and F) [1].

EXPERIMENTAL

To prepare the FRC the following materials were used:

- commercial Portland cement "425";
- alluvial aggregates with maximum size 15 mm, whose characteristics appear in table 1;
- steel fibres 50 mm in length and 0.50 mm diameter;
- commercial superplasticizer composed of a sulphonated naphthalene-formaldehyde condensate.

To carry out the factorial programme of experiments, cement content, water/cement ratio and fibre content have been examined at 2, 3 and 4 levels respectively; the tests were repeated twice. In particular, levels $C_1 = 400$ and $C_2 = 475$ kg/m^3 of aggregate were chosen for cement content; for water/cement ratio the levels were $(W/C)_1 = 0.4$, $(W/C)_2 = 0.5$ and $(W/C)_3 = 0.6$ by wt; while for fibre content the levels were $F_1 = 0$, $F_2 = 0.5$, $F_3 = 1.0$ and $F_4 = 1.5\%$ in volume.

The superplasticizer dosage was varied in order to obtain, for all the mixtures, the same workability.

In order to carry out this factorial design, it was necessary to make 2x3x4x2=48 mixtures.

A laboratory pan mixer with a vertical axis was employed for the mixing.

Four cubic (15x15x15 cm), two prismatic (15x15x30 cm) and two prismatic (15x15x50 cm) specimens were made from each mixture according to the Standards for the compressive strength (UNI 6132), secant modulus of elasticity (UNI 6556) and flexure test (AICAP technical advice), respectively, and successively cured in a room conditioned at 22°C and 98% RH.

The 28 day compressive strength, tensile strength, secant modulus of elasticity,

TABLE 1 - Characteristics of aggregate.

Grading of aggregates

Gravel 5-15 mm		Sand	
Sieve size (mm)	Percentage by weight passing	Sieve size (mm)	Percentage by weight passing
20	100	5	100
15	97	3	87.1
10	44	2	68.9
7.1	17	1	41.1
5	4	0.2	12.9
3	0.2	0.074	1.6

Combined aggregate characteristics

Proportions	Gravel 5-15 mm 45% Sand 55%
Bulk density (kg/m^3)	2047
Absorption (%)	1.35

1st crack moment and failure bending moment have been measured according to the Standards. The tensile test has been carried out on the test crop ends of the flexure test (UNI 6135 - indirect tensile test).

During the flexural test, strains and deflections were also measured in order to evaluate the tenacity index and ratio of the fibrous concrete strain work to the matrix strain work (AICAP technical advice) [4, 5]. The flexural test geometry and strain-gauges and deflectometers arrangement are shown in fig. 1.

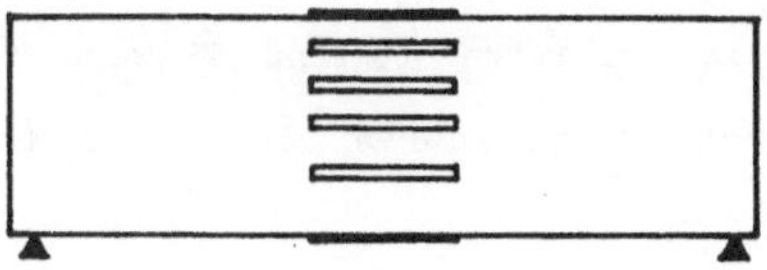

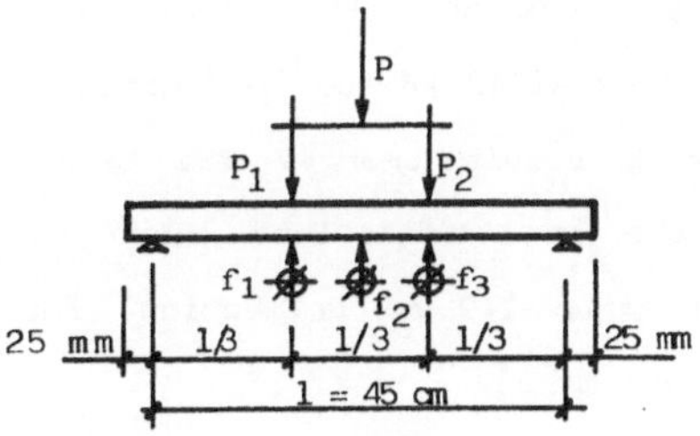

Fig. 1 - Flexural test geometry and strain-gauges and deflectometers arrangement.

RESULTS

The results obtained by experimental tests are presented in tables 2, 3 and 4.
Table 2 shows the values of 28 day compressive strength, tensile strength and secant modulus of elasticity; table 3 presents the values of the first crack moment and failure bending moment; table 4 gives the values of tenacity index and ratio of the fibrous concrete strain work to the matrix strain work.

TABLE 2 - Experimental results: 28 day compressive strength (kg/cm^2), tensile strength (kg/cm^2) and secant modulus of elasticity (kg/cm^2).

		F_1		F_2		F_3		F_4	
		C_1	C_2	C_1	C_2	C_1	C_2	C_1	C_2
I_1	$(W/C)_1$	573 41 427000	586 38 402000	582 49 412000	600 50 423000	555 61 395000	588 67 412000	592 78 390000	613 73 400000
	$(W/C)_2$	464 33 346000	504 33 389000	470 39 387000	503 44 388000	450 53 368000	530 64 381000	552 75 409000	565 66 384000
	$(W/C)_3$	368 34 320000	411 31 340000	362 34 304000	399 40 333000	380 42 320000	397 43 332000	369 54 312000	394 65 322000
I_2	$(W/C)_1$	582 33 433000	581 34 403000	619 56 426000	623 54 428000	560 62 440000	668 79 425000	570 78 405000	619 81 418000
	$(W/C)_2$	464 32 379000	502 38 380000	467 41 351000	500 44 373000	467 51 371000	535 63 366000	517 73 378000	527 74 388000
	$(W/C)_3$	360 28 309000	400 27 343000	373 34 317000	402 42 324000	489 52 325000	430 55 344000	400 60 337000	450 66 343000

Tenacity index and (FRC strain work)/(matrix strain work) ratio are parameters proposed by AICAP advice to characterize the capacity of a fibre reinforced concrete to absorb strain energy. The tenacity index is calculated by dividing the area below the load-deflections curve in midspain, relative to the flexure test, up to a maximum 1.9 mm deflection, for the area of the curve limited to the first crack moment [2].

The ratio between the FRC strain work and the matrix strain work is determined by dividing tha area below the load-deflections curve, up to limit configuration, for the analog diagram area relative to a fibre less sample, made from the same mix as the tested fibrous specimen [2]. The deflections are measured during the flexure test in the load application points (f_1 and f_3 in fig. 1).

TABLE 3 - Experimental results: 1st crack moment (kgm) and failure bending moment (kgm).

		F_1		F_2		F_3		F_4	
		C_1	C_2	C_1	C_2	C_1	C_2	C_1	C_2
I_1	$(W/C)_1$	225 225	225 225	225 338	233 315	263 450	278 473	281 653	285 645
	$(W/C)_2$	203 203	195 195	169 278	165 300	225 473	233 488	255 593	255 548
	$(W/C)_3$	188 188	195 195	150 323	169 278	206 360	225 465	225 525	244 525
I_2	$(W/C)_1$	218 218	218 218	188 360	225 368	270 458	293 600	285 615	308 653
	$(W/C)_2$	165 165	206 206	188 319	195 345	206 450	225 495	244 581	248 638
	$(W/C)_3$	173 173	176 176	173 323	180 356	218 431	225 473	236 525	244 540

TABLE 4 - Experimental results: tenacity index and ratio between the FRC strain work and the matrix strain work.

		F_2		F_3		F_4	
		C_1	C_2	C_1	C_2	C_1	C_2
I_1	$(W/C)_1$	31.0 45	39.8 53	47.0 67	62.6 90	70.8 172	73.6 181
	$(W/C)_2$	37.1 72	38.4 87	50.6 84	62.8 110	70.9 160	69.0 200
	$(W/C)_3$	38.2 91	34.0 111	48.0 129	63.5 167	62.5 165	66.8 185
I_2	$(W/C)_1$	33.0 47	41.4 51	45.6 69	62.1 88	72.3 166	72.4 181
	$(W/C)_2$	37.0 78	39.2 85	50.1 85	64.3 114	71.7 155	67.5 199
	$(W/C)_3$	38.6 88	34.1 110	47.5 135	63.2 168	63.4 163	66.7 179

Tables 5 and 6 present the results of the variance analysis, carried out to verify if the main effects and interactions among the considered factors have been estimated at an acceptable significance level (α) [1, 6].

TABLE 5 - Analysis of variance.

Source of variation	$F_{theoretical}$			$F_{calculated}$				
	α = 0.01	α = 0.05	α = 0.10	28 day compressive strength	tensile strength	secant modulus of elasticity	1st crack moment	failure bending moment
C	7.82	4.26	2.93	18.57	7.66	3.80	9.84	5.18
F	4.72	3.01	2.33	3.66	185.15	0.06	88.89	346.42
W/C	5.61	3.40	2.54	250.31	50.96	177.91	73.40	14.55
CxF	4.72	3.01	2.33	0.23	2.34	0.19	0.22	2.32
Cx(W/C)	5.61	3.40	2.54	0.53	0.08	2.04	0.18	0.01
(W/C)xF	3.67	2.51	2.04	2.01	2.11	1.75	0.80	1.82
Cx(W/C)xF	3.67	2.51	2.04	0.94	0.95	0.69	0.23	0.20

TABLE 6 - Analysis of variance.

Source of variation	$F_{theoretical}$		$F_{calculated}$	
	α =0.01	α =0.05	tenacity index	(FRC strain work)/(matrix strain work)
C	8.29	4.41	156.65	2069.69
F	6.01	3.55	1555.12	15247.43
W/C	6.01	3.55	10.81	2408.00
CxF	6.01	3.55	89.57	99.34
Cx(W/C)	6.01	3.55	9.92	82.89
(W/C)xF	4.58	2.93	10.52	810.17
Cx(W/C)xF	4.58	2.93	4.69	23.19

The comparison between F-theoretical and F-calculated values indicates that all the main effects of the various factors (excepting the effect of the fibre content on the secant modulus of elasticity) are resulted well estimated, while only some of the interactions among the factors have been sufficiently estimated.

Fig. 2 shows the main effects [1] of the factors C, W/C and F on 28 day compressive strength (fig. 2a), tensile strength (fig. 2b), secant modulus of elasticity (fig. 2c), 1st crack moment (fig. 2d), failure bending moment (fig. 2e) and tenacity index (fig. 2f), while fig. 3 shows the main effects of these factors on (FRC strain work)/(matrix strain work) ratio.

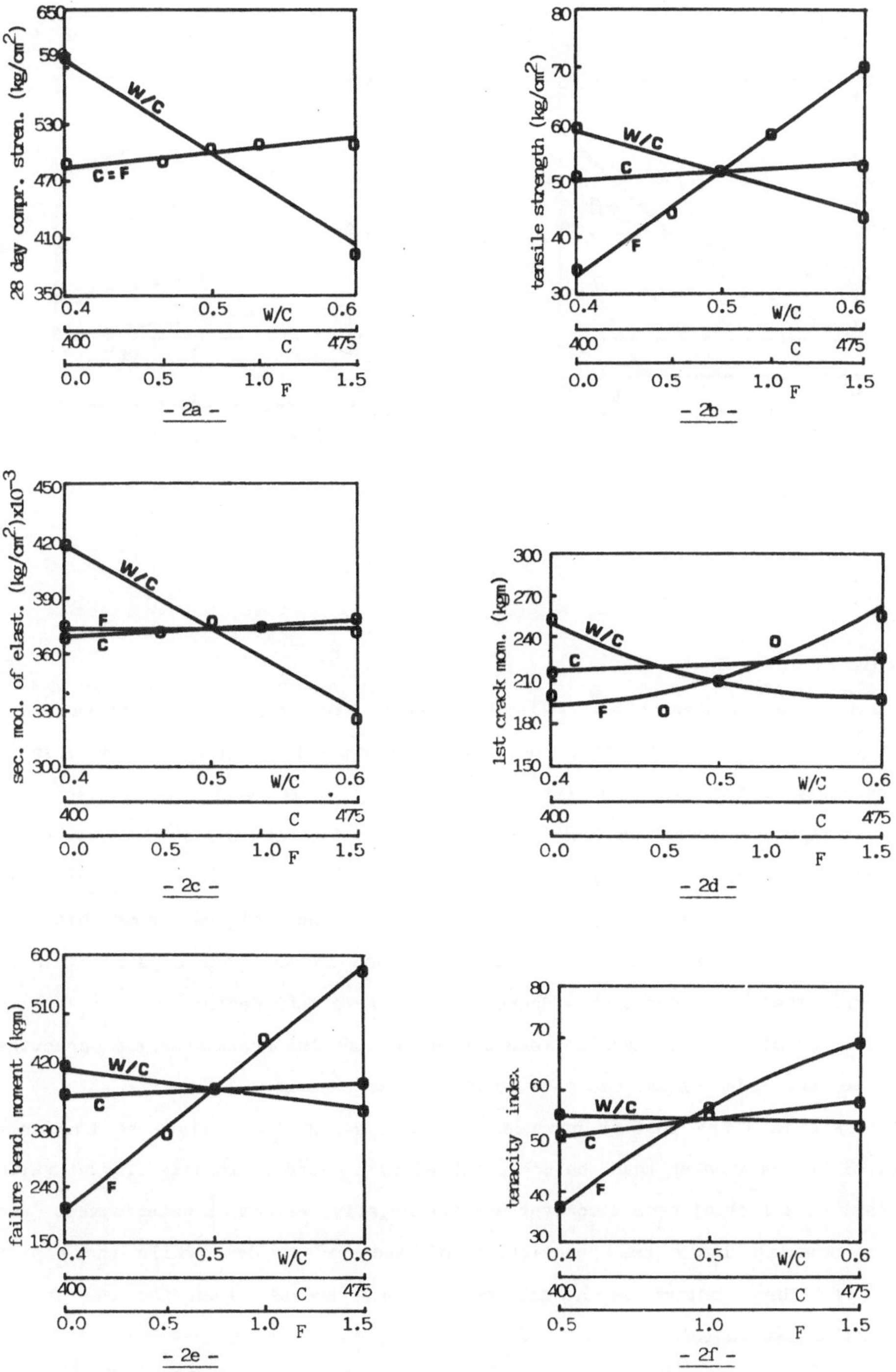

Fig. 2 - Main effects of the factors C (kg/m^3 of aggregate), W/C and F (% in volume).

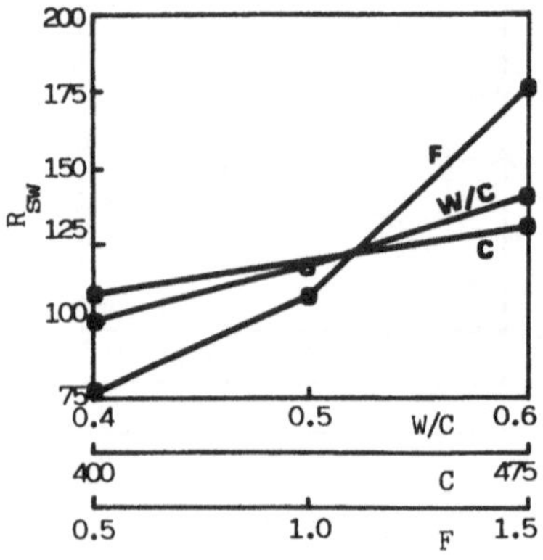

Fig. 3 - Main effects of the factors C, W/C and F.

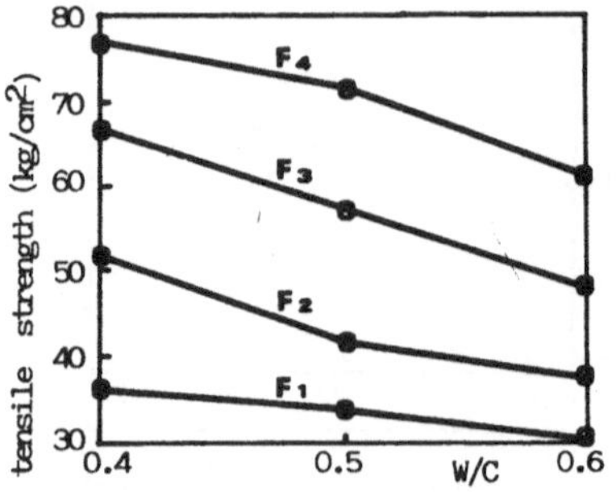

Fig. 4 - Interaction effects between the factors F and W/C on the tensile strength.

DISCUSSION

By examining the figures 2 and 3 it is possible to note that the fibre content negligibly influences the 28 day compressive strength and secant modulus of elasticity, moderately influences the 1st crack moment, while strongly influences the tensile strength, failure bending moment, tenacity index and ratio between FRC strain work and matrix strain work.

The analysis of gradients carried out on the curves of these figures shows that:

- an increase in fibre content produces an inrease in the tensile strength almost double than the one produced by a decrease in water/cement ratio;
- the influence of the increase in fibre content on the 1st crack moment is practically of the same order of the decrease in water/cement ratio;
- an increase in fibre content produces an increase in the failure bending moment almost six times greater than the one produced by a decrease in water/cement ratio;
- the positive effect of cement content and the negative effect of water/cement ratio on the tenacity index are practically of the same order, while the positive effect of fibre content is almost seven times greater than the one produced by the other two factors;
- the positive effect of fibre content on the ratio between FRC strain work and matrix strain work is almost double than the one produced by water/cement ratio and almost four times greater than that one produced by an increase in cement

content.

All these considerations are, obviously, valid within the variation limits studied for the considered factors.

The conclusions deriving from the statistical analysis of the results obtained are in agreement with the specific literature [7-10]. In fact, for the same fibre content, the performances of such a concrete depend on the resistence of the fibres to pull-out. This resistence, for the same fibre geometry and type, depends on the fibre-matrix interfacial bond and, therefore, on the concrete porosity, which is a function of water/cement ratio for a normal concrete.

In confirmation, fig. 4 shows the interaction effects between the factors fibre content and water/cement ratio on the tensile strength. Each curve of this figure has, as its parameter, a fibre content; the effect of cement content is "averaged". Fig. 4 shows that the increase in tensile strength, produced by an increase in fibre content, is a function of water/cement ratio, that is it increases as water/cement ratio decreases.

MATHEMATICAL MODELS

On the basis of the results obtained and of the considerations on the main effects and interactions of the factors C, W/C and F, an attempt was made to set up mathematical models, on a statistical basis, of the flexure behaviour of a steel fibre reinforced concrete.

In particular correlations between tensile strength, failure bending load, tenacity index and (FRC strain work)/(matrix strain work) ratio and the factors C, W/C and F, were obtained.

The statistical-mathematical method Stepwise regression [11] was adopted for the choice of the best model.

For each property characterizing the flexure behaviour, it was chosen, as the most satisfactory, the model in correspondence of which the minimum value of the mean square and the standard error of estimate was obtained.

The best resulting model for the tensile strength (R_t) is the following:

$$R_t = 0.0564F\times C - 70.94W/C + 68.3$$

in which the order of the variables is the one generated by the application of the Stepwise method; that is, the first term which enters into the regression and produces the highest coefficient of partial correlation is FxC, while the second is W/C.

The multiple regression coefficient is equal to 0.975; the standard error of estimate is equal to 3.7.

The best model for the failure bending moment (P_f) results the following:

$$P_f = 3485F - 3953W/C + 4625$$

The multiple correlation coefficient is equal to 0.988; the standard error of estimate is equal to 331.

The best model for the tenacity index (I_t) is the following:

$$I_t = 0.0729F\times C + 21.84$$

The multiple correlation coefficient is equal to 0.948 and the standard error of estimate is equal to 4.7.

For the ratio between FRC strain work and matrix strain work (R_{sw}), the model which better represents the experimental results is the following:

$$R_{sw} = 0.4333F\times W/C\times C$$

The multiple correlation coefficient is equal to 0.901; the standard error of estimate is equal to 22.

The introduction of all the variables was significant in each model at significance level $\alpha < 0.01$ [6].

Fig. 5 shows the interpolations of the experimental results with the models proposed.

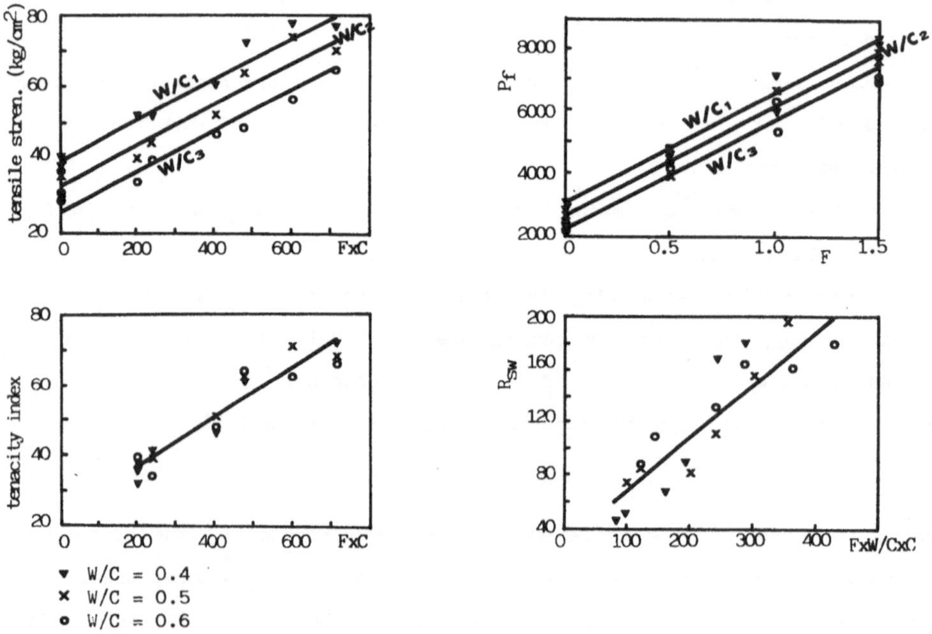

Fig. 5 - Mathematical models.

CONCLUSIONS

1 - The influence of fibre content on 28 day compressive strength and secant modulus of elasticity is practically negligible; a moderate positive influence on the first crack moment is noted, while it is very strong the influence of fibre content on tensile strength, failure bending moment, tenacity index and ratio between FRC strain work and matrix strain work.

2 - The adopted method of experimentation and elaboration of the data confers a high confidence level on the obtained results.

3 - The proposed models may be generalized by introducing generic constants, which are functions of the characteristics of the fibres, of the cement type, etc.

4 - The obtained results and the proposed models may be validly used as a testing basis for the construction of models for the design of structural elements in steel fibre reinforced concrete.

REFERENCES

1. Davies,O.L. and Goldsmith, P., Statistical Methods in Research and Production, ICI, 1972, p 247.
2. Raccomandazioni Tecniche AICAP per l'impiego del conglomerato cementizio fibroso, 1980.
3. Corradini, G., Pajewski, L.A., Scoccia, G. and Volpe, R., Influence of mix proportioning in the design and application of a fibre reinforced concrete, 21st Annual Meeting of Society of Engineering Science, Virginia Polytechnic Institute and State University, Blacksburg, 15-17 october 1984, Abstracts, p 83.
4. Corradini, G., Pajewski, L.A., Scoccia, G. and Volpe, R., Valutazione sperimentale del comportamento a flessione di un calcestruzzo fibrorinforzato, 25° Convegno AITEC, Parma, 17-18 ottobre 1985, pp 239-247.
5. Babut, R. and Brandt, A.M., The method of testing and analysing of steel fibre reinforced concrete elements in flexure, RILEM Symposium, 1979, 7.11.
6. Himmelblau, D.M., Process Analysis by Statistical Methods, John Wiley, New York, 1968.
7. Swamy, R.N. and Mangat, P.S., A theory for the flexural strength of steel fibre reinforced concrete, Cement and Concrete Research, Vol 4, 1974, pp 313-325.
8. Stroeven, P., Structural modelling of plain and fibre-reinforced concrete, Composites, Vol 13, april 1982.
9. Laws, V., Micromechanical aspects of the fibre-cement bond, Composites, Vol 13, april 1982.

10. Argon, A.S. and Shack, W.J., Theories of fibre cement and fibre concrete, RILEM Symposium, London, 1975.
11. Buzzi Ferraris, G., Analisi e identificazione dei modelli, CLUP, Milano, 1975.

Proceedings of the International Symposium on Fibre Reinforced Concrete, December 16–19, 1987, Madras, India

MODELING OF IMPACT ON FIBER REINFORCED CONCRETE CLOSED THICK CYLINDRICAL SHELLS

by

Geetha Rao
Department of Ocean Engineering
Massachusetts Institute of Technology
Cambridge, Massachusetts, U.S.A.

D.V. Reddy
Department of Ocean Engineering
Florida Atlantic University
Boca Raton, Florida, U.S.A.

ABSTRACT

The crack response of a cylindrical fiber reinforced concrete thick shell to high local concentrated impact loading is computed using the general purpose finite element program, ADINA (Automatic Dynamic Incremental Analysis). Based on previous experimental findings, only material nonlinearity is considered and the impact zone modeled as a system of thick shell elements. The Drucker-Prager yield surface is used for the elastoplastic analysis taking into account the tension cut-off and strain hardening effects. The effect of the reinforcement is incorporated into the general material properties.

INTRODUCTION

Impact loads on marine structures, nearshore and offshore, can be caused by collision, hurricanes, gusts, cavitation, blast, or earthquakes. As fiber reinforcement considerably enhances the impact strength and improves the resistance to cracking, it can be very effective in reinforced concrete coastal and ocean structures. The significantly reduced crack widths decrease the seawater access to the rebar and enhance the service life. Also, the damaged impact area does not spall suddenly as in the case of conventional reinforced concrete.

This paper addresses the analysis of the single impact behavior of fiber reinforced thick closed shells using the finite element program, ADINA (Automatic Dynamic Incremental Analysis).

BACKGROUND

In view of the large amount of literature on shell impact, only references relevant to marine applications and fiber-modification will be listed. The principal contributions are those of Skov and Olsen [1], Bhargava and Rehnström [3], Schrader, et al. [4], ACI Committee 544 [5], Olsen [6], Brakel and Oostlander [7], Suaris and Shah [8], Suaris et al. [9], Hibbert and Hannant [10], Emrich, Herter, and Puffer [11],

Hughes and Speirs [12], Hughes and Beeby [13], Knab and Clifton [14], Clifton and Knab [15], Gopalaratnam et al. [16], Reddy et al. [17], Geetha Rao and Reddy [18], Arockiasamy et al. [19, 20], and Reddy et al. [21].

An experimental investigation of cylindrical concrete shells, subject to impact, was carried out by Furnes and Amdahl [22], and Dellaripa and Reddy [23]. The experimental shells were similar to the oil storage caisson of the 'CONDEEP' type oil platform. In the latter study, a total of three shells were constructed from the same mix proportions except for different fiber volumes. A pendulum type machine was used for constant energy impact loading at the midsection of the shell wall. The instrumentation included deflection and strain gages and the response was recorded on a high speed strip chart. The number of blows to cracking and failure were determined and the failure strains examined. From these a comparative measure of the capacity of the shells was obtained.

ANALYTICAL INVESTIGATION

Cylindrical fiber reinforced closed concrete shells used in the present experiment are geometrically similar to the 'CONDEEP' type oil storage tank, Olsen [6]. As seen in Fig. 1b, the experimental fiber reinforced concrete shell is one of many cells in a honeycombed arrangement. The scaled shell is 48 in. (1219.2 mm) high, with outside and inside diameters of 27 in. (685.8 mm) and 24 in. (609.6 mm) respectively, and a wall thickness of 1.5 in. (38.1 mm). Comparisons of the 'CONDEEP' and scaled experimental shell of Ref. 22 are given in Fig. 2. In the model shell, shown in Fig. 2, the spherical closures at the top and bottom in the 'CONDEEP' were replaced by rigid diaphragms in the model. This is because impact effects are confined to the midsection of the shell wall. The effects of the fiber modification were included in the material properties.

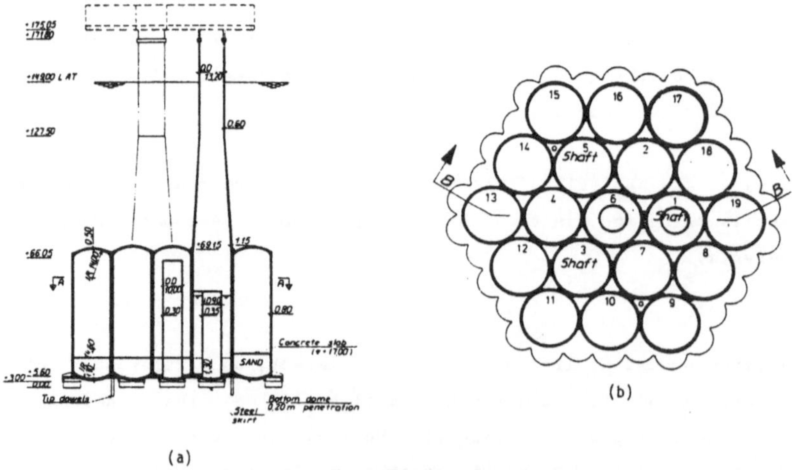

(a) (b)

Fig. 1 CONDEEP platform
(From Ref. 6, unit: m)

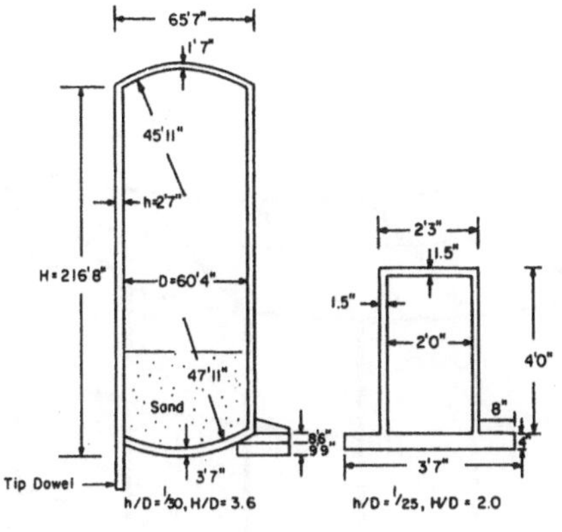

Fig. 2 Comparison of the CONDEEP cell and the experimental shell (1 in. = 25.4 mm)

Finite Element Analysis

The constitutive model used is an elastoplastic one with a Drucker-Prager yield surface. Tension cut-off and strain hardening considerations are included. The structural analysis package, ADINA (Automatic Dynamic Incremental Nonlinear Analysis) [24], is used to analyze the model.

Based on the experimental work of Ref. 23, Fig. 3, which indicated that most of the significant response was within a limited radius of the impacted region, this part is modeled as a system of thick-shell elements available in the ADINA element library, Fig. 4. Each element is described by eight nodes at the corners of a 3-D solid, all of which were used as integration points. All eight nodes of the interior elements had three translatory degrees of freedom. The rotating degrees of freedom were completely constrained. For the boundary elements, the internal nodes had three degrees of freedom while the external nodes were partially constrained with displacements

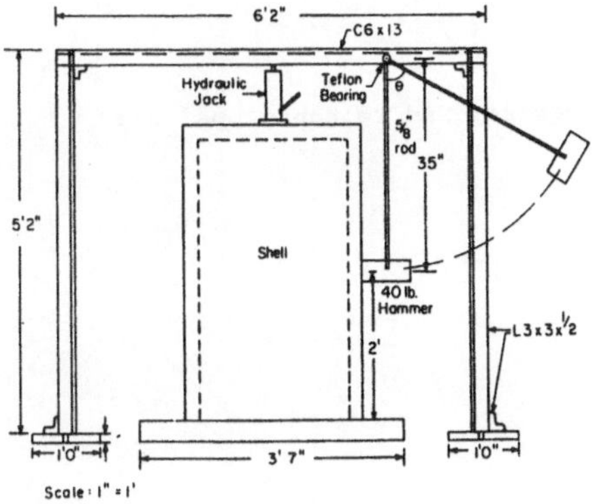

Fig. 3 Pendulum impact test set-up (From Ref. 23, 1 in. = 25.4 mm)

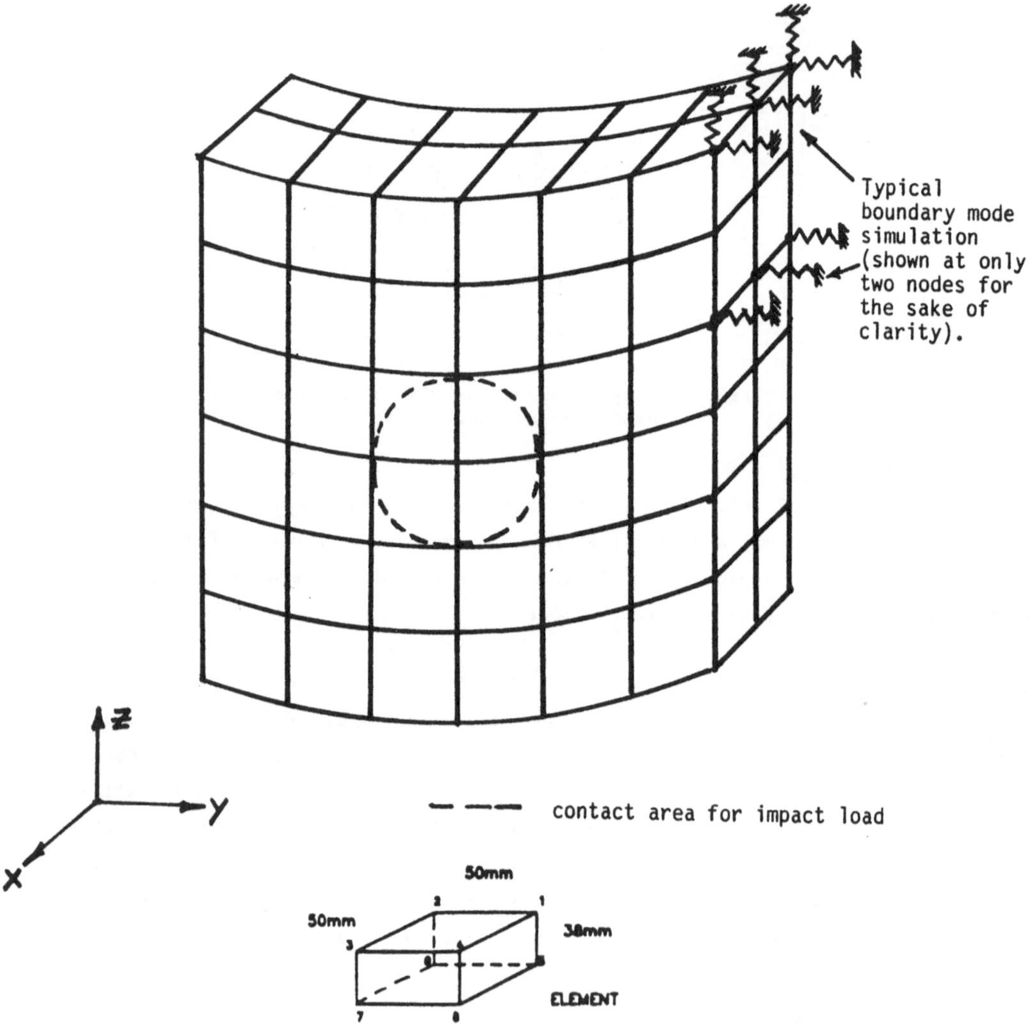

Fig. 4 Finite element model for the frequency and response analysis

proportional to the stiffness of the adjacent element. Although no dashpots were located at these points, artificial Rayleigh damping was introduced for the entire system to ensure stability. The impacted region consists of two layers of 36 elements in a square array for the frequency analysis, and a single layer for the stress analysis to reduce the number of degrees of freedom. For the frequency analysis, there were a total of 147 nodal points for the 72 elements. From these 357 free equilibrium equations and 84 prescribed displacement constraint equations were set up. The single element thickness simplification for the stress analysis reduced the number of nodal points to 98.

A lumped mass (material density = 2.4 x 10-7 kg/mm3) matrix is used and the eigen-frequencies and modes corresponding to the initial state of the structure are calculated, Fig. 5. These are then used in a mode superposition analysis to determine

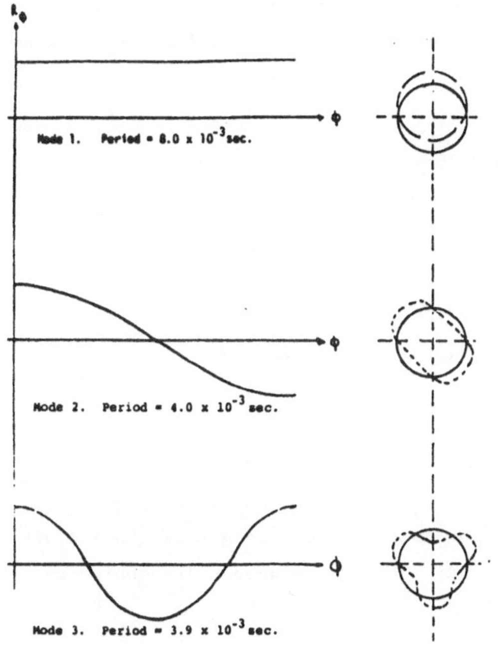

a) Panel mode b) Cylinder mode

Fig. 5 Mode shapes

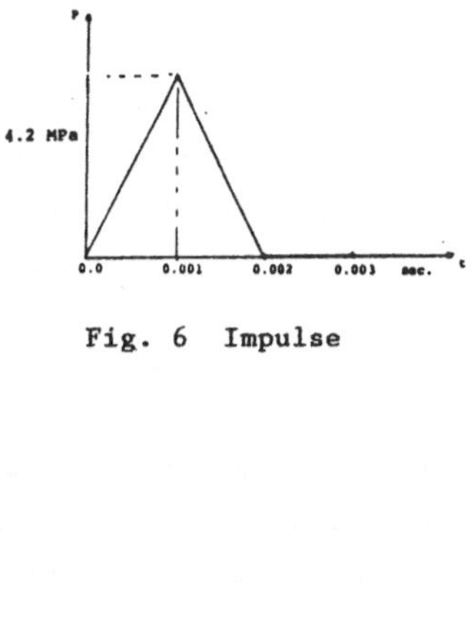

Fig. 6 Impulse

the final response. Although the natural frequencies change during the response solution, the mode superposition in ADINA uses the mode shapes corresponding to the initial state of the structure. This implies that only local nonlinearities are present, i.e. a slightly nonlinear system.

The Newmark method was used for the direct time integration with the following parameters:

$$\delta = 1/2$$

and

$$\alpha = 1/4,$$

which are defined in Eqs. 1:

$$^{t+\Delta t}\{\dot{U}\} = {}^{t}\{\dot{U}\} + [(1 - \delta)\ {}^{t}\{\ddot{U}\} + \delta\ {}^{t+\Delta t}\{\ddot{U}\}]\ \Delta t, \qquad (1a)$$

$$^{t+\Delta t}\{U\} = {}^{t}\{\dot{U}\} + {}^{t}\{\dot{U}\}\ \Delta t + [(1/2-\alpha)\ {}^{t}\{\ddot{U}\} + \alpha\ {}^{t+\Delta t}\{\ddot{U}\}\ \Delta t^2\ . \qquad (1b)$$

A maximum limit of 15 equilibrium iterations was set with a convergence tolerance of 1%.

The pressure load in time followed a triangular impulse shape, starting at zero, and reaching a peak value of 4.2 MPa at 0.001 sec. and dropping back to zero at 0.002 sec. Fig. 6. The lumped masses were calculated using a material mass density of 2.4 x 10^{-7} kg/mm^3. The material parameters for the Drucker-Prager yield function were taken as follows:

$$\alpha\ (\text{Drucker}) = \frac{2\ \sin\theta}{(3-\sin\theta)\sqrt{3}} = 0.2556$$

and

$$k = \frac{6c\ \cos\theta}{(3-\sin\theta)\sqrt{3}} = 0.2313\ ,$$

where

c and θ = cohesion and angle of friction of the material respectively.

The following cap hardening parameters, ADINA [25], were used:

$$W = -0.001$$

and

$$D = -3.5\ [\text{MPa}]^{-1}.$$

The initial cap position was taken at the compressive strength of 300 MPa with a tension cut-off value of 30 MPa. A Poisson's ratio of 0.3 was used. The entire analysis was repeated for a load of 10.0 MPa.

RESULTS AND DISCUSSION

The first mode was a very long-period mode with almost uniform displacement of the impacted region in the radial direction. The actual mode shape of the entire cylinder was complicated since the ends were fixed in a circle and the center deflected in the first mode shape. The second and third mode shapes were higher order shape functions of the first mode.

It must be mentioned here that the triangular impulse shape of the impact decomposes into an infinite number of Fourier components; hence, to get a reasonably accurate value of the response, a large number of modes needs to be considered.

Figs. 7 shows the resulting crack locations obtained in the stress analysis. The material model was assumed to be in the plastic state for a long time; hence regions of excessive displacement were obtained. These remain displaced even after the impact load has dropped off. The stresses in these regions reached plastic states and appear as cracks in the physical model.

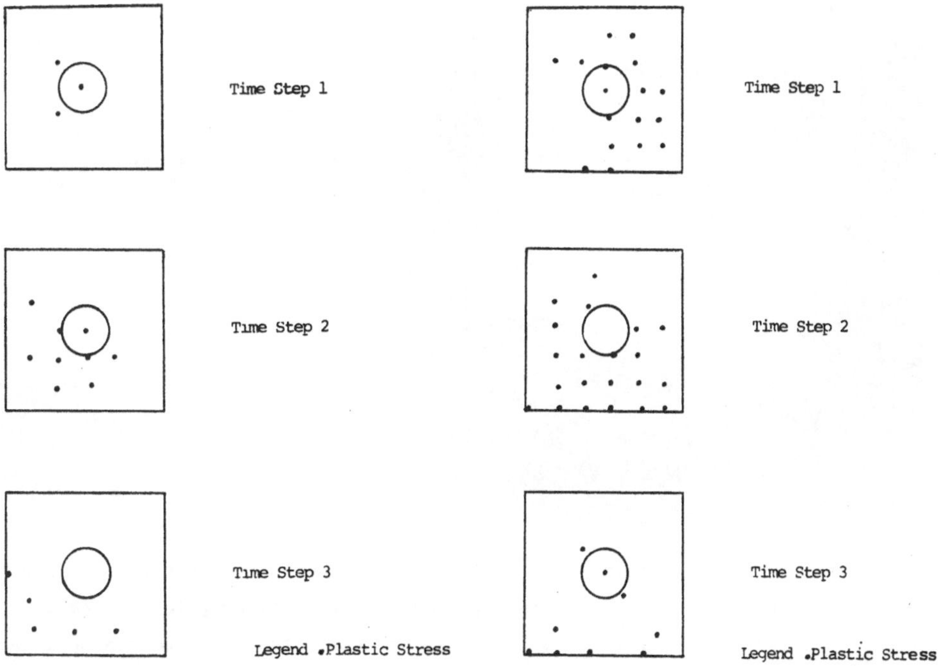

a) Load = 4.2 MPa b) Load = 10 MPa

Fig. 7 Computed crack locations

For lower levels of the impact loading, a stress-pulse was generated which propagated radially outward. For higher impact values, there was extensive cracking in the region immediately surrounding the region of contact. This also propagated outward, but the stresses remained plastic even after the stress pulse had passed. Thus, in the case of brittle materials, the energy absorbed will not be radiated but will instead cause fracture and shearing of the impacted region as demonstrated by Ref. 23.

Fig. 8 shows the crack patterns of an experimentally impacted shell. The shell was reinforced with glass fibers. There is a central collapse region with radiating cracks which appear to have propagated considerably. The crack points obtained from computation have a similar pattern. Although the direction of cracking cannot be obtained directly, a crack once initiated will extend outward along other crack points.

Fig. 8 Crack pattern of a glass fiber concrete shell.

Initially, the analysis was formulated to include the entire structure. But, since the significant response was limited to the impacted region, a finer grid was required in this area. This led to instabilities in the region where the fine and coarse grids met. However, as the relatively small value of the response, outside the impacted region, did not warrant the measures required to remove the instabilities, analysis of the panel was considered adequate.

ACKNOWLEDGEMENT

The authors would like to thank Dr. S.E. Dunn, Professor and Chairman, Department of Ocean Engineering, Dr. T.E. Shoup, Dean of Engineering, Florida Atlantic University, and Dr. H. Wang, Professor and Chairman, Department of Coastal and Oceanographic Engineering, University of Florida for their interest and encouragement. Appreciation is expressed to Dr. A.S.J. Swamidas of the Faculty of Engineering, Memorial University of Newfoundland and Dr. M. Arockiasamy, Department of Ocean Engineering, Florida Atlantic University, for valuable discussions. Thanks are due to the MIT Student Information Processing Board for supporting the computing costs. The work described was part of a Florida Sea Grant funded project No. R/C-7 entitled, "Strength and Durability of Fiber and Polymer-Modified Concrete for Marine Structural Applications". This support is gratefully acknowledged. Thanks are due to Ms. Shirley Behar for the patient and excellent typing of this manuscript.

REFERENCES

1. Skov, K. and Olsen, S.O., Impact Resistance of Reinforced and Prestressed Concrete Members, Materials and Structures, Vol. 8, No. 44, 1974, pp. 117-124.

2. Sorensen, K.A., Behavior of Reinforced and Prestressed Concrete Tubes under Static and Impacted Loads, Proc. First Int. Conf. on the Behavior of Offshore Structures (BOSS), Norwegian Inst. of Tech., Trondheim, Norway, 1976, pp. 798-8B.

3. Bhargava, J. and Rehnström, A., Dynamic Strength of Polymer Modified and Fiber-Reinforced Concretes, Cement and Concrete Research, Vol. 7, 1977, pp. 199-208.

4. Schrader, E.K., Brandshaug, T., Ramakrishnan, V., and Coyle, W.V., A Comparative Evaluation of Concrete Reinforcement with Straight Steel Fibers and Collated Fibers with Deformed Ends, Report SDSM&T-CBS 7801 submitted to Bekaert Steel Wire Corporation, 1978, pp. 1-52.

5. ACI Committee 544, Measurement of Properties of Fiber Reinforced Concrete, J.ACI, July 1978, pp. 283-289.

6. Olsen, O., Implosion Analysis of Concrete Cylinders Under Hydrostatic Pressures, Journal of the American Concrete Institute, Vol. 75, No. 3, 1978 pp. 82-85.

7. Brakel, J. and Oostlander, L.J., Concentrated Loading on Thick-walled cylinders, Proc. Second Int. Conf. on the Behavior of Offshore Structures (BOSS), London, 1979.

8. Suaris, W. and Shah, S.P., Strain-Rate Effects in Fiber-Reinforced Concrete Subjected to Impact and Impulsive Loading, J. of Composites, April 1982, pp. 153-159.

9. Suaris, W., Gokoz, U., Youngquist, O., and Shah, S.P., Analysis of Inertial Effects in the Instrumentation Impact Testing of Fiber Reinforced Concrete, U.S. Army Research Office, University of Illinois at Chicago Circle, August 1980.

10. Hibbert, A.P. and Hannant, D.J., Impact Resistance of Fibre Concrete, Tech. Rept., Transport and Road Research Laboratory, Crowthorne, England, 1981.

11. Emrich, F., Herter, J., and Puffer, G., Nonlinear Finite Element Analysis of Reinforced Concrete Beams Under Impact Load in Comparison with Experimental Results, BAM Conference, Berlin, June 1982, pp. 455-471.

12. Hughes, G. and Speirs, D.M., An Investigation of the Beam Impact Problem, Cement and Concrete Association, Technical Report 546, April 1982.

13. Hughes, G. and Beeby, A.W., Investigation of the Effect of Impact Loading on Concrete beams, The Structural Engineer, Vol. 60b, No. 3, September 1982, 45-52.

14. Knab, L.I. and Clifton, J.R., Cumulative Damage of Reinforced Concrete Subjected to Repeated Impact, Cement and Concrete Research, Vol. 12, 1982, pp. 359-370.

15. Clifton, J.R. and Knab, L.I., Impact Testing of Concrete, Cement and Concrete, Cement and Concrete Research, Vol. 14, 1983, pp. 541-548.

16. Gopalaratnam, V.S., Shah, S.P., and Reji John, A Modified Charpy Test for Cement Based Composites, U.S. Army Research Office, Northwestern University, Department of Civil Engineering.

17. Reddy, D.V., Gopal, K.R., Nomikos, G.G., and Thiel, D.T., Strength and Durability of Fiber and Polymer-modified Concrete for Marine Structural Applications, Florida Sea Grant Rept., 1984, 195 pages.

18. Geetha Rao and Reddy, D.V., Modeling of the Impact on Concrete Shells, Proc. Eighth International Conf. on Port and Ocean Engineering under Arctic Conditions (POAC), Narssarssuag, Greenland, September, 1985, pp. 589-602.

19. Arockiasamy, M., Swamidas, A.S.J., Hamlyn, D., and Munaswamy, K., Response Studies of Concrete Shell Panel Models to Simulated Bergy Bit Impact, Proc. Int'l. Assn. of Hyd. Research (IAHR) Symp., Hamburg, W. Germany, August, 1984.

20. Arockiasamy, M., Swamidas, A.S.J., Cheema, P., and Munaswamy, V., Impact Strength of Fibre Reinforced Concrete Cylindrical Panels, Proc. Eighth Int. Conf. on Struct. Mech. and Reactor Tech. (SMIRT), Brussels, Belgium, August 1985.

21. Reddy, D.V., Hartt, W.H., and Dunn, S.E., et al., Effect of Marine Exposure Upon Service Life of Reinforced Concrete, Florida Sea Grant Rept., 1986, 197 pages.

22. Furnes, O., and Amdahl, J., Computer Simulation Study of Offshore Collisions and Analysis of Ship-Platform Impacts, Applied Ocean Research, Vol. 2, No. 3, 1980, 119-127.

23. Dellaripa, F., and Reddy, D.V., Impact Capacity of Fiber Reinforced Concrete Closed Shell Structures, Proc. Int. Symp. on Fibre Reinforced Concrete, Madras, India, December 1987.

24. ADINA Engineering, ADINA, a Finite Element Program for Automatic Dynamic Incremental Nonlinear Analysis, Rept. AE 81-1, Watertown, Massachusetts, September 1981.

25. ADINA, System Theory and Modeling Guide, Rept. AE 83-4, Watertown, Massachusetts, September 1983, 3.5-8 to 3.5-9.

Proceedings of the International Symposium on Fibre Reinforced Concrete, December 16-19, 1987, Madras, India

THEORIES FOR SHRINKAGE AND CREEP OF STEEL FIBRE REINFORCED CONCRETE

P.S. Mangat, M. Motamedi Azari

Department of Engineering, University of Aberdeen, Marischal College, Broad Street, Aberdeen, Scotland, AB9 1AS

SYNOPSIS

The paper presents theoretical models to predict shrinkage and creep strains of cement matrices reinforced with aligned, long (continuous) steel fibres. These theories are derived using the concepts which were previously applied to composites reinforced with randomly oriented short steel fibres. These assume that fibres restrain the shrinkage and the flow component of creep of cement matrices through interfacial bond stress.

A wide range of experimental data on shrinkage and creep are given for composites reinforced with randomly oriented short fibres and with aligned long fibres. These results show a significant influence of fibre volume on these properties. The theoretical expressions derived in the paper show satisfactory validity to the experimental results.

INTRODUCTION

The restraint provided by aggregate particles to the shrinkage of cement matrices is primarily due to the relative difference in the stiffness of the two phases. Theoretical models have been developed which consider idealised spherical aggregate inclusions to restrain radical deformation of the matrix by maintaining strain compatibility (1). Although there are similarities between the influence of steel fibres and aggregates on the deformation characteristics of cement matrices (2), the long, thin shape of fibres leads to some basic differences. The contact area at the tip of a fibre is too small to permit significant stress transfer through strain compatibility. The shrinking matrix in the vicinity of a fibre tends to slide past the fibre length and any restraint is possible only through the fibre-matrix interfacial bond strength.

Theories for shrinkage and creep of cement matrices reinforced with randomly oriented discrete steel fibres were derived previously on the basis of the above concepts (3, 4). In this paper the concepts are extended to composites reinforced with long (continuous), aligned steel wires which were arranged in a square array within a mortar matrix. The theories are supported by experimental data on composites reinforced with continuous and discontinuous fibres.

SHRINKAGE THEORY - CONTINUOUS FIBRES

A model for shrinkage and creep of cement matrices reinforced with continuous steel fibres is shown in Figure 1. The element within the dotted area is enlarged in Figure 2, which represents a continuous fibre that reinforces a cylindrical zone of the matrix of diameter S and length L. The symbols S and L represent fibre spacing and fibre (and specimen) length respectively. The shrinkage (or creep) of the cylindrical zone of the matrix is restrained by the interfacial bond strength. The bond, in turn, is a function of the radial deformation of the matrix cylinder caused by shrinkage (and also the lateral deformation due to sustained stress in the case of creep).

In order to derive expressions for shrinkage, a knowledge of bond stress distribution along the fibre is necessary. From existing literature (5, 6), a typical distribution under tension is given in Figure 3. This distribution is assumed for the shrinkage theory owing to lack of information on bond stress distribution under a compressive state of stress. The bond stress in Figure 3 is maximum at the ends of a long fibre and gradually reduces to zero beyond the transfer length, ℓ_t. This implies that by increasing the fibre length, the restraint against shrinkage is not necessarily increased since the large portion of fibre beyond the transfer length does not provide any restraint through bond. Based on the above considerations, and assuming elastic behaviour, the following expression for free shrinkage can be derived:-

$$\varepsilon_{fs} \frac{L}{2} = \varepsilon_{os} \frac{L}{2} - \tau(\pi d\ \ell_t)\ \left(\frac{L/2}{AE}\right) \qquad \ldots (1)$$

where ε_{fs} = shrinkage strain of fibre reinforced composite.

ε_{os} = shrinkage strain of control matrix.

ℓ_t = transfer length.

τ = average fibre-matrix bond strength.

d = fibre diameter.

E = elastic modulus of matrix.

A = area of matrix cylinder surrounding a fibre, which equals $\frac{\pi s^2}{4}$.

Equation (1) simplifies to the following expression:

$$\varepsilon_{fs} = \varepsilon_{os} - \frac{4\tau d\ \ell_t}{(s^2 E)} \qquad \ldots (2)$$

The interfacial bond stress, τ, is a function of the radial pressure, P, exerted by the shrinking matrix at the fibre-matrix interface and the coefficient of friction, μ. It is given by the expression:-

$$\tau = \mu P \qquad \ldots (3)$$

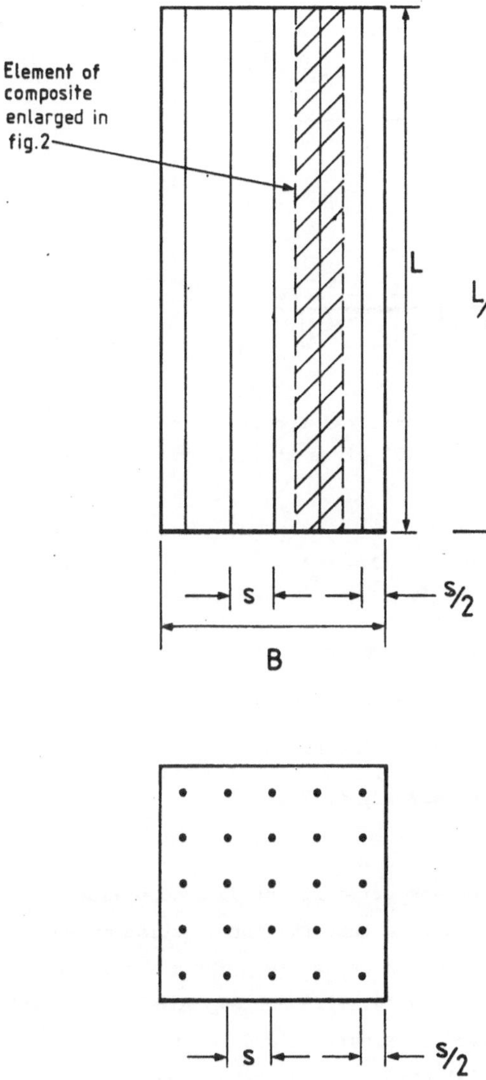

Figure 1 Wire distribution in test specimens.

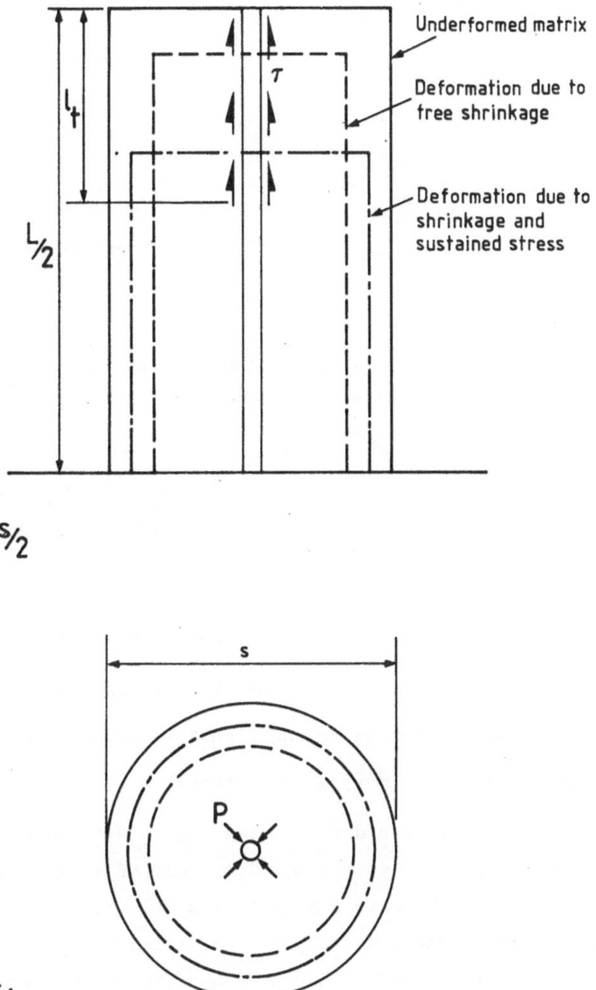

Figure 2 Shrinkage and creep model for fibre reinforced matrices.

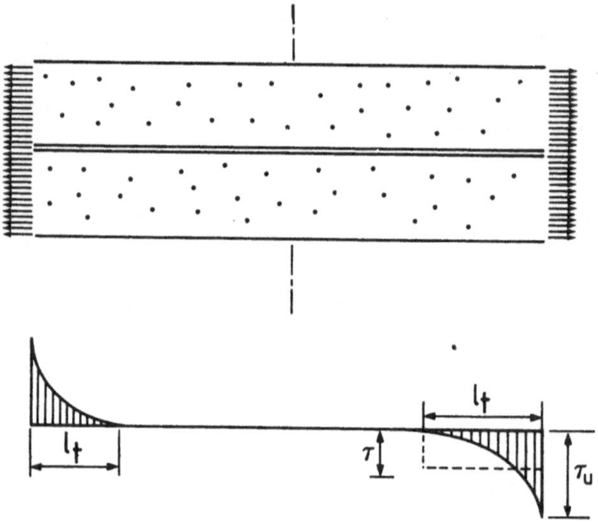

Figure 3 Bond distribution in an uncracked matrix.

Applying the classical equations for thick cylinders (7) gives the following expression for P:-

$$P = \frac{\varepsilon_{os}(s - d)E}{d} \left(\frac{(s^2 + d^2)}{(s^2 - d^2)} + \nu + \frac{E}{E_s}(1 - \nu_s) \right)^{-1} \qquad \ldots (4)$$

where ν = Poisson's ratio of the matrix which is assumed to equal 0.2.

ν_s = Poisson's ratio of steel fibre which is assumed to equal 0.3.

E_s = elastic modulus of steel fibre assumed to be 200 kN mm^{-2}.

A detailed derivation of equation 4 is given elsewhere (3, 4, 8).

CREEP THEORY - CONTINUOUS FIBRES

Creep of concrete under uniaxial compression comprises primarily of delayed elastic deformation and flow creep. The basis of this theory is that the introduction of steel fibres in cement matrices restrains only the flow component of creep while the delayed elastic component of creep is unaffected by the fibres (4, 9). A creep model for cement matrices reinforced with continuous fibres is given in Figures 1 and 2. In this case the radial deformation which results in interfacial bond is a function of both matrix shrinkage, ε_{os}, and the lateral strain, $\varepsilon_{o\ell}$, induced by the sustained creep stress. The following expression can be derived for the model in Figure 2 in a similar manner as adopted previously for discontinuous fibre reinforced composites (4, 8):-

$$\varepsilon_{fc}\left(\frac{L}{2}\right) = \varepsilon_{oc}\left(\frac{L}{2}\right) - \tau(\pi d\, \ell_t)\left(\frac{L/2}{AE_c}\right) \qquad \ldots (5)$$

where ε_{oc} = creep strain of control matrix.

ε_{fc} = creep strain of composite

and E_c = stiffness of cement matrix against flow creep, which equals the ratio of the sustained stress to the flow creep strain, ε_{op} (4).

The sustained stress and flow creep are linearly related especially at relatively low stress-strength ratios (4, 9).

Equation 5 can be simplified to give:-

$$\varepsilon_{fc} = \varepsilon_{oc} - \frac{4\tau d\, \ell_t}{s^2 E_c} \qquad \ldots (6)$$

Applying the classical equations for thick cylinders (7), the following expression for the radial pressure, P, acting on a fibre is derived:-

$$P = \frac{(\varepsilon_{os} - \varepsilon_{o\ell})\ (s - d)E}{d} \left[\frac{(s^2 + d^2)}{(s^2 - d^2)} + \nu + \frac{E}{E_s}(1 - \nu_s) \right]^{-1} \qquad \ldots (7)$$

where $\varepsilon_{o\ell}$ = lateral strain of the control matrix due to the sustained creep stress.

Equation 7 is similar to equation 4 for the shrinkage case except for the additional term $\varepsilon_{o\ell}$. This term can be calculated by multiplying Poisson's ratio of the matrix by its total strain under stress (sum of elastic and creep strains) (4).

SHRINKAGE AND CREEP THEORY - DISCRETE FIBRES

These theories for cement matrices reinforced with randomly oriented short fibres have been derived in detail elsewhere (3, 4, 10) and are briefly introduced in this section. Based on an idealised fibre distribution, the spacing between fibres, s, is given by the following expression (10):-

$$s^3 + (0.41\ell)s^2 - \left(\frac{\pi d^2}{4}\right)\frac{\ell}{v_f} = 0 \qquad \ldots (8)$$

where v_f = volume of fibres

ℓ = fibre length.

An expression for free shrinkage of the composite, ε_{fs}, in terms of the shrinkage of the control matrix, ε_{os}, is given as:-

$$\varepsilon_{fs} = \varepsilon_{os} - \frac{0.3362\ \tau\ d\,\ell^2}{s^2 E(0.41\ell + s)} \qquad \ldots (9)$$

The coefficient of friction at the fibre interface, μ, is given by (3, 10):-

$$\mu = \frac{\tau}{P} = \frac{(\varepsilon_{os} - \varepsilon_{fs})\ s^2(0.41\ell + s)}{0.3362\ \ell^2\ \varepsilon_{os}(s - d)} \left[\frac{(s^2 + d^2)}{(s^2 - d^2)} + \nu + \frac{E(1 - \nu_s)}{E_s} \right] \qquad \ldots (10)$$

Expressions for creep are also derived on a similar basis (4, 9).

EXPERIMENTAL PROGRAMME

Details of the experimental programme are given in Table 1. The mix proportions used were 1:2:0:0.55 (Mortar), 1:2.5:2.2:0.58 (Concrete), 1:2.75:0:0.58 (Mortar) and 1:0:0:0.35 (Cement paste). The materials used were Ordinary Portland Cement, washed sand conforming to zone 2 of BS 882 and coarse aggregate of 10 mm nominal size.

Continuous and discontinuous steel fibres were used for reinforcement. The continuous fibres had a plain surface and were either annealed or hard drawn. Four types of discontinuous steel fibres were used namely hard drawn (plain), melt extract, crimped and hooked. Details of the fibres are given in Table 1.

Series	Mix	Mix Proportions	Fibre Details				Specimen Size (mm)
			ℓ (mm)	d (mm)	v_f (%)	Fibre Type	
I	S_o	1:2:0:0.55 (Mortar)	-	-	-	-	50x50x250
	$S_{AC}5$		250	0.56	0.98	Annealed (Plain)*	
	$S_{AC}2.5$				3.56		
	$S_{HC}2.5$			0.41	1.90	Hard Drawn (Plain)*	
	S_H1		25	0.4	1		
	S_H3				3		
II	A_o	1:25:1.2:0.58 (Concrete)	-	-	-	-	100x100x500
	$A_m1.5$		22.5	0.40	1.5	Melt Extract*	
	A_m3				3		
	A_c1		48.7	1.14	1	Crimped**	
	A_c2				2		
	A_c3				3		
III	M_o	1:2.75:0:0.56 (Mortar)	-	-	-	-	
	M_m3		22.5	0.40	3	Melt** Extract	
	M_h3		28.0	0.48	3	Hooked*	
IV	P_o	1:0:0:0.35 (Cement paste)	-	-	-	-	
	P_m3		22.5	0.40	3	Melt** Extract	

* Continuous aligned fibres of spacing 2 mm and 5 mm.

** Discontinuous short fibres.

Table 1 Experimental Programme

Two sizes of prism specimens, 50 x 50 x 250 mm and 100 x 100 x 500 mm were used. These were cast in three layers, each layer being compacted on a vibrating table. The prisms were then covered with a polythene sheet for 24 h. Subsequently all the specimens were transferred to the temperature and humidity controlled room when a temperature of 20^{0}c and relative humidity of 55 per cent was maintained throughout.

Two prisms specimens per mix were tested for each test. The shrinkage and creep measurements were made by means of a Demec extensometer over a gauge length of 200 mm on the four faces of each prism. The creep specimens were loaded after 28 days of curing for a period of up to 90 days at sustained stress-strength ratios of 0.3 and 0.55 as indicated in Figure 5. These specimens were then unloaded and creep recovery measurements taken for a further period of up to 60 days. The shrinkage measurements started 24 h after casting and regular measurements were made thereafter up to 360 days age.

RESULTS AND DISCUSSION

Free Shrinkage

Some typical free shrinkage data of the mixes listed in Table 1 are presented in Figure 4. The relationship between fibre volume and shrinkage at different ages of up to 360 days is plotted. The results in Figures 4a to 4d are for composites reinforced with discontinuous, randomly oriented fibres. The results in Figures 4e and 4f include data for both continuous and short fibres. All the graphs in Figure 4 indicate significant reduction of shrinkage in cement matrices with increasing fibre volume. The effect of continuous fibres is not markedly different from short fibres.

Compression Creep

Creep data of mixes listed in Table 1 are shown in Figure 5. The results in Figures 5a to 5d are for composites reinforced with randomly oriented short fibres. The results in Figure 5e are for composites reinforced with continuous, annealed steel fibres. Figures 5a and 5e represent data at a stress-strength ratio of 0.3. The remaining figures are for data at 0.55 stress-strength ratio. The results in Figures 5a and 5e show that creep decreases with increasing volume of both short and continuous fibres. For example at 60 days age the reduction in creep at v_f of 3% is 29% and 30% with short and continuous fibres respectively. This indicates a similar restraint to creep by short or continuous fibres thus confirming that only the transfer length of continuous fibres is effective in providing restraint.

A comparison of Figures 5a and 5b shows that fibre restraint to the creep of concrete matrix is more effective at a stress-strength ratio of 0.3 (Fig. 5a) as compared to 0.55 (Fig. 5b). This is due to the greater magnitude of the lateral strain, $\varepsilon_{o\ell}$, at stress-strength ratio of 0.55 thus leading to a reduction in radial pressure P and hence in the restraining bond stress τ.

In most graphs of Figure 5, the slope of the lines representing 90 days age data are steeper than at earlier ages. This indicates greater efficiency of fibre restraint at later ages. This is due to the fact that the flow component of creep is dominant at later ages, which is effectively by fibre reinforcement (9). At earlier ages the

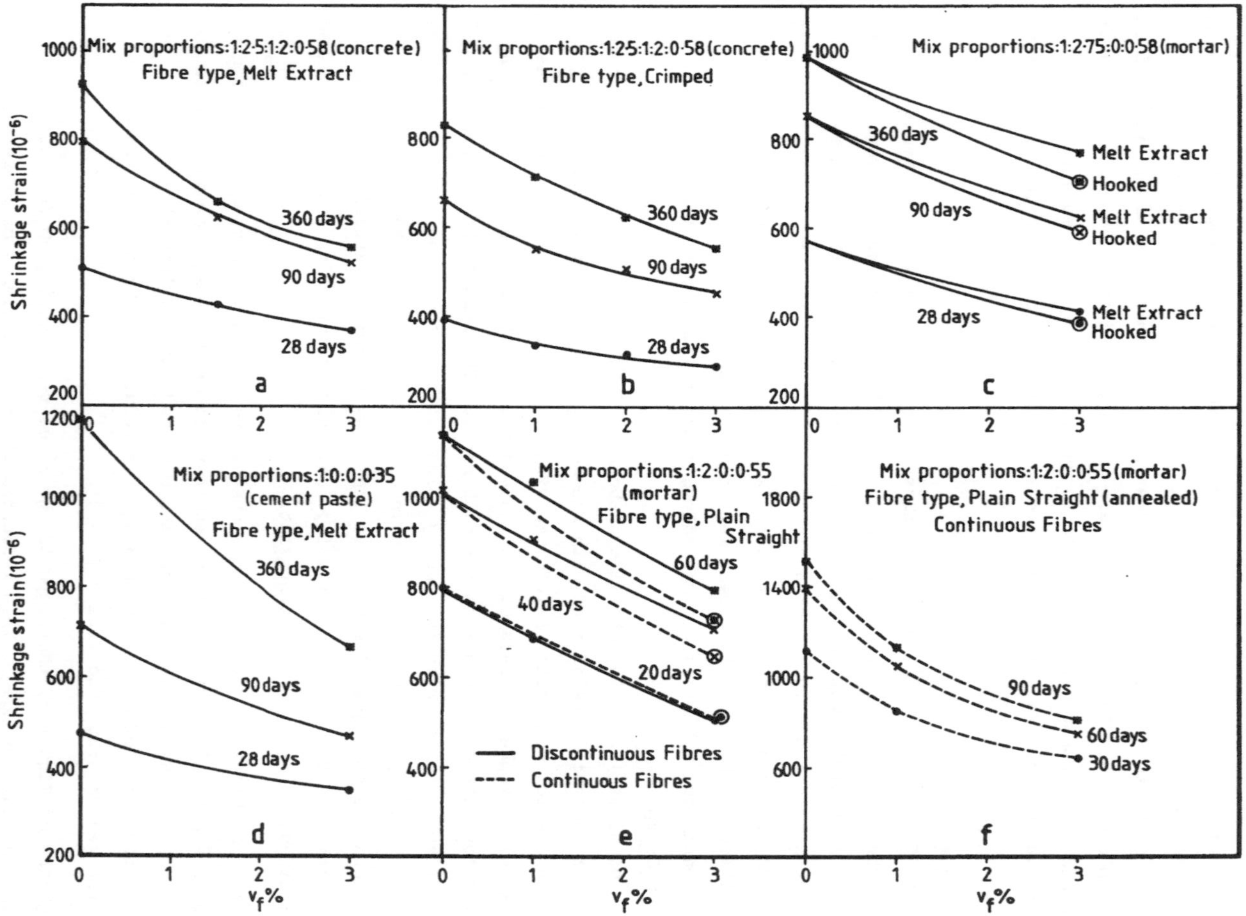

Figure 4 Influence of steel fibres on free shrinkage.

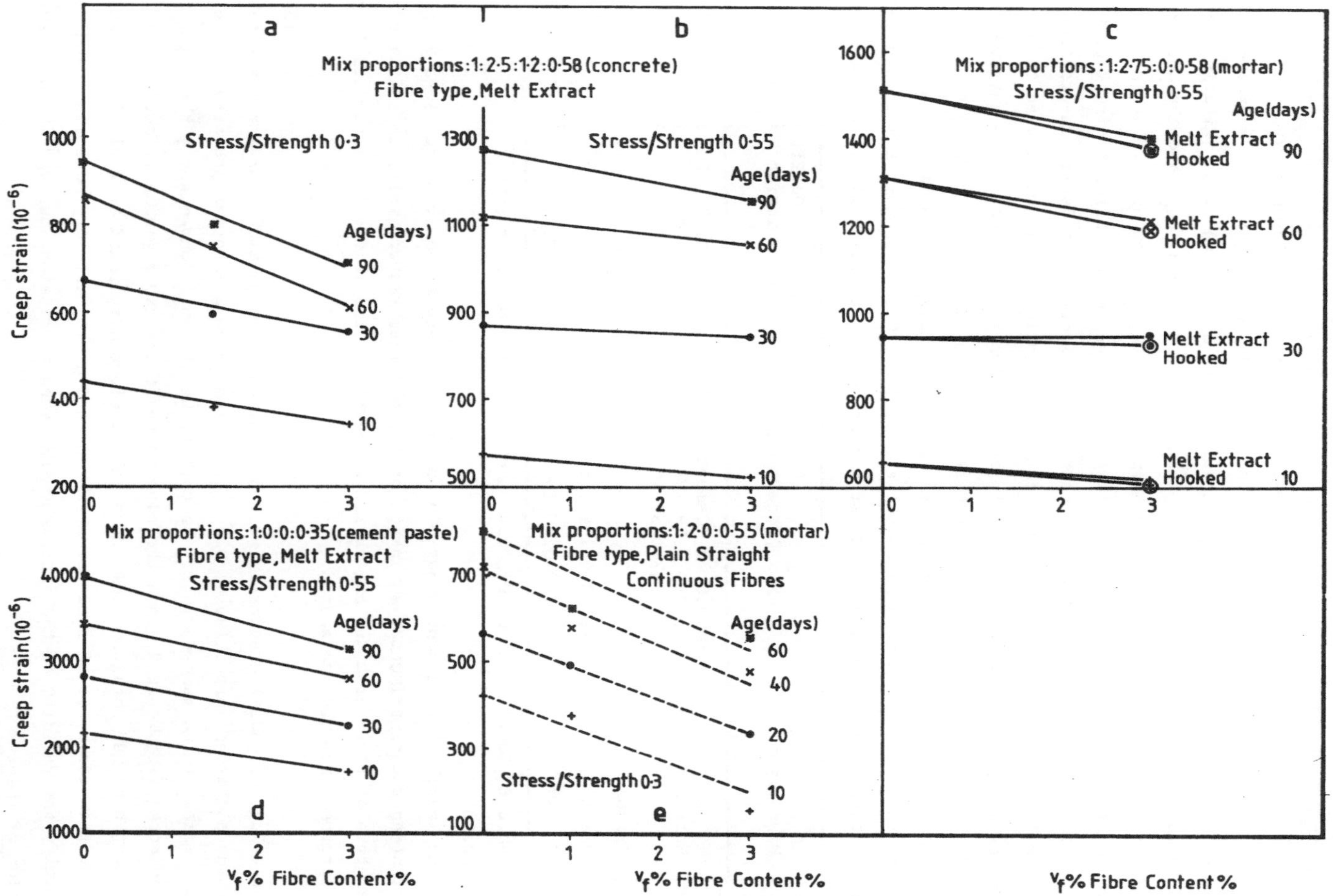

Figure 5 Influence of steel fibres on creep.

delayed elastic creep component is more significant and fibres are ineffective in restraining this component of creep (9).

Coefficient of Friction μ

The values of coefficient of friction, μ, for cement matrices reinforced with randomly oriented discrete steel fibres were calculated from equation 10 using the experimental data on shrinkage shown in Figure 4. The μ values were determined by substituting in equation 10 the free shrinkage strains of plain control concrete, ε_{os}, and the corresponding strains, ε_{fs}, of fibre reinforced matrices at various ages of cure up to 520 days. It was found that age of cure and fibre content has no significant effect on the values of the coefficient of friction. This has been discussed in detail in previous publications (3, 4). Therefore, the coefficient of friction, μ, is considered to be a material property of the fibre-matrix interface, which is governed by the surface texture and shape of steel fibres and the constituents of the matrix. The average μ values for the various types of steel fibres and cement matrices are listed in Table 2. The μ values in a concrete matrix range between 0.10 and 0.12, for mortar matrices the values range between 0.07 and 0.09, and for cement paste matrices μ is 0.12.

Fibre Type	Mix Proportions	Average Coeff. of Friction (μ)
Melt Extract	1:0:0:0.35 (Cement paste)	0.12
	1:2.75:0:0.58 (Mortar)	0.07
	1:2.5:1.2:0.58 (Concrete)	0.10
Crimped	1:2.5:1.2:0.58 (Concrete)	0.12
Plain	1:2:0:0.55 (Mortar)	0.07
Hooked	1:2.75:0:0.58 (Mortar)	0.09

Table 2 Coefficient of Friction μ at interface of short fibres.

Transfer Length

The transfer length, ℓ_t, of continuous fibres has been determined from equations 2 and 6 by substituting the experimental shrinkage or creep values respectively, in addition to the parameters of fibre geometry and matrix properties. The values for interfacial bond, τ, were obtained from the expression.

$$\tau = \mu P \qquad \text{... (3)}$$

where the values of radial pressure P were determined from equation 4 or 7 for the case of shrinkage and creep respectively; μ is considered to be a material property of the interface, which is the same for short or continuous fibres. Consequently the value of 0.07 is selected from Table 2 since the continuous fibres had a plain surface. The transfer length values obtained from these calculations are listed in Table 3. These values, for the wide range of data for shrinkage and creep, are very consistent and hence justify the validity of the shrinkage and creep expressions derived in this paper

Application of Theory

The theoretical expressions (equations 2, 6 and 9) can be used to predict the shrinkage and creep of steel fibre reinforced cement matrices. The information required to do so is the shrinkage and creep properties and stiffness of the plain concrete matrix and parameters of fibre geometry and volume. Additional information required is the coefficient of friction which can be obtained from Table 2 for the appropriate matrix and fibre combination. In the case of composites reinforced with continuous or long fibres only, the value of transfer length can be taken as the average of those listed in Table 3. The resulting calculated values of shrinkage or creep of the composite are within an acceptable margin of error for engineering purposes.

Test	Shrinkage									Creep					
Fibre type	Hard drawn			Annealed						Annealed					
Fibre spacing (mm)	2.5			2.5			5.0			2.5			5.0		
Age (days)	90	60	30	90	60	30	90	60	30	60	40	20	60	40	20
Transfer length ℓ_t (mm)	5.3	5.2	5.2	7.5	7.4	7.0	6.8	6.7	6.3	5.0	5.4	6.5	5.9	5.2	3.1

Table 3 Transfer Length of Continuous Fibres

CONCLUSIONS

1. Shrinkage and creep of cement matrices decreases with increasing volume of steel fibre reinforcement. The influence of short randomly oriented fibres and long (continuous) aligned fibres is not significantly different.

2. Restraint to shrinkage and creep is provided by the transfer length of long (continuous) fibres through the interfacial bond strength.

3. The bond strength is a function of radial pressure exerted on a fibre by the matrix and coefficient of friction which is a constant property of the fibre-matrix interface.

4. The theoretical expressions given in the paper can be used to predict shrinkage and creep strains of composites reinforced with short or long (continuous) fibres.

ACKNOWLEDGEMENTS

The authors gratefully acknowledge undergraduate students, Judith Mason and Zulkiflee B.A. Rahman, for some of the experimental work reported in this paper.

REFERENCES

1. Pickett, G., *Effect of Aggregate on Shrinkage of Concrete and Hypothesis Concerning Shrinkage,* Journal of American Concrete Institute, Vol. 52, 1965, pp. 581-590.

2. Mangat, P.S. and Motamedi Azari, M., *Influence of Steel Fibre Reinforcement on the Fracture Behaviour of Concrete in Compression,* The International Journal of Cement Composites and Light Weight Concrete, Vol. 6, No. 4, November 1984, pp. 219-232.

3. Mangat, P.S. and Motamedi Azari, M., *A Theory for the Free Shrinkage of Steel Fibre Reinforced Cement Matrices,* Journal of Material Science, Vol. 19, 1984, pp. 2183-2194.

4. Mangat, P.S. and Motamedi Azari, M., *A Theory for the Creep of Steel Fibre Reinforced Cement Matrices Under Compression,* Journal of Materials Science, Vol. 20, 1985, pp. 1119-1133.

5. Swamy, R.N., Mangat, P.S. and Rao, C.V.S.K., *The Mechanics of Fibre Reinforcement of Cement Matrices,* Journal of American Concrete Institute, SP-44, 1974, pp. 1-28.

6. Bartos, P., Review Paper: *Bond in Fibre Reinforced Cements and Concrete,* International Journal of Cement Composites and Light Weight Concrete, Vol. 3, No. 3, 1981, pp. 159-177.

7. Timoshenko, S., *Strength of Materials, Part II,* Advanced Theory and Problems, Third Edition, D. Van Nostrand Co. Int., New York, 1956.

8. Motamedi Azari, M., *Fracture Behaviour and Long Term Deformation Characteristics of Steel Fibre Reinforced Concrete Compression Members,* Ph.D. Thesis, University of Aberdeen, 1984.

9. Mangat, P.S. and Motamedi Azari, M., *Compression Creep Behaviour of Steel Fibre Reinforced Cement Composites,* Materiaux et Constructions, Vol. 19, No. 113, 1986, pp. 361-370.

10. Mangat, P.S. and Motamedi Azari, M., *Shrinkage of Steel Fibre Reinforced Cement Composites,* Materiaux et Constructions, In Press.

Proceedings of the International Symposium on Fibre Reinforced Concrete, December 16-19, 1987, Madras, India

MODELLING STRUCTURAL RESPONSE OF REINFORCED CONCRETE BEAMS WITH STEEL FIBRES

R.N. Swamy, R. Jones and T.P. Chiam, University of Sheffield, England

SYNOPSIS

A nonlinear 3-dimensional finite element model is presented to predict the complete structural response of reinforced concrete beams, made with plain or fibre concrete, and subjected to short term loading. The model incorporates material nonlinearities arising from cracking or yielding; it is applicable to beams of any arbitrary shape but of uniform cross-section; it is valid for lightweight and normal weight concrete, takes into consideration shear transfer across cracks, and recognises both shear and flexural failures. Tension steel is represented by bar elements; concrete by 8-node isoparametric hexahedra but the model can be easily modified to incorporate higher order elements. The accuracy of the analytical model is checked by tests on full scale reinforced concrete beams of rectangular, I and T sections, and involving a wide range of parameters of fibre volume, tension steel, moment-shear ratio, and type of concrete. Within its limitations, the model is shown to predict very satisfactorily the deformation characteristics during loading, and the mode of failure.

INTRODUCTION

Finite element modelling is recognised as a powerful tool of analysis, and is applicable to a wide range of structural problems. This paper describes the development of a finite element model and the associated computer program to predict the structural behaviour of reinforced concrete beams without and with steel fibres.

The model presented is a nonlinear 3-dimensional finite element model, which is capable of predicting the entire structural response up to failure, whether shear or flexural, of both plain and fibre reinforced conventional reinforced concrete beams subjected to short term monotonically increased loading. In the model, the concrete is represented by 8-node isoparametric hexahedra, and the tension steel by bar elements. The model takes into account material nonlinearity arising from tensile cracking of concrete, yielding of concrete, crushing of concrete, or yielding of main reinforcement, or a combination of two or more of these effects. A comprehensive test programme is also reported to check the accuracy of the model, and it is shown that very good correlation is obtained between the model and the results of lightweight concrete I beams, and normal weight rectangular and T beams without or with steel fibres, and failing in shear or flexure. One of the advantages of the form of the computer program developed is that it is very general, and that it can be easily modified to incorporate higher order isoparametric elements, and other failure criteria for concrete.

FINITE ELEMENT MODEL

Finite Element Approximations

The model used here integrates concrete and steel into a single element with separate stress-strain relationships for concrete and steel reinforcement. This is achieved by determining separately, the stiffness matrix of the steel reinforcement and that of the parent concrete element. The stiffness matrix of the reinforced concrete element is then obtained by adding up the stiffness matrix of each component. The advantage of this approach is its flexibility for different arrangements of the reinforcement and its computational simplicity.

Three Dimensional Concrete Elements

The 8-node isoparametric hexahedron has been adopted in this study. This hexahedron is a linear element and for exact integration requires 2 x 2 x 2 Gauss points. To assess how well normal and shear stresses are represented at the Gauss points using exact integration, a preliminary elastic finite element analysis of one of the test beams was carried out using three discretization layouts and 2 x 2 x 2 Gauss point integration. Form this study, a model of 7 x 15 elements with 1 x 2 x 2 Gauss point integration was selected as giving a far better representation of stresses.

Bar Elements

The reinforcing bars in the present study are represented by separate 2-node bar elements, and are allowed to lie within the isoparametric hexahedral elements, such that new reinforcement can be easily added to an existing model without altering the original finite element topology.

Steel-Concrete Bond

Bond stress-slip characteristics in reinforced concrete are complex, and when incorporated in analytical models, require not only greater computational efforts but also greater computer time. Obviously, better prediction of structural behaviour can be obtained by incorporating the steel-concrete bond interaction. In this paper, the simpler approach of perfect bond between steel and concrete is assumed.

Mathematical modelling of Concrete

The stress-strain and failure behaviour of concrete is characterized in this analysis by an elastic-perfectly plastic fracture model. Fig. 1 shows schematically the actual and idealised stress-strain relations assumed in this analysis for plain and fibre concretes in uniaxial compression and for plain and fibre concretes in uniaxial tension respectively. The post-cracking tensile strength of fibre concrete σ_{cu} is

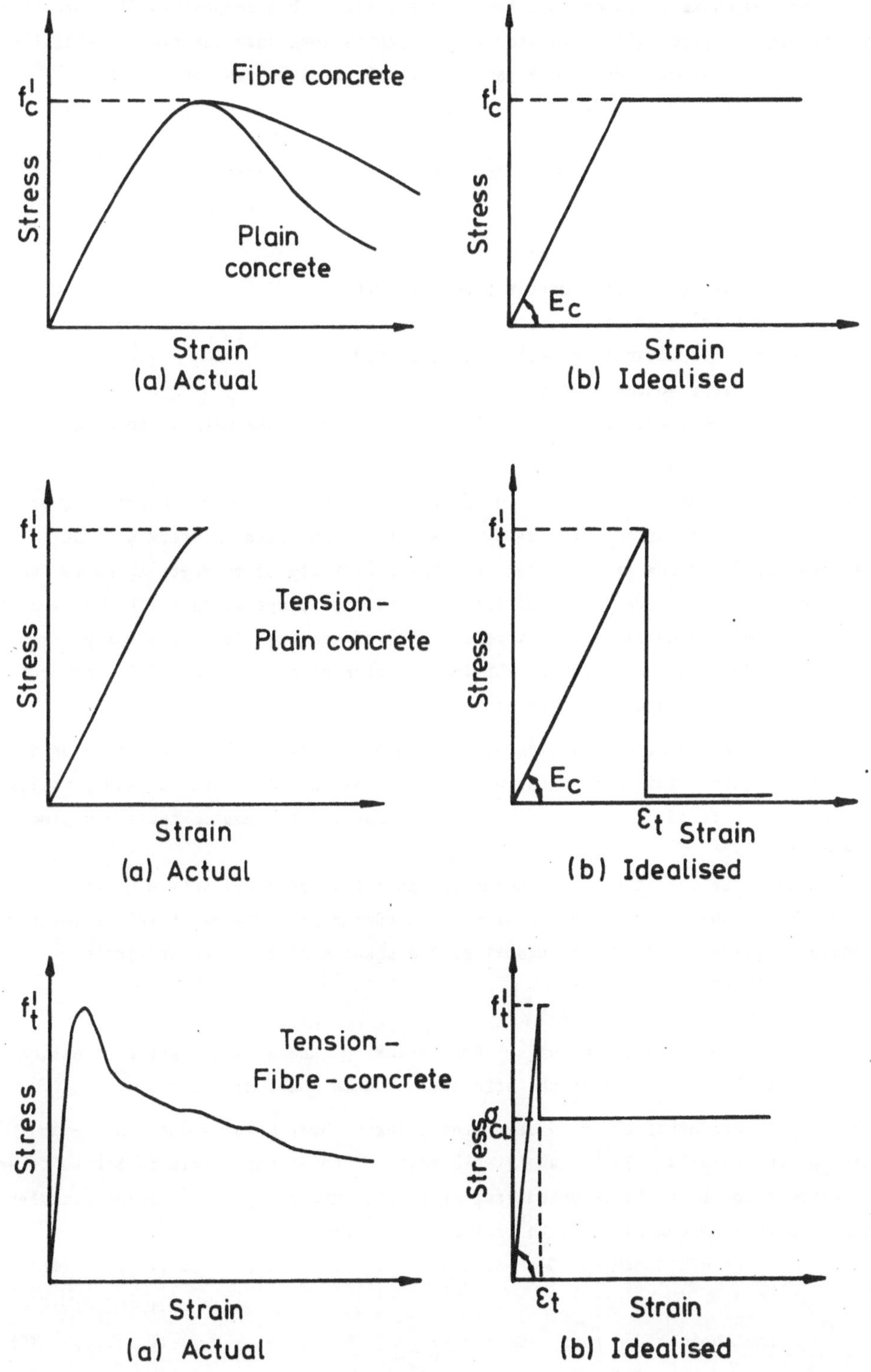

FIG.1 Schematic Uniaxial Stress - Strain Relations for Concrete

based on the composite mixtures rule for discontinuous fibre composites [1], and Laws' length efficiency factor [2]. Thus, for fibre lengths less than and greater than the critical length, the post-cracking tensile strength is given respectively by

$$\ell_f < \ell_c : \sigma_{cu} = 0.41\ \tau(\ell_f/d_f)\ V_f \quad \ldots\ldots\ldots\ldots\ldots\ldots\ldots\ldots \quad (1)$$

$$\ell_f > \ell_c : \sigma_{cu} = 0.41\ (1 - \frac{\sigma_{fu}}{4\tau}\frac{d_f}{f})\ \sigma_{fu}\ V_f \quad \ldots\ldots\ldots\ldots\ldots\ldots \quad (2)$$

where ℓ_f = fibre length

ℓ_c = critical fibre length

τ = average fibre - matrix bond strength

d_f = fibre diameter

σ_{fu} = fibre ultimate tensile strength, and

V_f = fibre volume fraction

The interfacial bond stresses were calculated according to the method proposed by Swamy et al [1, 3, 4].

Concrete is assumed to be linearly elastic and isotropic prior to cracking or yielding. The yield surface of both plain and fibre concrete is assumed to be approximated by Von Mises yield criterion. The plasticity of concrete is caused not by plastic flow but by the cumulative effect of microcrack propagation. The addition of steel fibres improves ductility and strain softening, so that full plasticity assumption is adequately confirmed. The failure strains of plain and fibre concrete are taken to be 0.0035 and 0.0045 respectively.

Cracking under shear-compression or tension-compression stress states is predicted by fracture criteria similar to those proposed by Calzona [5]. The maximum principal tensile stress criterion is used to predict cracking in the tension-tension regime.

<u>Shear Retention Factor</u>

To account for the effects of aggregate interlock and dowel action of the reinforcement, a shear retention factor was introduced into the constitutive relations of cracked concrete. Thus, after cracking, the shear stress τ - shear strain γ relation is defined by

$$\tau^* = \beta\ G\gamma^* \quad \ldots\ldots\ldots\ldots\ldots\ldots\ldots \quad (3)$$

where G is the material shear modulus before cracking, and β is the shear retention factor. βG may be considered as the effective cracked shear modulus.

In a real reinforced concrete beam, the cracked shear modulus decreases with increasing crack widths. In the analytical model proposed here, this factor is taken into consideration by tests reported earlier by the authors [6]. The shear transfer stiffness is then approximated by the following equations:

$$\text{Gravel concrete} : G_s = 3.0\left(\frac{1}{w}\right)^{1.75} \quad \ldots\ldots\ldots\ldots\ldots\ldots\ldots\ldots \quad (4)$$

$$\text{Lightweight Concrete} : G_s = 2.5\left(\frac{1}{w}\right)^{1.75} \quad \ldots\ldots\ldots\ldots\ldots\ldots\ldots\ldots \quad (5)$$

where G_s is the shear transfer stiffness in N/mm² and w is the crack width in mm. In the present study, both constant and variable shear retention factor was used.

Modelling of Steel Reinforcement

The uniaxial stress strain behaviour of steel was idealised by an elastic-plastic model with an elastic modulus of 200 kN/mm².

Nonlinear Solution Procedure

In the present study, only time independent material nonlinearity due to monotonically increased loading is considered. Geometrical nonlinearity, creep and shrinkage effects are not considered. The nonlinear solution was carried out by using a combined incremental and iterative scheme.

The computer program developed in this study can analyse plain or fibre reinforced lightweight and normal weight concrete beams of any arbitrary shape but of uniform cross-section throughout. To minimise computational time, the present program allows the user to terminate the analysis temporarily after a specified number of load increments so that a check of the results obtained can be carried out.

EXPERIMENTAL WORK

Test Details

To examine the accuracy of the finite element model developed earlier, an extensive test programme consisting of 22 simply supported beams tested under a symmetrical four point loading system was carried out. Series 1 consisted of 18 lightweight concrete beams of I section; Series 2 consisted of 2 beams of rectangular section made of normal weight concrete whilst Series 3 had 2 beams of T-section of the same normal weight concrete. All the beams in any series had the same cross-section as shown in Figs. 2, 3, and 4.

Series 1

Three major parameters which significantly affect the shear strength of reinforced concrete beams were investigated in this series:-

fibre volume	-	0% and 1%
tension steel (based on rib width)	-	1.55%, 2.76% and 4.31%
shearspan - effective depth ratio	-	2.0, 3.43 and 4.91

All beams were 3.5 m long and tested over a span of 3.0 m. The beams were made with a lightweight aggregate concrete, having a 28% cement replacement by fly ash, and an average 28 day compressive strength of about 40 N/mm². The steel fibres were crimped, 0.5 x 50 mm. All the steel bars were of high tensile steel.

Series 2 and 3

The four beams in these two series (Figs. 3 and 4) were made of gravel concrete with 30% by weight of cement replaced by fly ash. The concrete mixes were designed to have a 28 day compressive strength of 40 to 45 N/mm². The beams in each series had a plain concrete beam, and a fibre concrete beam containing 0.8% fibre volume of 0.5 x 50

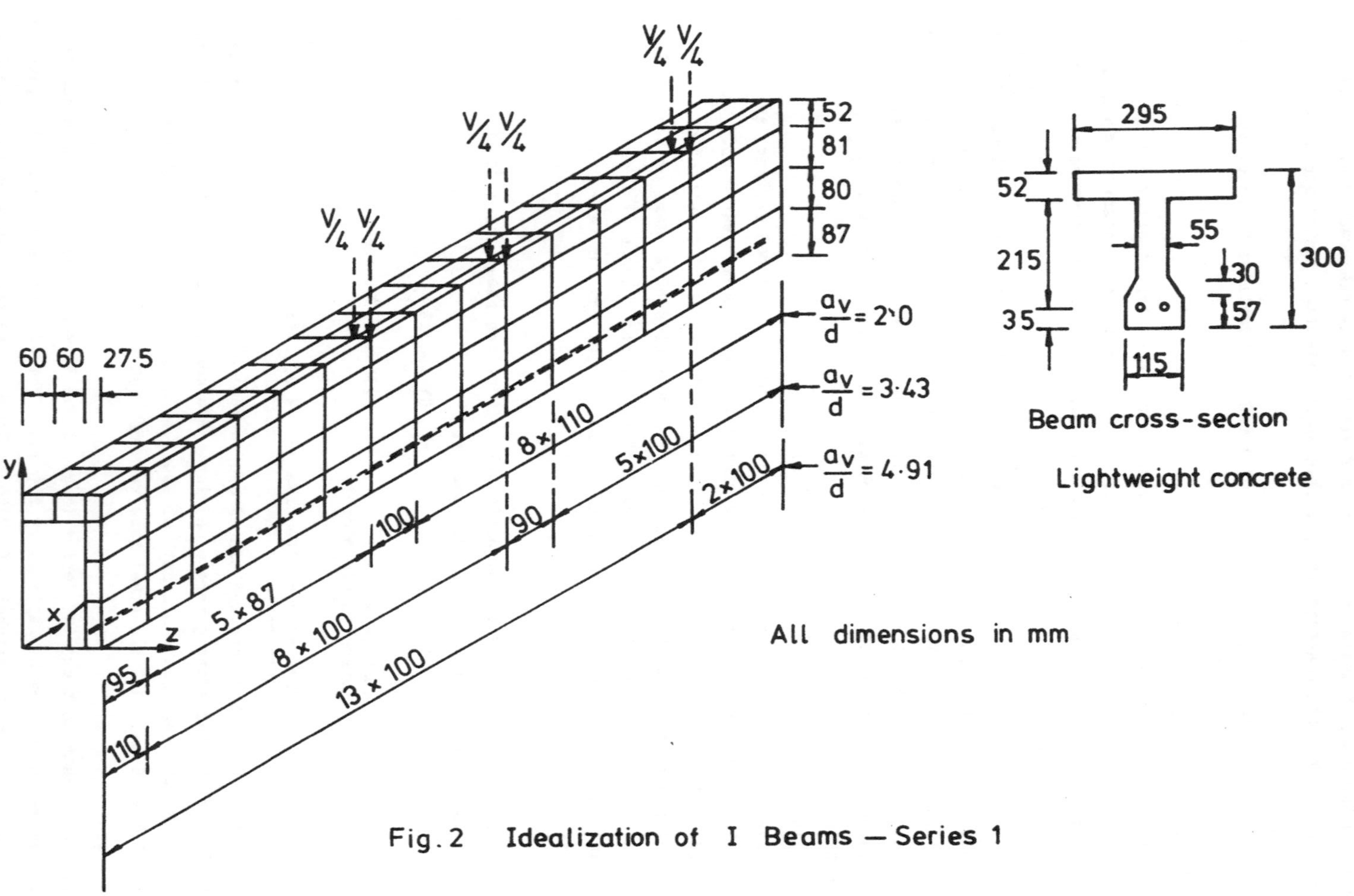

Fig. 2 Idealization of I Beams — Series 1

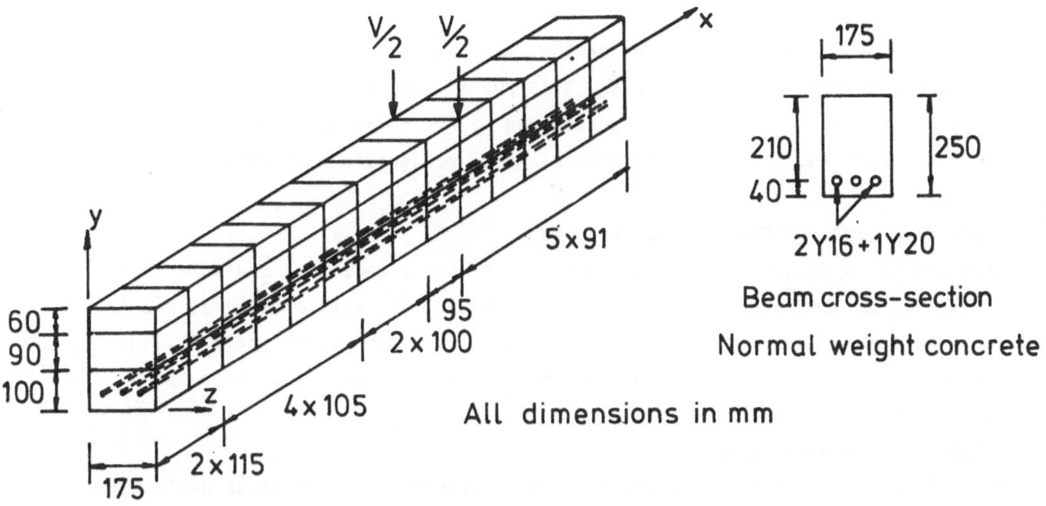

Fig. 3 Idealization of Rectangular Beams – Series 2

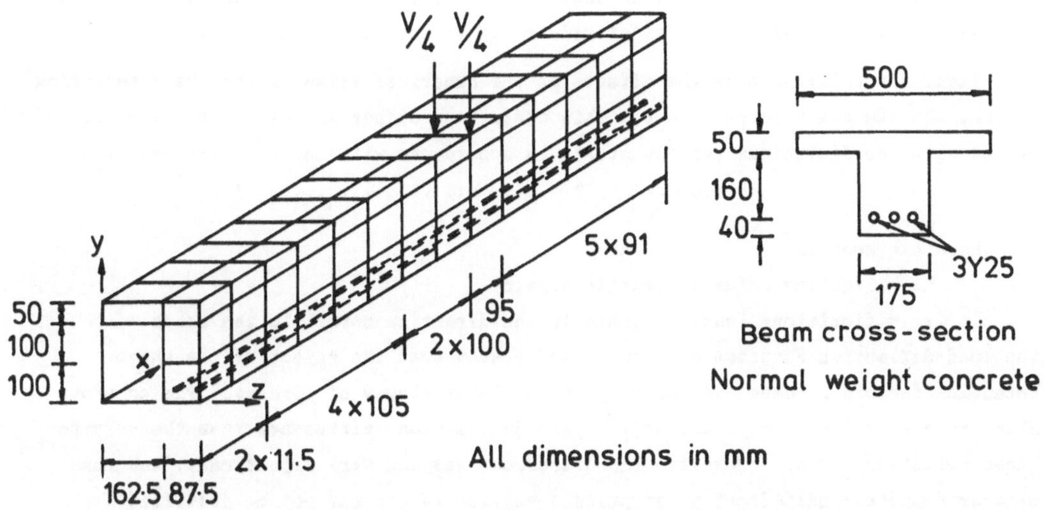

Fig. 4 Idealization of T – Beams __ Series 3

mm crimped fibres (as in Series 1). The tension steel was high tensile deformed bars, the percentage of steel being 1.95% in Series 2 and 4% (based on rib width) in Series 3. All the beams were 3.4m long and tested over an effective span of 2.8 m under a constant shear span-effective depth ratio of 4.5 [7].

Beam Discretization

Due to symmetry, only one quarter of the beam was (generally) solved analytically. Each beam was idealised as shown in Figs. 2, 3 and 4. The dimension of the discretization in the x-direction was determined by the shear span effective depth ratio of the beam. The concrete and steel material properties were obtained from control tests.

COMPARISON OF FINITE ELEMENT AND TEST RESULTS

From the large amount of data obtained for the three series of tests of this study, only typical information on load-displacement, load-tensile steel strain and load-concrete strain relation are presented here. In the numerical analysis, both constant and variable shear retention factors were used. Their effects on the analytical results were evaluated.

Load-Midspan Displacement : Series 1

Figs. 5 and 6 show typical analytical and experimental load-midspan deflection relations for two plain and two fibre concrete beams of series 1. All these beams had a longitudinal steel ratio of 2.76%; beams 2TL-1 and 2TLF-1 were tested at a shear span - effective depth ratio of 2.0 whilst beams 2TL-2 and 2TLF-2 were tested at a ratio of 3.43. Beams 2TL-1 and 2TL-2 were made without fibres; beams 2TLF-1 and 2TLF-2 had 1% fibre volume incorporated in the concrete.

The results show reasonably good correlation, although the stiffness of the fibre concrete beams was generally overestimated in the later stages of loading. This is possibly due to the assumption of a constant post-cracking tensile stress; in reality, this value will decrease as the crack widens due to increasing fibre pull-out.

Figs. 5 and 6 also show the effects of the numerical value of the shear retention factor. In the study reported here, this factor was either maintained constant or varied with the fictitious tensile strain in accordance with the following equation:

$$\beta = \alpha \left(\frac{E_t}{E_1} \right)^{1.75} \qquad (6)$$

where α = constant

E_t = limiting uniaxial tensile strain

E_1 = fictitious tensile strain in the direction normal to the crack.

The load-deflection diagrams of Figs. 5 and 6 show that the effect of the shear retention factors becomes evident only at the later stages of loading. The constant shear retention factor gave slightly higher deformation stiffnesses than the varying shear retention factor. However, the difference was not very significant, and the greater computational effort in adopting a varying factor may not be justified.

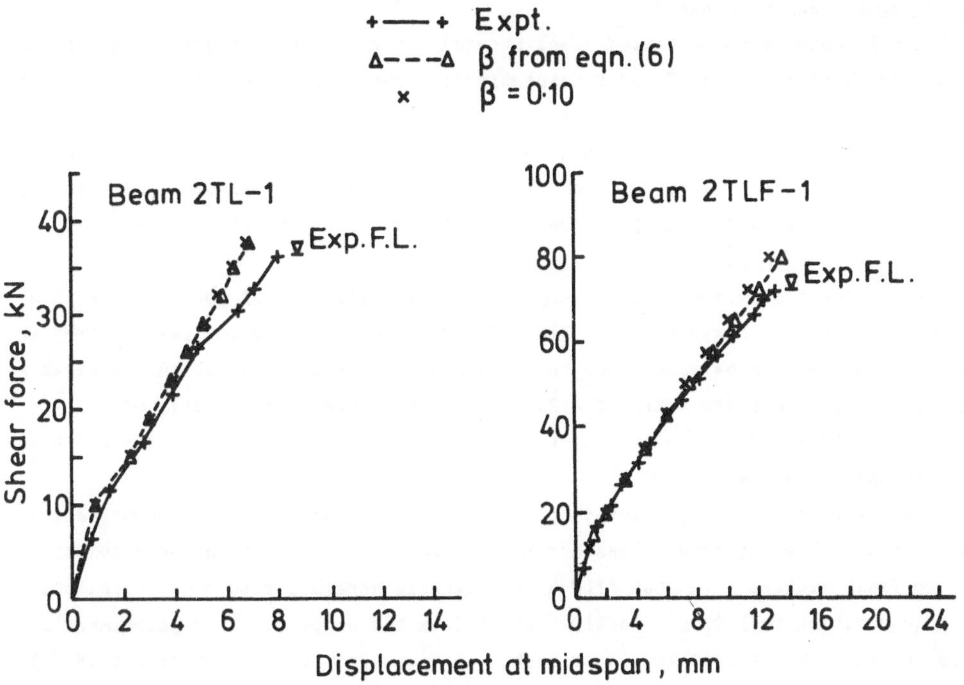

FIG. 5 LOAD - MIDSPAN DISPLACEMENT CURVES

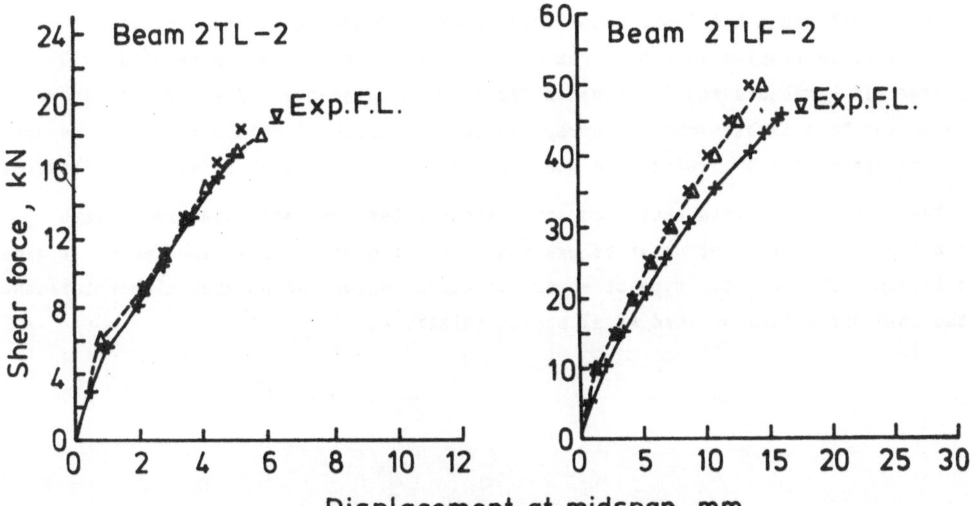

FIG. 6 LOAD — MIDSPAN DISPLACEMENT CURVES

Steel Strains : Series 2 and 3

Typical computed and measured steel strains for the normal weight rectangular and T beams are shown inn Figs. 7 and 8 respectively. The analytical predictions are seen to be in close agreement with test results for both plain and fibre reinforced beams. The theory predicted that the reinforcement had yielded in the rectangular fibre reinforced concrete beam (Fig. 7), and this was confirmed by the test. The value of the shear retention factor had practically no effect on the computed strains (Fig. 7).

Longitudinal Concrete Strain Profile : Series 1

The predicted and experimental longitudinal distributions of concrete strain at 10 mm from the top surface are shown in Figs. 9 and 10 respectively for beams 1TLF-3 and 2TLF-3. Both these beams were fibre reinforced with 1% fibre volume and tested at a shear span-effective depth ratio of 4.91 - beam 1TLF-3 had a steel ratio of 4.31% whilst beam 2TLF-2 had a steel ratio of 2.76%. Beam 1TLF-3 failed in shear and beam 2TLF-3 in shear-flexure.

Both the measured and predicted strain profiles showed diminishing compressive strains towards the supports; these strains became tensile at regions near the support prior to final failure. In beam 1TLF-3, the tensile strain spread over a distance of 270 mm (equivalent to 1.01 x effective depth) from the support. This phenomenon is consistent with theoretical and experimental observations reported by others [8, 9]. Figs. 9 and 10 also show that the top fibre strain adjacent to the point load was generally higher than that at midspan after the formation of shear cracks - again a feature observed in shear failures and reported extensively in literature. The results show generally good correlation with the predicted values.

CONCLUSIONS

The data presented in this paper show that the nonlinear 3-dimensional finite-element model developed predicts adequately the entire structural response up to failure of both plain and fibre reinforced concrete beams, subjected to short term monotonically increasing loading. The concrete in the model is represented by 8-node isoparametric hexahedra and the tension steel by bar elements. The model is shown to be valid for both lightweight and normal weight conventional reinforced concrete beams made from plain or fibre reinforced concrete, and failing either in shear or flexure.

The results indicated that a constant shear retention factor generally gave marginally higher deformation stiffness than a varying shear retention factor at high load levels. However, the type of shear retention factor had no appreciable influence on the load deflection or load steel strain relations.

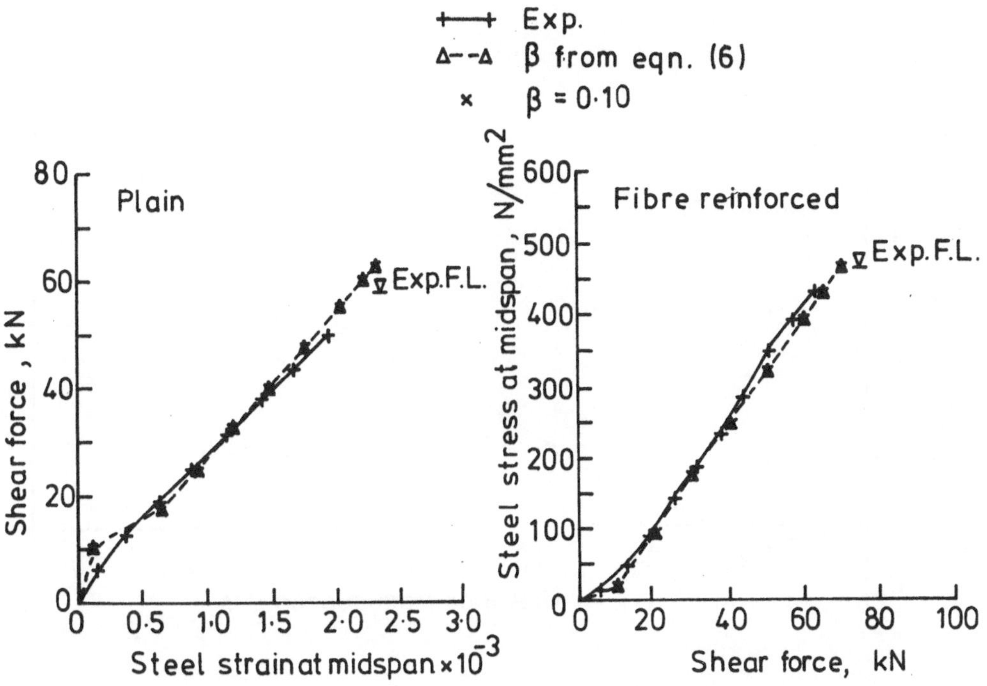

FIG. 7 LOAD-STEEL STRAIN CURVES FOR RECTANGULAR BEAMS

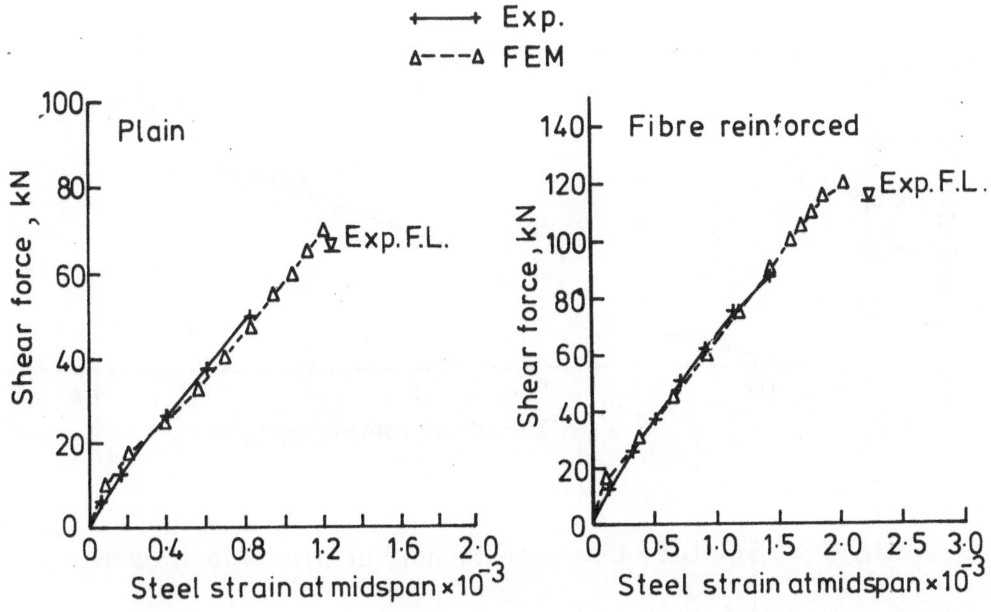

Fig. 8 Load-Steel Strain Curves for T-Beams

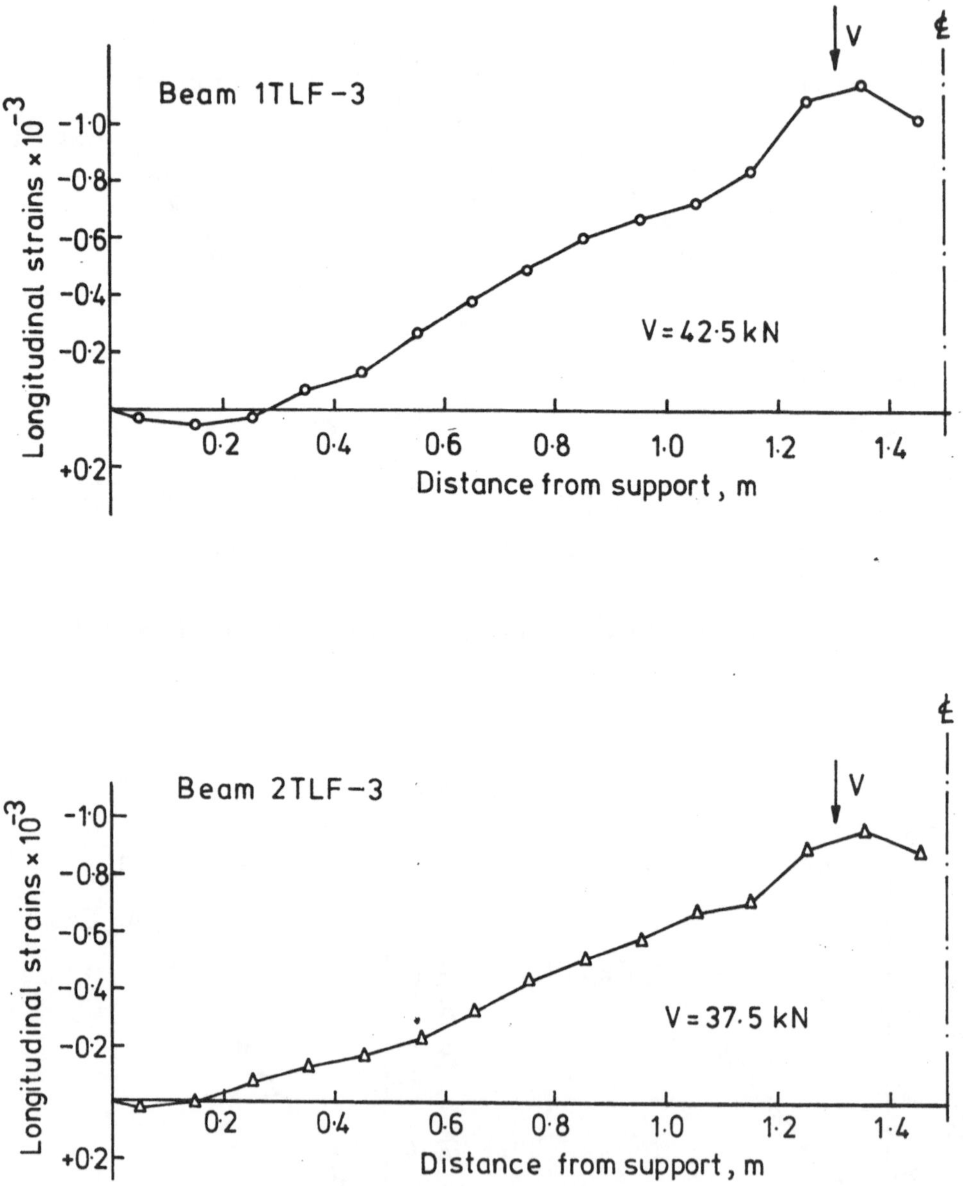

Fig. 9 Predicted Concrete Strain Variation Along Span

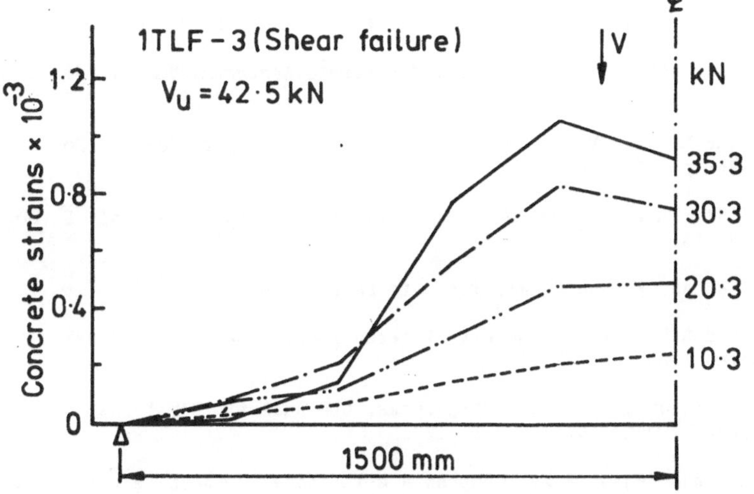

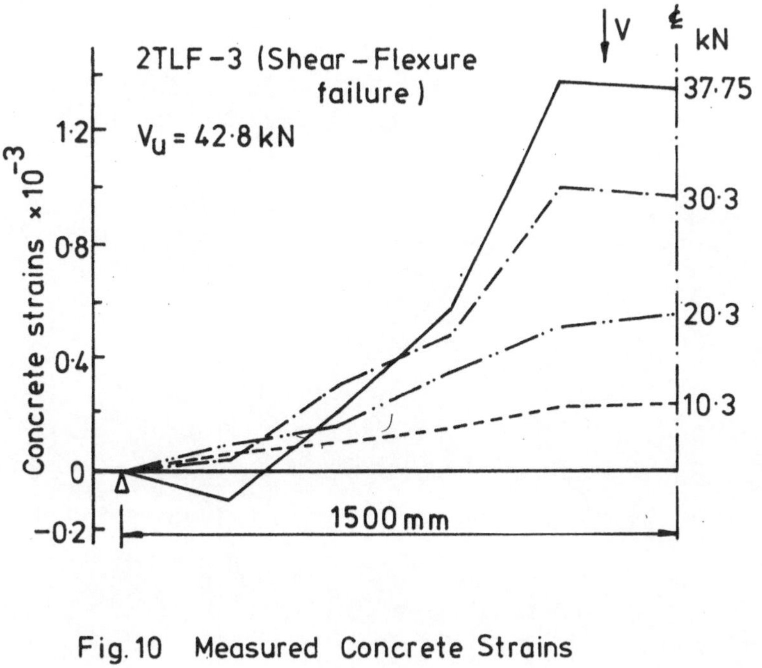

Fig. 10 Measured Concrete Strains

REFERENCES

1. Swamy, R.N., Mangat, P.S. and Rao, C.V.S.K., In : *Fibre Reinforced Concrete*, ACI Publication SP-44, 1974, pp. 1-28.

2. Laws, V., *Journal of Applied Physics*, Vol. 4, 1971, pp. 1737-1746.

3. Swamy, R.N. and Mangat, P.S., *Cement and Concrete Research*, Vol. 4, No. 2, March 1974, pp. 313-325.

4. Swamy, R.N. and Mangat, P.S., *Cement and Concrete Research*, Vol. 6, No. 5, September 1976, pp. 641-650.

5. Calzona, R., *The International Journal of Lightweight Concrete*, Vol. 2, No. 2, June 1980, pp. 105-113.

6. Swamy, R.N., Jones, R. and Chiam, T.P., To be published.

7. Swamy, R.N. and Bahia, H.M., Concrete International - Design and Construction, Vol. 7, No. 3, March 1985, pp. 35-40.

8. Leonhardt, F., Shear in Concrete Structures, CEB Bulletin D'information No. 126, Shear and Torsion, June 1978, pp. 67-124.

9. Muhidin, N.A., and Regan, P.E., Chopped steel fibres as shear reinforcement in concrete beams, Fibre Reinforced Materials : Design, Engineering and Applications, Institution of Civil Engineers, London, 1977, pp. 149-163.

CHAPTER II

PROPERTIES AND BEHAVIOUR OF STEEL FIBRE REINFORCED CONCRETE

Proceedings of the International Symposium on Fibre Reinforced Concrete, December 16-19, 1987, Madras, India

MATERIALS AND PROPERTIES OF FIBRE REINFORCED CONCRETE

V. Ramakrishnan
Professor of Civil Engineering
South Dakota School of Mines and Technology
Rapid City, SD 57701, U.S.A.

SYNOPSIS

Extensive research has established that the addition of various types of fibres in concrete improves the static flexural strength, flexural fatigue strength, impact strength, shock resistance, ductility and fracture toughness. The time dependent properties like creep, shrinkage, and durability are also affected by the inclusion of fibres. The degree of alteration of fresh as well as hardened concrete properties depends on the type, size, shape, quantity and aspect ratio of these fibres. This paper presents a comprehensive review on the types of fibres and their influence on fresh and hardened concrete properties. Emphasis is given to the recent research results with particular reference to steel fibres.

INTRODUCTION

The most promising materials used in construction today are the concrete fibre composites. To increase the fracture resistance of cementitious materials, fibres are frequently added, thus forming a composite material. Although the concept of reinforcing brittle materials with fibers is quite old, the recent interest in reinforcing portland cement based materials with randomly distributed fibres was spurred by pioneering research on steel fibre reinforced concrete conducted in United States in the 1960's. Since then there has been substantial research and development activities throughout the world.

Concrete fibre composites (shotcrete and concrete) have been found more economical for use in airport and highway pavements, bridge deck overlays, curtain walls, sewer pipes, cavitation/erosion resistance structures, tunnel linings, rock slope stabilization, dome structures, security concrete, refractory concrete, precast concrete products, earthquake-resistance structures and explosive resistance structures (like MX missile silos) (1 to 25). The behavioral efficiency of these composites is far superior to that of plain concrete and many other construction materials of equal cost.

High strength concrete [with compressive strength higher than 41.0 MPa (6000 psi)] is being used in the construction of buildings, and other reinforced and prestressed concrete structures. One major drawback of high strength concrete is that it is brittle. The failure will be sudden and catastrophic, particularly in structures which are subjected to earthquake, blast, or suddenly applied loads. An ideal solution to overcome the serious disadvantage of high strength concrete is to add steel fibres in the concrete to convert it into a ductile material and avoid sudden failures.

The constructed facilities of the world have been deteriorating due to the effect of the natural environment, excessive use beyond the original design, aging of the materials and general obsolescence. Fibre reinforced concrete composites are almost ideal materials for repair, rehabilitation, retrofit, and renovation of the world's deteriorating infrastructure. Extensive research findings on FRC have been reported all over the world. It is not the intent of this paper to review all the published papers. Highlights of the recent research in regards to the types of fibres and their influence on the fresh and hardened concrete are presented with specific examples, mostly from the author's work (1 to 23).

TYPES OF FIBRES

A great variety of fibre materials in various sizes and shapes, have been developed for use in FRC. They are commercially available for use in the construction industry. Currently steel, glass, polymeric, and carbon fibres are commonly used. There are also limited applications of natural (mineral or organic) fibres, such as bamboo akwara, armid, asbestos and cotton.

A convenient numerical parameter describing a fibre is its aspect ratio, defined as the fibre length divided by diameter (or equivalent diameter in the case of non-round fibers). Typical aspect ratios used range from 30 to 150 for length dimensions of 6.0 to 76mm (0.25 to 3.0 in).

Steel Fibres: Steel fibres for use in concrete are available in a number of shapes, sizes and metal types. Currently they are produced by three different processes. 1. A sheet of metal is cut or slitted, producing a square or rectangular fibre. 2. Cold-drawn wire is chopped to specific length. Some are collated with water-soluble glue into bundles of 10 to 30 fibres to facilitate handling and to increase their bond and anchorage parameters. 3. Melt-extracted fibres are produced by a process whereby a rotating, cooled disc with indentations of the size of fibre is dipped in the surface of a molten pool of high quality metal.

Many different types of fibre, with round, rectangular and cresent shaped cross sections, are commercially available. They range in ultimate strength from 345 to 2070 MPa (50,000 to 300,000 psi). Fibre sizes range from 13 x 0.25 mm to 64 x 0.76 mm (0.5 x .01 in. to 2.5 x 0.03 in.). However, most successful shotcrete applications have utilized fibres with length of 13 to 30 mm (0.5 to 1.2 in.). In one investigation, brass plated cut steel wire tire cord with a tensile strength of 2758 MPa (400,000 psi) has been successfully used (6). Fibres with hooked or deformed ends could be used in smaller quantities because they develop higher pullout resistance. Fibres with large surface area, square or rectangular as compared to round, have more concrete bonding area. And fibres with pitted surface have a greater surface area.

For high temperature application up to 1650°C (3000°F), like in refractory shotcretes, stainless steel fibres are used. Corrosion of steel fibres in concrete with a high water-to-cement ratio may cause deterioration. The free moisture in wet concrete provides an aqueous medium which facilitates transport of chlorides toward the metal. It also increases the electrical conductivity of material, thus aiding any tendency for electrochemical corrosion. However, in actual field cases, corrosion of carbon steel fibres have been found to be minimal. When fibres were exposed on a surface, they showed evidence of corrosion; however, internal fibres showed no corrosion. Fibre loadings in construction projects have typically ranged from 0.5 to 2.0 percent by volume: 47 to 157 kg/m^3 (66 to 265 lb/yd.^{3}). Higher percentages of fibres have been generally used with straight fibres.

Most of the early applications of FRC consisted of using relatively high fibre loadings of straight steel fibre, of relatively small diameter and high aspect ratios. These large quantities were needed to obtain higher flexural strengths. This caused serious problems, namely "balling" of fibres during the mixing operation and "pull-out" of the straight fibre during loading due to lack of adequate anchorage. Balling of fibres in the mixer prevents uniform distribution and also causes problems when the concrete is placed. Because of the pull-out of the fibres, the SFRC did not achieve the required ductility and post-crack load-carrying capacity. The addition of straight fibres had to be done by special vibrating sieves or by manual sprinkling. The introduction of a new type of fibre (with hooked ends and bundled together with a water soluable glue) had eliminated these problems (8). Collation of the fibres had eliminated balling and the fibres could be added to the mixer along with the aggregate all at one time eliminating the need for special devices and additional labour for addition of fibres. The improved anchorage (hooked ends) has made it possible for considerably smaller quantities (40 percent less) of fibres to produce the desired properties (8). The later development of other deformed fibres (corrugated, crimped, paddeled, etc.) had similar results of eliminating balling and they had improved anchorage (19).

Glass Fibres: Commonly used glass fibres are round and straight and have diameters of 0.005 to 0.015 mm (0.0002 to .0006 in.), but these fibres may be bonded together to produce glass fibre elements with diameters of .013 to 1.3mm (.0005 to .050 in.). The strength of glass fibre is comparable to that of steel fibre, its density is lower, and its elastic modulus is about one third that of steel. A major disadvantage of glass fibre is its high vulnerability in the alkaline cementitous environment which seriously damages the long-term properties of glass fibre reinforced concrete. Attempts have been made in the recent years to develop alkali resistant glass fibres or low alkali cements.

The major application for glass fibre has been the spray-up process in which the glass fibres and a cement-rich mortar are sprayed simultaneously on a surface.

Polymeric Fibres: Synthetic polymeric fibres are derived from organic polymers and the following have been tested in portland cement concrete: acrylic, aramid, kevlar, nylon, polyester and polypropylene. Among these the polypropylene has the most successful commercial application so far. The common forms of these fibers are smooth-monofilament, twisted, fibrillated and tridimentional mat. The monofilaments were of a geometry, size and shape, similar to that which were commercially available in steel and glass; approximate diameter 0.25 mm (0.01 in.) and 12 to 50 mm (½ to 2 in.) long, with an aspect ratio of 50 to 100. Dosage rates varied from 0.1 to 2.0 volume percent. Polypropylene has a low density and is also chemically inert. It has a hydrophobic surface which does not absorb part of the mixing water.

The major short comings of polymeric fibres are low modulus of elasticity, poor bond with cement matrix, combustibility and low melting point. Their bond to cement matrix is improved by twisting several fibres together or by treating the fibre surface. The long term properties of polymeric fibre concrete are not available.

Carbon Fibres: Carbon fibres produced from petroleum pitches in large volumes are low-cost and low-modulus. These fibres are very small in diameter, and are generally used in shorter lengths. They are also manufactured as continuous mats and continuous straight fibres. They can be manufactured in strength as high as steel with a density only one-fifth that of steel. Carbon fibres are inert in aggressive environments, abrasion-resistant and stable at high temperatures, with relatively high stiffness.

The uniform dispersion of carbon fibres in concrete is more difficult than the other fibre types.

ADDITIVES AND ADMIXTURES

A variety of additives and admixtures are added to FRC, particularly in the wet mix to improve strength, adhesiveness, cohesiveness, and freezing and thawing and abrasion-resistance characteristics, to increase workability, to inhibit corrosion and to reduce permeability. Major additives and admixtures used are briefly discussed.

Silica Fume: Condensed silica fume and ferrosilicon dust (essentially the same material but with slightly different chemical compositions) are by-products of silicon metal and ferrosilicon alloy manufacturing. The light grey and extremely fine powder was once discarded as waste material; research and testing have turned it into a valuable product, sometimes costing more than that of cement. Silica fume is highly pozzolonic and can be used in FRC as a cement replacement or to reduce permeability and to increase the strength. (4,17). When silica fume is added, a superplasticizer has to be added to achieve the required workability.

Water Reducers: Water reducers are used to improve the wet-mix workability, cohesiveness in the plastic state, and pumpability.

Accelerators: Mainly used in fibre shotcretes.

Superplasticizers: Known as high-range water reducers, superplasticizers are chemically distinct from normal plasticizers or water reducers. They are mainly sulfonated melamine formaldehyde condensates or sulfonated naphthalene formaldehyde condensates or modified lignosulfonates. They are used either to increase the strength or to increase the workability considerably without reduction in strength.

A major difficulty encountered in the early use of FRC, was the considerable reduction in workability caused due to the addition of fibres especially when large quantities of fibres were needed to achieve the required increase in the flexural strength. A high cement content and additional water were added in order to restore the workability, with a consequent increase in shrinkage strains. This impeded the widespread use of FRC. The author has shown (9) that this shortcoming can be eliminated by the use of superplasticizers. Steel fibre reinforced (SFRC) "flowing" concretes were produced with as low as 0.25 water to cement ratio and a high slump of 230 to 280 mm (9 to 11 in.). Thus the potential for economical utilization of FRC has been tremendously increased with use of superplasticizers.

Fly Ash: The most important consideration in the use of fly ash is its ability to decrease the water demand of concrete mixes without affecting their workability characteristics. Fly ash can reduce bleeding, and result in lower heat of hydration and improved cohesiveness. It also facilitates easier transportation, pumping, compaction and finishing of the material.

Polymer Latex Additives: These additives could be added to FRC to impart special desired properties, like adhesion improvement, permeability reduction, resistance to chloride attack, increased durability in freezing and thawing conditions, impact resistance, steel protection, and strength improvement.

Admixtures containing calcium chlorides, fluorides, sulfites, sulfates, and nitrates should never be used in SFRC, because these chemicals could promote corrosion of the steel fibers.

Properties of Fibre Reinforced Concrete

The properties of fibre concrete in the fresh and hardened states are influenced by the type, geometry, volume fraction and material properties of fibres, the matrix proportion, and the fibre-matrix interfacial bond characteristics.

Fresh Concrete Properties: The significant problem with fibre concrete mixes is a practical one that of ensuring adequate workability (flowability and compactability) that will facilitate the concrete to be placed, compacted, and finished with ease and also ensuring a uniform fibre distribution. The balling of fibres, segregation of the mix, and excessive bleeding during placing and compaction should also be avoided. For a given mix proportion, the degree of compaction seriously affects the strength and other properties. Therefore a knowledge of the fresh concrete properties is essential for proper design of the mix. Adequate information is available about the performance characteristics of fibre reinforced concretes with and without the use of superplasticizers (5, 7, 9, 23, and 25).

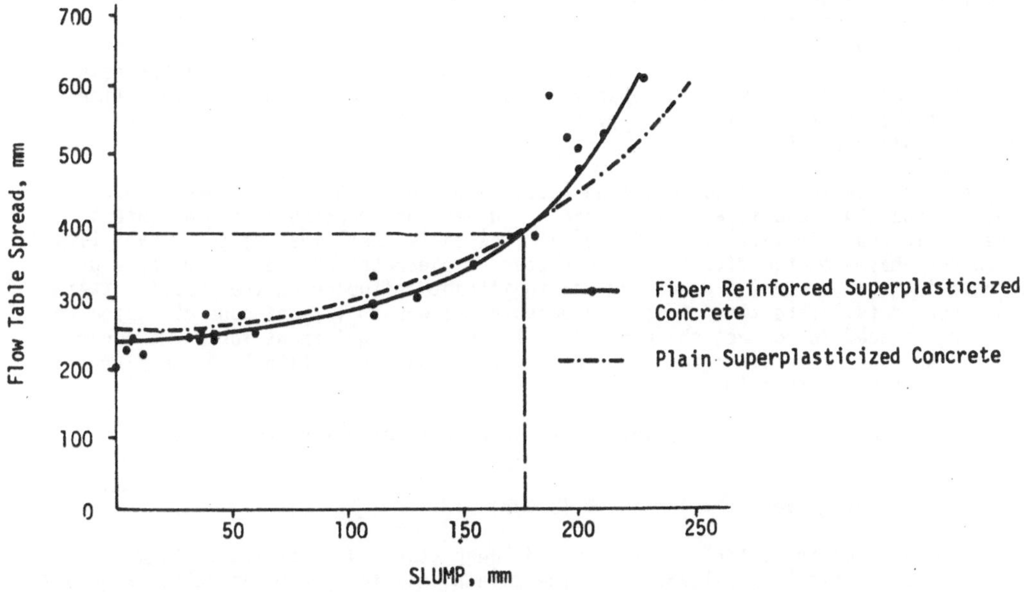

Fig. 1 Relationship Between Slump and Flow Table Spread

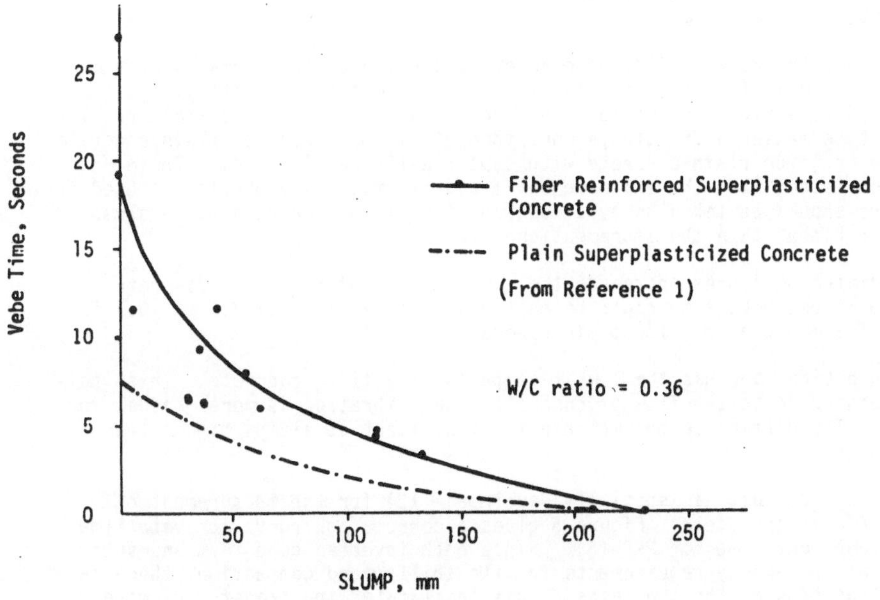

Fig. 2 Relationship Between Vebe Time and Slump

The relationship between slump and flow table spread for plain and steel fibre reinforced superplasticized concretes is shown in Fig. 1. These two curves show only a small difference between them; the relationship between slump and flow table spread is approximately the same for both superplasticized concretes, with and without fibres (5). In Figure 2, the continuous curve and broken curve show the relationship between slump and vebe time for superplasticized concretes with and without fibres respectively. For any slump the vebe time for plain concrete is much less compared to fibre reinforced concrete (5). It was also observed that the relationship between vebe time and flow table spread is linear for plain concretes, whereas it is curvilinear for fibre concretes (9).

The increase in fibre content from 32.63 kg/m^3 to 56.36Kg/m^3 had negligible effect on the slump and flow table spread. However, an increase in fibre content caused an increase in vebe time. Slump and flow table tests measure the workability of concrete based on the flowability character of concrete. Hence, an increase or decrease in fibre content did not have a significant influence on the results attained from these tests. Vebe time is used to measure the workability of concrete based on the energy needed to compact the concrete. The energy requirement for fibre concrete is more than for plain concrete. The energy needed was proportional to the fibre concrete in the concrete (5).

It was shown that slump and vebe time can be related using the general form of the equation.

$$\log_{10} \text{(Vebe time)} = a - b \text{ (slump)}.$$

This form resulted in better correlation and lower standard error (23). Figure 3 provides the graphical visualization of the data points for plain concrete. Figure 4 presents the comparative behavior of the three types of concretes.

Based on an analysis of extensive experimental results, the following conclusions were made (23).

1. The relationship between vebe time and slump was independent of the air-content.

2. The addition of superplasticizer increases the cohesiveness of the mixture when subjected to vibration.

This indicates that the rate of flow of concrete under vibration increases at a much slower rate for the mix with the superplasticizer. Hence the possibility of segregation is delayed to a higher slump for the concrete with the superplasticizer. The vebe time cannot be measured for slumps more than 89 mm (3.5 in.) for plain concrete and 163 mm (6.4 in.) for plain concrete with superplasticizer (Fig. 4). These numbers were consistent with the experimental measurements. If vibration is used for compaction, care should be taken to avoid segregation if the concrete has a slump comparable to or higher than the aforementioned values.

3. The addition of fibres increases the cohesiveness of the fresh concrete. Based on the equation, vebe time could be measured up to a slump of 221 mm (8.7 in.) as compared to 163 mm (6.4 in.) for plain concrete (Fig. 4).

4. The regression line has the lowest slope for the fibre concrete. This indicates that the increase in the flow of concrete under vibration is more gradual for fibre concrete. The difference becomes highly significant at slumps higher than about 100 mm (4 in.).

The inverted cone test was specially developed (12) for the measurement of workability of FRC in the field. Figure 5 gives a correlation curve for vebe time and inverted slump cone time for FRC (8). Since both inverted cone test and vebe test are based on the energy requirements for flowability and compaction, there is a linear correlation between the two tests. This facilitates the transfer of more accurate laboratory test results to field practice.

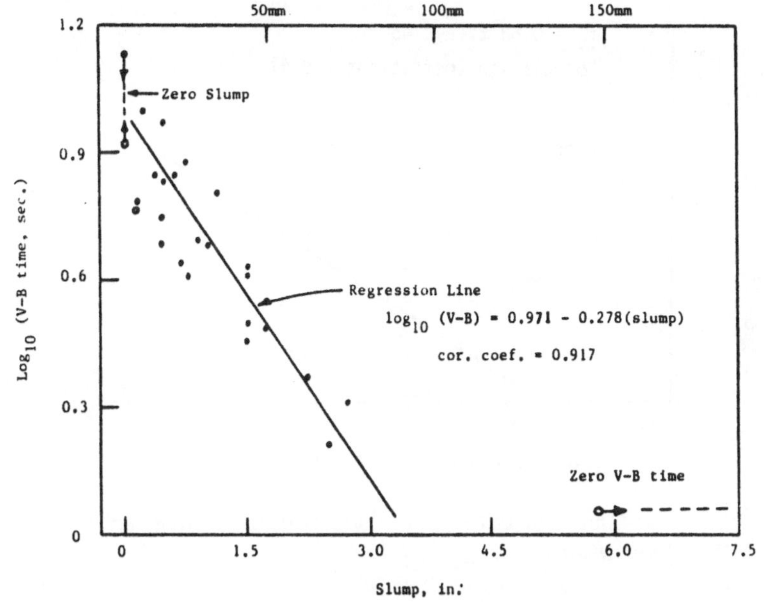

FIG. 3—*Variation of V-B time with respect to slump—plain concrete.*

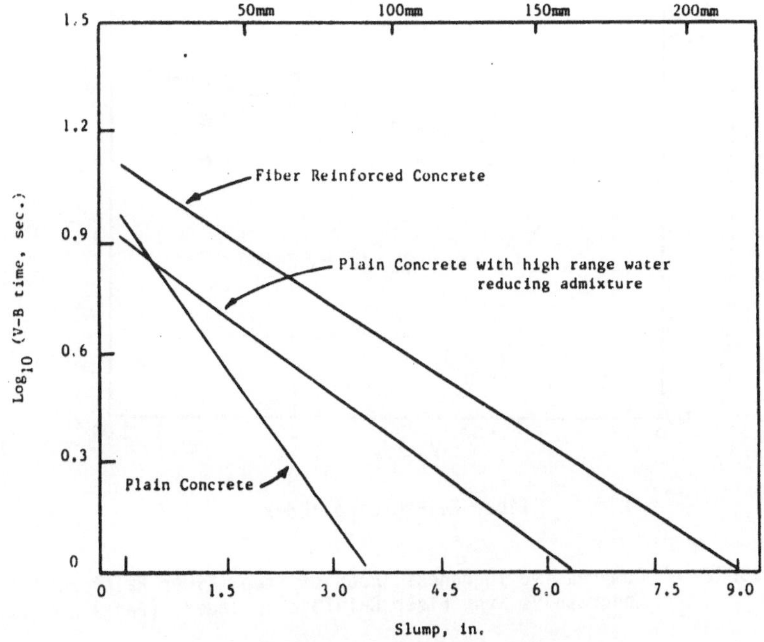

FIG. 4—*Comparison of the three types of concrete.*

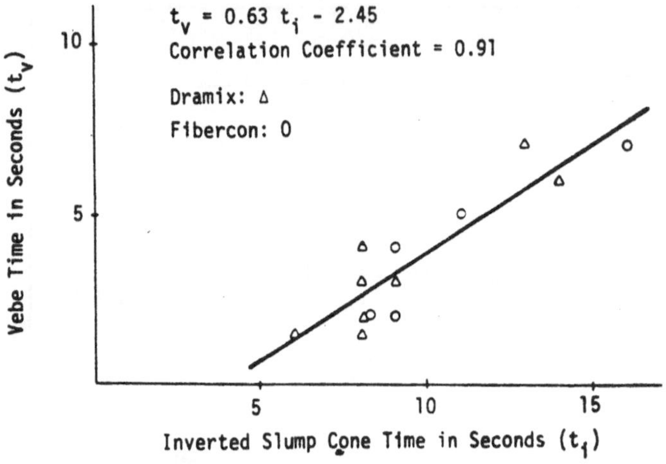

Fig.5 —Correlation curve for Vebe time and inverted slump cone time

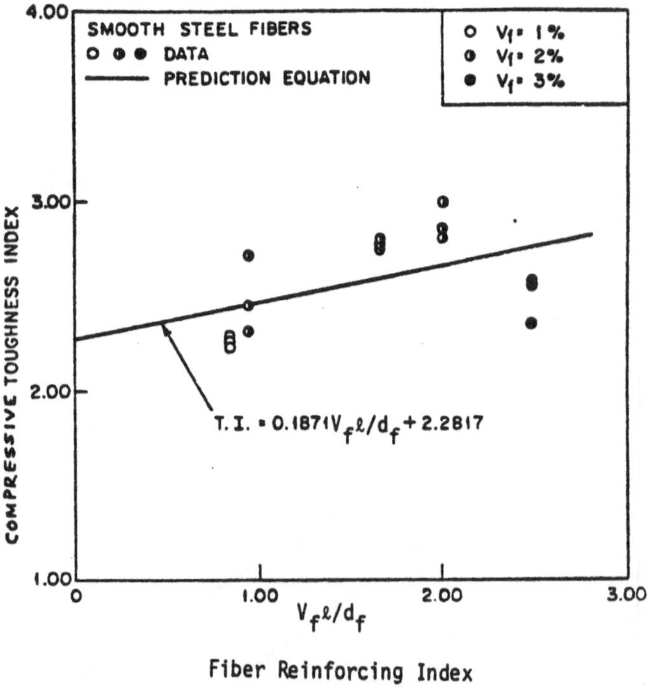

FIGURE 6 Compressive Toughness Index of Steel Fiber Reinforced Concrete vs. the Fiber Reinforcing Index. (Fanella & Naaman)

Properties of Hardened Concrete:

The most significant influence of the incorporation of fibres in concrete is to delay and control the tensile cracking of the composite material. Thus an inherently unstable tensile crack propagation in concrete is transformed into a slow controlled crack growth. The fibres provide a ductile member in a brittle matrix, and the resultant composite has ductile properties which are significantly different from plain concrete. This is the predominent feature of FRC, since it is the dynamic properties, like energy absorption and fracture toughness, that distinguish the material from plain concrete.

All modes of failure are affected by fibres. The strengthening mechanism of the fibres involves transfer of stress from the matrix to the fibre by interfacial shear, or by interlock between the fibre and matrix if the fibre surface is deformed. The fibre and the matrix share the tensile force until the matrix cracks, and then the total force is transferred to the fibres. This change in the mechanism of failure causes significant improvements in the following properties: ductility, toughness, impact resistance, tensile and flexural strengths, fatigue life, abrasion resistance, shrinkage, durability, and cavitation resistance.

Behavior in Compression: The increase in compressive strength due to the addition of fibres is variable ranging from negligible to 20 percent. However, there is a substantial change in the compressive stress-strain response. This change is generally characterized by a noticeable increase in strain at peak load and a significant increase in ductility resulting in substantially higher toughness. This increased toughness is useful in preventing sudden and explosive failure under static loading and in absorbing energy in dynamic loading. Typical increases in the compressive toughness index varies from 200 to 300 percent (Fig. 6). The toughness increases with the volume fraction of fibres and the aspect ratio. The available test data indicate that steel fibres lead to a higher toughness than glass or polypropylene fibres. Because of the mixing difficulties at higher fibre content, causing a harsh mix and increased air entrapment, a deterioration in the stress-strain response is obtained. This is accompanied with a lower material density.

Matrix properties also have influence on the effectiveness of fibres in improving the compressive behavior of concrete. The improvements due to fibre addition are relatively more significant at lower matrix compressive strengths. Addition of silica fume increases the compressive strength and toughness of steel fibre reinforced concrete through increase of the fibre-matrix interfacial bond (4).

Extensive research (2, 4 to 9) has indicated that there is no appreciable change in the linear part of the stress-strain curve in compression when randomly oriented fibres are added to concrete in different volume fractions. The modulus of elasticity values calculated were almost the same for plain and FRC specimens with different types of fibres and for various volume fractions.

Behavior in Flexure:

A significant difference in the behavior of plain and fibre reinforced concrete is found in the flexure test (8). When the fibre concrete beams are loaded in flexure, two stages of behavior have been generally observed in the load-deflection curve (Fig. 7). The behavior is more or less linear up to the first crack and then the curve is significantly non-linear and reaches its peak at the ultimate strength or at the maximum sustainable static load.

Two factors that significantly influence the flexure test are the fibre type and fibre volume. Figure 7 prevents the flexural load-deflection curves for two types of steel (straight and hooked) fibres (8), and Figure 8 shows the comparison of load-deflection curves for different volume fractions of fibres (5).

The hooked fibre proved its capacity as a crack arrester. The cracks are prevented from propagating until the composite ultimate stress is reached. The mode

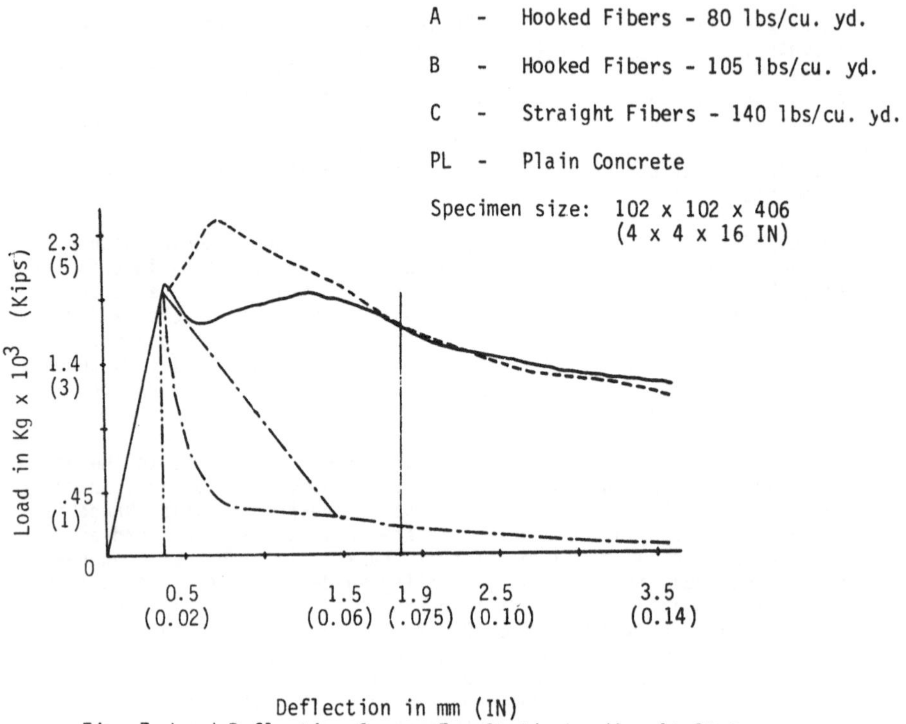

Fig. 7 Load-Deflection Curves For Static Loading At 28 Days

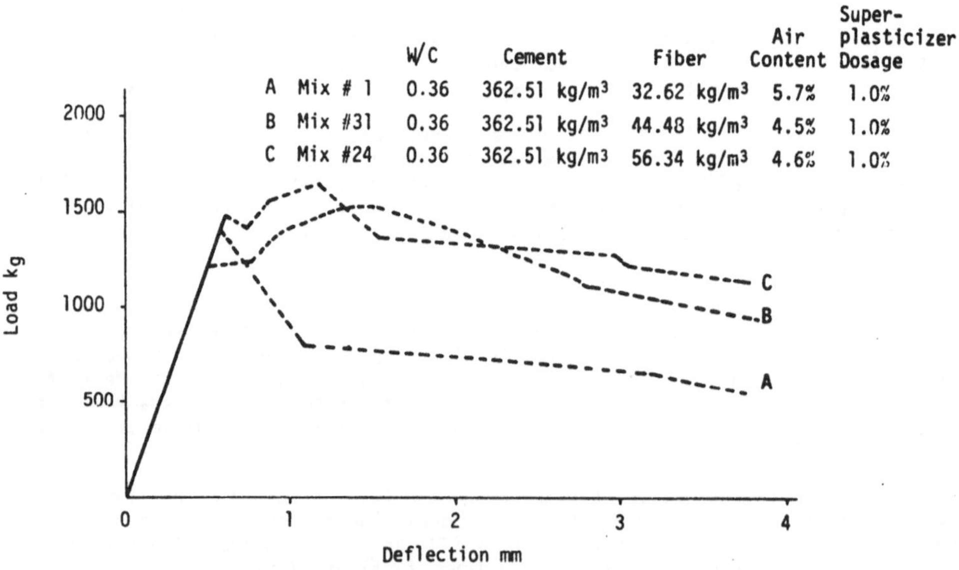

Fig. 8 Comparison of Load Deflection Curves for Mixes with Different Fiber Contents (7-Day)

of failure was a simultaneous yielding of the fibres and the matrix. During the test one could actually hear the popping sound of the fibers failing in tension. It seems obvious that the deformed ends of these fibers contribute significantly to the increase in the bond between fibre and matrix. The significance of good bond can be seen from the load-deflection curves of Figure 7 which are typical curves. These curves indicate a ductile behavior and a large energy absorption. The two fibre concentrations used gave similar strength results. For the specimen reinforced with straight fibres due to the sudden failure, the load dropped from the maximum value at Point A to the load indicated by Point C with a considerable increase in deflection, as shown in Figure 7. Therefore Line ABC is an apparent load-deflection curve whereas the actual curve may be closer to Line ADC as shown in Figure 7. Observation of the failure and an inspection of the failed specimens have shown that the mode of failure for the straight fibre specimens was strictly a bond failure. Compared to the load-deflection curve of the hooked fibre reinforced concrete, the straight fibre reinforced concrete has a lower load-carrying capacity. The load-deflection curves also indicate the advantage of fibrous concrete versus plain concrete in obtaining higher load carrying capacity, higher ultimate strength, and higher toughness.

All types of fibres, when added in sufficient volume fractions, increase the first crack and ultimate flexural strengths of concrete. The effect of steel fibres on the ultimate flexural strength is significant and with its better pull-out performance, it is especially effective at large deformations, and crack widths. Polymeric fibres, having relatively low modulus of elasticity, slightly reduce the initial stiffness and ultimate strength but their better extensibility results in a appreciable post-peak performance and toughness. Glass and carbon fibres have relatively small diameters; therefore for the same volume fraction, there are a large number of fibres with closer spacings. Hence they are more effective in limiting the size of the cracks and arresting the microcracks. Glass and carbon fibres can improve the pre-peak performance of the material and increase the ultimate strength.

Flexural Toughness (Energy Absorption):

Toughness or energy absorption of concrete is increased considerably by the addition of fibers. The toughness index is a measure of the amount of energy required to deflect the 4 in. beam used in the modulus of rupture test, a given amount compared to the energy required to bring the beam to the point of first crack (12). It is calculated as the area under the load-deflection curve up to 0.075 in. (1.9 mm), divided by the area under the load-deflection curve of the fibrous beam up to the first crack strength (proportional limit, defined as first deviation from linear). A flexural toughness index may be also calculated as the ratio of the the toughness of the steel fibre concrete at specified deflection to the toughness up to first crack as shown in ASTM C1018-86 or up to maximum stress or to the toughness of the unreinforced matrix.

The first crack load, the maximum load, and the area under the load-deflection curve upto 1-9 mm (0.075 in.) center point deflection are shown in Fig. 7. All specimens made of plain concrete failed immediately after the first crack and hence the toughness index for these specimens is equal to 1. In general the toughness index for FRC varied greatly depending on the position of the crack, the type aspect ratio, and volume fraction of the fibre and the distribution of fibres. However, the average toughness index for beams with hooked fibres was two to three times greater than for those beams reinforced with straight fibres (8).

Relationship Between 28 Day Compressive and Flexural Strengths:

Based on the experimental results an empirical equation (similar to the one in ACI 3/8/87) had been proposed in the form: $f_r = k(f'_c)^{1/2}$ in which f'_c = 28 day compressive strength, f_r = 28 day flexural strength and k is a constant. Figure 9 shows the influence of fibre content on k. For a one percent probability of underestimation of f_r, the constant k is equal to 0.76 expressed in MPa (9.3 expressed in psi) (22).

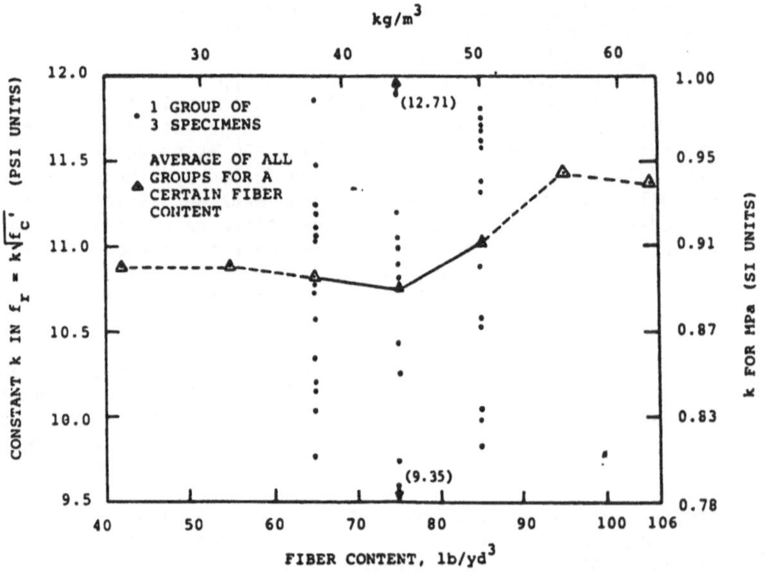

Fig. 9 —Variation of constant k with respect to fiber content

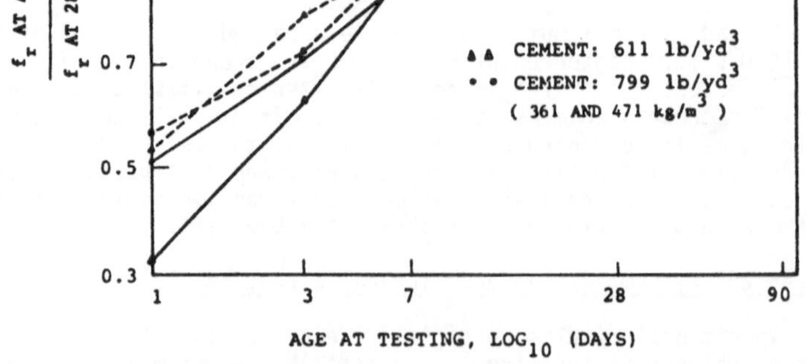

Fig. 10—Comparison of flexural strength gains for fiber reinforced and plain concretes

Time Dependent Variation of Flexural Strength and Toughness Index:

As in the case of reinforced concrete design, the design of FRC is based on 28 day strengths. However, in field applications the properties at both later ages becomes important. For example if FRC is to be used for overlay repair and the road has to open to traffic, say in one day, the engineer has to know the strength and other properties at that age in order to make an informed decision. Similarly one has to know the rate of increase of flexural strength with time and any possible changes in toughness in order to evaluate the long term behavior of e.g. airport pavements.

The results of an experimental investigation (9) are presented in Figures 10 and 11. The following observations are made from these figures (22): 1. the flexural strength varies linearly with the logarithm of time, 2) flexural strength gain with time is slower than compressive strength gain, 3) higher cement content in combination with lower water-cement ratio results in higher early strengths, and 4) the toughness index slightly decreases with an increase in maturity. Additionally it was shown that first crack and ultimate impact resistance increase with maturity (22).

Direct Tension: There is no standard test to determine the stress-strain curve of FRC in direct tension. Various parameters, like the size of the specimen, method of testing, stiffness of the testing machine, gage length, and the unpredictable cracking pattern, will influence curves and the tensile strength. The strength in direct tension is generally the same as that of plain concrete. However, the toughness of FRC is one of the two orders of magnitude higher due to the large frictional energy developed during fibre pullout.

Shrinkage and Creep: A knowledge of shrinkage, creep, and creep recovery characteristics of structural concrete is essential for the analysis and design of concrete structures. The total shrinkage and creep strains measured according to the ASTM procedures over a period of more than a year had shown that these strains were less than those obtained for the generally used structural concrete (9). Similar to plain concrete, the shrinkage and specific creep strains of FRC increased with an increase in the water cement ratio, and the concrete that exhibited higher shrinkage had also higher creep.

A study of steel fibre shotcrete (6) has shown that the addition of fibres reduces the drying shrinkage values and the reduction is proportional to the quantity of fibres added (Fig. 12).

A comparison of creep strains for FRC with two different cement contents is shown in Figure 13.

Thermal Conductivity: A limited study (9) has shown that the addition of randomly distributed steel fibres had no significant effect on the thermal conductivity of superplasticized concrete.

Impact Strength: A simple impact test proposed by ACI Committee 544 (12) consists of a 4.54 Kg (10.0 lbs) hammer dropped onto a steel ball resting on the test specimen which is 64mm (2½ in.) high and 152mm (6 in.) in diameter. The hammer is dropped consecutively and the number of blows required to cause the first crack and ultimate failure are recorded. First crack is defined as the first visible crack. Ultimate failure is is reached when the cracks have opened sufficiently to make the specimen touch three of the four positioning lugs at the base plate. Test results from a comparative study of different fibre types (8) is shown in Figure 14. Fibres with good anchorage (hooked fibres) provide a superior impact resistance to the concrete (500 percent more than straight fibres). The higher energy required to pull the fibres out of the matrix provides the impact strength.

Steel fibre reinforced beams have been subjected to impact loading in instrumented drop-weight and Charpy-type systems by Suaris, Shah, Naaman, and Gopalaratnam. They reported that the total energy absorbed (measured from the load-deflection curves) by SFRC beams can be as much as 40-100 times that for unreinforced beams.

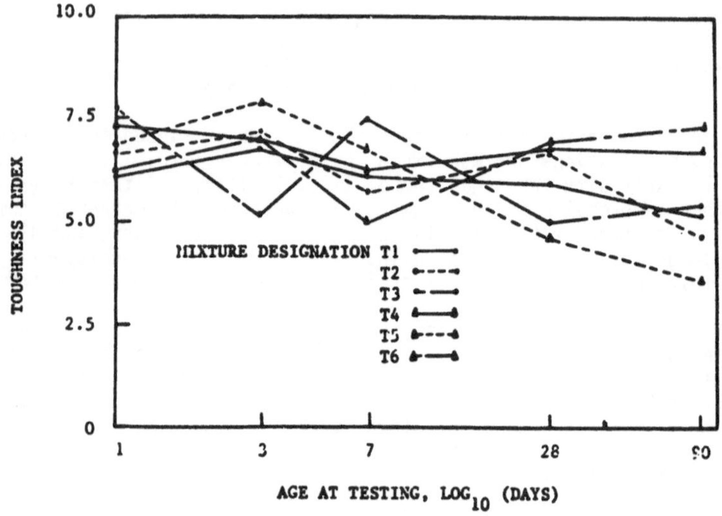

Fig.11—Toughness index versus age at testing

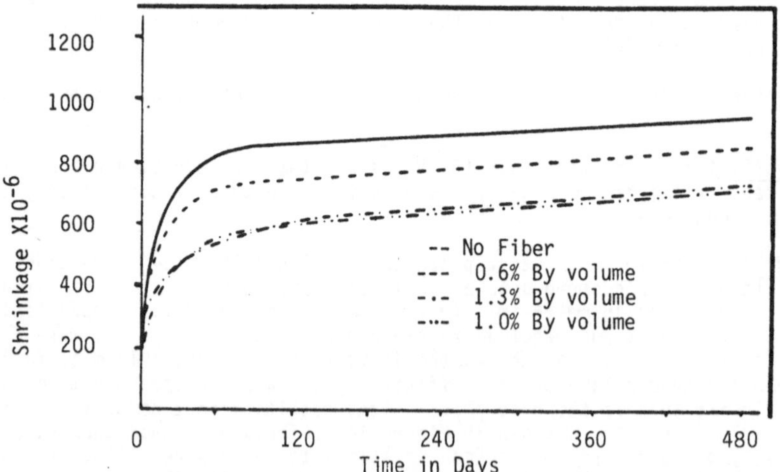

Fig.12 -Shrinkage Curves - 38 mm (1½ in) Beams. Time Reckoned Since End Of Wet Curing At The Age Of 28 Days.

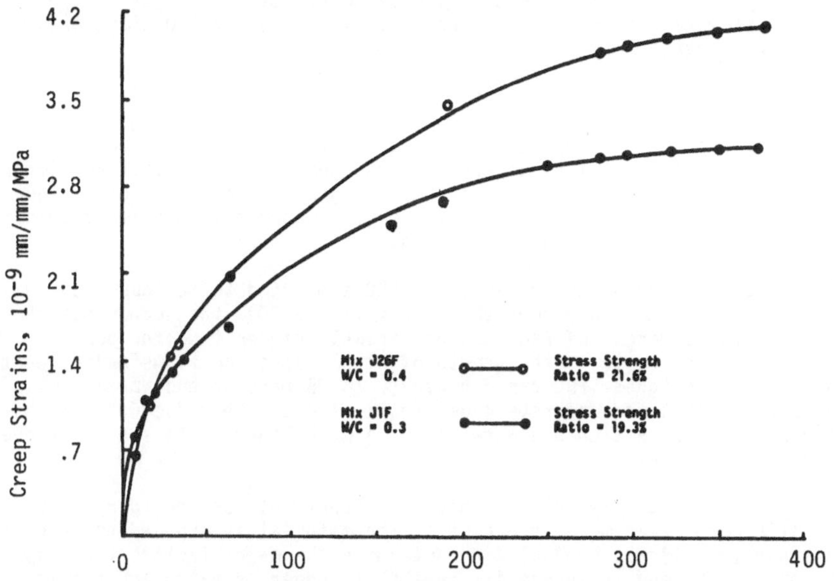

Fig. 13 Comparison Of Creep Strains For Two Different Mixes Of Superplasticized Fiber Reinforced Concrete.

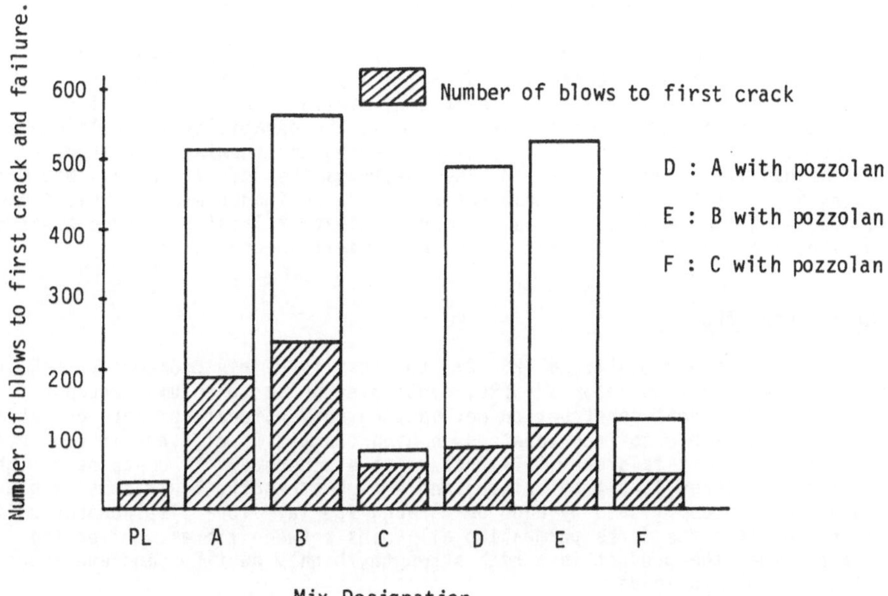

Fig. 14 Impact Resistance (28 Days)

In general the dynamic strength of FRC subjected to explosive charges, dropped weights, and dynamic flexural, tensile and compressive loads is 3 to 10 times greater than that for plain concrete.

Fatigue Strength: Fatigue strength is an important property of FRC because it is the behavior of the material under dynamic loading that clearly distinguishes the material from plain concrete. In many applications, particularly in pavements and bridge deck overlays, the flexural fatigue strength and the endurance limit are important design parameters because these structures are designed on the basis of fatigue load cycles. The greatest advantage in adding fibres to concrete is the improvement in flexural strength in both static and fatigue loading.

An extensive investigation of the behavior of FRC subjected to fatigue loading, using various types of steel and polypropylene fibres (18 to 20), the author has shown that the flexural fatigue strength of FRC is significantly higher than for non-reinforced concrete. Fibres restrain the growth of microcracks and flaws under repetitive stress. A properly designed SFRC can achieve 90 to 95 percent endurance limit. With a higher endurance limit the concrete cross sections could be reduced. Alternatively, using the same cross section would result in a much longer life span or higher load carrying capacity or both.

Freeze-Thaw Resistance: Durability is an essential property of most building materials and components. It becomes even more important if the material is exposed to freezing and thawing. A study (9, 21) was conducted to determine the major factors that may affect the behavior of SFRC and influence its durability under freezing and thawing conditions. Figure 15 presents the variation of relative dynamic modulus expressed in percentage with respect to the number of freeze-thaw cycles. The principle variable is the air content which varied from 1 to 10.8 percent. Figure 16 presents the comparison of plain and FRC for low air content of 2.5 percent. A better performance was shown by FRC in three out of five cases. However the difference is not significant, considering the variation that may be possible in experimental results.

In general it is concluded that, as in the case of plain concrete, air content is the most significant factor. Comparison of FRC and plain concrete indicates that their behavior under freeze-thaw loading is essentially similar. An increase of cement content and a reduction of water-cement ratio improves the durability. The toughness index of FRC does not change appreciably with freezing and thawing. An inspection of the specimens had indicated (9) that in SFRC specimens the severity of visible cracking and spalling as a result of freeze-thaw was reduced when inadequate air-void system was present. Based on the results, it was recommended that at least 8.0 percent air content should be specified for FRRC exposed to severe freeze-thaw conditions.

Highly Reinforced SFRC

In 1982, Lankard had published (26, 27) the results of his pioneering work in the development of highly reinforced SFRC, containing up to 20 volume percent of steel fibres. Using a different construction method than conventional concrete or fibre concrete, it was possible to incorporate such high quantities of fibre in concrete. Designated as SIFCON, it is slurry infiltrated fibre concrete. It contains a high quantity of steel fibres (6-24% by volume) and a high strength, low viscosity slurry. Fibre placement is accomplished by hand or through special fibre displacement units. Slurry is poured into the fibre permeating all voids between fibres. After the slurry has hardened the product is a high strength, highly ductile concrete especially resistant to dynamic loads.

Results of author's tests (19) on SIFCON beams (102 x 102 x 356 mm) for flexural strength, toughness index and flexural fatigue strength are given in Fig. 17 and Table 1.

Six beams were prepared with the same cement/fly ash slurry and their replicate specimens of three compositions identified as follows: two beams (X1 and X2) with

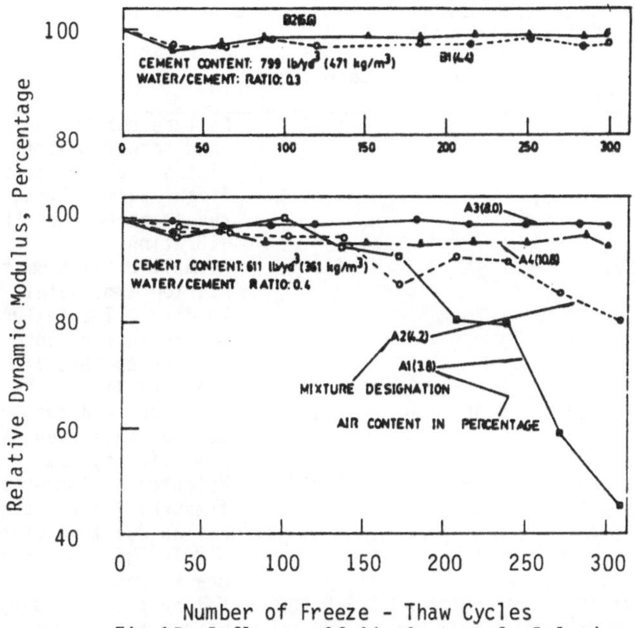

Fig.15 Influence Of Air Content On Relative Dynamic Modulus, Fiber Content 45.5 Kg/m3 (75 lb/yd3)

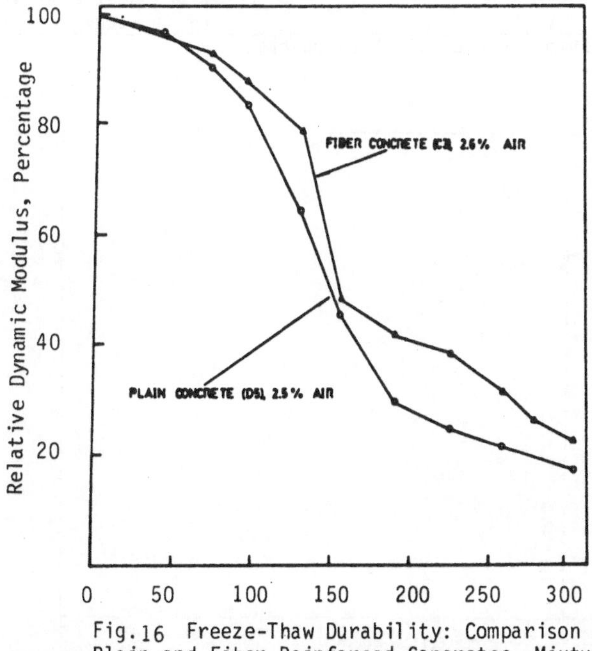

Fig.16 Freeze-Thaw Durability: Comparison Of Plain and Fiber Reinforced Concretes, Mixtures 05-C3

Table 1 - Results of Flexural Fatigue Testing of SIFCON Beams

Beam	Age	Flexural Fatigue Stress Minimum MPa	Flexural Fatigue Stress Maximum MPa	% of Static Flexural Stress	Number of Cycles to Failure	Observations
X1	12	3.43	27.42	80	1400	Failure due to inadequate bond between matrix and fibers. Undamaged clean fibers were falling from the specimen when it was vibrating
X2	77	3.43	25.73	75	4400	A small crack 50mm high started immediately after loading. The failure was due to lack of bond similar to that of specimen X1.
XS2	72	3.43	17.15	50	2,000,000	No cracks and the beam did not fail even after 2 million cycles
XS2*	74	3.43	25.73	75	52,400	Retested at higher maximum flexural fatigue stress. Failed due to a within the middle third of large crack span
XF1	73	3.43	27.42	80	1,175,000	Failed due to a single wide crack at the mid-span
XF2	76	3.43	25.73	75	1,935,800	Failed due to a 10mm wide crack at the middle third of the beam

Beam size 102 X 102 X 356mm. Loaded at third points.

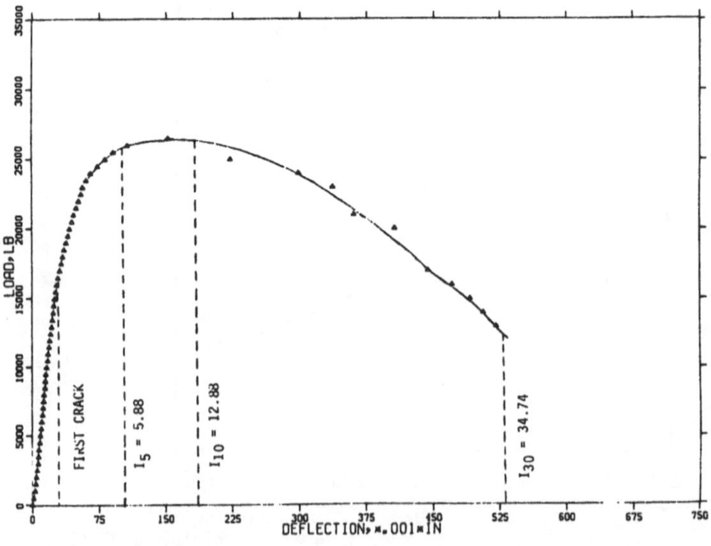

Fig. 17 Load-Deflection Curve for SIFCON Beam - XS1

13.5 volume percent, 50 mm (2 in.) long Xorex 1 steel fibre. Two beams (XS1 and XS2) with 11.0 volume percent, 50 mm (2 in.) long Xorex II steel fibres. Two beams (XF1 and XF2) with 16.0 volume percent of a 50/50 blend of 50 mm (2 in.) long Xorex 1 and 19 mm (3/4 in.) long Flexten slit sheet fiber. All of the beams were cured in water until they were tested.

Beam XS1 was tested in static flexure after 71 days of curing in water. The load-deflection curve, the first crack strength and the toughness index values are shown in Fig. 17. The beam failed at an ultimate flexural stress of 32.8 MPa (4756 psi). There seemed to be adequate bond between the fibres and the slurry matrix. We could hear a few fibres breaking in tension with a popping sound. The beam was very ductile with a high energy absorption capacity and post-crack load carrying capacity. It had a very large deflection of 13 mm (0.525 in.) before failure and it had a very large strain to failure. The toughness index values calculated according to ASTM-C1018 were I_5 = 5.88, I_{10} = 12.80 and I_{30} = 34.74.

Fatigue Strength of SIFCON: Fatigue test results are given in Table 1. All fatigue tests were done in a 45.3 metric ton (50 ton) dynamic capacity testing machine which had different ranges and the load could be adjusted accurate to 0.5 Kg (1 pound) at the load range at which the tests were done. All tests were done at a frequency of 20 cycles per second and the machine was adjusted so that the machine will automatically stop either when the specimen failed or after 2 million cycles of loading. Beams with Xorex I fibres failed at low cycles due to inadequate bond development between the matrix and the fibres. Beams with Xorex II fibres and beams with a blend of Xorex I (corrugated) and Flexten (straight) fibres had high fatigue strengths. The endurance limit for 2 million cycles was 75 percent of the ultimate static flexural strength.

REFERENCES

1. Ramakrishnan, V., "Steel Fiber Reinforced Shotcrete-A State-of-the-Art Report" proceedings, "Steel Fiber Reinforced Concrete", US-Sweden joint seminar (NSF-STU), Stockholm, June 3-5, 1985, pp 7-24. (Published by Elsevier Science Publishing Co., Inc. New York)

2. Ramakrishnan, V., "Superplasticized Fiber Reinforced Concrete for the Rehabilitation of Bridges and Pavements" Transportation Research Record 1003, Transportation Research Board, National Research Council, Washington, D. C., 1985, pp. 4-12.

3. Schrader, E. K., Paxton, James, and Ramakrishnan, V., "Composite Concrete Pavements with Roller Compacted Concrete" Transportation Research Record 1003, Transportation Research Board, National Research Council, Washington, D. C., 1985, pp. 50-56.

4. Ramakrishnan, V., and Srinivasan, V., "Performance Characteristics of Fiber Reinforced Condensed Silica Fume Concretes" ACI Special Publication SP-79, Vol. 11, American Concrete Institute, Detroit, Mich. 1983, pp. 797-812.

5. Ramakrishnan, V., Coyle W., Kopac, Peter, A., and Parko, Thomas J., "Performance Characteristics of Steel Fiber Reinforced Superplasticized Concrete," ACI Special Publication SP-68, American Concrete Institute, Detroit, Michigan 1981, pp. 515-534.

6. Ramakrishnan, V., Coyle, W. V., Dhal, Fowler, Linda, and Schrader, E. K., "A Comparative Evaluation of Fiber Shotcrete", Concrete International: Design and Construction, Vol. 3, No. 1, Jan. 1981, pp 56-69.

7. Ramakrishnan, V., Coyle, W. V., Kulandaisamy, V., and Schrader, "Performance Characteristics of Fiber Reinforced Concrete with Low Fiber Contents", ACI Journal, Proceedings, V. 78, NO. 5, Sept.-Oct., 1981, pp 384-394.

8. Ramakrishnan, V., Brandshaug, T., Coyle, W. V., and Schrader, E. K., "A Comparative Evaluation of Concrete Reinforced With Straight Steel Fibers and Fibers with Deformed Ends Glued Together into Bundles", ACI Journal, Proceedings, V. 77, No. 3, May-June, 1980 pp 135-143.

9. Ramakrishnan, V., and Coyle W. V., "Steel Fiber Reinforced Superplasticized Concrete for Rehabilitation of Bridge Decks and Highway Pavements", Report DOT/RSPA/DMA-50/84-2, Office of University Research, U. S. Department of Transportation, November 1983, p. 410. (Available from the National Technical Information Service, Springfield, Virginia 22161).

10. ACI Committee 506, "State-of-the-Art Rèport on Fiber Reinforced Shotcrete". Concrete International: Design and Construction, V. 6, No. 12, December, 1984, pp 15-27.

11. ACI Committee 544, "State-of-the-Art Report on Fiber Reinforced Concrete", Report ACI 544 1R-82, Concrete International: Design and Construction, May, 1982.

12. ACI Committee 544, "Measurement of Properties of Fiber Reinforced Concrete", ACI 544.2R.78, ACI Manual of Concrete Practice, Part 5, 1982.

13. "Fiber Reinforced Concrete - International Symposium" American Concrete Institute Special publication, SP-81, ACI Detroit, 1984.

14. Ramakrishnan, V., "Performance Characteristics of Steel Fiber Reinforced Superplasticized Concrete", Proceedings, Symposium L. "Advances in Cement-Matrix Composites", Materials Research Society, University Park, Pa., Novermber, 1980.

15. Ramakrishnan, V., "Authors closure, A comparative Evaluation of Concrete Reinforced with Straight Steel Fibers and Fibers with Deformed Ends Glued Together into Bundles", ACI Journal Proceedings, V. 78, No. 2, March-April, 1981, p 151.

16. Ramakrishnan, V., "The Role of Superplasticized Fiber Reinforced Concrete and Fiber Shotcrete in the Rehabilitation of Bridges", Proceedings, V. 1, International Symposium on Rehabilitation of Structures, Bombay, India, December 21-23, 1981, pp 111/21 to 111/28.

17. Ramakrishnan, V., and Srinivasan, V., "Silica Fume in Fiber Reinforced Concrete". The Indian Concrete Journal, December, 1982, pp 326-334.

18. Ramakrishnan, V., "Flexural Fatigue Strength of Dramix Steel Fiber Reinforced Concrete" a Report submitted to Bekaert Steel Wire Corporation, Nile, Illinois, August, 1985.

19. Ramakrishnan, V., "Flexural Fatigue Strength of SIFCON and Concrete Reinforced with Zorex Steel Fibers", A Report submitted to Ribbon Technology Corporation, Ohio, June, 1985.

20. Ramakrishnan, V., "Flexural Fatigue Strength of Polypropylene Fiber Reinforced Concrete", A report submitted to Forta Corporation, Grove City, PA. 16127, July, 1985.

21. Ramakrishnan, V., and Balaguru, P. N., "Freeze-Thaw Durability of Fiber Reinforced Concrete", ACI Journal, Proceedings, V. 83, No. 3, pp 374-481.

22. Balaguru, P., and Ramakrishnan, V., "Mechanical Properties of Superplasticized Fiber Reinforced Concrete Developed for Bridge Decks and Highway Pavements", American Concrete Institute, special publication SP-93, Concrete in Transportation, ACI, Detroit, 1986, pp 563-584.

23. Balaguru, P., and Ramakrishnan, V., "Comparison of Slump Cone and V-B Tests as Measures of Workability for Fiber-Reinforced and Plain Concrete," Cement, Concrete and Aggregates, CCAGDP, vol. 9, No. 1, Summer 1987.

24. Balaguru, P. and Ramakrishnan, V., "Properties of Lightweight Fiber Reinforced Concrete" to be published in American Concrete Institute Symposium. Vol. (SP) in 1987.

25. Balaguru, P., and Ramakrishnan, V., "Properties of Fiber Reinforced Concrete: Workability, Behavior Under Long Term Loading and Air-Void Characteristics" submitted for publication.

26. Lankard, D. R., and Lease, D. H., "Highly Reinforced Precast Monolithic Refractories", Bulletin American Ceramic Society, Volume 61, No. 7, 1982, pp 728 - 732.

27. Lankard, D. R., and Newell, J. K., "Preparation of Highly Reinforced Steel Fiber Reinforced Concrete Composites", ACI publication SP-81, Detroit, Michigan, 1984, pp 286-306.

ACKNOWLEDGEMENTS

The author gratefully acknowledges the financial support given by the United States National Science Foundation under Grant No. MSM-8612082.

Proceedings of the International Symposium on Fibre Reinforced Concrete, December 16-19, 1987, Madras, India

STATE-OF-THE-ART ON
FRACTURE OF CEMENT CONCRETE WITH AND WITHOUT FIBRE REINFORCEMENT

Prakash Desayi, Professor, Civil Engineering Department, Indian Institute of Science, Bangalore, INDIA.
N. Ganesan, Lecturer, Civil Engineering Department, Regional Engineering College, Calicut, INDIA.

SYNOPSIS : Many attempts have been made in the recent years to determine the fracture properties of cementitious materials. In this paper, the various investigations relating to fracture of hardened cement-paste, mortar, concrete and fibre reinforced concrete (with continuous and discontinuous fibres) are briefly reviewed and the current developments taking place in the application of fracture mechanics to such materials are described.

INTRODUCTION

In the recent years many attempts have been made to determine the fracture parameters which characterize the fracture behaviour of cementitious materials. Kaplan [1] was the first to apply Linear Elastic Fracture Mechanics (LEFM) to determine the fracture toughness of concrete. Since then several experimental and theoretical studies have been made to apply the concepts of LEFM and Elastic Plastic Fracture Mechanics (EPFM) to determine the fracture parameters of hardened cement-paste, concerete and fibre reinforced concrete (with continuous and discontinuous fibres). The effect of various factors such as:

i) material features which include water-cement ratio, air content, fine aggregate content, shape, type and volume fraction of coarse aggregate etc.,

ii) geometrical effects which include the size of specimen , pre-crack length etc., and

iii) type of tests like single edge notched beam, double centilever beam, double torsion beam, compact tension tests etc.,

on the fracture of cementitious materials have been studied. The aim of this paper is to review the various developments in the application of fracture mechanics to cementitious materials. Since extensive reviews [2 to 6] on fracture of hardened cement paste, mortar, concrete and fibre reinforced concrete are available, they are discussed briefly and the current developments taking place in the application of fracture mechanics to cementitious materials have been reviewed in detail. The literature is divided into three portions namely (i) hardened cement paste, mortar and concrete, (ii) fibre reinforced concrete with discontinuous fibres and (iii) fibre reinforced concrete with continuous fibres.

HARDENED CEMENT PASTE, MORTAR AND CONCRETE

The first experimental study of the applicability of fracture mechanics to concrete was conducted by Kaplan [1] who carried out bending tests on single edged notched beams of mortar and concrete and calculated the critical strain energy release rate (G_c). He showed that G_c depends on the size of specimen as well as loading technique. Even though Kaplan observed slow stable crack growth prior to fracture instability he did not make any correction for slow crack growth in his clacu- lations. He concluded that "Griffith [7] concept of a critical strain energy release rate being a condition for rapid crack propagation and consequent fracture is applicable to concrete". Subsequently, Glucklich [8] developed the Griffith theory and applied to concrete in greater detail. He was the first to study the dissipative effects occuring in concrete at a crack tip and charged them to micro-cracking.

Romualdi and Batson [9] tested large concrete plates with centrally located slots in tension and found that G_c increases with slot length. They also pointed out that no correction was made for the slow crack growth prior to fracture and hence the computed values of G_c based on original crack length were "somewhat in error". In 1969 Naus and Lott [10] carried out extensive research on portland cement concretes to obtain K_c. The variables included in their study were (i) water-cement ratio, (ii) the air-content, (iii) fine aggregate content, (iv) the curing time and (v) the maximum size of coarse aggregate. From their study they showed that the fracture toughness varies with the above variables.

Based on their tests on 102x102x508 mm notched beams, Welch and Haisman [11, 12] concluded that even though the methods of test and assumptions used in the calculations of G_c and K_c appear to influence their magnitude significantly, the fracture toughness values may however be considered to be material constants for a particular concrete in the same way as ultimate strength, dynamic elastic modulus or even vibrational damping coefficients.

A number of studies [10, 13 to 16] showed that for normal weight concrete, where the aggregate was stronger than the cement matrix, the fracture toughness increased with increasing volume. However, for light weight aggregates which are weaker than the cement matrix, an increase in aggregate volume leads to a decrease in fracture toughness [6, 17, 18, 19]. It was also shown that fracture toughness increased as the maximum size of coarse aggregate increased [10, 16, 19, 20]. However, Peterson [15] did not find any change in the fracture toughness with the increase in maximum size of coarse aggregate. Many investigators found that K_c and G_c increased as the water-cement ratio decreased [6, 19, 21, 22]. However, some investigators [10, 23] found that this was true only for cement paste and mortar, while the fracture toughness of concrete was not sensitive to the water-cement ratio.

A number of investigators found that the fracture toughness parameters K_c and G_c appear to be independent of crack length while some others have produced completely different results. Higgins and Bailey's [22] experiments on beam specimens with variable overall depth yielded constant K_c value for hardened cement paste over a wide range of crack depths. Based on the tests of 38x38x250 mm notched beams and 50x100x350 mm double cantilever beams, Brown [24] and Brown and Pomeroy [21] found that K_c was independent of crack growth for cement paste. Using 105 mm long double torsion specimens, Evans et. al [25] found that fracture mechanics parameters were independent of crack length for crack lengths greater than 20 mm in the case of mortars. Similarly from bending tests on 76x76x292 mm beams Kim et. al [26] found that K_c was independent of the relative notch depth ratio. In 1977, Hillemeier and Hilsdorf [27] conducted an analytical and experimental investigation on hardened cement paste compact-tension specimen. Their results showed that K_c values decreased with increasing crack length, finally approaching a constant value. The constant value was independent of both the initial crack length and the method by which the crack had been prepared. Gjorv et. al [13] tested 50x50x500 mm notched beams in bending and found that K_c for cement paste and light weight concrete were independent of crack length. However, for the case of mortar and concrete, K_c decreased with the increase of crack length. Kesler et. al [28] tested cement plates in tension and concluded that the fracture toughness K_c of cement paste, mortar and concrete decreased with increasing crack length. Hence, they concluded that the concepts of linear elastic fracture mechanics are not directly applicable to cement paste, mortar and concrete. They also proposed a formula for predicting the microcracking zone in front of a crack.

Watson [29] from tests on single edge notched hardened cement paste beams of dimensions 20x20x250 mm, obtained different G_c values for different span/depth ratios. For beams with the same span/depth ratio but with varying width of crack front (beam thickness) from 50.8 to 254 mm, Mindess and Nadeau [30] found that K_c was almost constant. From their studies on 80x340x600 mm concrete plates in tension, both with central cracks and with single edge cracks, as well as 100x100x150 mm notched beams in bending, Mazars [31] found that G_c was essentially independent of both crack length and the type of specimen.

In most of the above studies the linear elastic fracture parameters G_c and K_c were calculated from the maximum load and the initial notch depth; the slow crack growth that might have taken place during the test was not considered. As slow crack growth occurs during testing, these studies underestimated the effective crack length at the peak load. Some investigators tried to account for slow crack growth using compliance measurements. However, the compliance of the specimen with a cast or sawn notch cannot be compared to that of a "naturally" cracked specimen [6]. One of the reasons is that there may be interaction between the two rough surfaces of a fractured crack while there is no such effect in an artificially introduced crack [5].

One more fundamental problem in the fracture toughness testing is to arrive at a valid specimen size. It appears that Walsh [32, 33] was the first to study the problem of specimen size on the observed fracture toughness of concrete. He suggested that the specimen must be large enough so that the assumed zone of disturbance at the crack-tip is small compared to the specimen dimensions. Based on the beam tests, he showed that for concrete a minimum specimen size should be 230 mm, or generally, the requirement can be expressed as

$$(K_c/\sigma_t)^2 > 50 \text{ mm} \tag{1}$$

Where σ_t is the nominal failure stress in flexure. However, Cook and Crookham [34] and Brown [35] pointed out that LEFM could be applied to relatively small concrete beams as long as slow crack growth was accounted for. According to Bazant [36] for true Griffith behaviour, the cross sectional dimensions of the member should be atleast 100 times the maximum aggregate size. Hillerborg et. al [37] proposed a tied crack model for concrete and used finite element analysis to study crack propagation. They reported that for a test to be valid based on LEFM, notched beams should have a depth 10 times a quantity called characteristic length, l_{ch}, where

$$l_{ch} = (K_c/\sigma_t)^2 \tag{2}$$

According to the above equation one needs a depth of beam equal to 2 or 3 meters for concrete. From tests on large double cantilever beams, Sok et. al [38] concluded that for valid fracture tests the concrete specimens must be large enough so that more than 0.5 m of crack propagation can take place. Based on the tests on double cantilever beams, Visalvanich and Naaman [5] showed that fracture toughness K_c of mortar could be predicted analytically provided that the test specimen is large enough to accommodate crack growth larger than 200 mm. Wecharatana and Shah [39] found that for double torsion specimens of mortar unstable crack propagation occured only at a crack length of 305 to 308 mm.

It is well known that concrete exhibits a non-linear load-deflection behaviour because of its heterogeneous composition and formation of microcracks during loading. Hence attempts made to apply LEFM lead to the above contradictions and have not been very successful. Some investigators have attempted to apply the concepts of elastic-plastic fracture mechanics to characterize the fracture behaviour of cementitious materials. Three extensions of linear elastic fracture mechanics into the elastic-plastic region [40] namely (i) J-integral approach (ii) critical crack opening displacement method and (iii) R-curve analysis have been applied to concrete and fibre reinforced concrete.

Mindess et. al [41] and Halvorsen [42] suggested that J-integral appeared to be an useful fracture criterion for concrete. Carrato [43] showed that J-integral was independent of the notch geometry and that in notched beam tests the notch depth became important only when the maximum aggregate size approached the size of the uncracked ligament. Some investigators used R-curve analysis to characterize the fracture behaviour of mortar and concrete. It appears that Brown [24] was the first to apply R-curve concept to mortar beams. He determined K_R curves for mortar and used the compliance technique to find the actual crack depth. Sok et. al [38] used large double cantilever specimens to determine K_R curves for concrete. The arms were prestressed to prevent the crack propagation perpendicular to the notch due to the high bending stresses. Based on the tests on double cantilever beams and double torsion beams, Wecharatana and Shah [44] showed that R-curves for mortar and concrete appear to be independent of test configuration. Recently Wecharatana and Shah [45] proposed a method to calculate fracture resistance of concrete (in terms of R-curves) which incroporates the non-linear effect of process zone. Bazant and Cedolin [46] recently showed that an approximate linear elastic fracture analysis using an equivalent crack length instead of the actual crack length and considering the fracture energy as a function of crack extension (R-curve) fits the existing fracture data (obtained with three point and four point bend specimens) satisfactorily for concrete. A number of investigations on fibre reinforced concrete based on R-curve are also available and are discussed in the next section.

Movenzadeh and Kuguel [14] and Harris et. al [47] have proposed a direct method for determining the fracture characteristics of brittle materials. In this method the area per unit corss-section under stable load-deflection of plot of a three point bend specimen is taken to represent the fracture energy of concrete. Petersson [48] compared both the indirect method (in which G_c is determined using LEFM knowing the maximum load of a three point bend test on a notched beam) and direct method (as explained above) of determining the fracture energy. He concluded that, (i) the indirect method is size dependent and to obtain relevant results, large size specimens (approximately 2 to 3 meters deep) are required and (ii) this method allows much smaller specimens to be used than the indirect method. Peterson [15] also showed that the fracture energy did not seem to be affected by the quality of aggregate. A strong aggregate gave high value of G_c while a weak aggregate gave a low value of G_c. G_c increased when water-cement ratio decreased and the age of concrete increased. Also G_c did not appear to be affected by the maximum particle size.

Recently Mindess [40] studied the effect of specimen size on the fracture energy of concrete and concluded that the values of G_c increased considerably for the largest specimens tested in their study. Nallathambi and Karihaloo [50] and Nallathambi et. al [51] showed that the fracture energy (G_c) increased with increasing maximum coarse aggregate size (which does not support the results reported by Petersson [15]) and depth to span ratio but decreases with increasing pre-crack size. Recently Jenq and Shah [52] proposed a two parameter fracture model for concrete which takes into account the non-linear slow crack growth.

Critical stress intensity factor and crack tip opening displacement were proposed as the two fracture criteria. From the experimental results of notched beams of varying dimensions, they showed that the proposed fracture criteria were size independent. Wittman and Metzener-Gheorghita [53] carried out tests on large specimens with four different geometries and found that failure of large size concrete elements could be predicted realistically on the basis of LEFM. They also suggested that for comparatively small specimens, an approach which takes total fracture energy into consideration is more appropriate.

FIBRE REINFORCED CONCRETE (DISCONTINUOUS FIBRES)

Harris et. al [47] tested three point bend specimens of both glass and steel fibre reinforced concrete to determine the work of fracture (or surface energy) and the critical stress intensity factor, K_C. They found that the work of fracture could be increased by two orders of magnitude by adding as little as 2 per cent by volume steel fibres and K_C increased by a factor 2. Also their investigation showed that the effective volume fraction of randomly distributed fibres is about 55% of the nominal volume fraction of fibres. Brown [54] assessed the resistance to crack propagation in glass fibre reinforced concrete. He found that the pesudo-toughneess (or apparent stress intensity factor) of this material increased linearly with crack growth at a rate proportional to fibre content. Naaman et. al [55] made an attempt to include the effect of fibre debonding in the fracture mechanics approach. They showed that the process of debonding in fibre reinforced concrete develops a pseudo-plastic zone at tip of the crack. From their studies on steel fibre reinforced concrete, Beaumont and Aleszka [56] showed that the fracture energy varies linearly with the volume fraction and aspect ratio of fibres. Based on the tests on notched fibre concrete beams reinforced with varying volume fractions of steel, alkali-resistant glass and poly-propylene fibres, Swamy [57] showed that the slow crack growth cannot be neglected in any characterization of the fracture resistance of fibre composites.

J-integral was first applied to fibre reinforced concrete by Mindess et. al [41]. They determined the critical value of J-integral at the maximum load (J_{CM}) using notched versus unnotched beam technique. From their tests they found that J_{CM} values are greater for fibre reinforced concrete than for plain concrete. Also they showed that J_{CM} was much more sensitive to the fibre content than G_C. Subsequently, Halvorsen [42] carried out experimental studies to determine J_{CM} values for steel fibre reinforced concrete beams. He found that (i) J_{CM} values increase significantly above a fibre content of 0.75% by volume and (ii) J_{CM} is size dependent. In these tests the effect of relative notch depth was also examined and no significant difference was obtained J_{CM} values determined on beams with different relative notch depths. In their studies on steel fibre reinforced concrete beams, Velazco et. al [4] found that the values of J_{CM} increased as the relative notch depth decreased. This indicates that J_{CM} is notch dependent which is contrary to the results of Halvorsen [42]. Velazco et. al [4] also used crack opening displacement approach (COD) to steel fibre reinforced concrete beams. They determined COD_C at fracture load and found that the values are not independent of notch depth. Recently Agarwal et.al [58] investigated the fracture behaviour of fibre reinforced epoxy resin using single edge notched (SEN) specimens. They evaluated J-integral using energy rate interpretation and found that the value was independent of crack length when the ratio of crack length to specimen width (a/w) is larger than 0.35. Also they found that critical value of J-integral agreed well with the critical stress intensity factor obtained using R-curve approach in their earlier study.

From the fracture tests on cementitious materials, it has been observed by many investigators that the fracture of such materials is accompained by extensive slow stable crack growth prior to unstable crack propagation. Hence, in order to account for slow crack growth attempts have been made to apply the concept of R-curve approach. Mai [59] studied the fracture behaviour of asbestos cement composite using both three point notched bend and double cantilever beam (DCB) specimens. It is found that crack initiation toughness (K_i) was independent of the geometry of the two specimens used. However, opposite trends were observed for the total fracture energy versus crack length curves. The fracture energy decreased with crack length for notched bend beams but increased for DCB specimens. They used quasi-static fracture energy method to evaluate fracture energy of DCB specimens. Lenain and Bunsell [60] found that crack resistance of sheet abestos cement could be characterized by R-curves. They also proposed a model to determine the stress intensity factor required for crack propagation in the fibre reinforced concrete. Also they investigated the effect of initial notch on R-curves from tests conducted on a series of DCB specimens and found that R-curves are not identical. K_R curves for an asbestos cellulose cement composite have been determined by Mai et. al [61]. They studied the influence of beam depth on the K_R-curves and used beam depths between 25 mm and 200 mm. They found that K_R-curve is a useful approach to describe the fracture behaviour of asbestos cements and it appears to be independent of both size and starter notch depth. However, they suggested that a minimum size of the notched bend beams (depth 200 mm and relative notch depth 0.3) is required to obtain complete K_R curve. Velazco et. al [4] experimentally determined K_R curves for different relative notch depths and fibre volume for steel fibre reinforced concrete beams. Recently Visalvanich and Naaman [62] and Wecharatana and Shah [63] proposed models for predicting fracture resistance of fibre reinforced concrete and compared their theoretical values with the experimental data of double cantilever and notched beam tests. Wecharatana and Shah [44] proposed a method in which the concept of strain energy release rate based on LEFM was modified in order to account for both the elastic and inelastic energy absorbed during crack growth. Using the test results obtained from the Double Cantilevel Beam (DCB) and Double Torsion Beam (DTB) specimens, R-curves were plotted for mortar, concrete and fibre reinforced concrete. While R-curves were independent of testing configuration for both mortar and concrete, the R-curves for fibre reinforced concrete obtained using DCB specimens were higher than those for DTB specimens. For all the DCB specimens, the intial notch depth was the same and the effect of initial notch depth was not investigated.

Using compact tension specimens Bentur et. al [64] studied the crack patterns developed in specimens reinforced with parallel and randomly oriented steel fibres. They found that (i) in the case of specimens with parallel fibres, the cracks propagating from the precast notch did not run in a strai ght path across the fibres (which was assumed by many models); in most instances the cracks seemed to be arrested in the matrix ahead of the fibre and turned 90 degrees on a new course parallel to the fibre and continued to propagate across the fibre. This resulted in a 'pseudo-debonding' event since the separation was entirely within the matrix and the debonding effect was not actually felt at the interface of the steel, and (ii) in the case of specimens reinforced with inclined fibres, the cracks tended to by pass such fibres by propagating parallel to them to the end of the fibre and then continued again in the forward direction.

FIBRE REINFORCED CONCRETE (CONTINUOUS FIBRES)

In all the studies, described in the previous Sections, the materials considered were hardened cement paste, mortar, concrete and fibre reinforced concrete having discontinuous fibres. Only a few investigation have been come across on the application of fracture mechanics to material reinforced with continuous fibres. Romualdi and Bstson [9, 65] appear to be the first to apply fracture mechanics to concrete beams reinforced with continuous fibres. The continuous steel wires were aligned on the tension side of the beam and they found that the fracture arrest could be achieved by reducing the spacing of reinforcement to a suitable scale. Also from theoretical studies they showed that the tensile cracking strength increased proportionately to the inverse square root of the spacing of reinforcement. Weiss [66] carried out theoretical studies on the crack development in concrete with closely spaced reinforcement and developed a series of formulae to predict the effect of reinforcement on the formation and development of cracks. Bowling and Groves [67] carried out experimental studies on rectangular double cantilever and single edge notched bend specimens. The material considered was epoxy resin reinforced with nickel wires. From their studies they showed that the specimen geometry had an important effect on the fracture behaviour of composite. Also they proposed a model to predict the stress intensity factor required for crack propagation in the fibre reinforced composite taking into consideration the effect of bridging of fibres. Bazant and Cedolin [68] used a 'smeared' crack to model cracking in concrete. They showed that the smeared representation of cracking and an adequate representation of bond-slip phenomenon consititutes an efficient and realistic method for taking into account cracking in finite element analysis of reinforced concrete. Ingraffea et. al [69] proposed a model for bond-slip between steel reinforcing bar and the surrounding concrete using non-linear fracture mechanics and finite element method. Recently Desayi and Ganesan [70, 71] studied the fracture behaviour of ferrocement using Notched Beam and Double Cantilever Beam specimens. From the tests on notched beam specimens it has been found that the apparent J-integral appears to be a useful fracture criterion for ferrocement. Using the test results of DCB specimens, crack growth resistance curves (R-curves) were obtained using different methods available in literature and compared. From these studies it has been found that (i) R-curves obtained taking into account the effect of permanent deformation were dependent on the initial notch depth, (ii) the R-curves were influenced by the percentage of mesh reinforcement and (iii) the values of strain energy release rate in R-curve increased with the increase of percentage of mesh reinforcement.

CONCLUSION

Research on the application of the concepts of fracture mechanics to the fracture, failure and cracking of comentitious materials with and without fibre-reinforcement is being actively pursued by a number of investigators. These studies have thrown considerable light on the qualitative nature of the fracture behaviour and the various parameters affecting the same. It is however, felt that a lot more research is required towards directing the existing knowledge as well as conducting additional investigation in developing the fracture mechanics approach for use in the designs of cementitious structural elements.

NOTATIONS

COD	-	Crack opening displacement
G_c	-	Critical strain energy release rate
J_{CM}	-	Critical value of J-integral at the maximum load
K_c	-	Critical stress intensity factor
K_i	-	Crack initiation toughness
l_{ch}	-	Characteristic length
σ_t	-	Nominal failure stress in flexure.

BIBLIOGRAPHY

1. Kaplan, M.F., "Crack propagation and the fracture of concrete", Journal of American Concrete Institute, Proc. Vol. 58, No.5, November 1961, pp.591-610.
2. Carpinteri, A., "Application of fracture mechanics to concrete structures", Journal of the structural division, Proceedings of the American Society of Civil Engineers, Vol. 108, No. ST4, April 1982, pp.833-848.
3. Swamy, R.N. "Fracture mechanics applied to concrete", Development in Concrete Technology-1, F.D. Lydon, Ed. Applied Science Publishers Ltd., Barking, 1979, pp.221-281.
4. Velazco G., Visalvanich, K., Shah, S.P. and Naaman A.E., "Fracture behaviour analysis of fibre reinforced concrete beams", Progress Report, Grant ENG 77-23534, National Science Foundation, Washington, DC, March 1979, 115p.
5. Visalvanich, K. and Naaman, A.E., "Fracture behaviour and Analysis of fiber-reinforced concrete using double cantilever beam technique", Progress Report for National Science Foundation ENG 77-23534, May 1985, 173 pp.

6. Mindess, S. "The application of fracture mechanics to cement and concrete: a historical review", prepared for RILEM Technical Committee 50-FMC.

7. Griffith, A.A., "The phenomena of rupture and flow in solids", Philosophical Transactions, Royal Society of London, Series A, V.221, 1920, pp.163-198.

8. Glucklich J. "Fracture of plain concrete", Journal of Engineering Mechanics Division, Proc., American Society of Civil Engineers, Vol. 89, (EM6) December 1963, pp.127-138.

9. Romualdi, J.P. and Batson, G.B., "Mechanics of crack arrest in concrete", Journal of Engineering Mechanics Division, Proceedings of the American Society of Civil Engineers, Vol. 89, EM3, June 1963, pp.147-168.

10. Naus, D.J. and Lott, J.L. "Fracture toughness of portland cement concrete", Journal of American Concrete Institute, Vol. 66, June 1969, pp.481-489.

11. Welch, G.B. and Haisman, B., "Fracture toughness measurements of concrete", Uniciv Report, No. R.42, University of New South Wales, Kensington, N.S.W. Australia, January 1969, 17pp.

12. Welch, G.B. and Haisman, B. "The application of fracture mechanics to concrete and the measurement of fracture toughness", Materials and Structures (RILEM), Vol. 2, No. 9, May-June 1969, pp.171-177.

13. Gjorv, O.E., Sorensen, S.I. and Arnesen, A., "Notch sensitivity and fracture toughness of concrete" Cement and Concrete Research, Vol. 7, No. 3, May 1977, pp.333-344.

14. Moavenzadeh, F. and Kuguel, R. "Fracture of concrete", Journal of Materials, American Society for Testing and Materials, Vol. 4, No. 3, September 1969, pp.497-519.

15. Petersson, P.E., "Fracture energy of concrete: practical performance and experimental results", Cement and Concrete Research, Vol. 10, No. 1, January 1980, pp.91-101.

16. Shah, S.P. and McGarry, F.J., "Griffith fracture criterion and concrete", Journal of Engineering Mechanics Division, Proceedings of the American Society of Civil Engineers, Vol. 97, No. EM6, December 1971, pp.1663-1676.

17. Koyanagi, W. and Sakai, K. "Observations on the crack propagation process of mortar and concrete", Review of the Twenty Fifth General Meeting, The Cement Association of Japan, TOKYO, 1971, pp.153-157.

18. Okada, K. and Koyanagi, W. "Effect of aggregate on the fracture process of concrete", in Mechanical Behaviour of Materials, Vol.IV, Proceedings of the International Conference on Mechanical Behaviour of Materials, Kyoto, 1971, The Society of Materials Science, Japan, 1972, pp.72-83.

19. Togawa, K., Sajoh, T. and Araki, K. "Parameters on the fracture toughness of mortar and concrete", Review of the Twenty Seventh General Meeting, The Cement Association of Japan, TOKYO, 1973, pp.117-120.

20. Strange, P.C. and Bryant, A.H. "Experimental Tests on Concrete Fracture", Journal of the Engineering Mechanics Division, Proceedings of the American Society of Civil Engineers, Vol. 105, No. EM2, April 1979, pp.337-342.

21. Brown, J.H. and Pomeroy, C.D. "Fracture toughness of cement paste and mortars", Cement and Concrete Research, Vol. 3, 1973, pp.475-480.

22. Higgins, D.D. and Bailey, J.E., "Fracture measurements on cement paste", Journal of Materials Science, Vol. 11, No. 11, November 1976, pp.1995-2003.

23. Lott, J. and Kesler, C.E., "Crack propagation in plain concrete", Symposium on Structure of Portland Cement Paste and Concrete, Special Report 90, Highway Research Board, Washington, D.C., 1966, pp.204-218.

24. Brown, J.H., "Measuring the fracture toughness of cement paste and mortar", Magazine of Concrete Research, Vol. 24, No. 81, December 1972, pp. 185-196.

25. Evans, A.G., Clifton, J.R. and Anderson, E. "The fracture mechanics of mortar", Cement and Concrete Research, Vol. 6, No. 4, 1976, pp.535-548.

26. Kim, M.M., Ko, H.-Y. and Gerstle, K.H. "Determination of fracture toughness of concrete", Fracture in Concrete, Eds. W.F. Chen and C.E. Ting, American Society of Civil Engineers, New York, 1980, pp.1-14.

27. Hillemeier, B. and Hilsdorf, H.K. "Fracture mechanics studies on concrete compounds", Cement and Concrete Research, Vol. 7, No. 5, September 1977, pp.523-536.

28. Kesler, C.E. Naus, D.J. and Lott, J.L., "Fracture mechanics-its applicability to concrete," Proceedings of the International Conference on Mechanical Behaviour of Materials, Vol. 4, Japan, 1971, The Soceity of Materials Science, Japan, pp.113-124.

29. Watson, K.L., "The estimation of fracture surface energy as a measure of the toughness of hardened cement paste", Cement and Concrete Research, Vol. 8, No. 5, September 1978, pp.651-656.

30. Mindess, S. and Nadeau, J.S., "Effect of notch width on K_{IC} for mortar and concrete", Cement and Concrete Research, Vol. 6, No. 4, 1976, pp.529-534.

31. Mazars, J. "Existence of a critical strain energy release rate for concrete", in Fracture 1977, Vol. 3, ICF4, Waterloo, pp.1205-1209.

32. Walsh, P.F. "Fracture in plain concrete", The Indian Concrete Journal, Vol. 46, No. 11, November 1972, pp.469-470, 476.

33. Walsh, P.F. "Crack initiation in plain concrete", Magazine of Concrete Research, Vol. 28, No. 94, March 1976, pp.37-41.

34. Cook, D.J. and Crookham, G. "A discussion of the paper 'Notch sensitivity and fracture toughness of concrete' by O.E. Gjrov, S.I. Sorensen and A. Arnesen", Cement and Concrete Research, Vol. 8, No. 3, 1978, pp.387-388.

35. Brown, J.H. "Reply to P.F. Walsh in Discussion of measuring the fracture toughness of cement paste and mortar, by J.H. Brown", Magazine of Concrete Research, Vol. 25, No. 85, 1973, pp.220-221.

36. Bazant, Z.P., "Material behaviour under various types of loading", High Strength Concrete Ed. S.P. Shah, Proceedings of a Workshop held at University of Illinois at Chicago Circle, Chicago, 1979, pp.79-92.

37. Hillerborg, A. Modeer, M. and Petersson, P.E., Analysis of crack formation and crack growth in concrete by means of fracture mechanics and finite elements", Cement and Concrete Research, Vol. 6, No. 6, 1976, pp.773-782.

38. Sok, C., Baron, J. and Francois, D. "Mechanique de la rupture applique au beton hydraulique", Cement and Concrete Research, Vol. 9, No. 5, 1979, pp.641-648.

39. Wecharatana, M. and Shah, S.P. "Double torsion tests for studying slow crack growth of portland cement mortar", Cement and Concrete Research, Vol. 10, No. 6, November 1980, pp.833-844.

40. Mindess, S. "The fracture of fibre reinforced and polymer impregnated concretes", The International Journal of Cement Composites, Vol. 2, No. 1, February, 1980, pp.3-11.

41. Mindess, S., Lawrence, F.V. and Kesler, C.E. "The J-integral as a fracture criterion for fibre reinforced concrete", Cement and Concrete Research, Vol. 7, No. 6, November 1977, pp.731-742.

42. Halvorsen, G.T. "J-integral study of steel fibre reinforced concrete", The International Journal of Cement Composites, Vol. 2, No. 1, February 1980, pp.13-22.

43. Carrato, J.L. "Experimental evaluation of the J-integral", M.S. Thesis, University of Illinois at Urbana-Champaign, 1980.

44. Wecharatana, M. and Shah, S.P., "Slow crack growth in cement composities", Journal of Structural Division, Proceedings of the American Society of Civil Engineers, Vol. 108, No. ST6, June 1982, pp.1400-1413.

45. Wecharatana, M. and Shah, S.P., "Prediction of non-linear fracture process zone in concrete", Journal of Engineering Mechanics, Proceedings of the American Society of Civil Engineers, Vol. 109, No. 5, October 1983, pp.1231-1246.

46. Bazant, Z.P. and Cedolin, L., "Approximate linear analysis of concrete fracture by R-curves", Journal of Structural Engineering, American Society of Civil Engineers, Vol. 10, No. 6, June 1984, pp.1336-1355.

47. Harris, B., Varlow, J. and Ellis, C.D., "The fracture behaviour of fibre reinforced concrete", Cement and Concrete Research, Vol. 2, 1972, pp.447-461.

48. Petersson, P.E., "Fracture energy of concrete: method of determination", Cement and Concrete Research, Vol. 10, No. 1, January 1980, pp.79-89.

49. Mindess, S., "The effect of specimen size on the fracture energy of concrete", Cement and Concrete Research, Vol. 14, No. 3, May 1984, pp.431-436.

50. Nallathambi, P. and Karihaloo, B.L. "Size dependent effects of fracture toughness of concrete", Sixth International Conference on Fracture, New Delhi, India, Vol. 1, December 4-10, 1984, pp.2785-2794.

51. Nallathambi, P., Karihaloo, B.L. and Heaton, B.S. "Various size effects in fracture of concrete", Cement and Concrete Research, Vol. 15, No. 1, January 1985, pp.117-126.

52. Jenq, Y. and Shah, S.P. "Two parameter fracture model for concrete", Journal of Engineering Mechanics, American Society of Civil Engineers, Vol. 11, No. 10, October 1985, pp. 1227-1241.

53. Wittmann, F.H. and Metzener-Gheorghita, "Fracture toughness of concrete-determination on large specimens", RILEM, Materials and Structures, March-April 1985, No. 104, pp.93-95.

54. Brown, J.H. "The failure of glass fibre-reinforced notched beams in flexure", Cement and Concrete Research, Vol. 25, No. 82, March 1973, pp.31-38.

55. Naaman, A.E., Argon, A.S. and Moavenzadeh, F. "A fracture model of fiber reinforced cementitious materials", Cement and Concrete Research, Vol. 3, No. 4, 1973, pp.397-411.

56. Beaumont, P.W.R. and Aleszka, J.L. "Cracking and toughning of concrete and polymer-concrete dispersed with short steel wieres" Journal of Materials Science, Vol. 13, No. 8, August 1978, pp.1749-1760.

57. Swamy, R.N. "Influence of slow crack growth on the fracture resistance of fibre cement composites", The International Journal of Cement Composites, Vol. 2, No.1, February 1980, pp.43-53.

58. Agarwal, B.D., Patro, B.S. and Prashant Kumar, "J-integral as a fracture criterion for short fibre composites: an experimental approach", Engineering Fracture Mechanics, Vol. 19, No. 4, 1984, pp.675-684.

59. Mai, Y.W., "Strength and fracture properties of asbestos-cement mortar composites", Journal of Materials Science, Vol. 14, No. 9, September 1979, pp.2091-2102.

60. Lenain, J.C. and Bunsell, A.R. "The resistance to crack growth of asbestos cement", Journal of Materials Science, Vol. 14, No. 2, February 1979, pp.321-332.

61. Mai, Y.W., Foote, R.M.L. and Cotterell, B. "Size effects and scaling laws of fracture in asbestos cement", The International Journal of Cement Composites, Vol. 2, No. 1, February 1980, pp.23-34.

62. Visalvanich, K. and Naaman, A.E., "Fracture model for fibre reinforced concrete", Journal of the American Concrete Institute, Vol. 80, No. 2,

March-April 1983, pp.128-138.

63. Wecharatana, M. and Shah, S.P. "A model for predicting fracture resistance of fiber reinforced concrete", Cement and Concrete Research, Vol. 13, No. 6, November 1983, pp.819-829.

64. Bentur, A., Diamond, s. and Mindess, S. "Cracking process in steel fibre reinforced cement paste", Cement and Concrete Research, Vol. 15, No. 2, March 1985, pp.331-342.

65. Romualdi, J.P. and Batson, G.B., "Behaviour of reinforced concrete beams with closely spaced reinforcement", Journal of the American Concrete Institute, Vol. 60, No. 6, June 1963, pp.775-789.

66. Weiss, V. "Crack development in concrete with closely spaced reinforcement and in similar materials", Cement and Concrete Research, Vol. 3, No. 2, March 1973, pp.189-205.

67. Bowling, J. and Groves, G.W. "The propagation of cracks in composites consisting of ductile wires in a brittle matrix", Journal of Materials Science, Vol. 14, No. 2, February 1979, pp.443-449.

68. Bazant, S.P. and Cedolin, L. "Fracture mechanics of reinforced concrete", Fracture in Concrete, Eds. W.F. Chen and E.C. Ting, American Society of Civil Engineers, New York, 1980, pp.28-35.

69. Ingraffea, A.R., Gerstle, W.H., Gergely, P and Saouma, V. "Fracture mechanics of bond in reinforced concrete", Journal of Structural Engineering, American Society of Civil Engineers, Vol. 110, No. 4, April 1984, pp.871-890.

70. Desayi, P. and Ganesan, N. "Fracture behaviour of ferrocement beams", Journal of Structural Engineering, American Society of Civil Engineers, Vol. 112, No. 7, July 1986, pp.1509-1525.

71. Desayi, P. and Ganesan, N. "Fracture properties of ferrocement using double cantilever beam specimens", The International Journal of Cement Composites and Light Weight Concrete (UK), Vol. 8, No. 2, May 1986, pp.121-132.

Proceedings of the International Symposium on Fibre Reinforced Concrete, December 16-19, 1987, Madras, India

EXPERIMENTAL STUDY OF SFRC UNDER COMPRESSION, INDIRECT TENSION & PURE BENDING

T.ABDUL HAFEEZ KHAN, JNT UNIVERSITY ,HYDERABAD, VISITING PROFESSOR, UNIVERSITY OF CONSTANTINE, CONSTANTINE, ALGERIA.
SAMAI MOHAMED LAID, UNIVERSITY OF CONSTANTINE, CONSTANTINE, ALGERIA.
RAMAKRISHNA, B., JNT UNIVERSITY, HYDERABAD, INDIA.

SYNOPSIS: This investigation provides a behavioral study of SFRC under compression , indirect tension and pure flexure. The fibre variables adopted were length, diameter and volume. Some test data was viewed in the light of the Composite Material Theory and the Spacing Theory. The spacing varied through the volume fraction of fibres was seen to have an effect over the tensile strength of SFRC while the spacing varied through fibre diameter was seen to have little effect. Test data was also perused in in view of some established relationships between different strengths of SFRC like the cube strength, split cylinder strength and modulus of rupture.

INTRODUCTION

The high technology composites based on steel are finding wider and wider use in Engineering applications. Hence the need to understand their behavior better . For example, two different theories have been propulgated to explain the tensile strength of SFRC like the Composite Material Theory and the Fibre Spacing Theory. There is need to establish and reestablish the validity of these theories and infact try to understand one in the light of the other.

The present investigation is aimed at the behavioral study of SFRC under compression, indirect tension and pure bending. The test results were viewed in the light of the two theories.The test data was also utilised to have a verification on some of the established relationships between different strengths of SFRC .

EXPERIMENTAL PROGRAMME

The behaviour of SFRC in compression was examined by testing a series of cubes. For the behaviour of SFRC in tension,series of tests were carried on cylinders. The split cylinder test was preferred over the direct tension test for the following reasons. The direct test is difficult to perform because of of stress concentration caused by machine stiffness, specimen alignment, size and shape of specimen and fibre orientation. A series of beams were tested to determine the behaviour of SFRC in pure bending.

A mix proportion of 1.0 part of ordinary Portland cement (conforming to Indian standard specifications), 2.4 parts of sand and 4.0 parts of crushed granite rock and Water cement ratio of 0.55 were adopted. The concrete used for the fabrication of test specimens was impregnated with steel fibres of varying lengths, diameters and volume fractions. The mixing of concrete was done in a rotary type mixer, care being taken while mixing to ensure uniform distribution of fibres in the entire mass of concrete. Plain concrete specimens were cast to serve as reference specimens. An Avery 50 tonnes Universal Testing machine was used for carrying out the tests.

EFFECT OF FIBRE VOLUME

(a) On Tensile Strength:

The tensile strength measurements were made on SFRC members containing 0.5, 0.75, 1.0 percent by volume of 0.195 cm, 0.125 cm and 0.091 cm diameter fibres and of lengths 4,5 and 6 cms. The split cylinder test was adopted to determine the indirect tensile strength of SFRC. The tensile strength was found to be a linear function of volume fraction of fibres.,while the other parameters were kept constant. The addition of fibres amply contribute to the tensile strength of SFRC. There is however a limit upto which volume fraction of fibres can be increased . At higher percentages balling effect will occur resulting in non-uniform distribution of fibres and resultant reduction in strength. The use of dispersers would obviate this difficulty. Earlier investigation has shown that by increasing the sand content higher volume fraction of fibres can be incorporated thus resulting in mixes having improved strength. Fig. 1 represents graphically the relationship between the relative tensile strength and volume percent of fibres.

(b) On Flexural Strength :

The flexural measurements were made on beams of size 140x10x15 cm. Steel wires of 20 gauge (0.09 cm dia.) in lengths of 3.81 cm , 5.01 cm and 6.35 cm were used for reinforcement. Volume percents of 0.5, 1.0 and 1.5 were used. The modulus of rupture of SFRC was seen to be a linear function of volume percent of fibres.

(c) On Compressive Strength :

The compressive strength measurements were made on standard SFRC cubes. The volume percents of 0.5 , 1.0 and 1.5 were adopted. Fibres of 0.09 cm dia. and of lengths 3.81 cm, 5.01 cm and 6.35 cm were used. In general the improvement in compressive strength due to fibre addition is small. The compressive strength of SFRC was seen to be a linear function of volume percent of fibres.

EFFECT OF ASPECT RATIO, lp/d RATIO

(a) Aspect Ratio: The volume fraction of fibres was kept constant and aspect ratio was varied through the length and diameter. It was seen that the increase of aspect ratio

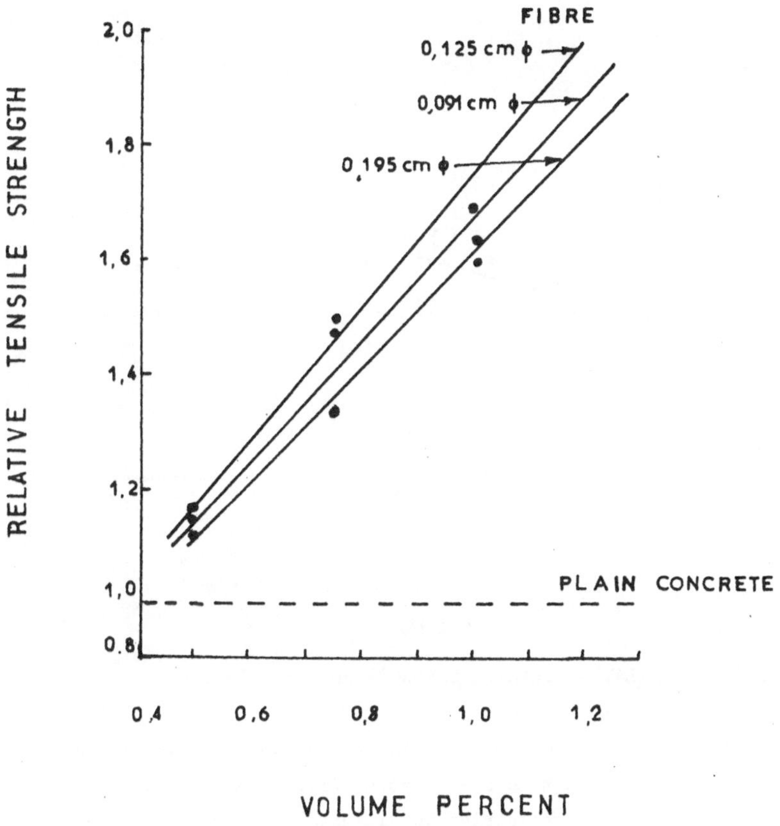

FIG. 1 : RELATIVE TENSILE STRENGTH VERSES VOLUME PERCENT

resulted in increase of strength. There is however a limit upto which strengths increase progressively. In the case of long fibres non-uniform distribution of fibres occur, putting a limit on the aspect ratio upto which strengths increase. This limit can be increased by the use of dispersers. The strengths were seen to be nearly linear functions of lp/d ratio, this being the combined effect of aspect ratio and volume percent. Fig. 2 shows one such relationship namely between relative tensile strength and lp/d ratio.

EFFECT OF FIBRE SPACING

Fibre spacings were computed adopting the following formula suggested by Romualdi and Mandel(2) :

$$s = 13.8 \; d\sqrt{\frac{1}{p}} \qquad (1)$$

As seen from the above formula, the length parameter has no effect on fibre spacing.

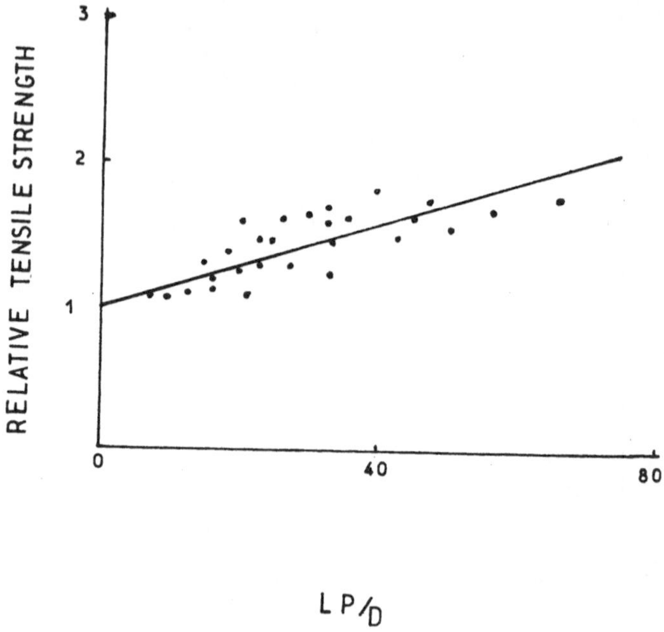

FIG. 2 : RELATIVE TENSILE STRENGTH AND LP/D RATIO.

In the present investigation the fibre spacing was varied through volume fraction keeping the fibre diameter constant as also through fibre diameter keeping the volume fraction constant. The specimen geometries and material were kept constant in this investigation. It was seen that the tensile strength of SFRC increases with decrease of spacing when the spacing was varied through fibre volume fraction,keeping the diameter constant. The relative tensile strength was seen to be inversely proportional to the square root of the spacing. This is represented in the Fig. 3. On the contrary when the fibre spacing was varied through diameter keeping the volume fraction constant, the effect on the tensile strength was seen to be marginal. Table(1) depicts this state of affair. This finding of the present investigation is in keeping with the findings of Shah S.P and Hannent,D.C reported in 1971,1974. Romualdi et al(1,2) who were the earliest to propose the Fibre Spacing Theory in 1963 had concluded from their investigation that the tensile strength of SFRC was only effected by the spacing, if the material & specimen geometries were kept constant. The findings of the present investigation and those of Shah are identical and are different from those drawn from the Fibre Spacing Theory.

As of date, two different theories namely the Composite Material Theory devoloped by Hannent,D.C.,Naaman and Swamy R.N. et al and the Fibre Spacing Theory initially proposed by Romualdi and Baston are available to predict the tensile strength of SFRC. According to the law of mixtures in the Composite Material theory, the tensile strength of SFRC(short steel fibrous concrete) is expressed as:

$$\sigma_u = \sigma_m (1 - v_f) + \eta_e \tau_s (\frac{l}{d}) v_f \qquad (2)$$

Based on the Fibre Spacing theory , the following formula has been proposed by Dr.Kuzusuke Kobayashi (3,4) which is similar to the one suggested by Romualdi et al :

$$\sigma_t = K (\frac{1}{\sqrt{s}} - \frac{1}{\sqrt{s_c}}) + \sigma_{mu} \qquad (3)$$

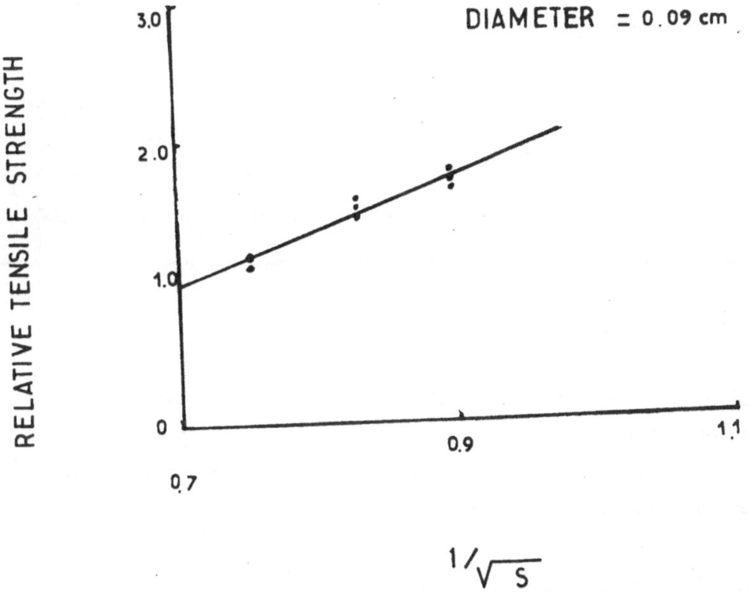

FIG. 3 RELATIVE TENSILE STRENGTH VERSES $1/\sqrt{S}$

From the present investigation it was seen that the fibre spacing do effect the tensile strength of SFRC but only when the variation in spacing was obtained through variation in fibre volume fraction and not through variation in fibre diameter. This indicates that there is some hidden relation between the Crack arrestor phenomenon, the very basis of the Spacing theory and the Composite Material theory. Perhaps this had led the investigators Guan Liqui et al(5) to adroitly seek a better understanding of these theories, to view one in the light of the other. They have derived equation (2) originally deduced from the Composite Material approach, using the crack arrestor concept. The basis of one theory has been utilised to arrive at the same equation given by the other theory. This then is the common ground and the meeting point for these two theories which were hitherto seen to be as opposed to one another. This indeed is a happy augury for the better understanding of this important high technology composite based on steel.

TABLE. 1 CALCULATED AVERAGE FIBRE SPACING AND TENSILE STRENGTH OF SFRC
(spacing varied through fibre diameter)

Specimen identification	Diameter in cm.	Calculated average fibre spacing in cm	length cm	Tensile strength kg/cm^2	Volume percent
BR 10	0.091	1.76	4	24.76	0.50
BR 01	0.125	2.44	4	26.17	0.50
BR 19	0.195	3.81	4	25.46	0.50
BR 14	0.091	1.43	5	35.37	0.75
BR 05	0.125	1.99	5	36.99	0.75
BR 23	0.195	3.11	5	32.26	0.75
BR 18	0.091	1.24	6	39.61	1.00
BR 09	0.125	1.73	6	40.32	1.00
BR 27	0.195	2.69	6	37.49	1.00

SOME ESTABLISHED RELATIONSHIPS - A REVIEW

The test data of the present investigation was utilised for a review of some established relationships connecting (1) cube strength of SFRC to cube strength of corresponding plain concrete (2) split cylinder strength to compressive strength of SFRC (3) split cylinder strength of SFRC to modulus of rupture.

Evaluation Of Cube Strength Of SFRC

R.Narayanan et al (6) have proposed the following formula connecting cube strength of SFRC to cube strength of corresponding plain concrete :

$$f_{cuf} = f_{cu} (1 + 0.1 F) \qquad (4)$$

where f_{cuf} = cube strength of SFRC

f_{cu} = cube strength of corresponding plain concrete

F = fibre factor $= (lp/d) \times d_f$

d_f = bond factor: the recommended value of bond factor for uncramped fibre=0.5 and this value has been adopted for the computations.

Adopting the above formula the cube strength of SFRC test specimens of the present investigation were computed and compared with the observed values. Table (1) shows this comparison. From the limited test data it was seen that the predicted values are conservative and in keeping with the observation of the proposers namely that the formula provides lower bound of gain in cube strength due to fibre addition

TABLE 2 : COMPRESSIVE STRENGTH OF SFRC

Beam identification	Bond factor d_f	volume percent p	lp/d	Fibre factor F	calculated compressive strength kg/cm^2	observed compressive strength kg/cm^2	ratio : observed value/calculated value
FC 1	0.5	0.005	21.20	0.106	253	265	1.04
FC 2	0.5	0.010	42.40	0.0212	255	282	1.10
FC 3	0.5	0.015	63.50	0.317	258	282	1.09
FC 4	0.5	0.005	28.20	0.140	254	273	1.07
FC 5	0.5	0.010	51.50	0.256	257	312	1.20
FC 6	0.5	0.015	84.60	0.423	261	303	1.16
FC 7	0.5	0.005	35.30	0.175	255	287	1.12
FC 8	0.5	0.010	70.60	0.350	259	296	1.14
FC 9	0.5	0.015	106.00	0.530	263	273	1.04

EVALUATION OF SPLIT CYLINDER STRENGTH AND MODULUS OF RUPTURE

R.Narayanan et al have proposed the following formula connecting the split cylinder strength and compressive strength of SFRC.

$$f_{spf} = \frac{f_{cuf}}{A} + B + C\sqrt{F} \qquad (5)$$

where A = A non-dimensional constant having a value of $20 - \sqrt{F}$

B = A non-dimensional constant having a value of $0.7\ N/mm^2$

C = A non-dimensional constant having a value of $1.0\ N/mm^2$

Using the above equation , the split cylinder strength of the test specimens of the present investigation were computed for comparison with the observed values. Table(3) shows this comparison. The predicted values are seen to be on the conservative side. The same authors have proposed the following formula for the prediction of modulus of rupture of SFRC from its split cylinder strength :

$$f_{spf} = 0.65\ f_{rf} \qquad (6)$$

The predicted values of modulus of rupture of the test specimens, using the above formula were compared with the observed values. The comparison is furnished in Table. (4).

The predicted values are seen to be conservative. As rightly pointed out by the authors the formulae are useful in arriving at a safe and quick estimate of the flexural strength of SFRC from the knowledge of cube strength of plain or fibrous concrete.

TABLE. 3 : SPLIT CYLINDER STRENGTH OF SFRC

Specimen identification	Fibre factor F	Compressive strength of SFRC kg/cm^2	Calculated value of split cylinder strength of SFRC kg/cm^2	Observed value of split cylinder strength kg/cm^2	Ratio of observed value and the calculated value
FC 1	0.105	253	23.16	22.86	0.99
FC 2	0.212	255	25.07	30.68	1.22
FC 3	0.317	258	25.76	31.62	1.22
FC 4	0.140	254	23.75	25.75	1.08
FC 5	0.256	257	25.18	30.66	1.21
FC 6	0.423	261	27.00	32.70	1.21
FC 7	0.175	255	24.23	25.75	1.06
FC 8	0.350	259	26.35	30.50	1.15
FC 9	0.530	263	28.00	32.81	1.17

TABLE. 4 : MODULUS OF RUPTURE OF SFRC

Specimen identification	Observedvalue of ultimate flexural moment kgcm	Calculated value of split cylinder strength kg/cm^2	calculated value of modulus of rupture kg/cm^2	Observed value of modulus of rupture kg/cm^2	Ratio of calculated values and observed values
FC 1	13650	23.16	35.63	36.40	1.02
FC 2	15800	25.07	38.57	42.13	1.09
FC 3	17500	25.76	39.64	46.66	1.17
FC 4	14500	23.75	36.54	38.66	1.06
FC 5	17250	25.18	38.74	46.00	1.18
FC 6	17898	27.00	41.54	47.73	1.15
FC 7	16400	24.23	37.27	43.73	1.17
FC 8	18200	26.35	40.54	48.53	1.19
FC 9	20200	28.00	43.10	53.80	1.24

CONCLUSIONS

The following conclusions may be drawn from the present study.

(1) The compressive, tensile and flexural strengths of SFRC are functions of volume fraction of fibres. However there is a limit upto which volume fraction can be increased. The use of dispersers help elevate this limit, but then this also upto a limit.

(2) The aspect ratio has an effect on the compressive, tensile and flexural strengths. There is a limit of aspect ratio upto which strengths progressively increase.

(3) The strengths were seen to be nearly linear function of lp/d ratio, this being the combined effect of aspect ratio and fibre volume fraction.

(4) The tensile strength of SFRC increases with decrease of fibre spacing when the variation in spacing is obtained by varying fibre volume fraction, keeping the diameter of fibre constant. But when the spacings were varied through fibre diameter while keeping the fibre volume fraction constant, little effect on the tensile strength was discerned.

(5) The established relationships connecting

(a) the cube strength of SFRC to the cube strength of corresponding plain concrete.

(b) the split cylinder strength of SFRC to its cube strength.

(c) modulus of rupture of SFRC to its split cylinder strength.

were found to agree with the test results of the present investigation. These relationships can be utilised for arriving at a quick and safe estimate of the flexural strength of SFRC from the knowledge of its cube strength or that of the corresponding plain concrete.

(6) The fact that the fibre spacings varied through fibre volume fraction (keeping the diameter constant) alone effect the tensile strength of SFRC points out an existing common ground between the two theories namely the spacing theory and the Composite material theory which were hitherto thought as opposed to each other.

APPENDIX 1. REFERENCES

1. Romauldi, J.P. and Baston, G.B., Behaviour of Reinforced Concrete Beams with closely spaced Reinforcement, Am. Conc. Inst. Journal, Vol. No. 60, June 1963, pp. 775-790.
2. Romauldi, J.P. and Mandel, J.A., Tensile strength of concrete affected by uniformly distributed and closely spaced short lengths of wire reinforcement, Am. Conc. Inst. Journal, Vol. No. 61, June 1964, pp. 657-671.
3. Kazusuke Kobayashi and Ryokche, Strength and deformation of steel fibre reinforced concrete in uniaxial tension (in Japanese) , Proc. of the Japan Society of Civil Engineers, No. 257, Jan. 1977, pp. 85-94.

4. Kazusuke Kobayashi and Ryokche Cho, Reinforcement priciples and first crack strength of SFRC (in Japanese), Proc. of JSCE, No. 282, Feb. 1979, pp. 131-134.
5. Guanliqui and Zhao Guofan, A study of the mechanism of fibre reinforcement in short steel fibrous concrete, Symposium 1986, Sheffield, 13-17 July 1986, Vol.2.
6. Narayanan, R. and Kareen-Palanjan, A.S., Factors influencing the strength of fibre reinforced concrete, Rilem Symposium 1986, Sheffield, 13-17 july 1986, Vol. 2.
7. Shah, S.P. et al, Complete stress-strain curves for SFRC in uniaxial tension and compression, Rilem Symposium on Testing and Test methods of fibre cement composites, ACI. Ed. by Swamy, R.N., April 5-7, 1978, Paper 7.3, p. 399.
8. Swamy, R.N. and Manget, P.S., The interfacial bond stress in steel fibre cement composites, Cement & Research, Vol. 6, 1976, pp. 641-650.
9. Hannent, D.J., Fibre Cements and fibre concretes, John Wiley and Sons. Ltd., 1978, pp. 8-34. 219 P.
10. Shah, S.P. and Rangan, B.B.,Fibre reinfoced concrete properties, ACI Journal, Feb. 1971, pp.126-135.

APPENDIX. 2. NOTATIONS AND ABBREVIATIONS

The following symbols are used in this paper:

s = Average spacing of fibres.

d = Diameter of fibre.

p = Fibre content, percent by volume.

l = length of fibre.

F = Fibre Factor.

d_f = Bond Factor.

σ_u = Ultimate strength of SFRC.(tensile)

σ_{mu} = Ultimate tensile stress in the matrix.

v_f = Volume fraction of fibres.

K = A constant depending on bond between fibres and the matrix.

η_o = Orientation effeciency factor.

σ_t = Tensile strength of SFRC.

s_c = Max. fibre spacing capable of strengthening plain concrete with steel fibres.

f_{cu} = Cube strength of plain concrete.

f_{cuf} = Cube strength of SFRC.

f_{spf} = Split cylinder strength of SFRC.

f_{rf} = Modulus of rupture of SFRC.

τ_s = Bond stress at the moment when the tensile stress in the matrix is at ultimate stage.

Proceedings of the International Symposium on Fibre Reinforced Concrete, December 16-19, 1987, Madras, India

FLEXURAL BEHAVIOUR OF REINFORCED FIBRE CONCRETE BEAMS

H V Dwarakanath - Professor and Head, Department of Civil Engineering, M.S. Ramaiah Institute of Technology, Bangalore, India.

T S Nagaraj - Professor and Chairman, Department of Civil Engineering, Indian Institute of Science, Bangalore, India.

SYNOPSIS

Several investigators have shown that the presence of steel fibres in beams reinforced with high strength deformed steel bars is beneficial in modifying the deformational behaviour of reinforced fibre concrete beams with particular reference to the deformational characteristics. The beams reinforced with high strength deformed bars, both under-reinforced and over-reinforced, with fibres disposed in two types of locations are considered in this investigation. The two types of locating the fibres are - (i) over the entire depth and (ii) over half the depth of the beams on the tension side. It is observed that partial inclusion of the fibres over half the depth, in the case of under-reinforced beams, is equally beneficial as the full depth inclusion, in controlling cracking and deflection and in increasing the stiffness of the beams right from the beginning of loading upto the ultimate. In the case of over-reinforced beams, fibres in small quantities, as used in this investigation, are not found to be effective in bringing about any appreciable modification in the deformational behaviour of the beams.

INTRODUCTION

A review of the existing reports reveals that various aspects of the improvements which can be brought about by the addition of steel fibres in conventionally reinforced concrete structural members have been investigated and studied. In the case of deep beams, inclusion of fibres was found to prevent shear failures and to increase ultimate flexural capacity and resistance to shear failure. The failure was observed to be more ductile and gradual [1]. In the case of ordinary beams, replacement of vertical stirrups and bentup bars by fibres was found to provide effective shear reinforcement [2]. Improvement in strength under pure torsion and combined bending and torsion of concrete members has been observed when fibres were included [3]. Control of cracking and deflection has been reported as the most significant contribution of the fibres in flexure of reinforced concrete beams, reinforced with high strength deformed bars [4].

Improvement in the serviceability behaviour of concrete structural members reinforced with high strength deformed bars has been observed to be more profound and substantial compared to the increase in flexure loads [5]. Reduced deformations at all stages of loading, increase in ultimate punching shear loads and occurrence of ductile shear failures in the case of slab column connections which were conventionally reinforced have been observed with the inclusion of fibres in the connection regions [6]. In the light of these findings, fibre reinforced concrete supplemented by additives such as superplasticizers, silica fumes and polymers has been predicted to become a panacea solution [7]. However, steel fibres form an expensive constituent material of steel fibre concretes. It is therefore vitally important to determine ways and means of using the fibres in an optimum way with utmost care consistent with economy for achieving best possible results in practical sense.

In this context, a continued investigation regarding the effect of steel fibres in concrete beams reinforced with high strength deformed bars has been undertaken and reported in this paper. The scope of the investigation covers a detailed experimental study of both under-reinforced and over-reinforced fibre concrete beams, conventionally reinforced with high strength deformed steel bars, in flexure. The steel fibres are disposed in two ways viz., over the full depth of the entire beam and over only half the depth on the tension side of the entire beam. Two volume fractions of the steel fibres with one aspect ratio have been used for each mode of disposition of fibres in the investigation.

EXPERIMENTAL INVESTIGATION

The experimental investigation mainly consisted in casting 20 nos of large size conventionally reinforced beams each measuring 1800 mm x 100 mm x 208 mm and conducting flexure tests on the beams after proper curing. Casting was carried out in a suitable mould which consisted of two mild steel channels kept back to back and properly secured to each other for ensuring stability and dimensional accuracy.

Fully steel fibre-reinforced concrete beams and fully plain concrete beams were cast in three layers, each layer as one batch. Half depth steel fibre-reinforced concrete beams were cast in three layers, two layers of steel fibre-reinforced concrete and one layer of plain concrete, each layer as one batch. Using mortar spacers etc. for providing the required bottom cover was totally avoided as it was felt that spacers might act as crack initiators. Instead, the required cover was obtained with sufficient accuracy by initially casting a thin layer of the mix and vibrating the same using plate attachment to the Kongo Hammer. This was done in such a way as to obtain the desired cover thickness which had been premarked on the inner vertical sides of the formwork. The reinforcement cage was carefully lowered into the formwork after casting the cover bed as described above. The alignment was quickly checked up and corrected for, if required, and casting continued. Side covers and end covers were ensured by placing wooden strips of thickness equal to the cover thickness and by carefully and slowly withdrawing the same as the casting continued. The sectional dimensions were checked after casting to ensure that the deviations in the dimensions of the beam were of negligible order. Four percentages of tension steel were included in the four series of beams respectively. These percentages included two under-reinforced and two over-reinforced steel contents. In each series, for one specified percentage of steel, there were five numbers of beams - two numbers of beams of fibre reinforced concrete for the entire depth (one beam for each of the two specified volume fractions of the steel fibres), two numbers of beams with half depth fibre-reinforced concrete and half depth plain concrete (one beam for each volume fraction of the steel fibres) and one beam of plain concrete for the entire depth. In all, there were twenty beams.

Batching of materials was done by weighing the constituent materials. Layerwise batching was intended to achieve better uniformity and better control over mixing and other operations involved in the process of mixing and casting.

It was ensured that minimum time was spent in casting the layers one by one. Compaction of each layer was effected by internal vibration using an electrically operated vibrator, Kongo Hammer, fitted with a needle meant for laboratory use. The actual thickness of the half depth fibrous concrete layer was determined by depth measurement from the top at a few selected positions before casting the remaining layer of plain concrete.

The specimens were finished to a smooth and level surface and left in the mould for about 24 hours. Demoulding involved only removal of the sides of the formwork. The specimens were watered well and covered all around with wet gunny bags and maintained in a wet condition for 3 to 4 days at the end of which they were transported to the curing tank. The tank was replenished with fresh water at the end of every week. After 28 days of curing, the specimens were removed to a protected place with shade till the time of testing. Average age of the specimens at the time of testing, varied from 14 to 18 weeks.

One specimen was cast for each variation in the content and disposition of the fibres for a given percentage of conventional steel content. The beams require white washing a day prior to testing. In general, the beams required various markings for positioning the loading rollers and for fixing the guages and deflection strips. The deflection strips of aluminium flat were fixed at the centre of the span on the tension face so that it projected sufficiently on either side beyond the sides of the beam. The projections helped in taking the deflection readings with the help of dial guages mounted on a magnetic base. The strips were fixed to the specimen using araldite adhesive after rubbing the surface with sand paper and cleaning with acetone where the strips were to be attached. The guages were held in a specially fabricated equipment while measuring axial elongations and compressions in tension and flexure tests. It prevented damaging of dial guages and transducers even in the case of occurrence of a total collapse of the specimen in a test. It avoided the problem of demec discs falling off due to cracking of concrete. Further, it saved time and labour involved in the fixing of demec discs and in the placing of demec guages each time a measurement is to be made. In addition, it had the same accuracy as the one obtained with demec discs.

The beams were tested in a load frame designed for a capacity of 50 tonnes. The self reacting loading frame was quite rigid and consisted of riveted double I-section vertical posts and I-section top and bottom horizontal members. The loading device was kept on the floor of the loading frame and the beams were tested in an inverted position being supported on the roller assemblies fixed to the bottom flange of the top horizontal member of the loading frame. Thus the upper face of the beam was in tension and the lower face was in compression during testing. This inverted arrangement of testing was due to the loading device which applied the loading resting on the floor. Longitudinal strains were measured using LVDTs at steel level, one LVDT on either side of the beam. LVDTs had a range of $\pm$ 10 mm with a least count of 10 microstrains and, at other levels towards the compression face, dial guages were used. The other levels were 20 mm and 80 mm from the compression face where the dial guages with a least count of 0.01 mm were used. On one side of the beam, the guage length was 100 mm and on the other side it was 200 mm for measuring the longitudinal displacements. Dial guages with a least count of 0.01 mm were used for measuring the central deflection one on each side of the beam at equal distances from the edge of the beam. The loading was applied continuously through a strain rate controlled loading device. The test lasted for 90 to 120 minutes and it took about one hour for prior preparation before commencing the test.

The following precautions were taken while testing the large size beams :

(i) The loading device was firmly seated on a level horizontal surface and truly aligned vertical.

(ii) The test beam was supported on the roller assemblies such that the markings of the beam span were just on the rollers, the horizontal axis of the beam passed through the centres of the rollers and the centre of the beam span was vertically just above the centre of the loading device.

(iii) The centre of the loading joist coincided with the centre of the loading device and the rollers provided on the loading joist were exactly on the markings corresponding to the load span on the rest of the beam. It was ensured that the web of the loading joist was aligned centrally and parallel to the horizontal axis of the beam.

(iv) The load cell was centred with respect to the centres of loading joist and of the test beam.

(v) The probes of the dial guages were positioned at equal distances from the axis of the beam symmetrically on the deflection strips on either side of the beam.

(vi) The beam axis was marked on both top and bottom faces of the beam.

(vii) A swivel was provided between the loading joist and the load cell to ensure a truly vertical concentrated load.

(viii) A strong rope was provided at both ends of the beam looped around the beam and the top horizontal girder of the reaction frame to prevent any mishap.

DISCUSSION OF TEST RESULTS

The results of the test observations have been shown in Tables 1 to 4. Table 1 shows the percentages of conventional steel used in the beams and the test loads realized in comparison with the theoretical loads arrived at by ACI method. Tables 2 and 3 pertain to the deformational characteristics of the under-reinforced series of the beams and Table 4 pertains to those of over-reinforced series of beams.

A comparitive study of the details given in Tables 2 and 3 reveals that both half depth and full depth fibre inclusions bring about modifications in the structural characteristics associated with the deformational behaviour of beams like curvature, strain at steel level and central deflection in a much desired manner right from the beginning upto ultimate. The order of improvement in both types of inclusions is

Table 1 : Flexural loads obtained by Theory (ACI Method) and by Test (Conventionally Reinforced Plain Concrete Beams)

Sl No.	Beam Series	p(%)	b x d (mm^2)	Flexural load (kN) Analytical	Exptl.	Comparison
1	2	3	4	5	6	5/6
1	1	0.7654	100x183.27	38.3	40.1	0.96
2	2	1.2848	100x181.90	60.3	60.0	1.01
3	3	2.2131	100x180.04	83.8	80.0	1.05
4	4	3.7416	100x168.93	81.2	80.3	1.01

Table 2 : Modified Deformational Characteristics in Under-reinforced Beams due to Addition of Steel Fibres; Series 1; p = 0.7654%

Sl No.	Deformational Characteristics	v_f (%) 0 FDPC	0.75 FDFC	1.5 FDFC	0.75 TZHDFC	1.5 TZHDFC	Remarks
1	Relative ϵ_s	1	0.57	0.57	0.78	0.67	
2	Relative $1/\phi$	1	1.50	1.50	1.36	1.36	At σ_s=240 N/mm^2
3	Relative $1/\delta$	1	1.71	1.78	1.46	1.52	
4	Relative ϵ_s	1	0.62	0.61	0.69	0.78	
5	Relative $1/\phi$	1	1.50	1.38	1.38	1.28	At σ_s=285 N/mm^2
6	Relative $1/\delta$	1	1.59	1.67	1.43	1.48	
7	Relative ϵ_s	1	0.61	0.63	0.67	0.67	
8	Relative $1/\phi$	1	1.32	1.80	1.21	1.66	At σ_s=355 N/mm^2
9	Relative $1/\delta$	1	1.60	1.66	1.50	1.53	
10	Relative ϵ_s	1	0.86	0.89	0.93	0.75	
11	Relative $1/\phi$	1	0.91	0.95	0.98	0.80	At ultimate
12	Relative $1/\delta$	1	0.88	0.93	0.84	0.81	

FDPC means Full Depth Plain Concrete; FDFC means Full Depth Fibre Concrete; TZHDFC means Half Depth Fibre Concrete on Tensile Side.
ϵ_s Strain at Steel level; ϕ = Section Curvature with respect to maximum Concrete strain; δ = central deflection; σ_s = stress in steel corresponding to strain ϵ_s

Table 3 : Modified Deformational Characteristics in Under-reinforced Beams due to Addition of Steel Fibres; Series 2; p = 1.2848%

Sl No.	Deformational Characteristics	v_f (%)					Remarks
		0 FDPC	0.75 FDFC	1.5 FDFC	0.75 TZHDFC	1.5 TZHDFC	
1	Relative e_s	1	0.97	0.77	0.94	0.89	
2	Relative $1/\phi$	1	1.17	1.21	1.01	1.11	At σ_s = 240 N/mm^2
3	Relative $1/\delta$	1	1.14	1.21	1.05	1.14	
4	Relative e_s	1	0.82	0.85	0.88	0.91	
5	Relative $1/\phi$	1	1.12	1.05	1.12	1.05	At σ_s = 285 N/mm^2
6	Relative $1/\delta$	1	1.11	1.15	1.05	1.15	
7	Relative e_s	1	0.98	0.93	0.98	0.99	
8	Relative $1/\phi$	1	1.12	1.11	1.12	1.11	At σ_s = 355 N/mm^2
9	Relative $1/\delta$	1	1.12	1.12	1.07	1.12	
10	Relative e_s	1	1.03	1.16	1.01	0.96	
11	Relative $1/\phi$	1	1.09	1.22	1.05	0.95	At ultimate
12	Relative $1/\delta$	1	0.98	1.03	0.86	0.80	

Table 4 : Modified Deformational Characteristics in Over-Reinforced Beams due to Addition of Steel Fibre; Series 3 and 4; Ultimate Condition

Sl No.	Deformational Characteristic	v_f (%)		
		FDPC 0	FDFC 0.75	FDFC 1.50
1.	$\phi(x10^{-5}/mm)$	1.00	1.48	1.65
2.	$e_c(x10^{-3})$	1.00	1.17	1.59
3.	δ (mm)	1.00	1.22	1.33
4.	$\phi(x10^{-5}/mm)$	1.13 (+)	1.59 (*)	2.46 (x)
5.	$e_c(x10^{-3})$	1.00	1.20	1.21
6.	δ (mm)	1.00	0.97	1.38
		1.28 (++)	1.63	2.61 (xx)

FDPC : Full Depth Plain Concrete; FDFC : Full Depth Fibre Concrete; (+) : at 0.91 P_u; (**) : at 0.88 P_u; (xx) : at 0.82 P_u; after the ultimate; P_u : Ultimate Load.

practically the same. The enhancement in moment capacity in the case of full depth fibre inclusion is larger and this enhancement is observed to be greater. From the test observations, it is found that full depth fibre inclusion results in increased ductility characteristics and in preserving the structural integrity at ultimate.

A comparative study of the findings, observed in the tests on over-reinforced beams and presented in Table 4 reveals that the presence of the fibres is found to increase the ductility characteristics only at the ultimate and only with full depth fibre inclusion. With the two volume fractions of the fibres studied in this investigation viz. : 0.75% and 1.5%, there is an indication that larger fibre contents are likely to enhance the ductility characteristics of over-reinforced beams more significantly. Also, full depth fibre inclusion is found to impart structural integrity to the beams at ultimate from the observations at the time of testing.

Comparison with the Findings of other Investigators :

A comparison, in the strict sense of the term, is not possible among the findings of several investigations because of the variations in the variables of these investigations. However, a qualitative comparison can still be made with a view to study whether there exists a general agreement in the trend of improvement in the deformational characteristics of beams due to the addition of steel fibres. As such, Table 5 include the improvements observed by Swamy and Al Taan [4] and Table 6 includes the observations by Kormeling et al. [8] in their respective tests. A critical study of the values of the deformational characteristics shown in these tables will reveal that there is a general agreement in the trend of improvements in the deformational characteristics, with the inclusion of steel fibres in under-reinforced beams. A critical analysis of test results pertaining to over-reinforced beams and shown in Table 4 has already been made and in continuation, the conclusions arrived at by Mangat and Motamedi [9] regarding the marked increase in the ductility of compression members by a combination of stirrups and fibres is worth investigating whether such a combination can bring about substantial improvements in the ductility of over-reinforced beams. Further, both steel fibres and conventional steel being reinforcing elements in the concrete mass, incorporating the fibres, with efficient geometry and shape, in sufficient quantities on the compression side is believed to impart better ductility characteristics to over-reinforced beams.

CONCLUDING REMARKS

In the case of conventionally under-reinforced and over-reinforced beams, the increase in ultimate flexural loads decreases as the

Table 5 : Test Data of Beams Tested by Swamy and Al Taan (1981) - Conventionally Reinforced Fibre Concrete Beams

Sl No.	Deformational Characteristic	v_f (%)					Remarks
		0 FDPC	0.50 FDFC	1.0 FDFC	0.50 TSFC	1.0 TSFC	
1	Relative ϵ_s	1	0.27	0.24	0.61	0.80	At first
2	Relative 1ø/	1	1.93	1.90	1.18	1.21	crack
3	Relative 1/δ	1	1.39	2.17	1.43	1.54	stage
4	Relative ϵ_s	1	0.91	0.83	1.17	1.04	At design
5	Relative 1/ø	1	1.16	1.77	0.99	1.08	service
6	Relative 1/δ	1	1.15	1.20	1.11	1.16	load
7	Relative ϵ_s	1	0.67	-	0.80	0.81	At
8	Relative 1/ø	1	1.27	1.08	1.11	1.17	failure
9	Relative 1/δ	1	0.90	1.26	1.02	1.02	stage

1, 2 and 3 using conventional steel with yield strength 460 N/mm^2; 4 and 9 using conventional steel with yield strength 617 N/mm^2; crimped steel fibres 0.50 mm x 50 mm TS means tensile skin surrounding the conventional steel.

Table 6 : Test Data of Beams Tested by Kormeling, Reinhardt and Shah (1980)- Conventionally Reinforced Fibre Concrete Beams

Sl No.	Deformational Characteristics	Without fibres	With hooked fibres	With straight fibres	With paddled fibres	Remarks
1.	1/ø at cracking	1	2.027	1.487	2.114	2 nos of 4mm bars with
2.	1/ø at ultimate	1	2.007	1.599	1.648	yield strength 716 N/mm^2 for
3.	1/δ at ultimate	1	2.233	1.944	1.758	conventional steel
4.	1/ø at cracking	1	0.971	0.848	0.905	4 nos of 10mm bars with
5.	1/ø at ultimate	1	1.433	0.848	1.110	yield strength of 460 N/mm^2
6.	1/δ at ultimate	1	1.595	0.847	1.549	

percentage of conventional steel provided is increased. The increase in fibre content alone will not bring about correspondingly commensurate improvements and hence, use of lesser volume fractions of fibres with improved geometry appears to be more effective. Half depth mode of inclusion of fibres for under-reinforced beams and full depth mode of inclusion for over-reinforced beams are recommended for desired improvements in the deformational characteristics.

REFERENCES

1. Roberts, T.M. and Ho, N.L., "Shear Failure of Deep Fibre Reinforced Concrete Beams", International Journal of Cement Composites and Light Weight Concrete, Vol. 4, No. 3, 1982, pp. 145-152.

2. Batson, G.B., Jenkins, E. and Spatney, R., "Steel Fibres as Shear Reinforcement in Beams", Proceedings, Am. Concrete Inst., Vol. 69, No. 9, 1972, pp.640-644.

3. Narayanan, R. and Green, K.R., "Fibre Reinforced Concrete Beams in Combined Bending and Torsion", Indian Concrete Journal, Vol. 55, No. 8,1981, pp. 222-228 and 232.

4. Swamy, R.N. and Al Taan, "Deformation and Ultimate Strength in Flexure of Reinforced Concrete Beams made with Steel Fibre Concrete", Proceedings Am. Concrete Inst., Vol. 78, No. 5, 1981, pp. 395-405.

5. Swamy, R.N., Al Taan, Saad, A. and Ali, A.R., "Steel Fibres for Controlling Cracking and Deflection", Concrete International, Vol. 1, No. 8, 1979, pp. 41-49.

6. Swamy, R.N. and Ali, S.A.R., "Punching Shear Behaviour of Reinforced Slab - Column Connecting Made with Steel Fibre Concrete", Proceedings Am. Concrete Inst., Vol. 79, No. 5, 1982, pp. 392-406.

7. Naaman, Antoine, E., "Fibre Reinforcement for Concrete", Concrete International Design and Construction, Vol. 7, No. 3, 1975, pp. 21-25.

8. Kormeling, H.A., Reinhardt, H.W., and Shah, S.P., "Static and Fatigue Properties of Concrete Beams Reinforced with Continuous Bars and with Fibres", Proceedings Am. Concrete Inst., Vol. 77, No. 1, 1980, pp. 36-43.

9. Mangat, P.S., and Motamedi Azari, M., "Influence of Steel Fibre and Stirrup Reinforcement on the Properties of Concrete in Compression Members", International Journal of Cement Composites and Light Weight Concrete, Vol. 7, No. 4, 1985, pp. 261-267.

Proceedings of the International Symposium on Fibre Reinforced Concrete, December 16–19, 1987, Madras, India

CRACKING CHARACTERISTICS OF R.C. BEAMS WITH STEEL FIBRES

Dr. H. Achyutha
Dept. of Civil Engineering
Indian Institute of Technology
Madras

Dr.P. Sabapathi
Dept. of Civil Engineering
Thiagarajar College of Engg.
Madurai

Results of an experimental investigation on the effects of inclusion of steel fibres (smooth round G.I. wires) in conventional reinforced concrete beams on their cracking characteristics are presented in this paper. Factorial experimental design principle has been adopted in arriving at the number of beam specimens to be tested. The influence of concrete strength, aspect ratio and volume of steel fibres and quantity of tensile reinforcement, on the cracking characteristics has been presented in detail. The beneficial effect of steel fibres in reinforced concrete beams in increasing the working loads at a particular crack width has been highlighted.

INTRODUCTION

Steel Fibre Reinforced Concrete (SFRC) has found field applications in bridge decks, highway and airfield pavements, repair works, industrial flooring etc. Although SFRC has found interesting applications, it may never replace conventional reinforced and prestressed concrete structural members of large size. Recently some investigations [1 to 6] have been carried on use of steel fibres in conventional concrete structural members. The results of these investigations are quite encouraging to structural engineers as substantial improvements have been achieved in the structural behaviour of these members. The aim of this paper is to present some of the results of an extensive investigation carried out on the flexural behaviour of reinforced concrete beams with steel fibres [7]. The scope of the paper is limited to the presentation of the cracking characteristics of these members. The results presented in this paper cover the influence of concrete strength, aspect ratio and volume of steel fibres and quantity of tensile reinforcement. Factorial experimental design principle has been adopted in arriving at the number of beam specimens to be tested [8,9].

MATERIALS, SPECIMENS AND TESTING DETAILS

Concrete mix proportion of 1 : 1.25 : 1.25 of cement, sand and coarse aggregate by weight with water-cement ratio of 0.48 was adopted for all the beams except two beams for which the proportion is 1 : 1.4 : 2.1 with water-cement ratio of 0.62 for studying the effect of concrete strength. High strength deformed beam of nominal diameter 8 mm, 10 mm, 12 mm, or 16 mm were used as main tension reinforcement in the beams. Galvanised Iron (G I) wire of 0.56 mm diameter (24 gauge) was used as fibres for the entire experimental work. These wires were available in the form of circular coils which were cut to required length (aspect ratio) using a table type hand operated shear cutter. The Galvanised Iron wire had yield and ultimate strengths of 669 and 717 N/mm^2.

The beam specimens were 2500 mm long, rectangular in cross-section and an over all dimension of 126 x 203.2 mm. A clear cover of 25 mm was uniformly adopted for all the beams irrespective of the bar diameter. Typical sketches showing the reinforcement details are given in Fig. 1. The other details like number of beams, actual area, yield and ultimate strengths are furnished in Table 1.

TABLE 1 : DETAILS OF BEAM SPECIMENS

Sl. No.	Beam No.	Concrete Cube Strength f_{cu} N/mm^2	Age at test days	Fibre details a_f	V_f	Details of tension reinforcement: No. and nominal size of bar	A_{st} mm^2	f_{sy} N/mm^2	f_{su} N/mm^2
1	2	3	4	5	6	7	8	9	10
				a) Effect of Fibre Parameters					
1	1	32.96	48	80	1.6	2 x 10 0	161.32	375	499
2	2	34.84	50	80	1.0	2 x 10 0	175.00	615	869
3	7	32.57	53	50	1.6	2 x 10 0	158.68	400	516
4	8	32.18	54	50	1.0	2 x 10 0	158.68	400	516
5	13	33.79	96	50	1.3	2 x 10 0	154.92	365	475
6	14	31.17	98	80	1.3	2 x 10 0	154.92	365	475
7	15	34.59	91	65	1.6	2 x 10 0	152.08	402	542
8	16	36.11	93	65	1.0	2 x 10 0	152.08	402	542
9	3	29.82	45	65	1.3	2 x 10 0	165.42	400	499
10	9	28.43	50	65	1.3	2 x 10 0	152.38	580	797

11	17	35.82	88	65	1.3	2 x 10 0	158.44	426	550
12	5	34.53	48	No Fibre		2 x 10 0	165.42	400	499
13	11	35.03	105	No fibre		2 x 10 0	152.38	580	797
14	19	35.53	86	No fibre		2 x 10 0	158.44	426	550

b) <u>Effect of Quantity of Tension Steel</u>

15	4	29.82	49	65	1.3	2 x 8 0	113.20	386	494
16	6	34.53	49	No fibre		2 x 8 0	113.20	386	494
17	10	28.43	51	65	1.3	2 x 12 0	237.82	432	564
18	12	35.02	113	No fibre		2 x 12 0	237.82	432	564
19	18	35.82	90	65	1.3	2 x 16 0	408.48	510	682
20	20	35.53	87	No fibre		2 x 16 0	408.48	510	682

c) <u>Effect of concrete strength and type of steel</u>

21	21	36.95	77	50	1.9	3 x 16 0	651.60	479	662
22	22	36.33	81	No fibre		3 x 16 0	651.60	479	662
23	23	26.60	75	50	1.9	2 x 16 0	427.60	395	551
24	24	26.09	77	No fibre		2 x 16 0	427.60	395	551
25	25	53:99	80	50	1.9	2 x 12 0	200.00	...	1560
26	26	64.64	81	No fibre		2 x 12 0	200.00	...	1560

d) <u>Effect of Fibres on Ductility (Confirmation tests)</u>

27	45	46.65	86	80	1.5	2 x 16 0	421.00	443	553
28	46	46.65	87	80	1.5	3 x 16 0	631.50	443	553

e) <u>Effect of Fibre around Tension Steel only compared to Equivalent Additional Bar Reinforcement</u>

29	32	33.05	40	No fibre		2 x 12 0	243.56	446	580
30	33	33.75	37	80	0.676	2 x 12 0	243.56	446	580
		*23.41							
31	34	33.75	38	No fibre		2 x 12 0	243.56	446	580
						+1 x 8 0	+53.86	368	469
32	35	33.66	37	72	1.137	2 x 12 0	243.56	446	580
		*22.56							
33	36	33.66	38	No fibre		2 x 12 0	243.56	446	580
						+1 x 10 0	+90.54	414	515

34	37	27.77 *24.94	39	64.	1.529	2 x 12 0	243.56	446	580
35	38	27.77	37	No fibre		2 x 12 0 +1 x 12 0	243.56	446	580

* Compressive strength of fibre concrete around tension steel.

The main beam specimens were tested in a loading frame under symmetrical loading. The load was applied through a 500 kN capacity hydraulic jack fixed centrally at the top of the loading frame. Load from the hydraulic jack was transmitted to the specimen through a distributor beam to give a central pure moment zone of 600 mm length. The beams were supported on hinge at one end and roller at the other end. A sketch of the loading arrangement is shown in Fig. 2. During the testing of the beam specimens, in addition to crack measurements, deflections, concrete surface strains across the cross-section and steel stress were measured at various load levels.

CRACK MEASUREMENTS

The first visible crack and further cracks were identified by carefully observing the concrete surface through magnifying glass. Crack spacings and heights were measured with ordinary scale having a least count of 1.0 mm. Only the maximum width of each crack within the constant moment region was measured using a hand microscope with a least count of 0.025 mm.

RESULTS AND DISCUSSIONS

<u>First Visible Crack</u> : The First visible crack appeared within the central 100 mm length of the constant moment zone in most of the beams. Values of the load at which the first crack was observed in (a) beams without fibres, (b) with fibres around the tensile steel only and (c) beams without fibres but with extra tensile reinforcement equivalent to the fibre content, are presented in Table 2 and 3. In most of the beams with fibres, the first visible crack appeared at moments greater than the corresponding value of the identical beams without fibres by 50% or more. In the case of over reinforced beams (No. 21, 46) increase up to 22% was observed. There is an increase of the order of 30% in the load at first visible crack due to the addition of fibres around the tensile steel only, over that of beams with additional bar reinforcement equivalent to the fibre content. This indicates that the fibres are more efficient than the equivalent bar reinforcement in the initial stages of cracking.

TABLE : 2 RATIO OF MOMENT AT FIRST VISIBLE CRACK TO THE WORKING MOMENT

Beam No.	A_{st}	a_f	V_f	M/M_w	Increase %
6	2 No. 8 mm	...	...	0.6182	...
4	2 No. 8mm	65	1.3	0.8613	39.3
5	2 No. 10 mm	...	...	0.4648	...
11	2 No. 10 mm	...	...	0.5384	...
19	2 No. 10 mm	...	...	0.4492	...
			Mean	0.4841	
1	2 No. 10 mm	80	1.6	0.5115	5.7
14	2 No. 10 mm	80	1.3	0.7103	46.7
2	2 No. 10 mm	80	1.0	0.8766	62.5
7	2 No. 10 mm	50	1.6	0.9353	93.2
13	2 No. 10 mm	50	1.3	0.8493	75.4
8	2 No. 10 mm	50	1.0	0.9011	86.1
15	2 No. 10 mm	65	1.6	1.0795	123.0
17	2 No. 10 mm	65	1.3	0.8290	71.2
3	2 No. 10 mm	65	1.3	0.8355	72.6
9	2 No. 10 mm	65	1.3	0.7247	49.7
16	2 No. 10 mm	65	1.0	0.6105	26.1
20	2 No. 16 mm	...	...	0.2724	...
18	2 No. 16 mm	65	1.3	0.4728	73.6
45	2 No. 16 mm	80	1.5	0.6218	128.3
22	3 No. 16 mm	...	...	0.3690	...
21	3 No. 16 mm	50	1.9	0.4515	22.4
46	3 No. 16 mm	80	1.5	0.4243	15.0
24	2 No. 16 mm	...	...	0.2933	...
23	2 No. 16 mm	50	1.9	0.4117	40.4

M_W = Working moment at permissible stress (230 N/mm^2) in steel excluding the effect of fibres.

TABLE : 3 COMPARISON OF OBSERVED FIRST CRACK LOAD OF BEAMS WITH FIBRES SURROUNDING TENSION STEEL AND BEAMS WITH EQUIVALENT BAR REINFORCEMENT

Beam No.	A_{st}	a_f	V_f	Load at first crack KN	Increase %
32	2 No. 12 mm	...	...	8.0	....
34	2 No. 12 mm	...	...	8.0	...
and	1 No. 10 mm				
33	2 No. 12 mm	80	0.0676	10.0	25.0
36	2 No. 12 mm				
and	1 No. 10 mm	...	...	8.5	...
35	2 No. 12 mm	72	1.137	11.5	35.3
38	2 No. 12 mm				
and	1 No. 12 mm	...	...	7.5	...
37	2 No. 12 mm	64	1.529	10.0	33.3

Crack Height : It ia quite evident that the crack height has to be smaller because of the extra tensile force offered by fibres spanning the cracks. The variation of maximum crack height with moment is presented in Fig. 3 for some of the beams with fibres over the whole section along with identical beams without fibres. A reduction of about 25% of the maximum crack height is seen from the graph.

Crack Spacing : Conclusion regarding the crack spacinig has to be based only on statistical considerations. The maximum and mean crack spacing values obtained experimentally for the beams with and without fibres are presented in Table 4 along with the statistical anlysis. It can be concluded from this analysis that the presence of fibres does not decrease the maximum or mean crack spacing. This conclusion is further confirmed from the results of beams having different amounts of tensile reinforcement (Table 5).

TABLE : 4 CRACK SPACING DETAILS OF BEAMS WITHOUT FIBRES AND WITH FIBRES OVER THE WHOLE SECTION

Beam	a_f	V_f	Crack spacing in mm			
			at working load		at steel yield	
			Maximum	Mean	Maximum	Mean
5	...	...	172.0	131.0	118.0	104.8
11	...	...	242.0	141.3	180.0	113.0
19	...	...	150.0	95.0	150.0	95.0
		Mean	188.0	122.4	149.3	104.3
		SD	48.0	24.3	31.0	9.0
		Cov. %	25.6	19.9	20.8	8.6
1	80	1.6	128.0	75.6	128.0*	84.2
14	80	1.3	170.0	88.0	170.0	61.1
2	80	1.0	144.0	84.7	107.0*	71.0
7	50	1.6	255.0	255.0*	247.0*	144.3
13	50	1.3	176.0	125.3	167.0	94.3
8	50	1.0	215.0	176.0	137.0	111.2
15	65	1.6	570.0*	570.0*	150.0	142.5
17	65	1.3	150.0	147.5	165.0	115.0
3	65	1.3	175.0	139.0	175.0	139.0
9	65	1.3	315.0*	154.0	147.0	117.8
16	65	1.0	268.0	138.3	162.0	79.0
		Mean	186.8	125.4	159.1	105.4
		SD	49.0	34.9	13.1	29.5
		Cov. %	26.2	27.8	8.2	28.0

Note : 1. * Neglected due to significant difference from other values.

2. All beams had 2 No. 10 mm dia. HSD bars as tension steel.

TABLE : 5 DETAILS OF CRACK SPACING OF BEAMS WITH DIFFERENT AMOUNTS OF TENSION REINFORCEMENT

Beam NO.	A_{st}	a_f	V_f	Crack spacing - mm			
				at working load		at steel yield	
				Maximum	Mean	Maximum	Mean
6	2 No. 8 mm	...	...	225.0	153.0	225.0	153.0
4	2 No. 8 mm	65	1.3	138.0	104.0	138.0	104.0
12	2 No. 12 mm	...	...	220.0	132.5	138.0	88.5
10	2 No. 12 mm	65	1.3	245.0	129.3	193.0	86.2
20	2No. 16 mm	...	...	95.0	69.4	85.0	62.5
24	2 NO. 16 mm	...	...	100.0	68.6	100.0	68.6
18	2 No. 16 mm	65	1.3	165.0	109.6	100.0	67.0
45	2 No. 16 mm	80	1.3	135.0	97.5	135.0	97.5
23	2 No. 16 mm	50	1.9	105.0	78.1	105.0	78.1
22	3 No. 16 mm	...	...	90.0	65.6	90.0	65.6
21	3 No. 16 mm	50	1.9	110.0	73.1	110.0	73.1
46	3 No. 16 mm	80	1.5	95.0	70.6	95.0	70.6

Crack Width

a) Beams with fibre over the whole seaction : The maximum crack width only has been discussed in detail in this section and not the minimum or mean width as they have no practical significance. In most of the beams the cracks which occured first have the maximum width. The crack width variation with moment are given in Fig. 4. This figure shows that at working load a reduction of 50 to 90% in the maximum crack width of beams with fibres over that of beams without fibres. Increase in service load at 0.3 mm crack width is seen to vary between 25 an 85%. Similarly the increase is between 50 to 146% and 73 and 212%, at 0.2 mm and 0.1 mm maximum crack width. Variations of percentage increase in service load due to inclusion fibres are presented in Fig. 5. The effects of volume percentage and aspect ratio of fibre on the reduction in crack width and increase in service load are presented in Fig. 6. The general trend is that the increase in fibre content increased the percentage reduction in crack width, but with respect to the aspect ratio there seems to exist an optimum value around 80/65 for a volume percentage of 1.6.

b) <u>Beam with fibre surrounding the tensile steel</u> : *Maximum crack widths in beams with fibres around the tensile steel only and without fibres but with additional reinforcements equivalent to the fibre content are plotted against moment in Fig. 7. It is seen from this figure that it is possible to improve the crack resistance performance of fibre reinforced concrete beams by addition of properly calculated volume steel fibres around the tensile steel only, instead of increasing the amount of tensile steel.*

c) <u>Effect of repeated loading</u> : *Limited studies on 3 beams (one with fibre, another without fibre and the third with additional bar reionforcements equivalent to the fibre content) were carried out and the results are presented in Fig. 8. It is seen that the maximum crack width was about 45% less for the beam with fibre compared to that of a beam without fibres. But the benefit in the reduction of maximum width due to fibre addition was comparable to that with equivalent extra tension steel initially only and not after 20 cycles of loading. Further extensive experimental work in this direction is necessary for studying the influence of repeated loads.*

CONCLUSIONS

1. Load at first visible crack increases by 50 to 128 % due to inclusion of steel fibres over the whole section of a reinforced concrete beam. But the increase is of the order of 30% only, in beams with fibres around the tension steel only, compared to beams without fibres and beams with extra reinforcement equivalent to the quantity of fibre around tension steel in lieu of fibres.

2. The presence of fibres reduces the crack height by about 25%

3. The effect of fibres on maximum or mean crack spacing has been found to be insignificant irrespective of the fibre aspect ratio and volume percentage.

4. There is a reduction of 50 to 90% in the maximum crack width at working load depending on the fibre aspect ratio.

5. The general trend is that an increase in the fibre content and aspect ratio increased the percentage increase in service load at a specified crack width.

6. Research work to study the effect due to repeated loading on reinforced concrete beams with steel fibres is necessary.

REFERENCES

1. Samarrai Muffid A., and Elvery Robert H., "The influence of fibres upon crack development in reinforced concrete subject to uniaxial tension", Magazine of Concrete Research, Vol. 26, No. 89, 1974. pp. 203-211

2. Swamy, R.N. and Al-Noori, K.A.,"Flexural behaviour of fibre concrete with conventional steel reinforcement", Fibre Reinforaced Cement and Concrete (RILEM Symposium, 1975), The Construction Press Limited, Lancaster, 1975, pp. 187-196.

3. Henager Charles H., and Doherty Terrance J., "Analysis of reinforced fibrous concrete beams", Proceedings ASCE, Vol. 102, No. ST1, 1976 pp. 177-188

4. Swamy, R.N., Al-Tann, Saad A., and Ali Samy, A.R., "Steel fibre for controlling cracking and deflection", Concrete International Vol.1, No. 8, 1979, pp. 41-49

5. Kormeling, H.A., Reinhardt H.W. and Shah S.P. "Static and fatigue properties of concrete beams reinforced with continuous bars and with fibres", Proceedings, American Concrete Institute, Vol. 77, No. 1, 1980, pp. 36-43

6. Swamy, R.N. and Al-Tann "Deformation and ultimate strength in flexure of reinforced concrete beams made with steel fibre concrete", Proceedings American Concrete Institute, Vol. 78, No. 5, 1981, pp. 395-405

7. Sabapathi, P. "Flexural behaviour of reinforced concrete beams with steel fibres", Ph.D. thesis, Indian Institute of Technology, Madras, October 1983.

8. Adler, Yu.P, et.al.,"The Design of Experiments to Find Optimal Conditions", Translated from the Russian, by G.Leib, Mir Publishers, Moscow,. 1975.

9. Fisher, R.A.," The Design of Experiments", Oliver and Boyd, Edinburgh, 1960.

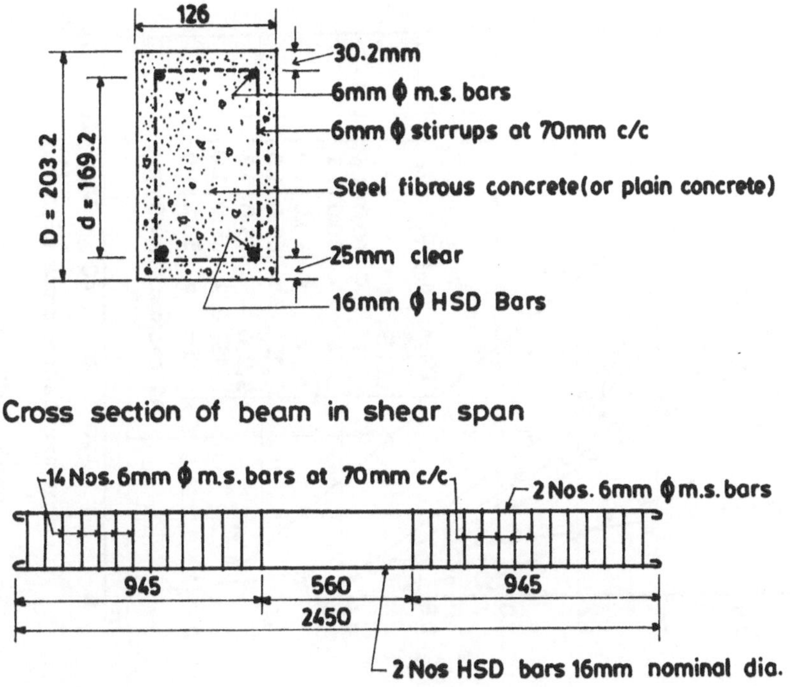

a. Cross section of beam in shear span

b. Longitudinal section of reinforcement grill

Fig. 1. Details of reinforcement

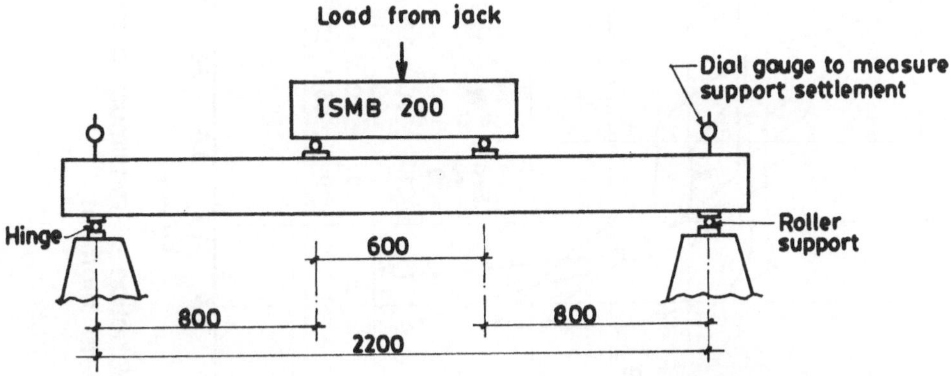

All dimensions are in mm

Fig. 2. Loading arrangement for test beam

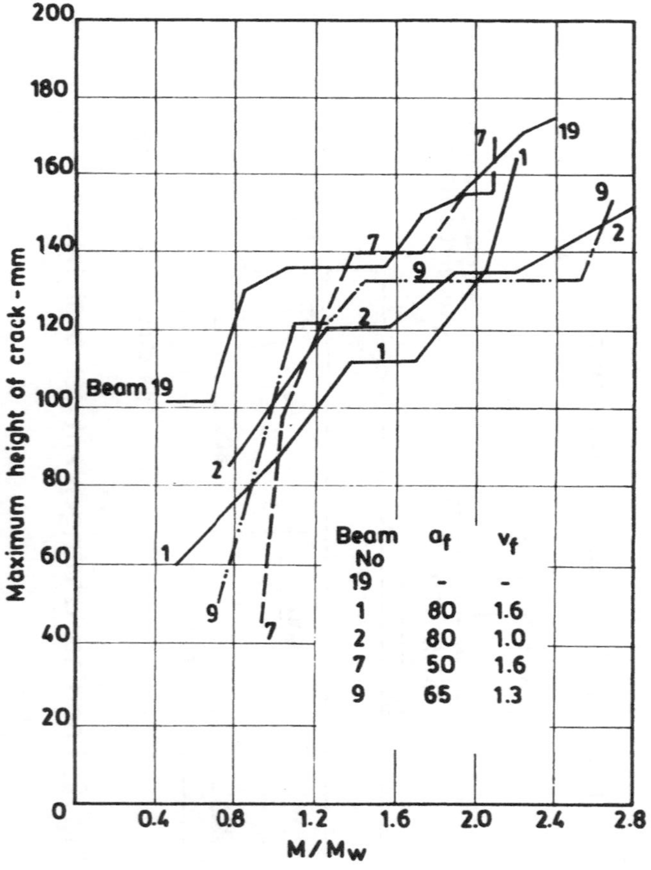

Fig. 3. Variation of maximum crack height with moment

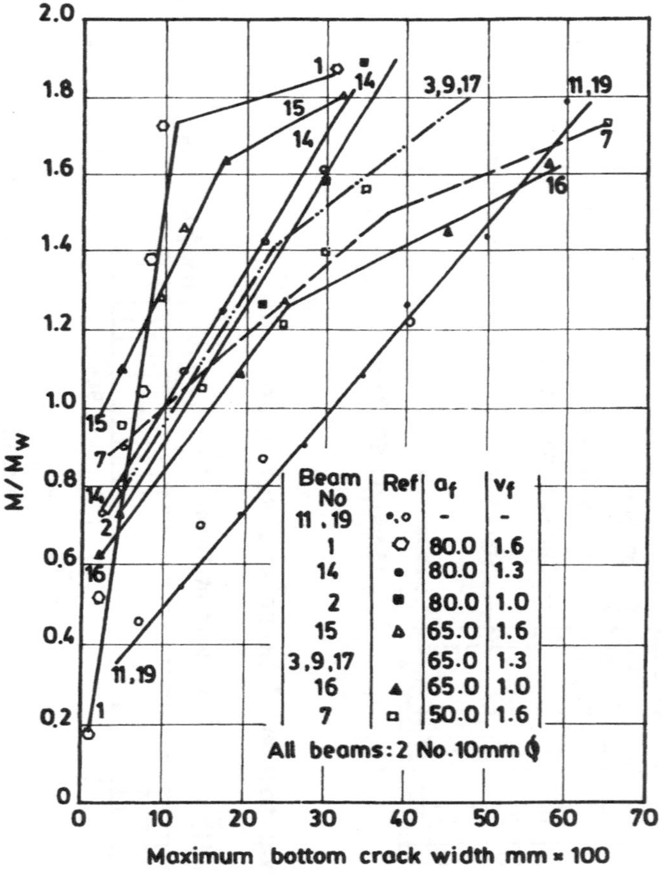

Fig. 4. Crack width in beams with fibres over the whole section and beams without fibres

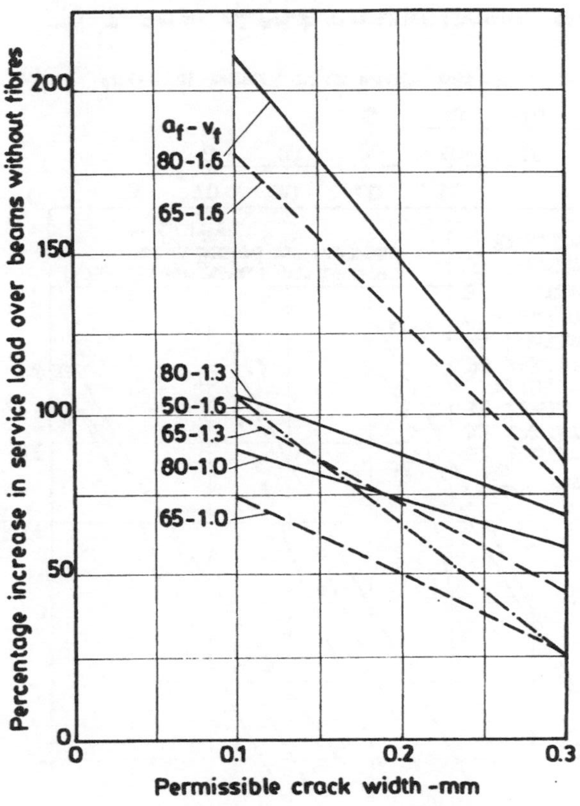

Fig. 5. Increase in service load at different crack widths due to fibre

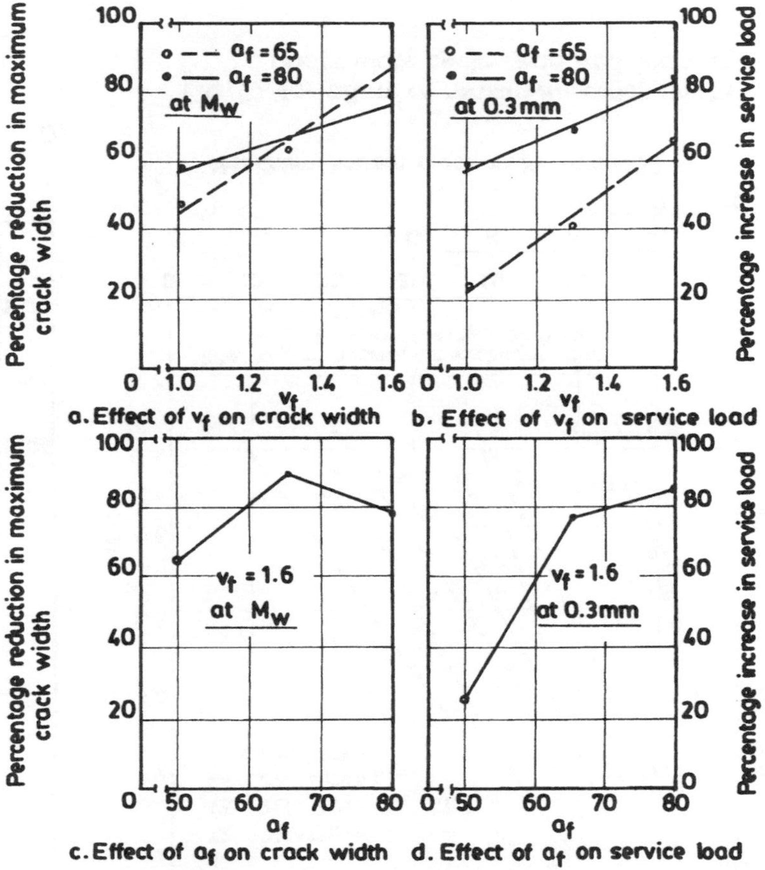

Fig. 6. Effects of v_f and a_f on percentage reduction in crack width and increase in service load

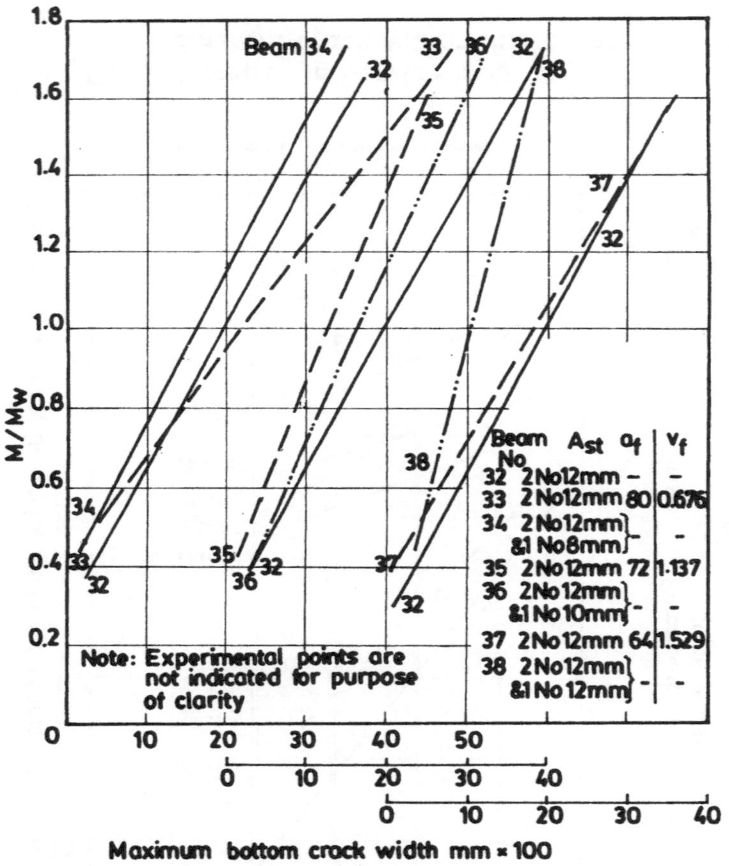

Fig. 7. Effect of fibre around tension steel on crack width

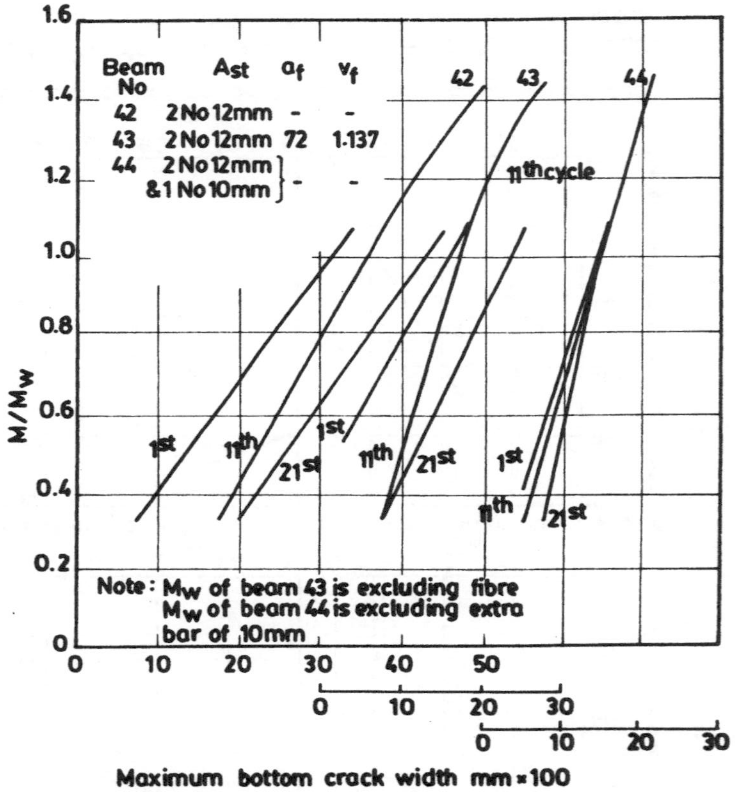

Fig. 8. Variation of maximum crack width under slow cycle repeated loading

Proceedings of the International Symposium on Fibre Reinforced Concrete, December 16-19, 1987, Madras, India

PERFORMANCE CHARACTERISTICS AND FLEXURAL FATIGUE STRENGTH OF CONCRETE STEEL FIBRE COMPOSITES

V. Ramakrishnan, Professor of Civil Engineering
South Dakota School of Mines and Technology
Rapid City, South Dakota 57701, U.S.A.
and
Charles Josifek, Vice President-Sales/Marketing
Ribbon Technology Corporation
Gahanna, Ohio 43230, U.S.A.

Synopsis: This paper presents the results of an experimental investigation to determine the flexural fatigue strength of concrete reinforced with deformed (corrugated) and melt-extract steel fibres. The test program included (1) flexural fatigue and endurance limit, (2) static flexural strength including load-deflection curve, determination of first-crack load, toughness index, (3) compressive strength, (4) static modulus, (5) pulse velocity, (6) unit weight, workability and finishability of fresh concrete.

There was no "balling" or tangling of the fibres during mixing and placing. Fibre reinforced concretes had better finishability and were easy to work with even at higher fibre concentrations. Due to the addition of fibres, the ductility and the post-crack energy absorption capacity were greatly increased. There was a tremendous increase in the static flexural strength and a very significant increase in the flexural fatigue strength. When compared to plain concrete, there was a considerable improvement in the endurance limit (for 2 million cycles).

INTRODUCTION

A review of previous research by the authors (1-10) and others (10 to 17) has well established that the addition of steel fibres in concrete considerably improves the static flexural strength, impact strength, shear and torsional strength, direct tensile strength, fatigue strength, shock resistance, ductility, and failure toughness. However the degree of these improvements depends on the type, size, shape and aspect ratio of the steel fibres. It has been shown that deformed steel fibres perform better in fresh concrete than straight steel fibres in improving the workability and eliminating balling. Deformed steel fibres are also more efficient than straight steel fibres in improving the desired hardened concrete properties.

Most of the previous work (1 to 10) was done using 0.50mm (0.02 in) diameter fibres. Where possible, it is more economical to use longer and larger fibres like 0.75mm (0.03 in.) diameter fibres because they are cheaper. However very little information is available about the performance of these longer and larger fibres in fresh and hardened concrete. This investigation will generate some data which will be useful for concrete design engineers.

Fibre reinforced concrete has been found more economical for use in airport and highway pavements. It is also used for thin section concrete applications such as sewer pipes, bridge overlays and curtain walls. Fibre concrete has been successfully used to minimize cavitation/erosion damage in structures such as sluiceways, spillways, navigation locks, and bridge piers where high velocity flows are encountered. It is also rapidly gaining acceptance as a suitable material for repair, rehabilitation and renovation of concrete structures. In many of these applications, particularly in pavements and bridge deck overlays, the flexural fatigue strength and the endurance limit are important design parameters because these structures are designed on the basis of fatigue load cycles. The greatest advantage in adding fibres to concrete is the improvement in flexural strength in both static and fatigue loading. Plain concrete has a fatigue endurance limit of about 50 to 55% of its static flexural strength (16, 17). A properly designed steel fibre reinforced concrete can achieve 90 to 95% endurance limit (4). With a higher endurance limit the concrete cross sections could be reduced. Alternatively, using the same cross section would result in a much longer life span or higher load carrying capacity or both.

However there is very little information available about the performance of steel fibre reinforced concrete in flexural fatigue loading. Therefore this investigation was undertaken to provide the needed information for concrete design engineers.

OBJECTIVES

The primary objective of this investigation is to determine the behaviour and endurance limit for concrete beams reinforced with deformed and melt extract steel fibers when subjected to non-reversed fatigue loading. The other main objectives of this investigation are as follows:

1. To determine the performance characteristics of four different types of steel fibres during mixing.

2. To determine the properties of the fresh and hardened concrete reinforced with these fibres and to record the post-crack behaviour under static flexure.

3. To compare the performance of fresh concrete, and the properties of the hardened concrete with and without fibers, using the same mix for both the concretes.

MATERIALS, MIXES, AND TEST SPECIMEN

Materials

Type I normal Portland Cement satisfying ASTM C150 was used for all mixes. The fine aggregate used was natural sand which had a saturated surface dry specific gravity of 2.64, an absorption of 1.64 percent and a fineness modulus of 3.0. The coarse aggregate used was crushed limestone with a maximum size of 3/4". The limestone had a saturated surface dry specific gravity of 2.68 and an absorption of 0.54 percent and a fineness modulus of 6.43. The water used was the tap water from the municipal supply system. It was maintained at 21.1°C to 22.20°C. The air-entraining admixture used was neutralized vinsol resin, and the superplasticizer used was "Mighty 150", a salt of naphthalene sulfonate formaldehyde condensate.

Fibres

Three of the steel fibres used were produced by the Ribbon Technology Corporation, Gahanna, Ohio. The XOREX fibres, are produced from a mild carbon steel with a tensile strength ranging from 860 to 1550 MPa (125,000 to 225,000 psi). They are available in two shapes. XOREX-1 is a flat, corrugated-ribbon type of fibre while XOREX-2 is a round corrugated-wire type fibre. They come in lengths of 25 to 75 mm (1 to 3 inches). The 50 mm (2 in.) fibres were used in this investigation. The diameter of the fibre (or equivalent diameter for the flat fibre) is .75 to 1.27 mm (0.03 in. to 0.05 in). This results in an aspect ratio of about 40 to 65 for the 50 mm (2 in.) fibre.

The melt extract fibres, produced by Ribbon Technology Corporation, are made from molten metal, as opposed to steel wires or sheets. The irregular shape improves bonding. The tensile strength is about 482 MPa (70,000 psi) and the equivalent diameter is about .75 mm (0.030 in.) for an aspect ratio of 67.

The FLEXTEN fibre is produced by D & C Supply Company, Inc. Bridgeville, Pennsylvania. It is a straight, smooth, rectangular carbon steel fibre with nominal dimensions of .25 x .56 x 25 mm (0.01 in. x 0.022 in. x 1.0 in.) and an aspect ratio of 60.

Mixes

Eleven mixes were made in this project. The same mix proportions as shown in Table 1 were used for all mixes. Each mix was made in one batch of .091m^3 (3.25 cu. ft.). Two mixes were control mixes and contained no fibres. The other mixes contained 59.33 Kg/m^3 (100 lb/cu.yd.) steel fibres. The mix designations, the types and quantities of fibres used are given in Table 2.

All mixing was done in a 0.168m^3 (six cubic feet) capacity drum mixer. The batching procedure used was as follows:

The fine and coarse aggregates were placed in the mixer in alternate layers; ½ fine, ½ coarse, ½ fine, ½ coarse. The 3/4 of the required mixing water was added and the contents were mixed for one minute to thoroughly wet the aggregates. Next the cement was added, followed by the remaining ¼ of the water with the air-entraining agent mixed in. The fibres were then sprinkled in by hand to insure even distribution. Finally the superplasticizer was poured in and the contents were mixed for three minutes. This was followed by a three minute rest period and a final mixing period of two minutes (ASTM designation C 192).

Specimens

The following specimens were cast from each mix: six 152 x 304 mm (6" x 12") cylinders for compression and static modulus tests. Nine 100 x 100 x 356mm (4" x 4" x 14") beams for static flexure and flexural fatigue tests.

The specimens were covered with a plastic sheet and cured for 24 hours at room temperature. They were then demolded and placed in a lime saturated water tank maintained at 22.2 degrees C (72°F). They remained in the water until they were tested at 7 or 28 days.

Tests For Plastic Concrete

The freshly mixed concrete was tested for temperature, slump (ASTM C 143), air content (ASTM C 231), and plastic unit weight. The results of these tests are given in Table 3.

Tests For Hardened Concrete

Cylinders were tested for compressive strength at 7 and 28 days (3 specimens each), according to ASTM C 39. At 28 days they were also tested for static modulus. Unit weight was calculated based on dimensions measured at both 7 and 28 days.

Beams were tested at 7 and 28 days (3 specimens each) for pulse velocity and static flexural strength (modulus of rupture). Load-deflection data from the test for static flexural strength was used to calculate the toughness index. At 28 days three or more beams were also tested for flexural fatigue strength.

In the test for static flexure the beams were loaded at the third points with a span of 304mm (12"). A dial gauge accurate to 0.00025mm (.0001 in.) was placed under the beam at the centerline. Load and deflection readings were taken

Table 1: Basic Mix Proportions

Fine aggregate	925.52 Kg/m^3	(1560 lbs/cu.yd.)
Coarse aggregate	925.52 Kg/m^3	(1560 lbs/cu.yd.)
Cement	390.38 Kg/m^3	(658 lbs/cu. yd.)
Water	156.03 Kg/m^3	(263 lbs/cu.yd.)
Steel fibers	59.33 Kg/m^3	(100 lbs/cu.yd.)
Air-entraining agent	312.17cc/m^3	(8.4 fl.oz/cu.yd.)
Water reducing agent	4162.23cc/m^3	(112 fl. oz/cu.yd.)

Table 2: Mix Designations

1. PL Control concrete mix without fibers.
2. OB-12A Replicate of above mix.
3. TX1 Trial mix with XOREX1 type fibers. 59.33 Kg/m^3 (100 lb/cu.yd.)
4. X1 Mix with XOREX1 type fibers. 59.33 Kg/m^3 (100 lb/cu.yd.)
5. OB-5 Replicate of above mix.
6. X2 Mix with XOREX2 type fibers. 59.33 Kg/m^3 (100 lb/cu.yd.)
7. OB-11 Replicate of above mix.
8. X1FL Mix with combination of two types of fibers: 39.55 Kg/m^3 (66.66 lb/cu.yd.) of XOREX1 and 19.77 Kg/m^3 (33.33 lb/cu.yd.) of FLEXTON
9. OB-6 Replicate of above mix.
10. ME Mix with MELT EXTRACT fibers. 59.33 Kg/m^3 (100 lb/cu.yd.)
11. OB-10 Replicate of above mix.

Table 3: Properties of Fresh Concrete

Mix	Dosage of Superplasticizer-cc	Concrete Temp °C	Slump cm	Air Content %	Plastic Unit Weight Kg/m^3
PL	400	22.1	21.0	7.6	2258.6
OB12A	200	24.0	10.2	5.4	2354.7
OB5	250	26.4	12.7	5.5	2402.8
OB6	250	26.8	8.9	4.3	2402.8
OB10	275	26.3	12.7	4.7	2418.8
OB11	270	26.2	12.7	4.4	2402.8
X1	400	22.7	12.1	5.0	2402.8
X2	400	20.4	20.3	6.9	2386.8
X1F1	400	19.8	12.7	7.0	2338.7
ME	400	19.3	7.6	5.8	2418.8
TX1	600	21.6	20.3	6.0	2338.7

All mixes were made with the same water to cement ratio of 0.40.

at intervals of 90.6 Kg (200 lbs.) up to first crack and at 45.3 Kg (100 lbs.) intervals, when possible, thereafter. The load was applied at a rate of about 453 Kg/min. (1,000 lbs/min.). The maximum load reached was noted for each beam and an average maximum load was calculated for each mix.

Fatigue Strength

The main thrust of the investigation was to determine the endurance limit in flexural fatigue loading of the various mixes used. For this test the endurance limit was defined as the maximum load at which the specimen could withstand 2,000,000 cycles of non-reversed fatigue loading. The 2,000,000 cycle limit was chosen to approximate the life span of a structure that may typically be subjected to fatigue loading, such as a bridge deck or highway pavement. It was known from previous testing that if the specimen could withstand 2,000,000 cycles without failure it could last for all practical purposes, forever (4, 6, 17). The range of the cyclic loading was expressed as a percentage of the average maximum load in static flexure for each mix. The lower limit for all tests was 10% of this maximum load. The upper limit was varied depending on the particular mix, generally from 55% to 80% of the maximum load. For example, suppose three beams from a mix were tested in static flexure and the maximum loads were 3,800 lbs, 4,200 lbs and 4,000 lbs the average maximum load for that mix would be 4,000 lbs, the lower limit for the flexural fatigue tests would be 400 lbs and the upper limit, for 70%, would be 2,800 lbs to start with. Our procedure was as follows. If the beams failed before 2,000,000 cycles the upper limit was lowered, say from 70% to 60% (2,800 lbs to 2,400 lbs). If the beam did not fail the upper limit was raised, say from 70% to 80% (2,800 lbs to 3,200 lbs) and the beam was retested. This was repeated until we could determine approximately at what percent of maximum load the beam could withstand 2,000,000 cycles without failure. Similar to static flexure test, the third-point loading was used with a span of 304mm (12") in the test for flexural fatigue strength.

The frequency of loading used was 20 cycles per second (Hz) for all tests. It has been shown that frequency has little or no effect unless extremely high rates are used (16). Therefore, for purposes of expediency, a rate of 20 Hz was used instead of 5 to 10 Hz as is sometimes used. Even so, a test to 2,000,000 cycles would take almost 28 hours to complete.

The machine used for these tests has a dynamic load capacity of 25,000 Kg (55,000 pounds). The machine could be operated in any of three modes: load control (force applied to the specimen), strain control (Strain induced in the specimen), or stroke control (distance traveled by the ram, or deflection of the specimen). Since this test was concerned with stress levels load control was used. Each of the three modes also had four operating ranges 100%, 50%, 20% and 10% of maximum. The readout from the transducers was measured in volts, with a maximum of 10 volts, and the operating range controls defined the measured quantity in terms of voltage. Therefore a voltmeter was used to measure the load to an accuracy of 0.5 Kg (one pound) when the 20% range was used. It was also verified that the machine maintained its accuracy of loading at 20 Hz.

There was a choice of three waveforms that could be used: sine wave, square wave, and triangular wave. The sine wave was used because it was felt that this would most accurately model real world behavior.

There was also a counter built-in which kept track of the number of cycles to the nearest 100. When a beam failed this counter reading was recorded as the cycles endured by the specimen. The counter also served to turn the machine off when a specified number of cycles had been reached, 2,000,000 in this case. A mechanical cutoff switch was installed which would turn off the machine when the beam broke.

TEST RESULTS AND DISCUSSION

Plastic Concrete Properties

The results of the tests on fresh concrete are given in Table 3. The slump test results indicate that satisfactory workability can be maintained even with a relatively high fibre content. This was achieved by adjusting the amount of superplasticizer used, with the water and cement content and hence the strength remaining the same.

Another factor that was checked was whether there was any balling of the fibres. This can be a nuisance and waste of material, and can prove to be a major problem in a large scale project. Deformed and melt extract fibers performed very well. In mixes with 59.33 Kg/m^3 (100 lbs/cu.yd.) of fibres, there was no balling noticed. For the same concentration of fibres, using a smaller fibre with more fibres per pound of steel will tend to decrease the slump and workability. This can be noticed by comparing mixes OB5 and OB6, where a third of the 50mm (2 in.) long corrugated steel fibres were replaced with 25mm (1 in.) long straight steel fibers. This resulted in a loss of slump.

Based on the limited number of small samples tested in this investigation it was observed that the finishability for most mixes was satisfactory. The mix with melt-extract fibres was the easiest to finish. Even though there was an apparent reduction in slump in fibre concrete mixes, the concrete was much more workable than a plain mix of corresponding slump. The fresh concrete in general was more stable and exhibited a significant reduction of surface bleeding. The cohesion was greatly improved and increased cohesion reduced the segregation tendency.

In general the slump, air content, and concrete temperature all varied slightly, but these variations in the fresh concrete properties did not significantly affect the hardened concrete properties.

HARDENED CONCRETE PROPERTIES

Compressive Strength

The results of the tests for compressive strength are given in Table 4. At both 7 and 28 days, the addition of fibres seemed to have little affect on the compressive strength of concrete. The minor differences noticed are those that are normally expected during the experimental work.

Static Modulus and Pulse Velocity

The results for static modulus and pulse velocity tests are given in Table 4. An analysis of the results has shown that the values were reasonably consistent for all mixes with and without fibres and the addition of fibres had little or no effect on the static modulus of concrete.

At 7 days the average pulse velocity was 4703 m/s (15,420 fps) with a maximum of 4831 m/s (15,840 fps) (+2.7%) and a minimum of 4612 m/s (15,120 fps) (-1.9%). At 28 days the average was 4831 m/s (15,840 fps) with a maximum of 4895 m/s (16,050 fps) (+1.4%) and a minimum of 4764 m/s (15,620 fps) (-1.4%). This indicates a good degree of consistency and quality control, and that fibre content had little or no effect on pulse velocity.

Static Flexural Strength (Modulus of Rupture)

The static flexural strength test results are given in Table 4. The 7-day strength was about 80 to 90% of the 28-day strength. This increase was similar to that of the compressive strength increase for these concretes. Comparing all the fibre reinforced concretes with those of corresponding plain concretes it

Table 4: Hardened Concrete Properties

Mix	Compressive Strength, MPa		Flexural Strength, MPa		Static Modulus $MPa \times 10^4$	Pulse Velocity m/sec		Unit Weight Kg/m^3		Toughness Index	
	7 days	28 days	7 days	28 days	28 days	7 days	28 days	7 days	28 days	7 days	28 days
OB-12A	34.33	38.75	5.03	4.96	3.88	4745.7	4832.9	2402.8	2402.8	1.0	1.0
PL	35.58	37.51	4.07	5.00	3.28	4765.5	4767.1	2354.7	2290.6	--	1.0
OB-5	34.61	41.51	4.41	5.21	3.95	4659.5	4800.6	2466.8	2450.8	10.1	6.7
OB-6	33.85	42.06	4.69	5.52	3.83	4684.5	4835.3	2450.8	2466.8	9.9	8.3
OB-10	37.23	43.23	5.96	5.96	3.81	4753.4	4834.1	2466.8	2482.9	4.5	7.5
OB-11	37.37	43.99	5.72	6.24	3.92	4827.4	4802.7	2450.8	2482.9	7.8	10.3
X1	37.30	44.40	4.14	5.03	3.63	4710.7	4825.0	2466.8	2418.8	7.87	8.03
X2	35.92	45.57	4.83	5.83	3.97	4730.5	4821.9	2418.8	2418.8	11.1	11.68
X1FL	35.23	42.75	4.72	5.55	3.79	4661.9	4700.0	2434.8	2418.8	7.81	7.74
ME	47.71	49.22	5.48	6.52	3.92	4850.9	4927.1	2466.8	2450.8	6.07	7.24
TX1	37.23	45.92	4.28	5.27	4.01		4773.2	2466.8	2418.8	--	11.41

6.89×10^{-3} MPa = 1Psi

can be seen that there is a substantial increase in the static flexural strength due to the addition of steel fibres.

Two typical load-deflection curves are given respectively in Figures 1 and 2 for the specimens with Xorex 1 and the melt-extract fibres.

A significant difference in the performance of plain and fibre concrete was found in the static flexure test. The deformed fibres proved their ability as crack arrestors. The cracks were prevented from propagating until the composite ultimate stress was reached. The mode of failure was simultaneous yielding of the fibres and the matrix because during the test one could actually hear the popping sound of the fibres breaking in tension and not due to bond failure. It seems obvious that the deformations contributed significantly to the increase in the bond between fibre and matrix.

Close-up photographs show many broken fibres in the interior of the section. A few fibres near the surface have been pulled out due to lack of proper cover and embedment. A close inspection of the broken surface has indicated that a large percentage of the fibres broke rather than being pulled out.

The significance of good bond can be seen from the load-deflection curves. The curves indicate a ductile behaviour. These curves also show the advantage of fibre reinforced concrete versus non-fibre reinforced concrete in obtaining higher load carrying capacity, higher flexural strength, higher toughness, and excellent energy absorbing characteristics. The apparent energy absorbed is represented by the total area under the load-deflection curve.

It should be recognized that there is always more scatter and sometimes one or two odd results in testing fibre reinforced concrete because of the random orientation of the fibres and possible non-uniform distribution of fibres in small specimens.

Toughness Index

The results of the calculation for toughness index are given in Table 4. The load-deflection curves were used to calculate these values. Toughness index is defined as the area under the load-deflection curve up to a specified deflection, divided by the area under curve up to the point where the concrete first cracks. The area represents the energy required to cause the deflection of the beam. The value of this index depends on what specified deflection criteria is used. The toughness index used in this investigation is according to ACI Committee 554 recommended procedure, using the area under the load-deflection curve up to a deflection of 1.9mm (0.075") for 102 x 102 x 356mm (4" x 4" x 14") beams tested over a 305mm (12") span. The toughness index is a measure of the ductility of the specimen. Plain concrete fails immediately upon cracking, without further load carrying capabilities. This is a brittle failure and the toughness index is 1.0. Concrete beams reinforced with fibers will continue to deflect in a ductile fashion and carry loads well past the point of first crack. In this case the toughness index will be greater than 1.0.

The mixes reinforced with deformed steel fibres had very high toughness indices at both 7 and 28 days. These varied from 4.5 to 11.68, the lowest for the beams reinforced with melt extract fibres and the highest for beams with Xorex 2 fibers. The flexural strength of beams with melt extract fibres was the highest indicating an excellent bond between the fibers and the matrix. However, for these beams, the loads dropped rapidly after the initial cracking causing lower toughness index values (Figure 2). One beam failed immediately after initial cracking and other beams had relatively brittle failure.

Fatigue Strength

The results from the flexural fatigue tests are given in Table 5. Column number four shows maximum flexural stress. This was calculated using the actual

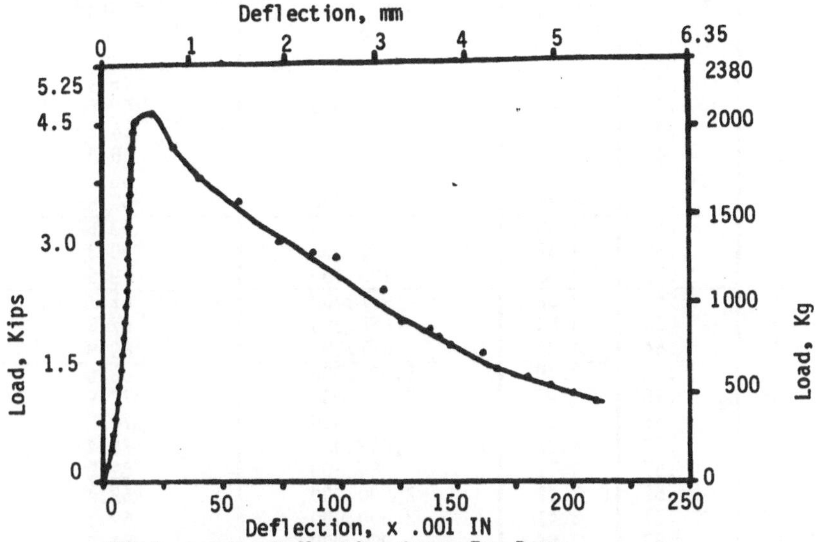

FIG 1: Load - Deflection Curve For Beam Reinforced With Corrugated Fibers (at 28 days)

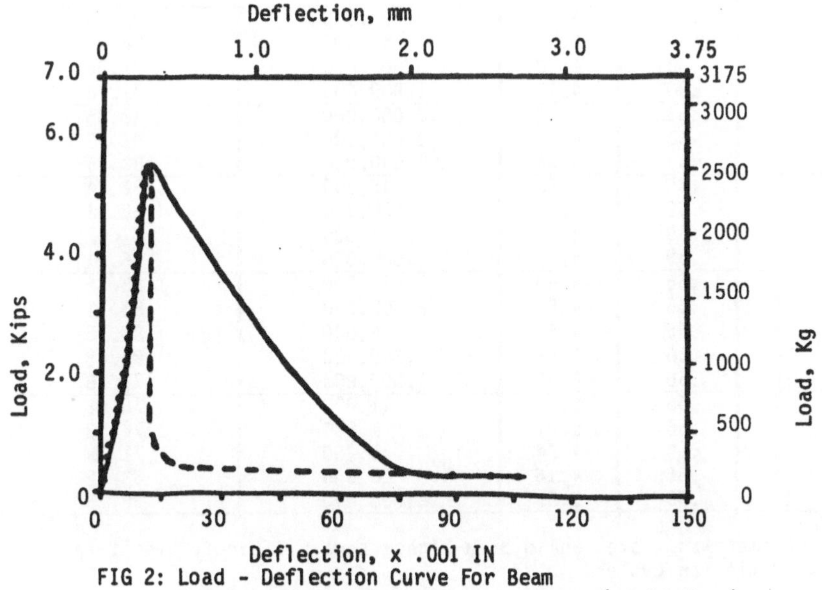

FIG 2: Load - Deflection Curve For Beam Reinforced With Melt Extract Fibers (at 28 days)

Table 5: Fatigue Test Results

Mix	Beam No.	Flexural Fatigue Stress, MPa		Number of Cycles to Failure	% of Static Flexural Stress
		Min.	Max.		
OB12	7	.49	2.45	2,000,000	49.4
OB12	12	.49	2.46	2,000,000	49.6
OB12	9	.49	2.71	2,000,000	54.6
OB12	10	.49	2.71	35,700	55.3
OB12	11	.51	2.80	13,400	56.4
OB12	8	.48	2.85	55,300	57.5
PL	4	.50	4.68	800	93.6
PL	10	.50	4.52	1,000	90.3
PL	12	.50	3.98	126,400	79.5
OB5	8	.50	3.52	907,400	67.7
OB5	9	.54	3.53	68,300	67.8
OB5	7	.54	4.32	517,100	82.9
OB6	8	.58	3.79	2,000,000	68.6
OB6	8*	.58	4.07	2,000,000	73.9
OB6	9	.56	3.93	148,800	71.3
OB6	8**	.59	4.38	1,249,800	79.4
OB6	7	.57	4.51	84,500	81.8
OB10	8	.57	3.12	2,000,000	52.4
OB10	8*	.57	3.41	2,000,000	57.1
OB10	8**	.57	3.69	1,085,900	61.9
OB10	7	.59	4.09	7,200	68.6
OB11	6	.59	4.17	2,000,000	66.9
OB11	6*	.59	4.47	2,000,000	71.6
OB11	9	.63	4.73	1,083,700	73.9
OB11	6**	.59	4.76	398,700	76.4
OB11	8	.63	5.09	429,700	81.6
X1	1	.50	4.72	97,900	93.70
	2	.50	4.47	59,300	88.90
	12	.50	4.38	2,000,000	87.00
	4	.50	4.12	1,909,600	81.78
TX1	4	.53	4.34	2,000,000	82.35
	6	.53	4.34	2,000,000	82.35
	8	.53	4.34	2,000,000	82.35
X2	12	.59	4.83	45,800	82.84
	6	.59	4.70	408,900	80.62
	4	.59	4.52	1,172,000	77.62
	7	.59	4.47	2,000,000	76.71
X1FL	1	.56	4.60	437,000	82.86
	9	.56	4.61	2,000,000	83.11
	3	.56	4.63	24,000	83.35
	4	.56	4.63	1,349,000	83.48
	10	.56	4.32	2,000,000	77.76
ME	1	.66	5.12	11,000	78.62
	2	.66	5.01	11,800	76.93
	3	.66	4.64	45,300	71.22
	4	.66	4.16	80,000	63.82
	12	.66	3.83	2,000,000	58.73

*This specimen was tested again at a higher stress after it didn't fail even at 2 million cycles.

**This specimen was tested for a third time at a higher stress. During the first and second tests, it didn't fail at 2 million cycles.

dimensions of the beam and the upper load limit used for the test. Column six shows maximum stress as a percentage. This actually indicates the upper load limit of the test as a percentage of the average maximum load from the static flexure tests for each mix. Column five shows the number of cycles at failure. The plain mix withstood a stress of only about 2.4 to 2.7 MPa (360 to 390 psi), and failed consistently around 2.7 MPa (400 psi). This represents an endurance limit of 50 to 55 percent of the modulus of rupture as determined from the static flexure test. This agrees with previous work done by the authors and others.

It is apparent from the test results that the deformed steel fibres provided a substantial improvement in fatigue strength. Beams with corrugated steel fibres safely withstood a stress of about 4.5 MPa (650 psi), 65% more than for plain concrete with an endurance limit of 72% of the flexural strength. For the beams with a mixture of corrugated and straight fibres, the endurance limit was higher at 75%. The beams with melt extract fibres also showed a marked improvement in fatigue strength, a 30% increase over plain concrete, with an endurance limit of about 60% of the modulus of rupture.

CONCLUSIONS

Based on the experimental investigation the following conclusions can be made:

1. Satisfactory workability can be maintained in fresh concretes reinforced with corrugated and melt-extract steel fibres even with the addition of about 1% by volume.

2. There was no balling of fibres even with higher concentration of fibres.

3. When compared to plain concrete, the fibre reinforced concrete had less bleeding, was more stable and had better finishability.

4. The addition of fibres did not produce any significant changes in the compressive strength, static modulus and pulse velocity values. A significant increase in the static flexural strength (modulus of rupture) was noticed in fibre concretes.

5. Due to the addition of steel fibres the mode of failure was changed from a brittle failure to a fully ductile failure with a great increase in the post-crack energy absorption capacity. There was a significant increase in the toughness index.

6. Flexural fatigue strength and the endurance limits were increased substantially with the addition of corrugated and melt extract steel fibres.

ACKNOWLEDGMENTS

The authors gratefully acknowledge that their research was conducted at the Civil Engineering Department of the South Dakota School of Mines and Technology, and was partly supported by the National Science Foundation under grant No. MSM-8612082.

REFERENCES

1. Ramakrishnan, V.; Brandshaug, Terje; Coyle, W. V.; and Schrader, Ernest K., "A Comparative Evaluation of Concrete Reinforced with Straight Steel Fibers and Fibers with Deformed Ends Glued Together into Bundles," ACI Journal, Proceedings V. 77, No. 3, May-June 1980, pp. 135-143.

2. Ramakrishnan, V.; Coyle, W. V.; Kulandaisamy, V.; and Schrader, E. K., "Performance Characteristics of Fiber Reinforced Concrete with Low Fiber Contents," ACI Journal, Proceedings V. 78, No. 5, Sept-Oct. 1981, pp. 384-394.

3. Ramakrishnan, V.; Coyle, W. V.; Dahl, Linda Fowler; and Schrader, E. K., "A Comparative Evaluation of Fiber Shotcrete," Concrete International: Design & Construction, V. 3, NO. 1, January 1981, pp. 56-59.

4. Ramakrishnan, V., "Flexural Fatigue Strength of Dramix Steel Fiber Reinforced Concrete" a Report submitted to Bekaert Steel Wire Corporation, Nile, Illinois, August, 1985.

5. Ramakrishnan, V., "Flexural Fatigue Strength of SIFCON and Concrete Reinforced with Xorex Steel Fibers", a Report submitted to Ribbon Technology Corporation, Ohio, June, 1985.

6. Ramakrishnan, V., "Flexural Fatigue Strength of Polypropylene Fiber Reinforced Concrete", A report submitted to Forta Corporation, Grove City, PA. 16127, July, 1985.

7. Shah, S. P. and Skarendahl, Ake, Editors, "Steel Fiber Concrete", Elsevier Applied Science Publishers, London, 1986, pp. 7-24.

8. Ramakrishnan, V. and Balaguru, P. N., "Freeze-Thaw Durability of Fiber Reinforced Concrete", ACI Journal, Proceedings, V. 83, No. 3, 1986, pp. 374-481.

9. Ramakrishnan, V., "Mechanical Properties of Superplasticized Fiber Reinforced Concrete Developed for Bridge Decks and Highway Pavements", American Concrete Institute, separate publication SP-93, Concrete in Transportation, ACI, Detroit, 1986, pp. 563-584.

10. Balaguru, P. and Ramakrishnan, V., "Comparison of Slump Cone and V-B Tests as Measures of Workability for Fiber-Reinforced and Plain Concrete," Cement, Concrete and Aggregates, CCAGDP, Vol. 9, No. 1, Summer 1987, pp. 3-11.

11. "State of the Art Report on Fiber Reinforced Concrete," ACI Committee 544, Concrete International, American Concrete Institute, May 1982.

12. ACI Committee 544, "Measurement of Properties of Fiber Reinforced Concrete," (ACI 544.2R-78), American Concrete Institute, Detroit, 1978.

13. Hannant, D. J.; Fibre Cements and Fibre Concrete, John Wiley and Sons, Ltd., 1978.

14. Fiber Reinforced Concrete, SP-81, American Concrete Institute, Detroit, 1984, 452 pp.

15. Fiber Reinforced Concrete, SP-44, American Concrete Institute, Detroit, 1974, 544 pp.

16. Neville Adam M.; "Hardened Concrete: Physical and Mechanical Aspects," American Concrete Institute, Detroit, Michigan and Iowa State University Press, Ames, Iowa, 1971.

17. Brandshaug, Terje, "A Comparative Evaluation of Concrete Reinforced With Straight Steel Fibers and Collated Fibers With Deformed Ends," M. S. Thesis, South Dakota School of Mines and Technology, 1978.

Proceedings of the International Symposium on Fibre Reinforced Concrete, December 16-19, 1987, Madras, India

DYNAMIC BEHAVIOUR OF FIBRE REINFORCED CONCRETE BEAMS

P. Srinivasulu N. Lakshmanan
K. Muthumani B. Sivarama Sarma

Scientists
Structural Engineering Research Centre
Madras, INDIA

SYNOPSIS

The paper presents an experimental study on the dynamic properties of steel fibre reinforced concrete (SFRC) beams having varying percentages of high strength steel fibres as well as medium strength main reinforcing steel. Test beams having a cross section of 100 mm x 200 mm and equally reinforced on the tensile and compressive faces have been studied. These have been subjected successively to steadystate forced vibration tests after they have been initially loaded to predetermined static loads to simulate different levels of cracking in them. The properties studied are the dynamic flexural rigidity $(EI)_d$ and damping(ξ). The influences of the parameters varied in the test programme on these dynamic properties at different stages of preloading, before and after cracking are discussed.

1. INTRODUCTION

It is well known that the addition of fibres to a concrete matrix would improve significantly many of its engineering properties. The fibres that are currently being used are steel, glass and polypropylene. Even natural fibres have been experimented by earlier researchers and efforts are continuing to evolve new fibres [1].

The main advantages of fibre reinforced concrete are better resistance to cracking, crack propagation, less deflection and ductility. Henager, Craig and others have shown high ductility for SFRC beams [2,3]. Tests at New Jersey Institute of Technology [4] have shown that addition of fibres significantly increases energy absorption capacity of structural joints.

The important dynamic applications of fibre reinforced concrete are machine foundations [5], blast and earthquake resistant structures [6] and structures exposed to impact forces [7]. Steel fibres are very popular and in most of the applications where they have been used, the fibre content generally varies from 0.2 to 2.0% by volume of concrete. For practical purposes, however, higher percentages of fibre may introduce problems of workability.

On the question of impact strength of fibre concrete, considerable literature is available [7,8,9]. The results available are, however, generally qualitative in nature. Impact is a complex dynamic phenomenon involving crushing, shear failure and tensile fracturing. It is also associated with penetration, perforation, fragmentation and scaling of the target being hit.

Although considerable amount of literature is available on the engineering properties of fibre reinforced concrete, such as flexural and shear strengths based on static tests, very little information is reported on the dynamic behaviour of this material. The variation in stiffness $(EI)_d$ before and after pre-determined cracking stages and the influence of fibre on damping capacity of SFRC structural elements needs to be carefully studied. For a rational analysis of the behaviour of SFRC under dynamic conditions, it is essential to have an accurate knowledge of the constitutive properties of the structural material when subjected to a wide range of strain rates and strain levels.

As part of a research programme leading to the identification of new industrial applications for fibre reinforced concrete, Structural Engineering Research Centre at Madras has undertaken static and dynamic tests on SFRC beams having varying percentages of fibre as well as tensile and compressive steel. The results presented in this paper form a part of this study. Earlier investigations carried out at this Centre on concrete elements [10] have shown convincingly that beams reinforced equally on compressive and tensile faces perform better under dynamic conditions and yield higher ductility. Hence similar reinforcement detailing has been adopted in the test beams during this investigation.

2. EXPERIMENTAL PROGRAMME

Six test beams designated as B1 to B6 and having a rectangular cross section of 100 mm x 200 mm have been tested under flexure over an effective span of 2.06 m. The steel fibres having the aspect ratio, (l/d), equal to 100 and characteristic yield strength of 1000 N/mm^2 have been used in all the test beams. High strength deformed bars having a characteristic yield strength of 415 N/mm^2 have been used as main reinforcing steel in all the test beams. The nominal grade of concrete used is M20. Table 1 contains the details of the main reinforcement and steel fibre used in these beams.

Ultrasonic wave velocity tests have been carried out on all test beams in order to evaluate dynamic modulus, E_d. This test is based on the principle of wave propogation in elastic media. Ultrasonic waves are passed between two fixed points on the test beam at which two probes - an emitter and a receiver are rigidly held. The time for the compressional wave (P-wave) to reach from one probe to another is indicated in digital form. Knowing the distance travelled and the time recorded, the wave velocity (V_p) is deduced. The dynamic modulus, E_d is obtained from the relation

$$E_d = \frac{\rho V_p^2 (1 + \nu)(1 - 2\nu)}{(1 - \nu)} \qquad \text{... (1)}$$

where ν is Poisson's ratio taken as 0.3 for concrete and ρ is the mass density. E_d is evaluated for all the test beams.

TABLE - 1

CHARACTERISTICS OF TEST BEAMS

Sl. No.	Beam Designation	Percentage of Volume of fibre, V_f	Percentage of Main Steel P_t	Steel at top and bottom	Stirrups
1.	B1	0.5	0.92	2 - 10*	2 legged 8 * at 150 mm centres
2.	B2	0.75	0.92	2 - 10*	- do -
3.	B3	1.0	0.92	2 - 10*	- do -
4.	B4	1.0	1.37	2 - 12*	- do -
5.	B5	1.0	2.55	2 - 16*	2 legged 8 * at 125 mm centres
6.	B6	0.0	0.92	2 - 10*	2 legged 8 * at 150 mm centres

* Bar diameter in mm

Static and dynamic tests have been conducted alternately on each test beam. After setting up the beam over the supports, the steady state dynamic test is conducted to determine the precracking dynamic response. Fig.1 shows schematically the dynamic test set-up used. An electro-dynamic shaker, a power amplifier and a signal generator have been used for exciting the test beams, while an electro dynamic pick-up, an amplitude measuring unit and a digital frequency tracking meter are used for collecting vibration response data from the test beams. The beam was excited at a constant force level, but at varying frequencies. The amplitude-frequency data obtained from each beam tested, is recorded.

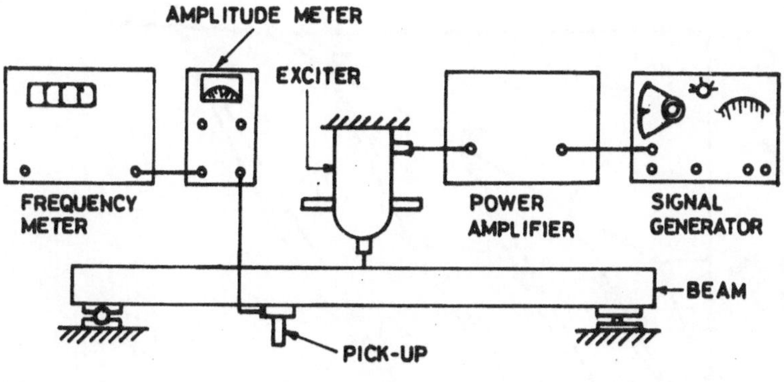

FIG. 1. SCHEMATIC DIAGRAM OF DYNAMIC TEST SET-UP

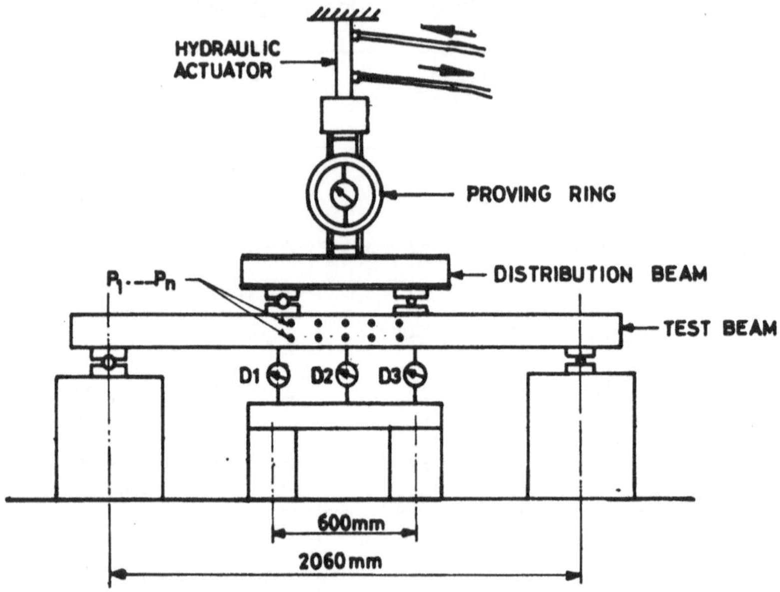

D1, D2, D3 – DIAL GAUGES

P_1... P_2.. P_n .– PELLETS FOR MECHANICAL STRAIN GAUGE

FIG. 2. SCHEMATIC DIAGRAM OF STATIC TEST SET-UP

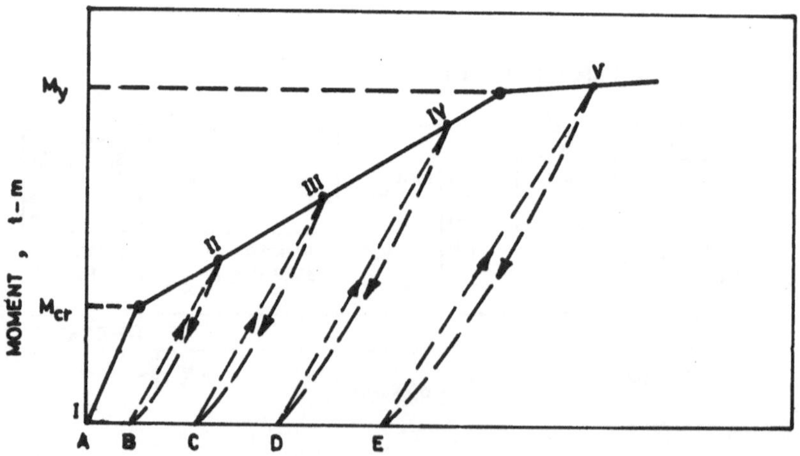

FIG. 3. IDEALISED MOMENT-CURVATURE RELATIONSHIP

Fig.2 shows schematically the test set-up used in static tests. During the static loading stage, strain and load deflections data are collected from the test beams. When the required load level is reached (to obtain a certain ratio of M/M_{cr}, where M and M_{cr} represent the applied static moment and cracking moment respectively), the static load is released and the static test set-up is replaced by the dynamic test set-up. Fig.3 shows the moment-curvature plot idealised as a tri-segmental curve. Points II to V in this figure represent the static load levels, while points B to E represent the stages at which dynamic tests have been conducted after cracking.

3. DISCUSSION OF RESULTS

3.1 Precracking Behaviour

3.1.1 Stiffness

Fig.4 shows response curves of two of the test beams B2 and B4 in the uncracked state. The frequency corresponding to the peak amplitude denotes the first resonant frequency of the test beam under flexure. The response plots are also used to evaluate damping in the test beams using band width technique [10]. The free and forced vibration tests in the uncracked state gave nearly the same frequency values. For example, free vibration test on beam B5 gave a value of 78 Hz (Fig.5) while the forced vibration gave a value of 76.5 Hz (Fig.4) for resonant frequency.

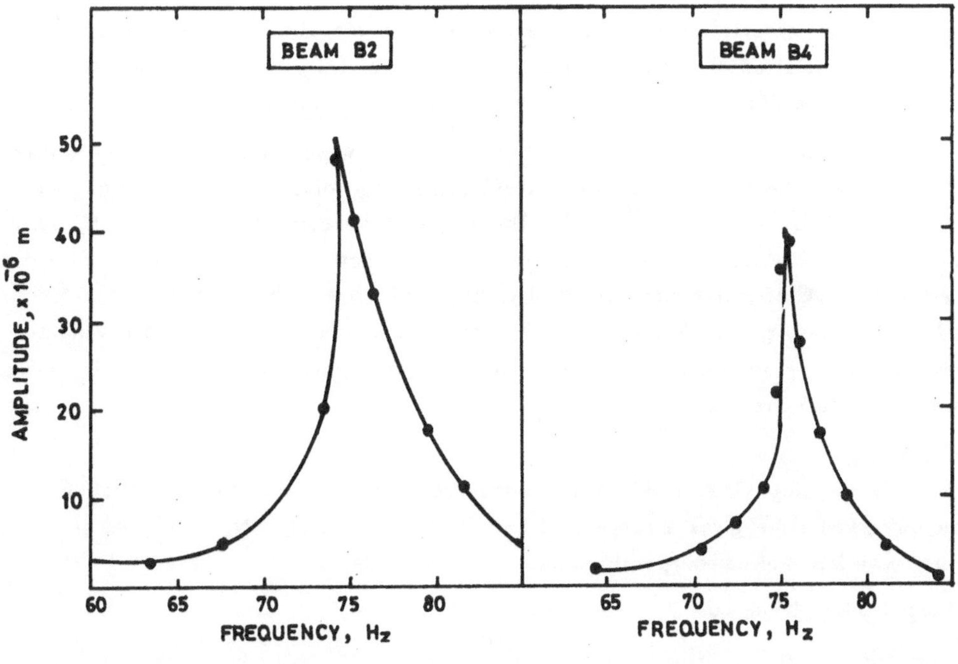

FIG. 4. TYPICAL RESPONSE OF UNCRACKED BEAMS (B2, B4) DYNAMIC TEST

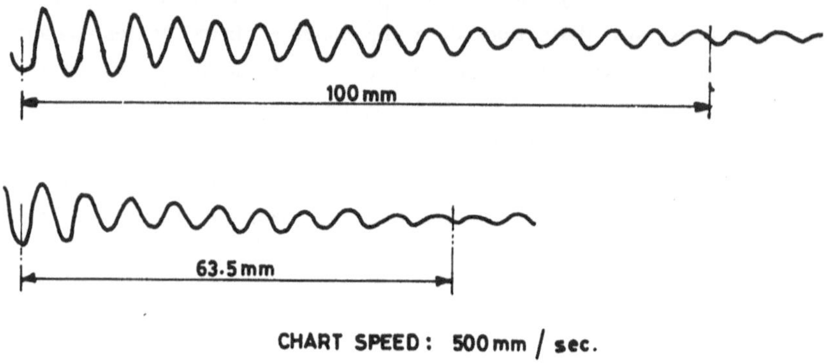

FIG. 5. TYPICAL FREE VIBRATION RECORDS
(BEAM B5 - UNCRACKED STAGE)

The resonant frequencies of all the test beams with varying parameters range from 72 to 78 Hz. The dynamic flexural rigidity $(EI)_d$ can be deduced for various test beams using the relation

$$(EI)_d = \frac{4}{\pi^2} \mu \ell^4 f_n^2 \qquad \text{... (2)}$$

where μ is mass per unit length of the test beam, l is the effective span and f_n is the experimentally evaluated frequency in Hz. The range of $(EI)_d$ is found to be from 175 to 215 tm^2 for the SFRC beams tested with varying fibre content in them ($0 < V_f \leq 1$).

The ultrasonic wave velocity tests gave a range of wave velocities from 3.7 km/sec to 4.0 km/sec for the test beams. The corresponding range of E_d values using Eq.1 works out to 2.6×10^6 t/m^2 to 3×10^6 t/m^2. This shows an increase of only 15% in E_d value for the test beams. Assuming that the gross section is effective for computation of moment of inertia (I_g), the effective flexural rigidity $(EI)_d$ works out to be in the range 175 tm^2 to 200 tm^2. This agrees closely with the experimental values obtained, thus validating the assumption of gross section for the computation of moment of inertia, and E_d value obtained from ultrasonic tests.

3.1.2 Damping

The damping value deduced from the free vibration and forced vibration test in the uncracked state gives a range of 0.8 to 1.6% of 'critical'. No consistant trend is seen in the variation of damping with variation of fibre content.

3.2 Post-Cracking Behaviour

3.2.1 Stiffness

Fig.6 shows the dynamic reponse of test beams B1 and B5 at various stages of cracking. Fig.7 shows a plot between (M/M_{cr}) and a factor α which is the ratio of $(EI)_d$ at a given preloading level to the uncracked $(EI)_d$ value.

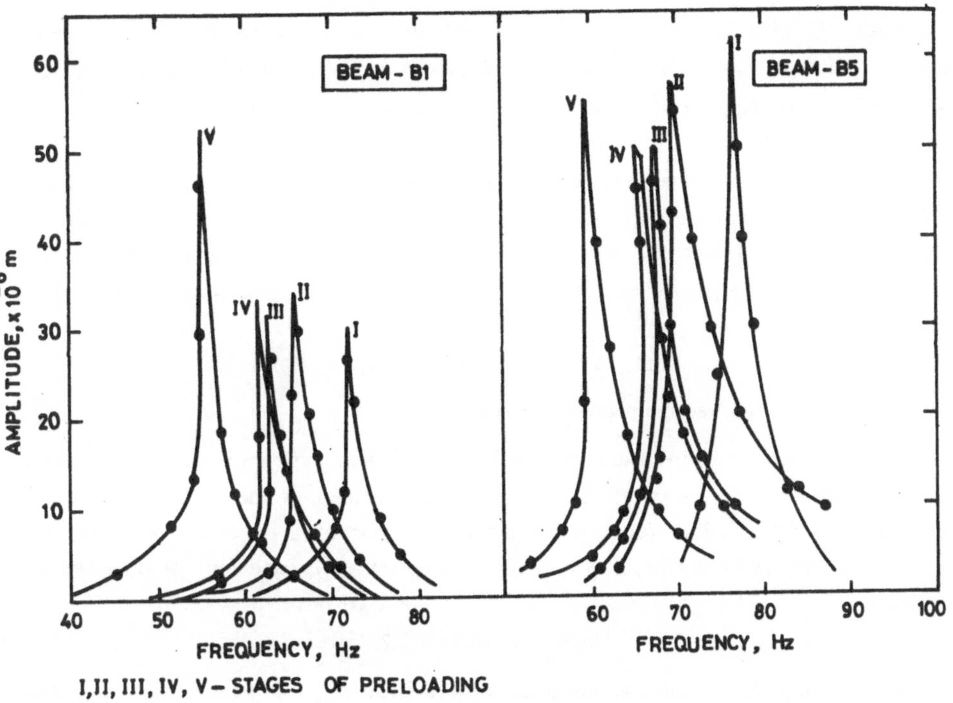

FIG. 6. DYNAMIC RESPONSE OF TEST BEAMS B1 and B5

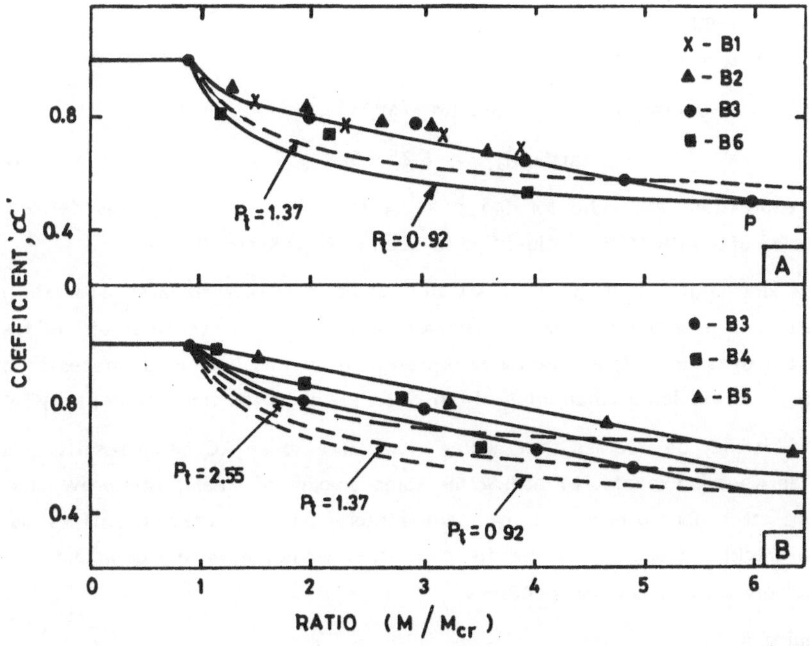

FIG. 7. VARIATION OF 'α' WITH (M/M_{cr})

Fig.7A shows the data of test beams B1, B2, B3 and B6 all having the same percentage of steel P_t equal to 0.92. It is seen that all points representing SFRC beams B1 to B3 follow a uniform decaying trend beyond the value of M/M_{cr} equal to 1, (representing cracked region) while beam B6 follows a similar trend but well below that of beams B1 to B3. Both curves, however, are seen to meet at a point P. Beyond this stage, the behaviour of SFRC and RC beams are expected to be identical. Fig.7B shows the variation of with (M/M_{cr}) for test beams B3, B4 and B5 all having V_f equal to unity, but varying p_t. The dotted lines in the figure represent the behaviour of RC beams deduced using the equation given below.

$$\alpha = [0.8\,(M_{cr}/M) + \{1 - 0.8\,(M_{cr}/M)\}\, I_{cr}/I_g] \qquad \ldots \quad (3)$$

where I_{cr} is equal to cracked moment of inertia.

As expected, the inclusion of steel fibres has significantly contributed to the increase in stiffness. However, it is seen that the maximum value of (M/M_{cr}) at which the behaviour of SFRC beam tends to be similar to that of a reinforced concrete beam decreases with increasing values of p_t. This limiting (M/M_{cr}) value may be obtained from the relation

$$M/M_{cr} = (6.25 - 0.25\,p_t) \qquad \ldots \quad (4)$$

It may be seen that the curvilinear variation of α with M/M_{cr} can be idealised as a multi-linear representation with ordinates of α defined thus:

i) $\alpha = 1$ for $0 < M/M_{cr} \leq 1$ ... (5.1)

ii) $\alpha = 0.83\,(p_t)^{0.15}$ at $M/M_{cr} = 1.5$ ... (5.2)
$(\alpha \neq 1)$

iii) $\alpha = [0.8\,(M_{cr}/M) + \{1 - 0.8\,(M_{cr}/M)\}\, I_{cr}/I_g]$
at $M/M_{cr} = 6.25 - 0.25\,p_t$ ... (5.3)

The above expressions are valid for $0.5 \leq V_f \leq 1.0$. Between the points defined above, linear variation of α with M/M_{cr} is justified as seen in Fig.7A and 7B.

The dotted line in Fig. 7A shows the variation of α with M/M_{cr} for the case of a RC beam with p_t equal to 1.37, an increase of nearly 0.5% over that of the test beam B6. Since the dynamic stiffness factor α represented by this line at all stages of cracking (upto $M/M_{cr} = 5$) is lower than that of an SFRC beam with fibre content V_f at least equal to 0.5, it may be inferred that adding steel fibre to an RC beam results in a higher stiffness than what is expected by adding the same amount of steel. This proves the advantage in using steel fibre over increasing main reinforcement if increased stiffness is desired in the post-cracking stage. Increase in main steel would, however, contribute to higher ultimate moment capacity of the specimen.

3.2.2 Damping

Using the band width method, the damping values of SFRC beams deduced from the forced response plots work out in the range of 1 to 2%, the higher limit being applicable to higher values of M/M_{cr} or increased levels of cracking. Considerable scatter is however seen in damping values deduced from this technique and no definite trend is seen with the parameters varied.

4. CONCLUSIONS

1. The dynamic stiffness of SFRC beams in the uncracked state is found to be marginally dependent in the fibre content. The maximum increse in stiffness is observed to be 15% for a value of V_f equal to 1.

2. The influence of cracking, defined by the static preloading level is found to be significant on stiffness as obtained from the response curves (Fig.6).

3. The increase in stiffness in the post-cracking stage is relatively larger for SFRC beams. This increase, however, is seen to be nearly the same for all the fibre volumes used ($0.5 \leq V_f \leq 1.0$).

4. To obtain increased stiffness for the same ultimate moment capacity, it appears more advantageous to use steel fibre than increasing an equal volume of main steel.

5. For application of SFRC in machine foundations where response amplitudes expected are low (in the range of 0 to 0.2 mm), the dynamic modulus for a M20 normal grade concrete may be taken as 3×10^6 t/m^2 and the gross section may be taken as effective for the purpose of moment of inertia calculations.

6. The damping values exhibited by SFRC beams show significant scatter. The average value in the uncracked state applicable for use in machine foundation designs may be taken as 1% critical.

7. The dynamic stiffness of SFRC beams in the post-cracking stage can be estimated on the basis of equations 5.1 to 5.3. These equations may prove useful in designs involving SFRC elements in blast or earthquake resistant designs.

Further tests are needed to quantify the variation of stiffness with fibre content in the range $0 \leq V_f < 0.5$. Tests are also needed to rationally evaluate the impact resistance of SFRC beams.

5. ACKNOWLEDGEMENT

This paper is published with the approval of Director, Structural Engineering Research Centre, Madras.

6. REFERENCES

1. Swamy, R.N., Wegstaffex, R.L., and Oakley, D.R., New fibres and new fibre cement composites, Section-2, Proc. of the Third International Symposium on the Developments in Fibre Reinforced Cement and Concrete, RILEM Symposium, FRC, July 1986.
2. Henager, C.H., Steel fibrous ductile concrete joint for seismic resistant structures, Reinforced Concrete Structures in Seismic Zones, SP - 53, ACI, Detroit, 1977, pp 371-386.

3. Craig, R.J., et.al., Behaviour of joints using reinforced fibrous concrete, FRC International Symposium, SP-81, ACI, Detroit, 1984.

4. Craig, R.J., Structural applications of reinforced fibrous concrete, Concrete International, December 1984, pp 28-32.

5. Jamrozy, Z. and Swamy, R.N., Use of steel fibre reinforcement for impact resistance and machinery foundations, International Journal of Cement Composites, Vol.1, No.2, July 1979, pp 65-75.

6. Fibre concrete cladding helping counter quake effects, Concrete International, Dec. 1984, pp 56-60.

7. Hibbert, A.P. and Henrant, D.J., Impact resistance of fibre concrete, TRRL Report No.654.

8. Snaris, W. and Shah, A.P., Test methods for impact resistance of fibre reinforced concrete, SP 81-12, ACI, Detroit, 1984.

9. Jerath, S. and Shibani, M., Dynamic stiffness and variation of reinforced concrete beams, Jl. of ACI, March-April 1985, pp 196-202.

10. Parameswaran, V.S., Lakshmanan, N., Srinivasulu, P., Krishnamurthy, T.S., Balasubramanian, K., Thandavamurthy, T.S., and Arumugam, M., Applications of laced reinforced concrete construction techniques to blast resistant structures, Report No.RCC-SR-86-1, SERC, Madras, October 1986, pp 1-107.

Proceedings of the International Symposium on Fibre Reinforced Concrete, December 16–19, 1987, Madras, India

BEHAVIOR OF FIBER REINFORCED CONCRETE PANELS SUBJECTED TO IMPACT LOADS

M. Arockiasamy,[+, ++] A.S.J. Swamidas,[++]
and K. Munaswamy[++]

[+]Department of Ocean Engineering
Florida Atlantic University
Boca Raton, Florida
U.S.A.

[++]Faculty of Engineering
and Applied Science
Memorial University of Newfoundland
St. John's, Newfoundland, Canada

ABSTRACT

This paper presents the experimental and analytical investigation on the impact behavior of fiber reinforced concrete cylindrical panels. It considers the impact forces and the structural responses of modeled concrete shell panels of a gravity platform to iceberg collision. The percentage increase in strength due to the addition of fibers was determined based on tests of plain reinforced and fiber reinforced concrete panels. Impact resistance tests were carried out on ten fiber reinforced and seven plain reinforced concrete shell panels. A three-dimensional nonlinear finite element program was developed, using eight-noded brick elements, for the static analysis of the concrete cylindrical shell panels; the nonlinearity was prescribed using the experimental stress-strain curves obtained from compression tests on fiber/plain reinforced concrete cylinders. The experimental results were also verified using other empirical formulae suggested by earlier investigators, in addition to the results of finite element analysis. While the failure modes of plain reinforced concrete panels are due to punching shear, those of the fiber reinforced panels could be attributed to the combined effects of punching shear, bending and elastic wave propagation.

INTRODUCTION

Concrete gravity structures have been proposed for oil and gas exploration in the offshore areas subjected to extreme winds, waves, ice sheet, annual and multi-year ice ridges, icebergs and bergy bits (such as Hibernia region in the eastern coast of Canada). These structures could experience large impact forces due to either accidental collisions with marine transport/supply vessels, icebreakers operating in the vicinity or ice-fields and icebergs. For large dynamic loads, fiber reinforced concrete structures are preferred because of their high resistance to crack formation, resistance to fatigue, seawater corrosion, freeze/thaw and abrasion. In this study, analytical and experimental procedures are used to determine the impact resistance of one of the panels of a concrete gravity platform proposed for the Hibernia region using fiber/plain reinforced concrete.

LITERATURE REVIEW

Most of the available studies deal with ship/supply boat collision with fixed/floating platforms. The discovery of oil in the iceberg-frequented waters of the Grand Banks of Newfoundland generated considerable interest in the problem of iceberg collision with gravity platforms.

In a supplementary role, steel fiber reinforced concrete is used to inhibit cracking, to improve resistance to impact or dynamic loading and to resist material disintegration. ACI Committee [1] has developed a guideline for specifying, mixing, formwork and reinforcing steel, placing and finishing of steel fiber reinforced concrete. Arockiasamy et al.[2] concluded from 4 impact tests on normal and fiber reinforced concrete cylindrical panels that fiber reinforced concrete panels were 1.65-1.70 times stronger than those of normal reinforced concrete. Based on their experimental investigations, Ramakrishnan et al.[3] and Schupack [4] concluded that fiber reinforced concrete had more durability and better resilience, larger ultimate modulus at rupture, toughness index, and impact (cracking and ultimate) resistance than plain reinforced concrete.

Jamrozy (5) observed a marked increase in the compressive and tensile strength at bending and increased resistance to fatigue and impact when steel fibers were added to concrete made of light weight industrial aggregate. Kavyrchine and Astruc (6) studied the impact tests on structural elements such as beams with two types of equipment and found that the resistance to local destruction of steel fiber concrete beams was better than that of ordinary concrete; but the energy necessary to destroy a beam by flexural failure after the first impact was lower with fiber reinforced concrete.

Extensive surveys of the published literature indicated that detailed studies on the impact response of offshore structures have been reported by FENCO and German and Milne[7], FENCO [8], Sorensen [9], Carlsen [10], Larsen and Engseth [11], Gerwick [12], Brakel et al.[13], Furness and Amdahl [14], Oliveira [15], Caldwell and Billington [16], Davies and Mavrides [17], Peterson and Pederson [18], Patterson and Johnson [19], Cammaert and Tsinker [20], Mavarikios and Oliveira [21], Arockiasamy et al.[22,23], Croteau et al.[24], and Swamidas and Arockiasamy [25]. In these studies, five analytical approaches have been utilized in dealing with the impact mechanics, one based on statical principles and the other four on dynamical principles. In the statical approach, the transient nature of the impact loads was replaced by an "apparently" equivalent static load; the dynamic effects were neglected. Hence, the maximum impact forces tend to be lower than actual forces.

Ellinas and Valsgard [26] presented a state-of-the-art paper of the methods used to determine the load-deformation characteristics of all major structural components of steel offshore structures and ships. They also presented a comprehensive review of the methods and principles for design against structural damage due to impact. The impact strengths of model fiber-reinforced/steel-reinforced concrete cylindrical panels of a gravity platform were measured by Arockiasamy et al. [27, 28, 29]. Geetha Rao and Reddy [30] studied the crack response of concrete closed thick cylindrical shells to impact loading. Dynamic response of fiber-reinforced and steel-reinforced plates and shells to impact loads and stimulated blasts has been investigated by Mayrhofer and Thor [31], Kavyrchine [32], Takeda et al. [33], and Kappler and Krings [34].

The present study deals with response of concrete gravity platforms to ice-mass impact. The modeled panels of a typical gravity platform were tested for its resistance to penetration against falling weight impact loads. Altogether 17 panels of fiber and plain reinforced concrete of 1/30 and 1/20 model scales were tested.

EXPERIMENTAL PROCEDURE

Model Fabrication

One of the outer wall panels of the gravity platform shown in Fig. 1 was chosen as the basic prototype [35]. Geometric scales of 1/30 and 1/20 were chosen for fabricating the model panels.

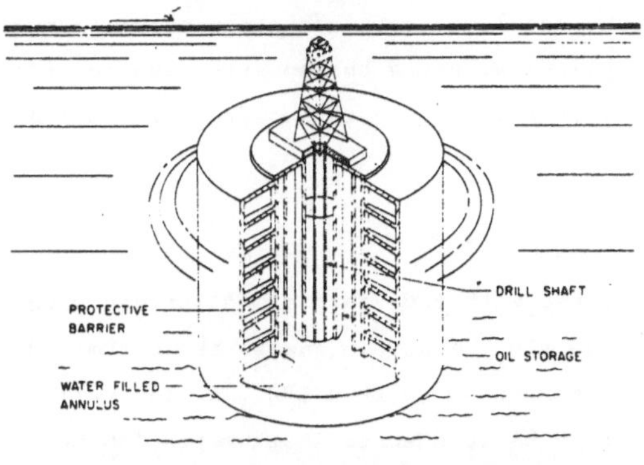

Fig. 1 Concrete gravity platform

Altogether, 21 panels were fabricated out of which seventeen were tested successfully: of the seventeen, ten were made of fiber reinforced concrete (FRC), and the other seven of plain reinforced concrete (PRC). The design compressive strength was 35 MPa, but the actual strength at 28 days was found to be between 43 to 50 MPa. An optimum fiber volume of 1% was used in this study based on the experimental work of Jamrozy and Swamy [36]. After a few trials, the mix proportion by weight was chosen as follows:

Coarse aggregate	1.77	Cement 1.0
Fine aggregate	1.83	Water 0.45
Superplasticizer	3 litres/100 kg	Fibers 0.17

The fibers used were hooked-end, brass-coated steel fibers, 30 mm in length and of nominal diameter 0.4 mm. A superplasticizer, Melment, was used to increase the workability of the fresh mix without increasing the water-cement ratio. The mix proportion for PRC panels were also the same save the super-plasticizer and fibers. Laboratory tests on fine and coarse aggregates and the steel fibers gave the following results:

Coarse aggregate (10 mm):
Dry rodded bulk density 1,832 kg/m^3.

Fine aggregate (6 mm):
Fineness modulus 3.06,
Relative density 2.55.

Steel fibers: Density 7.64 g/cm^3

Besides the steel fibers, conventional steel reinforcement was used in the edge beams, end traverses and the shell panel. While only one layer of reinforcement was provided (in the tensile zone) for the 1/30 scale model, two layers of reinforcement were provided (one each for tensile and compressive zones) in the 1/20 scale model. In addition, steel bars of 2.80 mm diameter were provided at 45° to the edges in all the four corners of the panel. After carefully mixing in the drum mixer so that no clumping of fibers took place, the panel was cast using the steel formwork. The panels were cured, along with the cylinders, using burlap saturated in ordinary water for 28 days, and then immersed in seawater at 0°C in the coldroom for a specified period before being tested.

Instrumentation and Testing

The models were instrumented with four sets of 45° strain gage rosettes, both on the top and at the bottom. In addition to strain gages, the dynamic displacement and the applied impact load were monitored (with a LVDT, a load cell and an accelerometer) automatically using a HP 3968A 16 Channel Instrumentation Recorder. The load was (i) applied hydraulically using a Tinius Olsen Universal Testing Machine of 1.335 MN capacity, and (ii) by falling weight impact hammers.

Analysis

Nonlinear, static, finite element analysis was carried out using 8-noded isoparametric elements with incompatible displacement formulation. The load vs. displacement time history was plotted and compared with experimental values.

RESULTS AND DISCUSSIONS

Of the 38, 152.4 x 304.8 mm cylinders tested, 22 were made of fiber reinforced concrete (FRC) and the other 16 were made of plain reinforced concrete (PRC). These were instrumented with two long vertical gages (gage length of 203.2 mm) placed on the opposite sides of the cylinder. In addition, the compressive displacements of the cylinders were also measured by a compressometer attached to the cylinder. Typical stress-strain curves for these cylinders are shown in Fig. 2. From these graphs it can be observed that the 28-day compressive strength of the FRC cylinders is considerably higher than the comparable PRC cylinders; after a long period of curing the difference between the two narrowed down. The average failure strengths after different periods of curing are given in Table I. Hence under axial loads, the fiber reinforced and plain reinforced concrete panels have almost the same strength. The variations of the Young's moduli under compressive strains are given in Fig. 3. The stress-strain behaviour of concrete under axial compressive load is nonlinear, and as evidenced from Fig. 3 a single Young's modulus cannot be used to describe load deformation behaviour of concrete.

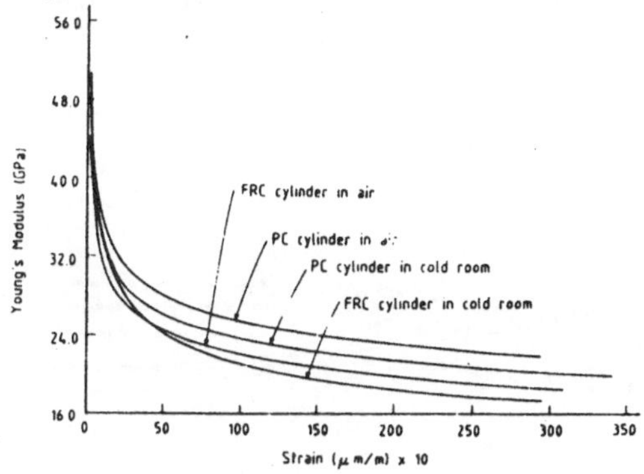

Fig. 2 Typical stress-strain curves for FRC and PRC cylinders

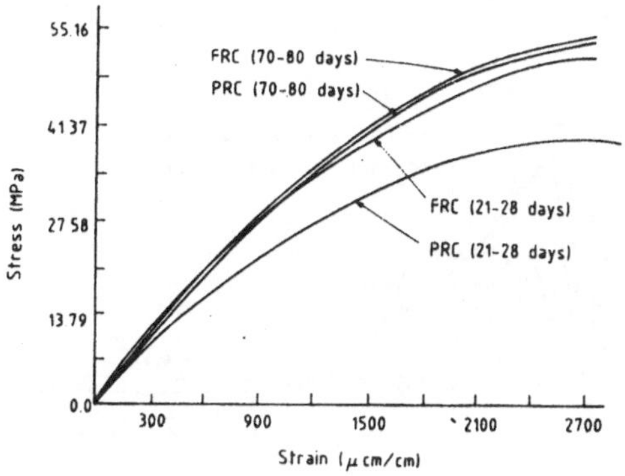

Fig. 3 Variation of Young's Modulus with axial compressive strain (40-45 days curing)

Table I: Compressive strength of concrete cylinders

Particulars		Average compressive strength (MPa) after curing/immersion for		
		21-28 days	40-45 days	70-80 days
Cylinders cured in ordinary water at room temperature	Fiber-reinforced concrete cylinders	50.72	52.06	56.21
	Plain concrete cylinders	43.68	53.30	54.83
Cylinders immersed in sea water at 0°C after 21-28 days curing in ordinary water	Fiber-reinforced concrete cylinders	-	46.40	-
	Plain concrete cylinders	-	51.30	-

The strength of the cylinders tested after immersion in seawater at 0°C was found to be lower than those tested in air, as shown in Table I. Fiber reinforced concrete seems to have lost more strength than plain reinforced concrete, probably due to corrosion of fiber reinforcement. The average failure strengths measured were 46.40 MPa and 51.30 MPa for FRC and PRC cylinders, respectively (after 40-45 days of immersion in seawater).

Impact tests on the FRC and PRC shell panels were carried out using two types of impact hammers, one 43.5 kg mass and the other 114.5 kg mass, the hammers were dropped from a height sufficient to puncture the panel. Earlier to these studies, the dynamic strengths of the shell panels were also measured by applying the loads suddenly (full-valve opening) in a UTM. The results of these tests in UTM are shown in Figs. 4 and 5.

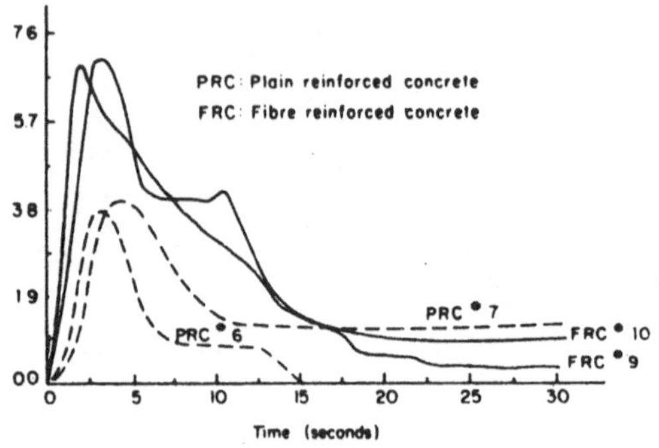

Fig. 4 Load-time history
(suddenly applied load)

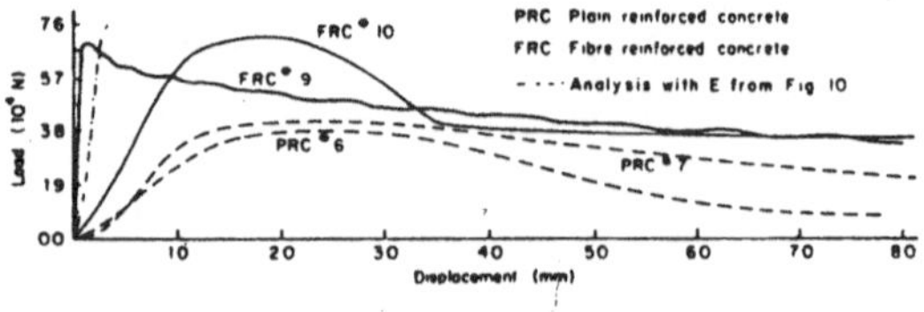

Fig. 5 Load vs. displacement
(suddenly applied load)

It is observed from these two figures that the fiber-reinforced concrete panels resisted much greater loads than plain reinforced concrete panels. Also, it is observed from Fig. 5 that the FRC panels absorb larger amount of energy than PRC panels before failure. From the falling weight impact tests, it was observed that FRC panels recorded the maximum strain slightly faster than the PRC panels indicating that the FRC panels had greater impact resistance and were more ductile. The impact resistance of the panels was measured by the accelerometer mounted on the falling weight hammer. The impact accelerations recorded on FRC and PRC panels are shown in Figs. 6 and 7 respectively.

Figs. 8 and 9 present the failure patterns of 1/20 and 1/30 scale models of panels, tested in air. From the maximum impact resistance of the panels values in Table II, it is seen that FRC shell panels can withstand a far greater impact force and absorb much more impact energy than PRC panels. It is also observed that the impact strength of panels immersed in seawater at 0°C gave highly varying results; this effect is being investigated further by immersing two more shell panels in seawater for a longer period of time.

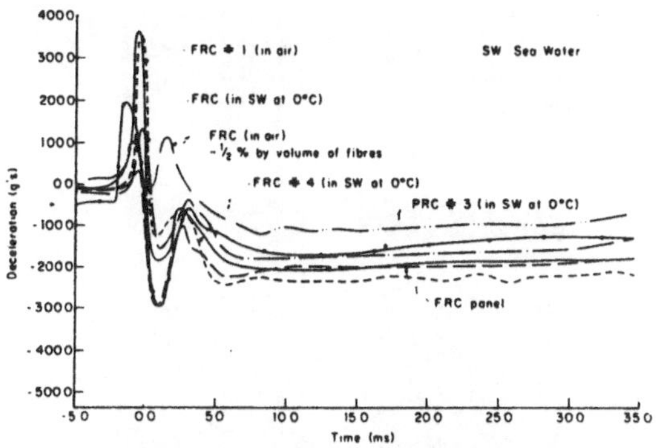

Fig. 6 Deceleration - time curve for shell panels (impacting mass of 43.5 kg)

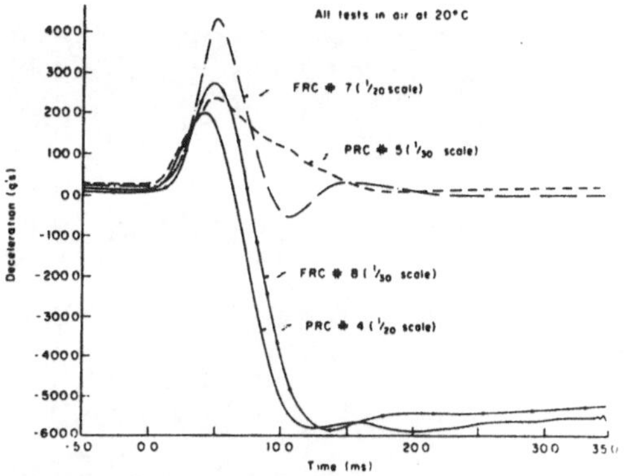

Fig. 7 Deceleration-time curve for shell panels (impact mass of 114.5 kg)

It is observed that the ratio of failure loads between the FRC and PRC panels (tested with 43.5 kg mass) ranges from 1.41 to 1.63 for falling weight impact loads while it ranges from 1.64 to 1.87 for suddenly applied loads. It is also observed that the ratio of failure loads between the falling weight impact loads (for 43.5 kg mass) and suddenly applied loads ranges from 2.16 to 2.36 for FRC panels and from 2.38 to 2.87 for PRC panels. Hence it can be generalized that the impact strength of fiber reinforced concrete panels are more than 2.5 to 3.0 times its static strength (which is assumed to be less than its strength under suddenly applied loads).

Table II: Measured impact resistance of shell panels

Nature of applied load		Panel model	Scale ratios	Impact resistance (N)
Falling weight impact	43.5 kg.	FRC #1	1/30	156,760
		FRC #2		150,230
		FRC #3		157,470
		FRC #4 (in seawater at 0°C)		152,771
		FRC #5 (in seawater at 0°C)		58,036
		FRC #6 (with 1/2% by volume of fibers)		50,786
		PRC #1	1/30	98,885
		PRC #2		103,420
		PRC #3		106,630
	114.5 kg	PRC #4	1/20	224,649
		FRC #7	1/20	477,379
		FRC #8	1/30	297,660
		PRC #5		157,254
Suddenly applied load (in a UTM)		FRC #9	1/30	66,750
		FRC #10		69,528
		PRC #6		37,233
		PRC #7		40,716

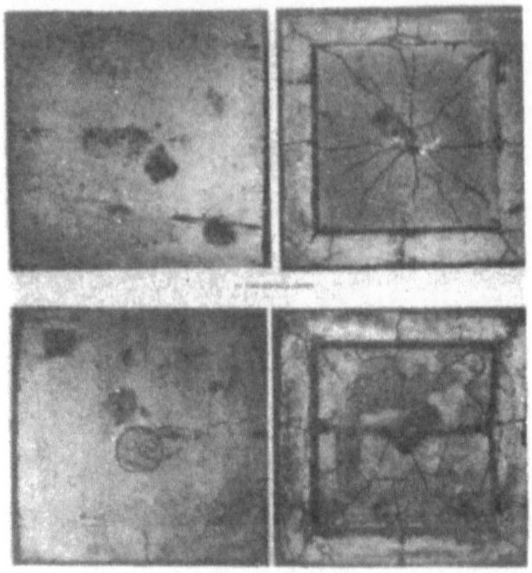

Fig. 8 Deceleration - time curve for shell panels (impacting mass of 114.5 kg.)

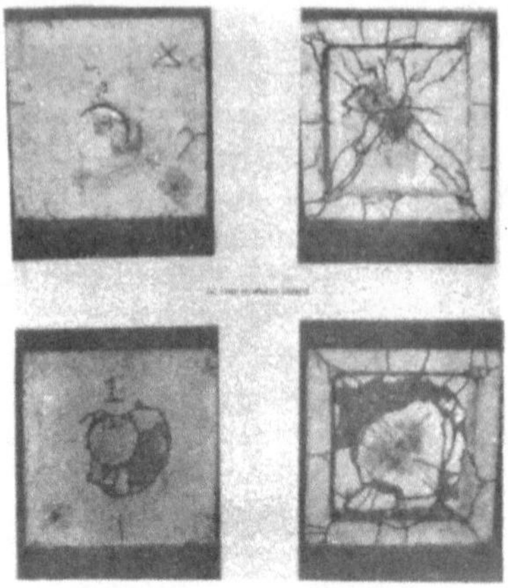

Fig. 9 Failure pattern of the 1/30 scale model

There is an obvious difference between the impact strengths of 1/30 model scale panels when impacted by 43.5 kg mass and 114.5 kg mass hammers, respectively. The impact strengths seem to be higher for the heavier mass compared to the lighter one. One possible reason that could be attributed is that for the lighter mass to puncture through, it requires a higher velocity, and at higher velocities the failure mode may be predominantly punching shear than flexure; whereas for the heavier mass, since the velocity required for failing the slab is lower, the failure mode seems to be governed by a combination of flexural and punching shear failures.

Table III compares the differences in the impact resistances obtained, using the various available empirical impact resistance relationships, and the experimental results. It is observed that the formula given by Brakel et al. [13] gives values closer to the plain reinforced concrete slabs tested in the UTM; since the tensile and shear strengths of fiber reinforced concrete are higher, the above equation must use higher tensile strength values for FRC panels. It is also observed that Jensen's formula [37] overpredicts the (falling weight) impact strength of plain reinforced concrete panels, while Sorensen's formula [9] underpredicts the same.

Table IV gives a comparison of the predicted impact resistance values for the prototype PRC and FRC panels (in air) of 1/30 and 1/20 scale ratios, using model-prototype relationships developed earlier in the study. It is observed that the results from both the scale models agree very well for plain reinforced concrete panels, while the results differ considerably for fiber reinforced concrete panels. One plausible reason is that while failure modes of plain reinforced concrete panels remain the same (i.e., punching shear), the failure modes of fiber reinforced concrete panels are combinations of punching shear, bending and elastic wave propagation failures; also the mode of laying the steel reinforcement mesh in the model must have influenced the impact resistance of the concrete panels considerably.

Table III: Comparison of failure loads from experiment with empirical formulae (1/30 scale models)

f_t = tensile strength of concrete = $0.62\sqrt{f_c'}$

f_c' = compressive strength of concrete

h_n = thickness of panel to the reinforcement

c = cohesion of concrete, bh = impact area

σ' = eff. normal stress at failure line

a = 0.25 x loaded perimeter

σ_m = eff, membrane stress

a_1 = width of load distribution

σ_{cp} = compressive prestress

D = diameter of cylinder

$\mu = [1 + \frac{\sigma_m}{f_t}]^{1/2}$, N_c = nondimensional constant, $\alpha_1, \alpha_2, \alpha_3$ = parameters dependent on load diameter and thickness of shell

$k_s = 0.65 + \frac{0.15}{(0.12+h_n)}$, $k_t = 0.50 + 50(\frac{h_n}{D})^2$, $k_p = 1.0 + 0.026\,\sigma_{cp}$

Particulars		Ultimate failure loads of concrete panels (N): Fiber reinforced concrete	Ultimate failure loads of concrete panels (N): Plain reinforced concrete
1. Experimental values	Falling weight with 43.5 kg mass	#1: 156,760 #2: 150,230 #3: 150,470	#1: 96,885 #2: 103,420 #3: 106,630
	Suddenly applied load (in UTM)	#9: 66,750 #10: 69,528	#6: 37,233 #7: 40,716
2. Formula due to Brakel et al. 1979 $P_U = 2\alpha_1\alpha_2\alpha_3 f_t h_n (a_1+h_n)$		45,299 (α_1=1.15)	43,769 (α_1=1.15)
3. Formula due to Caldwell and Billington (1981) $P_{br}=2.92 k_s k_t k_p h_n (a+h_n)\sqrt{f'_c}$ [Due to punching shear]		60,077	58,138
4. Formula due to Sorensen (1976) $P_U = \pi\mu h\,(a_1 + \mu h)\,f_t$		83,024	80,537
5. Formula due to Jensen (1979) $P_f = N_c(c+\sigma'\tan\phi)bh$		145,476	145,476

Table IV: Predicted impact resistance of prototype concrete panels using the experimental values

Particulars	From 1/20 model results	From 1/30 model results
Fiber reinforced concrete panels	190.952 MN	139.338 MN
Plain reinforced concrete panels	89.860 MN	92.081 MN

CONCLUSIONS

From the results of the above study, the following conclusions are made:

1. The model-prototype modeling and prediction seems to work out fairly well for plain reinforced concrete panels.
2. The impact resistance modeling has to take into account the following parameters while preparing a geometric model, viz., the mass, velocity and shape of impacting surface, the scaling of panel parameters (concrete strength, thickness, panel dimensions, reinforcement and fiber volumes) and the impact interface stiffness.
3. While the failure mode seems to remain the same for plain reinforced concrete panels it seems to vary considerably for fiber reinforced concrete panels.

ACKNOWLEDGEMENTS

The first author (M.A.) would like to thank Prof. S. Dunn, Chairman, Ocean Eng. Dept., and Dr. T. Shoup, Dean of Engineering, Florida Atlantic University for their continued encouragement and support. The support for this investigation to him by the Natural Sciences and Engineering Research Council of Canada grant, A 1799 and a Esso Resources of Canada grant is gratefully acknowledged. The authors thank the Dean, Dr. G.R. Peters and Associate Dean, Dr. T.R. Chari of the Faculty of Engineering and Applied Science, and Dr. I. Rusted, Vice-President, for their continued interest and active support. Grateful acknowledgement is due to Ms. Joan Varney and Ms. Shirley Behar for the careful typing of the manuscript.

REFERENCES

1. ACI Committee 544, Guide for specifying, mixing, placing and finishing steel fiber reinforced concrete, ACI Journal, March-April, 1984, pp. 140-148.

2. Arockiasamy, M., Swamidas, A.S.J., Hamlyn, D. and Munaswamy, K., Response studies of concrete shell panel models to simulated bergy-bit impact, IAHR Symp. on Ice, Hamburg, August 27-31, 1984, p. 12.

3. Ramakrishnan, V., Coyle, W.V., Kulandaisamy, V. and Schrader, E.V., Performance characteristics of fiber reinforced concretes with low fiber contents, ACI Journal, Technical paper No. 78-35, Sept.-Oct. 1981, pp. 388-394.

4. Schupack, M., Design of permanent seawater structures to prevent deterioration, Concrete International, March 1982, pp. 19-27.

5. Jamrozy, Z., Lightweight aggregate concrete with steel fiber admixture, RILEM Symp. on Testing and Test Methods of Fiber Cement Composites, The Construction Press, 1978, pp. 121-127.

6. Kavyrchine, M. and Astruc, M., Impact testing equipment for structural elements, RILEM Symp. on Testing and Test Methods for Fiber Cement Composites, The Construction Press, 1978, pp. 129-146.

7. FENCO, German and Milne, Iceberg impact on drillship, Project sponsored by Eastcoast Petroleum Operators Association, August 1971, pp. 40.

8. FENCO, Effects of ice floe impact on semi-submersible drilling vessels, Project Report submitted to BPOG Operations Ltd., Calgary, pp. 38.

9. Sorensen, K.A., Behaviour of reinforced and prestressed concrete tubes under static and impact loading, Proceedings of the First Int'l. Conf. on Behaviour of offshore structures, Norwegian Inst. of Tech., 1976, pp. 798-813.

10. Carlsen, C.A., Simplified collapse analysis of stiffened plates, J. of Norwegian Maritime Research, Vol. 5, No. 4, 1977, pp. 20-36.

11. Larsen, C.M. and Engseth, A.G., Ship collision and fendering of offshore concrete structures, Proc. of European Offshore Petroleum Conf. and Exhibition, EUR 17, Vol. 1, 1978, pp. 145-154.

12. Gerwick, Jr., B.C., Litton, R.W. and Reimer, R.B., Resistance of concrete walls to high concentrated ice loads, Proc. of XIII Offshore Tech. Conf., Houston, Texas, OTC 4111, May 4-7, 1981, pp. 425-436.

13. Brakel, J., Reinhardt, H.W. and Oostlander, L.J., Concentrated loading on a thick walled concrete cylinder, Proc. of the Int. Symp. on Offshore Structures, COPPE, Fed. Univ. of Rio de Janeiro, Brazil, October 1979, pp. 171-187.

14. Furness, O. and Amdahl, J., Computer simulation study of offshore collisions and analysis of ship-platform impacts, Norwegian Maritime Research, No. 1, 1980, pp. 2-12.

15. Oliveira, J.G., The behaviour of steel offshore structures and accidental collisions, Proc. of XIII Annual Offshore Tech. Conf., OTC 4136, Houston, Texas, May 4-7, 1981.

16. Caldwell, D. and Billington, C.J., Major ship collision damage to the prestressed concrete cylinder towers of offshore gravity structures, Proc. of II Int'l. Symp. on Integrity of Offshore Struct., Univ. of Glasgow, Scotland, July 1-3, 1981, pp. 339-361.

17. Davies, I.L.L. and Mavrides, A., Assessment of the damage arising from collision between ships and offshore structures, Proc. of II Int'l. Symp. on Integrity of Offshore Struct., Univ. of Glasgow, Scotland, July 1-3, 1987, pp. 363-380.

18. Petersen, M.J. and Pedersen, P.T., Collision between ships and offshore platforms, Proc. of Offshore Tech. Conf., OTC 4134, Houston, Texas, May 4-7, 1981.

19. Patterson, E. and Johnson, K., New non-linear methods for estimation of collision resistance of mobile offshore units, Proc. of Offshore Tech. Conf., OTC 4135, Houston, Texas, May 4-7, 1981.

20. Cammaeert, A.B. and Tsinker, G.P., Impact of large ice floes and icebergs on marine structures, Proc. of the 6th Int'l. Conf. on Port and Ocean Engg. under Arctic Conditions, Quebec, Canada, Vol. II, July 27-31, 1981, pp. 653-667.

21. Mavrikios, Y. and Oliveira, J.G., Design against collision for offshore structures, Report No. MITSF 83-7, Sea Grant College Program, M.I.T., Cambridge, Massachusetts, April 1983, pp. 164.

22. Arockiasamy, M., El-Tahan, H., Swamidas, A.S.J., Russell, W., and Reddy, D.V., Semi-submersible response to transient ice forces, J. of Ocean Engg., Vol. II (5), 1984, pp. 463-490.

23. Arockiasamy, M., Reddy, D.V. and Muggeridge, D.B., Dynamic response of moored semi-submersibles in an ice environment, Preprint SC-12, ASCE Structures Congress, Oct. 17-19, 1983, pp. 16.

24. Croteau, P., Rojansky, M., and Gerwick, M., Summer ice floe impacts against caisson-type exploratory and production platforms, Proc. III Int'l. Symp. on Offshore Mech. and Arctic Engg., New Orleans, Louisiana, U.S.A., Feb. 12-16, 1984, pp. 228-237.

25. Swamidas, A.S.J. and Arockiasamy, M., Iceberg impact forces on gravity platforms, III Int'l. spec. Conf. on Cold Regions Engg., Edmonton, Alberta, April 4-6, 1984, pp. 32.

26. Ellinas, C.P. and Valsgard, S., Collisions and damage of offshore structures: State-of-the-Art, Proc. of OMAE'85, Dallas, Texas, Feb. 17-21, 1985, pp. 475-495.

27. Arockiasamy, M., Swamidas, A.S.J., Hamlyn, D. and Munaswamy, K., Response studies of concrete shell panel models to simulated bergy-bit impact, IAHR Symp. on Ice, Hamburg, August 27-31, 1984, p. 12.

28. M. Arockiasamy, Swamidas, A.S.J., Munaswamy, K. and Hamlyn, D., Response of concrete shell panels to simulated bergy-bit impact, Arctic '85 "Civil Engg. in the Arctic Offshore" Conf., San Francisco, California, March 24-27, 1985, p. 18.

29. Arockiasamy, M., Swamidas, A.S.J., Cheema, P.S. and Munaswamy, K., Impact strength of fiber reinforced concrete cylindrical panels, Eighth Int. Conf. on Struct. Mech. in Reactor Tech., Brussels, Belgium, Vol. H., Concrete Structures, Aug. 19-23, 1985, pp. 125-130.

30. Geetha Rao and Reddy, D.V., Modeling of ice impact on concrete shells, Proc. eighth Int. Conf. on Port and Ocean Engineering under Arctic Conditions (POAC), Narssarssuag, Greenland, Sept. 1985.

31. Mayrohofer, C. and Thorn, H.J., Dynamic response of fiber and steel reinforced concrete plates under simulated blast-load, Proc. Conf. on Concrete Struct. Under Impact and Impulsive Loading, Berlin (West), June 2-4, 1982, pp. 279-288.

32. Kavyrchine, M., Punching strength of cylindrical concrete shells, Proc. of the Conf. on Concrete Struct. under Impact and Impulsive Loading, Berlin (West), June 2-4, 1982, pp. 304-314.

33. Takeda, J.I., Tachikawa, H., and Fujimoto, K., Fracture of reinforced concrete structural members and structures subjected to impact or explosion, Proc. of the Conf. on Concrete Struct. under Impact and Impulsive Loading, Berlin (West), June 2-4, 1982, pp. 289-295.

34. Kappler, H. and Krings, W., Calculations of concrete plates and shells under impact load, Proc. of the Conf. on Concrete Struct. under Impact and Impulsive Loading, Berlin (West), June 2-4, 1982, p. 15.

35. Weir, F.V., The comparative environmental risks associated with fixed platforms and floating platforms, and with tankers and pipelines, Proc. Symp. on Production and Transportation Systems for the Hibernia Discovery, Pet. Dir., St. John's, Nfld. 1981, pp. 337-355.

36. Jamrozy, Z. and Swamy, R.N., Use of fiber reinforcement for impact resistance of machinery foundations, The Int. J. of Cement Composites, Vol. I, (2), July, 1979, pp. 65-75.

37. Jensen, J.J., Impact of falling loads on submerged concrete structures, Proc. Int. Symp. on Offshore Struct., Brazil, Vol. I, Oct. 1979, pp. 215-231.

Proceedings of the International Symposium on Fibre Reinforced Concrete, December 16-19, 1987, Madras, India

STRUCTURAL APPLICATIONS OF FIBRE REINFORCEMENT IN REINFORCED CONCRETE MEMBERS

D.L.N. RAO
ANDHRA UNIVERSITY
WALTAIR, INDIA

M.R.A. KADIR
UNIVERSITY OF SALAHADDIN
ARBIL, IRAQ

KAWA TAHA ABU AL AWFFA
UNIVERSITY OF SALAHADDIN
ARBIL, IRAQ

SYNOPSIS

This investigation was designed to obtain experimental data on the deformation characteristics and strength of reinforced concrete beams made with steel fibre, in pure bending. A total of fourteen beams 75 x 150 x 1850 mm were tested to destruction to evaluate the effect of such variables as grade of bar reinforcement, length, diameter, aspect ratio, volume fraction and location of steel fibre in the concrete. The inclusion of steel fibres in the concrete significantly increased the post cracking stiffness at all stages upto failure. Narrower crack widths, substantially less deformations were recorded. Increases in ultimate strength were marginal.

INTRODUCTION

Inclusion of steel fibres in a conventionally reinforced concrete member is primarily not for strength increase; this strength increase can be more easily and economically obtained by bar reinforcement. The deficiencies of ordinary reinforced concrete members that can not be corrected easily by bar reinforcement, can be remedied to a significant extent by fibre reinforcement. Crack widths have to be limited to acceptable limits in reinforced concrete members for a variety of reasons, such as control of deflections, use of high grade reinforcing steel, impermeability, durability, corrosion prevention and maintenance of integrity and appearance of the structure. However, only limited data are available on the reinforcing effect of steel fibres when they are used in combination with bar reinforcement.

Batson et al [1] described beam tests to investigate the possibility of using fibres in place of shear reinforcement. Shah and Rangan [2] have shown that ductility of over-reinforced beams can be increased by steel fibres though to a lesser extent than stirrups. Swamy and Al-Noori [3] showed that the presence of steel fibres enables high strength steel with yield strength 700 N/mm^2 to be used with both crack width and deflection being controlled to within acceptable limits. It was also shown by them that with the use of steel fibres, over-reinforced beams could be made to behave in a ductile manner. Henager and Doherty [4] and Henager [5] have presented an analytical method to predict the ultimate flexural strength of steel fibre concrete beams with bar reinforcement. In their study, the bond stress, fibre stress, fibre aspect ratio and volume fraction of fibres were taken into account. While these tests related to mortar matrix rather than a concrete matrix which is generally used for structural members, Swamy and Saád A. Al-Ta'an [6] studied comprehensively the deformation characteristics and ultimate strength of conventionally reinforced concrete beams having bar reinforcement made with steel fibre concrete. These beams had 20 mm maximum size of aggregate and were under moment gradient. Maximum fibre content used was one percent.

The present paper reports the results of a test programme to study the influence of steel fibres on deformational behaviour and ultimate strength of conventionally reinforced concrete beams under pure flexure.

SCOPE OF INVESTIGATION

Details of Test Beams

The experimental programme involved flexural tests on fourteen reinforced concrete beams. All beams were of the same size 75 x 150 x 1850 mm. The principal variables in the tests were grade of bar reinforcement, length, diameter, aspect ratio, volume fraction of steel fibre and location of steel fibre in the concrete. Fibres were used either over the whole depth of the beam, upto three fourths height from the bottom or upto half height from bottom. To force the failure in the test zone, the portions of the beam beyond the test zone that were shear regions were reinforced with 6 mm stirrups to resist the transverse shear force. The nominal details of the beam cross-sections and the arrangement of reinforcement are shown in Fig. 1. A complete listing of the specimens and their properties are presented in Table - I.

Details of Materials

The concrete was made from ordinary portland cement; 10 mm maximum size natural gravel and local sand were used. The mix proportions by weight for cement, sand and gravel were respectively 1 : 1.4 : 2.8 with a water cement ratio of 0.6. Regular tap water was used for mixing. The steel fibres used in this study were round and of low carbon having a yield stress of 618 N/mm^2, which was also approximately ultimate stress. Two aspect ratios of fibre, 100 and 83.33 were used. Two volume fractions of fibre, 0.5 and 1.0 were employed. Flexural bar reinforcement was either plain mild steel or high yield strength deformed steel. The details are furnished in Table - II.

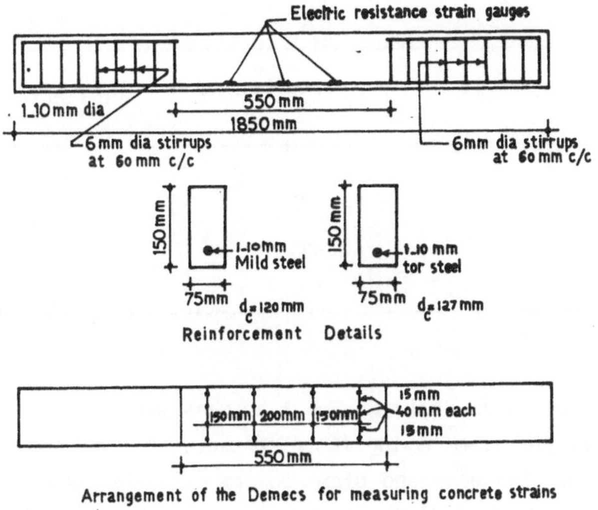

Fig. 1 : Details of Test Specimens.

Table - I : Details of Test Specimens.

Beam No.	Fibre Details				Cube strength (N/mm^2)	split cylinder (N/mm^2)	Age at test (days)	Tension reinforcement	
	diameter (mm)	length (mm)	aspect ratio	volume (%)				No.of 10 mm mild steel bars	No.of 10 mm deformed bars
A1	-	-	-	0.0	35.45	2.60	105	-	1
A2	-	-	-	0.0	35.45	2.60	105	1	-
A3	0.4	40	100	1.0	36.00	3.34	114	-	1
A4	0.4	40	100	1.0	36.00	3.34	114	1	-
A5	0.4	40	100	0.5	34.00	4.21	103	-	1
A6	0.4	40	100	0.5	34.00	4.21	103	1	-
A7	0.4	40	100	0.5 bottom 0.75 depth	34.00 41.76	3.28 4.21	105	-	1
A8	0.4	40	100	0.5 bottom 0.75 depth	34.00 41.76	3.28 4.21	102	1	-
A9	0.4	40	100	0.5 bottom 0.5 depth	34.00 41.76	3.28 4.21	99	-	1
A10	0.4	40	100	0.5 bottom 0.5 depth	34.00 41.76	3.28 4.21	113	1	-
A11	0.3	25	83.33	1.0	41.39	3.50	107	-	1
A12	0.3	25	83.33	1.0	41.39	3.50	113	1	-
A13	0.3	25	83.33	0.5	41.00	4.59	115	-	1
A14	0.3	25	83.33	0.5	41.00	4.59	112	1	-

Table - II : Physical Properties of Bar Reinforcement

Type of bar	Diameter (mm)	Yield stress (N/mm^2)	Ultimate stress (N/mm^2)
Mild steel	10 mm	375	487
Deformed steel	10 mm	533	836

Test Details

Beams were cast in steel moulds. The specimens were compacted uniformly on a table vibrator. For the beam having fibre concrete in the bottom zone upto 0.5 height or 0.75 height of the beam and plain concrete in the remaining top portion, first fibre concrete was poured and compacted upto the required height. Then immediately plain concrete was poured, upto the top of the beam in continuity of fibre concrete, compacted and finished. The age at test of the beams varied from 99 to 115 days.

All the test specimens were extensively instrumented to obtain the measurements rotation, curvature, concrete compressive strain, vertical deflection and crack width through out the loading range. A general view of the test set-up and typical locations of the various measurements are shown in Fig. 2.

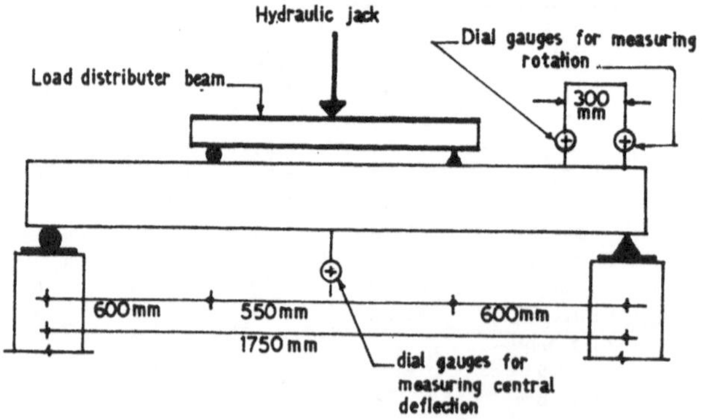

Fig. 2 : Test set-up

DISCUSSION OF TEST RESULTS

The principal test results are shown in Table - III. The effect of varying depths of fibre content in tensile zone on crack widths could be observed from Fig. 3. Beams A5, A7 and A9 constitute one group for comparison; these beams had deformed flexural bar. It can be seen that at the same crack width, the beam with fibres in

Table - III : Principal Test Results

Beam No.	Fibre content (%)	Aspect ratio	Cube strength (N/mm^2)	Type of steel	Moment at 1st crack (N. m.)	Moment at service load (N. m.)	Moment at ultimate (N. m.)
A1	0.0	-	35.45	deformed	880	3570	6070
A2	0.0	-	35.45	mild	1047	2207	3751
A3	1.0	100	36.00	deformed	1912	3750	6370
A4	1.0	100	36.00	mild	1730	2512	4269
A5	0.5	100	34.00	deformed	1471	3745	6434
A6	0.5	100	34.00	mild	1645	2158	3668
A7	0.5	100	34.00 / 41.76	deformed	1645	3911	6470
A8	0.5	100	34.00 / 41.76	mild	1645	2577	4380
A9	0.5	100	34.00 / 41.76	deformed	1471	3491	6193
A10	0.5	100	34.00 / 41.76	mild	1645	2189	3716
A11	1.0	83.33	41.39	deformed	1680	4231	7193
A12	1.0	83.33	41.39	mild	2151	2527	4295
A13	0.5	83.33	41.00	deformed	1477	3703	6295
A14	0.5	83.33	41.00	mild	2059	2492	4236

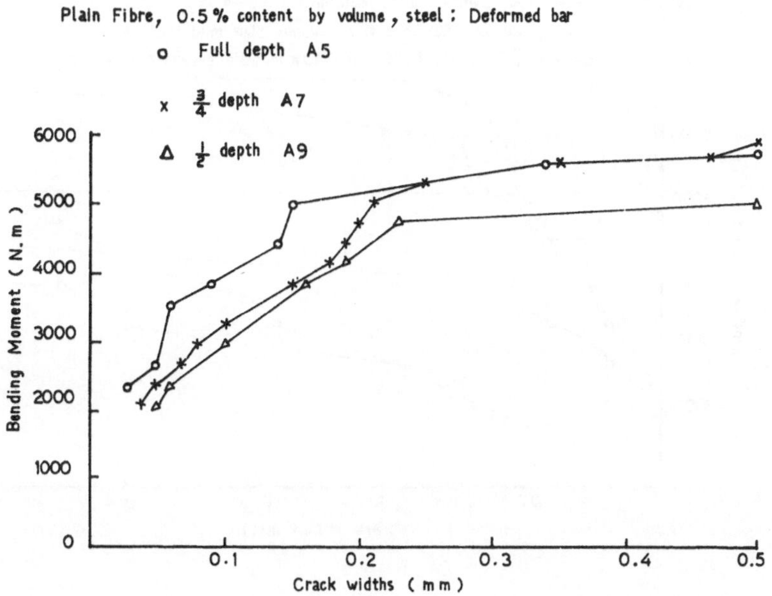

Fig. 3 : Effect of varying depths of fibre concrete

bottom 0.5 depth only, developed 25% less moment in initial stages of loading and 13% less moment in latter stages of loading than the beam with fibre in full depth. However for mild steel group (A6, A8 and A10) the beam with fibre in the bottom 0.75 depth registered more moment than the other two beams at the same crack width. Thus lack of fibres in the top half of the beam, resulted only in small differences in crack widths. As the fibre volume increased, the moment increased for all crack widths, but at different rates. The slopes of the curves for the smaller crack widths were much steeper than those for wider cracks. The effect of fibre concentration, on moments at different crack widths can be observed from Fig.4. A3 with 1% fibre developed nearly maximum moment at 0.2mm crack width. While A1 with no fibre and A5 with 0.5% fibre developed moments, as the crack widened; the latter far higher moments than the former at the same crack widths. The improvement in moment over the control specimen for A3 and A5 were respectively 250% and 235% at 0.1 mm, 232% and 191% at 0.2 mm and 123% and 106% at 0.3 mm. At service load, the number of cracks in the test zone for the fibre beams with deformed bar reinforcement were 4 to 6 and at ultimate they were 7 to 9. The corresponding beams with mild steel had fewer cracks, 2 to 4 at service load and 4 to 6 at ultimate.

DEFORMATION CHARACTERISTICS

The first crack load was defined as that load at which a maximum crack width of 0.01 mm was noted with the aid of a microscope. A load factor of 1.7 was used

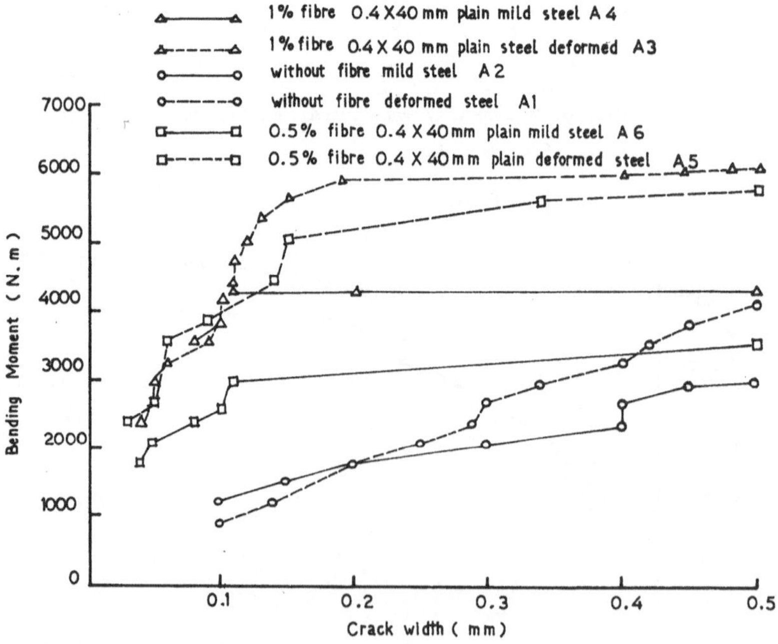

Fig. 4 : Effect of increasing fibre content on the crack width

to define experimental service load of each beam, the ultimate load was 1.7 times the service load of beam. The experimental flexural rigidity, EI was computed at first crack, service load and ultimate load stages, by dividing the moment in the test zone by curvature, obtained from concrete compressive strain readings. In the theoretical calculation of EI, uncracked moment of inertia was taken at first crack load stage, cracked moment of inertia at service and ultimate load stages.

DEFORMATIONS AT FIRST CRACK

The first crack moments of the beams varied from 0.15 to 0.28 of the ultimate moment and 0.25 to 0.47 of service moment for the beams without firbres. For the fibre beams it ranged between 0.23 and 0.5 of the ultimate moment and 0.39 and 0.85 of the service moment. The deformations at this stage of loading are only a fraction of those occurring at the design service loads. However the results showed that fibres are effective even when cracking is very slight.

DEFORMATIONS AT EXPERIMENTAL SERVICE LOAD

Effect of Fibre Concentration

Effect of fibre concentration on the deformations at service load can be studied from Table - IV. As the fibre content increased, for the beams with deformed bars the maximum reduction in deflection was 28%. Such reduction was 42% for mild steel beams. The rotation generally decreased as the volume content increased.

Table - IV : Effect of Fibre Concentration at Service Load

Beam No.	Fibre (%)	Moment (N.m)	Defl. (mm)	Span / Defl.	Rotation (rad x 10^3)	Concrete comp. strain (x 10^5 m/m)	N. A. depth	Curvature (rad/m x 10^3)	Test EI (kNm^2)	Theory EI (kNm^2)	EI test / EI theo.
A1	0.0	3570	8.04	218	11.86	84.00	27.20	30.88	116	168	0.69
A5	0.5	3786	6.48	270	9.88	27.96	35.10	7.97	470	177	2.66
A3	1.0	3747	5.76	304	8.26	70.99	37.20	19.08	197	182	1.08
A13	0.5	3703	6.48	391	8.17	71.93	32.90	21.86	169	195	0.87
A11	1.0	4231	6.14	285	12.60	61.14	31.40	19.47	211	196	1.11
A2	0.0	2207	4.23	414	7.59	45.06	35.30	12.76	173	168	1.03
A6	0.5	2158	3.13	559	4.51	89.63	38.00	23.59	92	177	0.52
A4	1.0	2512	3.66	478	5.18	32.49	47.80	6.80	369	182	2.03
A14	0.5	2492	2.67	655	5.16	61.45	27.40	22.43	111	195	0.57
A12	1.0	2527	2.45	714	5.90	42.35	46.60	9.09	278	196	1.42

The improvement in cracked flexural stiffness was substantial as the fibre content increased. Similar observations can be made with regard to the depth of neutral axis as the fibre concentration was augmented.

DEFORMATIONS AT ULTIMATE LOAD

Table - V shows the effect of fibre concentration on deformation characteristics. For the beams embedded with deformed bar, the maximum reduction in deflection

Table - V : Effect of Fibre Concentration at Ultimate Load

Beam No.	Fibre (%)	Moment (N. m)	Defl. (mm)	Span / Defl.	Rotation (Rad x 10^3)	Concrete comp. strain. (x 10^5 m/m)	N.A. depth (mm)	Curvature (rad/m x 10^3)	Test EI ($kN.m^2$)	Theory EI ($kN.m^2$)	EI test / EI theo.
A1	0.0	6070	47.35	37	69.50	358	20.10	178	34	168	0.2
A5	0.5	6434	31.22	56	42.00	330	1.95	169	38	177	0.22
A3	1.0	6370	44.23	40	59.57	410	23.00	178	36	182	0.20
A13	0.5	6295	26.30	67	40.80	374	19.50	192	33	195	0.17
A11	1.0	7193	38.99	45	52.90	470	16.00	294	25	196	0.13
A2	0.0	3751	18.47	95	30.04	250	22.60	114	33	168	0.20
A6	0.5	3668	33.20	53	41.47	300	25.00	120	31	177	0.18
A4	1.0	4269	9.33	188	13.57	100	25.00	40	107	182	0.39
A14	0.5	4236	30.86	57	60.00	320	17.00	188	23	195	0.12
A12	1.0	4295	9.60	182	10.63	470	16.00	294	15	196	0.08

for the fibre beams over the control specimens was 45%. This reduction was 50% for beams with mild steel reinforcement. The increase in flexural rigidity for fibre beams with deformed steel over the companion control specimens ranged between 32 to 53% with a few exceptions. The imporvement was substantially much for mild steel beams.

The increase in the fibre content resulted in the increase of neutral axis depth. The maximum increase was 6% for deformed steel beams. For mild steel beams it was upto 22%.

The amount of longitudinal reinforcement used was well below the balanced steel. The control beams and the fibre beams exhibited substantial ductile behaviour after reaching maximum load and before final collapse. The maximum concrete compressive strain just prior to ultimate load ranged between 300 to 470 x 10^{-5}m/m for fibre beams and 250 to 358 x 10^{-5}m/m for beams without fibre with a few exceptions.

ULTIMATE STRENGTH

For the sake of economy of fibres in principle, it is possible to provide fibre concrete only in the tensile zone of a beam. Such a provision will improve the deformation characteristics of the member and the absence of fibres in the compression zone have little influence on the strength. The effect of fibre concrete in varying depths in tensile zone of the beam can be examined from the results of the beams divided into the following groups. Group 1 A5, A7 and A9 and Group 2: A6, A8 and A10. Group 1 related to 0.5% and 0.4 x 40 mm fibre and deformed steel bar. A7 and A9 showed 3% more and 4% less strength than the control specimen, A5. Group 2 related to 0.5% and 0.4 x 40 mm fibre and mild steel bar. A8 and A10 were 20% and 2% stronger than the control specimen, A5. It shows lack of fibres in the compression zone did not adversely affect the strength.

To study the effect of fibre concentration on the flexural strength, the following groups may be examined. Group 1: A1, A5 and A3 and A1, A13 and A11. The percentage increase in strength of A5 and A3 over the control specimen A1 was 6 and 5 respectively. For A13 and A11 the percentage increase over the control specimen A1 was 4 and 19. For mild steel beams and deformed bar beams the percentage increase in strength with increase in fibre content is shown in Table - VI.

Table - VI : Effect of Fibre Content on Beam Strength

Beam No	% of fibre	Flexural strength (Nm)	% increase over control specimen
A1	0.0	6070	-
A5	0.5	6434	6
A3	1.0	6370	5
A1	0.0	6070	-
A13	0.5	6295	4
A11	1.0	7193	19
A2	0.0	3751	-
A6	0.5	3668	-
A4	1.0	4269	14

Generally the strength increase was more in beams with mild steel bar than those with deformed steel. In either case, the fibre contribution essentially remains the same; for a given volume fraction of fibre. When a fixed amount of fibre contribution is added to the smaller strength contribution of mild steel bar, the percentage increase appears larger than the corresponding beam with deformed bar. The aspect ratios used are 100 and 83.33; the difference being small, the effect on strength is not conclusive.

CONCLUSIONS

The following conclusions are drawn based on the findings of tests reported here.

1. The test results of this investigation clearly established that steel fibres permit substantial increase in bending moment to be developed, before a particular width of crack occurs.

2. The improvement of crack control due to the addition of fibres was more pronounced with deformed bars than with plain bars.

3. One of the very significant roles played by steel fibres is arresting advancing cracks and reducing the crack widths. Addition of even a small quantity of fibre of 0.5% caused significant reduction in crack widths.

4. The increased depth of neutral axis and the increased flexural stiffness of the fibre reinforced concrete beams at all stages of loading, reflected the ability of fibres in arresting the crack growth and crack widening.

5. At all stages, of loading, from first crack to ultimate, fibres were effective in resisting deformations.

6. The concrete compressive strains at the attainment of ultimate load ranged between 330 to 700 x 10^{-5} m/m.

7. Even at ultimate fibres were effective in reducing deformations and increasing flexural stiffness.

8. The influence of steel fibres in increasing the ultimate strength of an ordinary reinforced concrete beam is limited, in comparison with the substantial beneficial effects imparted to such beams, relative to reduction in crack width, and deformations and increase in stiffness upto ultimate stage.

REFERENCES

1. Batson, G.B., Jenkins, E. and Spatney, R. 'Steel Fibres as Shear Reinforcement in Beams,' *Journal of the American Concrete Institute*, October 1972, Vol. 69, pp 640-44

2. Shah, Surendra, P. and Rangan, B. Vijay, 'Effects of Reinforcements on Ductility of Concrete,' Proceedings ASCE, V 96, ST6, June 1970, pp 1167-1184.

3. Swamy, R.N. and Al-Noori, K.A, 'Flexural Behaviour of Fibre Concrete with Conventional Steel Reinforcement,' Proceedings, RILEM Symposium on Fibre Reinforced Cement and Concrete (London, Sept. 1975), The Construction Press Limited, Lancaster, 1975, pp 187-196.

4. Henager, Charles H. and Doherty, Terrence J. Analysis of Reinforced Fibrous Concrete Beams, Proceedings, ASCE, ST1 Jan. 1976, Vol. 102, pp 177-188.

5. Henager, charles. H. Ultimate Strength of Reinforced Steel Fibrous Concrete Beams, Proceedings, Conference on Fibre Reinforcd Materials: Design and Engineering Applications (London, Mar. 1977), *Institution of Civil Engineers*, London, 1977, pp 151-160.

6. Swamy, R.N. and Al - Ta'an, Sa'ad A. Deformation and Ultimate Strength in Flexure of Reinforced Concrete Beams Made with Steel Fibre Concrete, *Journal of the American Concrete Institute*, Title No. 78-36, Sept-Oct 1981, pp 395-405.

Proceedings of the International Symposium on Fibre Reinforced Concrete, December 16-19, 1987, Madras, India

ULTRAHIGH STRENGTH STEEL FIBERS FOR CONCRETE REINFORCEMENT

Y M SRINIVAS* K A GUPTE** R H G RAU***
* Research Metallurgist
** Sr. Research Metallurgist
*** General Manager (R&D)
Mukand Iron and Steel Works Limited
L B Shastri Marg, Kurla, Bombay 400 070.

ABSTRACT

Application of either unstressed or prestressed steel reinforcement is common for providing desired tensile strength values in cement concrete. This conventional reinforcement is susceptible to sudden crack propagation through regions of high-tensile loads. These high tension cracks are not only responsible to the loss of load bearing area but may also lead to enhanced corrosive attack on the exposed reinforcement bar surfaces. Use of steel fibers in the cement concrete matrix can improve toughness and flexural strength of structural members. These fibers when uniformly distributed in the concrete can arrest the propagation of the tensile cracks. In addition, they can also impart abrasion resistance to the concrete. The conventionally used fibers are made from 0.4 to 1.0 mm diameter wires of low carbon steel having tensile strengths of around 100 kg/mm^2. Recent research work on alloy design has brought out possible application of ultrahigh strength steel fibers having superior combination of strength and ductility. These high-strength high-ductility steel fibers are produced to dual phase ferritic-martensitic steel structures and have two-fold corrosion resistance over the conventional steel fibers. This paper discusses the basic principles of these ultrahigh strength (typically 165-200 kg/mm^2) steel fibers and their suitability for concrete reinforcement.

INTRODUCTION

Steel fiber reinforced concrete is a composite material with a uniform distribution of discontinuous small steel fibers in the conventional cement concrete. The technology of fiber reinforced concrete is by no means new. Attempts to use steel fibers in mortar and concrete date back to nearly half a century, though it is only in recent times that significant advance has been made in the utilization of this material. The steel fiber reinforced concrete (SFRC) is superior to conventional concrete because of its greater strength and crack arresting properties [1, 2].

The steel fibers used in most cases are either straight or with bent ends, as shown in Fig.1a, and have a tensile strength typically of the order of 80-100 kg/mm^2. As a result of poor mechanical anchorage the SFRC normally fails not by fracture of fibers but by the pull out of fibers from concrete matrix.

Fig.1a Two typical shapes of fibers commonly used for concrete reinforcement

Fig.1b Typical corrugated shape adopted for ultrahigh strength steel fibers

Recent work on material design has produced the ultrahigh strength steel fibers (UHSF) based on the ferritic-martensitic steel formula. These fibers can be produced in the size range of 1.0mm dia to 0.4 mm dia. with tensile strengths ranging from 165 kg/mm^2 to 200 kg/mm^2. As pointed out earlier, a premature failure of the bond between the fiber and matrix limits the possible improvements in composite properties from the increased strength levels of the fibers. By altering the shape of fibers so as to improve the mechanical anchorage in concrete, it may be possible to gain maximum advantage of UHSF in improving the performance of the composite. (Typical shape of corrugated UHSF fiber is as shown in Fig. 1b). Keeping these points in background, work was undertaken to :

a. compare the properties (flexural strength and toughness) of concrete beams reinforced with the fibers of mild steel and fibers of ultrahigh strength steel.
b. study the effect of shape and aspect ratio of these ultrahigh strength steel fibres on the flexural strength and toughness of the composite.

BACKGROUND ON UHSF

During last decade, extensive work has been carried out in the development and assessment of SFRC. Application of SFRC is now well established wherever the service calls for abrasion resistance, impact resistance or superior flexural strength levels. However, the work principally relates to low strength mild steel wires; for the obvious reason that higher strengths are associated with increased steel costs. Further, to derive benefits of the high strength, the mechanical anchorage needs to be improved. Emergence of high strength-high ductility ferritic-martensitic dual phase steel rods is showing an alternate economical route for the production of UHSF.

The dual phase steel is a composite of strong martensite and ductile ferrite. This microstructure imparts high strength as well as high ductility to the steel. The ferritic-martensitic steel (F-M steel) is characterized by low yield strength levels, continuous yielding phenomenon and high initial work hardening rates. These properties make the F-M steel wire rod an attractive candidate for drawing in a single multipass operation to fine wires with high strength and high ductility. This eliminates the need of intermediate heat treatment operation which is required in case of conventional carbon steels with ferritic-pearlitic micro structures. Details on wire drawing of F-M steels and associated benefits are discussed elesewhere [3]. The F-M steel fibers have following three distinct advantages over the conventional mild steel fibers of the same diameters:

a. Superior tensile strength
b. Superior ductility
c. Superior corrosion resistance.

In applications where the presence of continuous reinforcement is not essential to the safety and integrity of the structures eg.pavements,

Overlays and shotcrete linings, the improvements in flexural strength associated with the UHSF can be used to reduce section thickness or improve performance of the component or both.

EXPERIMENTAL WORK

Measurements of flexural strength and toughness were done by using the centre point loading system on 100 mm x 100 mm x 500 mm long beams. These cement concrete beams were reinforced with steel fibers of two different shapes and different sizes as detailed in Table 1.

Table 1. Details of Steel Fibers used for Reinforcement of Cement Concrete Beams.

Sr. No.	Steel Grade	Fiber Dia. d.mm	Aspect Ratio l*/d	Shape
1.	Mild steel (MS)	1.0	60, 80	Straight
2.	UHSF	1.0	40, 60, 80	Straight and Corrugated

* l = Length of the fiber

The specimens after casting were cured under water for 15 days; after which they were tested under centre point loading conditions.The test was terminated when the visible crack which initiated at the bottom reached the top of the beam as shown schematically in Fig.2.

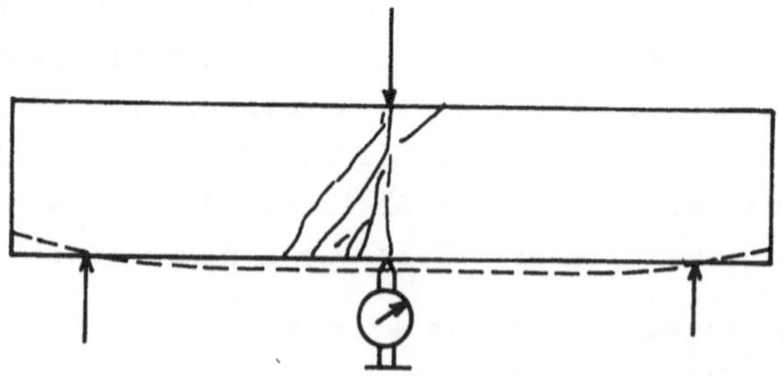

Fig.2 Schematic illustration of the fiber reinforced beam tested under centre point loading conditions

All other parameters were kept constant throughout the experimentation. The deflection of the beam under load was measured at the load line with a dial gauge. The load-deflection curves were plotted and the area under curve was taken as a measure of toughness. Flexural strength was calculated as per ASTM C-293-68. The cement concrete mix was kept constant corresponding to the grade M250. Coarse aggregate used in the mix had a max. size of 20 mm. The amount of fibres added was kept constant i.e. 1% by volume.

RESULTS AND DISCUSSION

The data generated from the testing was converted to load-deflection curves from which toughness and flexural strength were calculated. The results are summarised in Table 2. No deliberate effort was made to align the fibers in one direction during the preparation of the specimen. Hand mixing was done for the preparation of the mix, which was later compacted on a vibrating table.

Table 2. Toughness and Flexural Strength Values for Beams reinforced with UHSF and MS fibers of different aspect ratios.

Aspect Ratio	Fiber Type	Toughness Kg.m	Flexural Strength Kg/cm^2
40	UHSF - straight	1.0	45
	UHSF - corrugated	7.9	36
	MS - straight	5.1	48
60	UHSF - straight	6.5	48
	UHSF - corrugated	10.1	54
	MS - straight	1.5	54
80	UHSF - straight	16.2	50
	UHSF - corrugated	20.3	84

The load deflection curves for the aspect ratios of 40, 60 and 80 for the different types of fibers are shown in Fig. 3,4 and 5. Typical crack observed in a fibre reinforced concrete beam is shown in Fig. 6.

The results indicate that the UHSF corrugated fibers of small aspect ratio help in improving the toughness of the concrete (or post-cracking

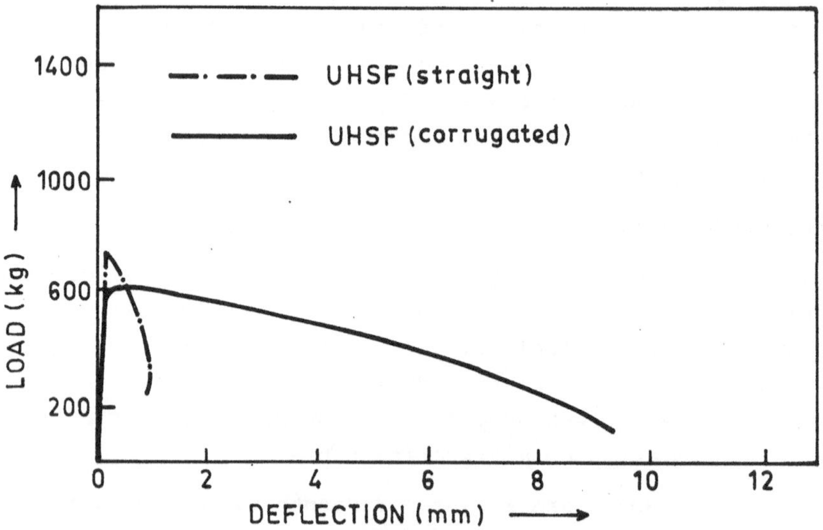

Fig.3 Load-Deflection curves for the beams reinforced with fibers of aspect ratio 40 (40AR)

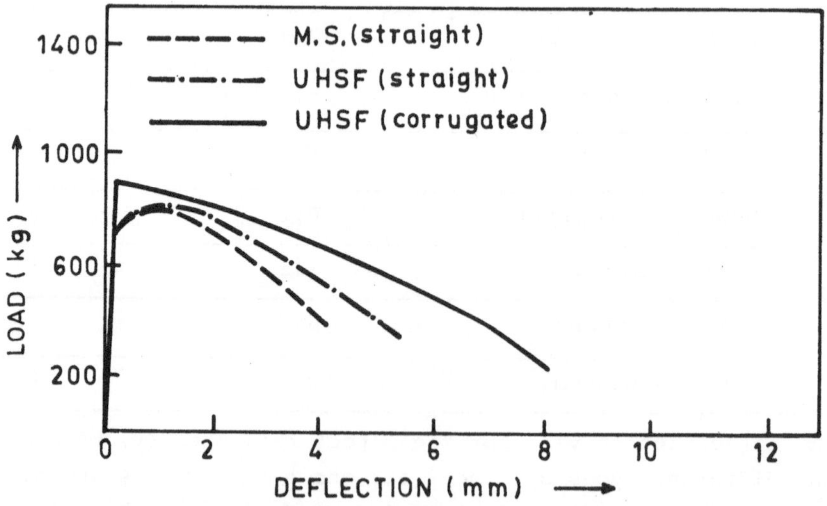

Fig.4 Load-Deflection curves for the beams reinforced with fibers of aspect ratio 60 (60AR)

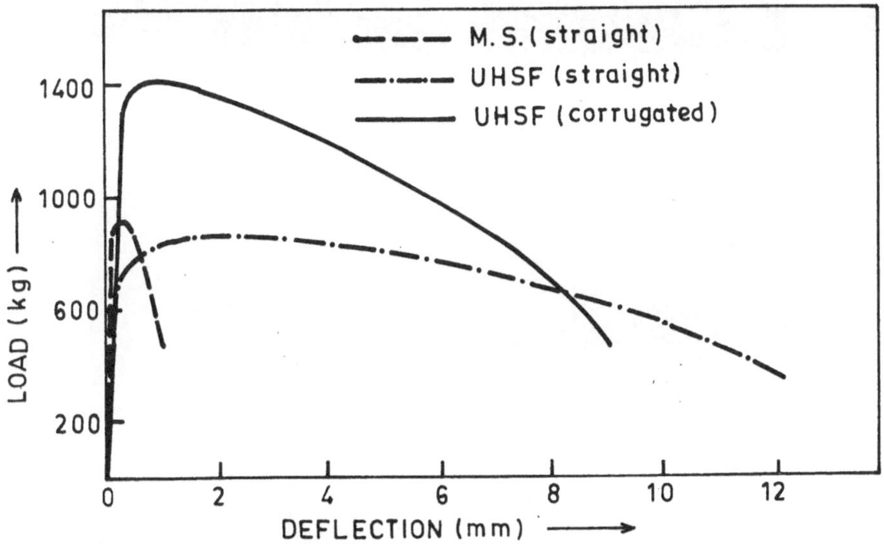

Fig.5 Load-Deflection curves for the beams reinforced with fibers of aspect ratio 80 (80 AR)

load carrying capacity of the concrete), and in case of the higher aspect ratio both the flexural strength as well as the toughness have improved. As can be seen from the Table 2; the beam cast with UHSF-corrugated fibers of 80 aspect ratio has shown best properties in both flexure as well as toughness. The aspect ratio, which influences both bond strength and orientation of fibers with respect to the crack propagation direction, appears to be the most significant factor affecting crack propagation resistance. Since the compaction of the specimens has been done on a vibrating table for the same period of time, it is quite probable that the random orientation of fibers was more in case of the lower aspect ratio fibers. Also it has been found that when short straight fibers are used with random orientation, the fibers do not contribute towards post crack load carrying capacity of the composite [4,5]. However with the use of corrugated fibers (40 AR) the toughness has increased, indicating that the bond has improved between the fiber and the matrix. A further confirmation of the improved characteristics of the beam reinforced with UHSF corrugated can be observed from the results of beam reinforced with 60 aspect ratio fibers and 80 aspect ratio fibers. Since the length of the fibers is increasing, the mechanical anchorage also consequently improves (bond strength being a function of surface area of contact) thus contributing to the improved properties of the composite. Further, due to the external vibrations which were used as means for compaction, it is

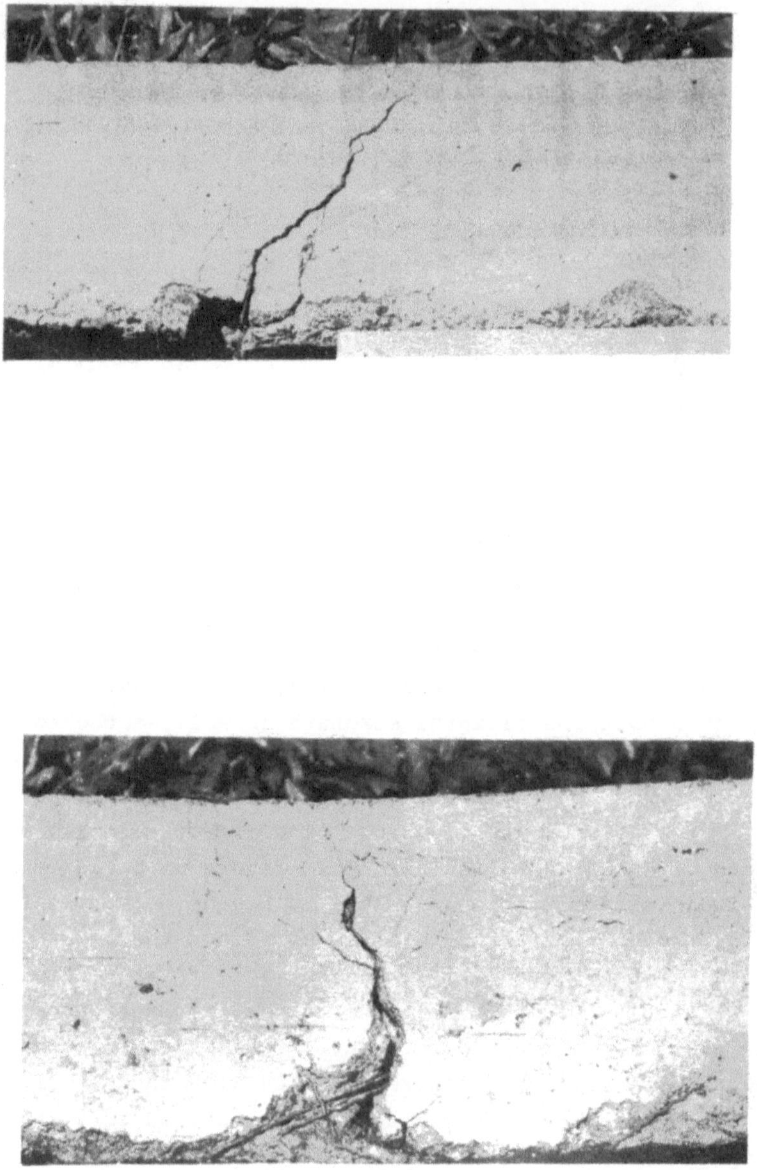

Fig.6. Typical crack patterns observed in fiber reinforced beams, tested under centre point loading conditions.

quite probable that the fibers tended to get aligned in one direction, which further enhanced the crack resisting properties. As can be observed, the best properties of the fiber reinforced concrete have been obtained for the beam having UHSF-80 aspect ratio corrugated. Increasing the aspect ratio would no doubt further enhance the anchorage but it would create not only problems during mixing of the concrete, but would also increase the risk of segregation, and thus leading to inconsistent properties of the composite. This could be one probable reason for the low toughness value obtained for the beam reinforced with straight mild steel fibers of 80 aspect ratio.

Studies on improvement in bond strength in steel fiber reinforced concrete indicated that the best way to improve bond strength would be to provide positive anchorages for the fiber in the matrix [6] and this can be done by adopting the corrugated shape as done in the case of the ultrahigh strength fibers.

CONCLUSIONS

1. For maximum utilisations of high strength levels of UHSF, the mechanical anchorage in the cement concrete matrix should be effective.
2. The bond strength of the fibers in concrete can be improved by shaping the fibers accordingly so as to give a positive anchorage- as in the case of corrugated fibers.
3. UHSF, having a corrugated shape and an aspect ratio 80, have shown considerable improvement in both flexural strength and toughness as compared to conventional fibers. This appears to be the optimum aspect ratio for UHSF of corrugated shape.
4. By a judicious use of the cement concrete mix (and employing a shape which contributes to a positive anchorage of the fiber with the matrix) and by ensuring uniform distribution of the UHSF corrugated fibers of 80 AR in the mix, it is possible to derive the maximum benefits of the UHSF characteristics, thereby enhancing significantly the properties of the fiber reinforced composite.

REFERENCES

1. Guide for Specifying, Mixing, Placing and Finishing Steel Fiber Reinforced Concrete. ACI 544. 3R-84

2. Gurusamy K, Mangat P S, Proc. of Symposium on 'Our World in Concrete Structures' Singapore, 1986, vol. V pp. 157.

3. Gupte K A, Rau R H G, Proc. of International Conf. on Advances in Metal Forming Technologies ICMF-87, ASM (to be published).

4. Bentur A, Diamond S, and Mindess S, Cement and Concrete Research, vol. 15, 1985, pp. 331-342.

5. Shah S P, and Rangan B V - American Concrete Institute Journal, vol. 62, no. 2, Feb. 1979, pp. 128-135.

6. Tattersall G H, Urbanuwicz C R - Magazine of Concrete Research, vol. 87, June 1974, pp 105-113.

Proceedings of the International Symposium on Fibre Reinforced Concrete, December 16-19, 1987, Madras, India

STUDY OF FIBRE REINFORCED CONCRETE

P.K.Nagarkar
Chief Engineer & Director

S.K.Tambe
Research Officer

D.G.Pazare
Asstt. Res. Officer

S.K.Deshpande
Jr. Scientific Assitt.

Maharashtra Engineering Research Institute, Nashik, Maharashtra, India.

SYNOPSIS

Fibre reinforced concrete is the two phase concept in which two materials are combined to produce a composite of superior properties which has been known since ancient time. Fibre reinforced concrete can be defined as a composite material consisting of cement based matrix containing an ordered or random distribution of fibres. Concrete is very good in compressive strength but weak in tension. It has very low durability and little resistance to cracking. These drawbacks of plain concrete can be removed by the addition of fibres in the concrete. The common fibres generally used in the concrete are Nylon, Coir, Jute and Steel. The fibre used in the study taken up by M.E.R.I., Nashik, were Nylon and Steel only, having different aspect ratio. The test results indicated that the compressive, split tensile and flexural strength increased by addition of fibres; however the workability of concrete was reduced.

INTRODUCTION

When concrete is reinforced by short closely spaced fibres, an ideal two phase composite is produced which is term as fibre-reinforced concrete.

The object of the study is to determine, in case the randomly oriented fibres, the percentage reinforcement, and the effect of aspect ratio on the properties of concrete i.e. compressive strength, flexural strength, split tensile, bond with steel, abrasion resistance and modulus of elasticity. The properties were assessed for fresh as well as hardened concrete.

SCOPE

The fibre reinforced concrete is essentially a composite system wherein the tensile strength is slightly improved. The additional improvement in the properties of fibre reinforced concrete is bond with steel alongwith other properties are observed.

The study was taken up to see the effect of aspect ratio on the various properties of fibre reinforced concrete. Aspect ratio is defined as the ratio between the length and diameter of the fibre.

MATERIALS

Cement :

Cement used in the study is tested as per I.S. 4031-1967. The cement conforming to I.S. limits of I.S. 269-1976 is used. Test results are given in Table 1.

Sand :

Locally available river sand is used. The grading and physical properties of sand are given in the Table 2.

Coarse Aggregate :

Locally available aggregate is used for this study. Physical properties of aggregate used in the study are given in Table 3. 20 mm M.S.A. aggregate is used. The combination of aggregate (i.e. 20 mm - 10 mm) is taken as 50:50 to get good grading.

Fibres :

The material used as fibre in this study is steel or nylon. These fibres are cut into 20 mm and 60 mm lengths. Their physical properties are given in Table 4.

EXPERIMENTAL PROCEDURE

Design of Mix :

The concrete mixes were designed as per I.S. 10262-1982 by absolute volume concept. The concrete mixes were run with 350 and 450 kg/m^3 cement content. The unit water content were kept constant for plain concrete as well as fibre reinforced concrete.

The nylon and steel fibres were used. The addition of fibres in the concrete was in the percentage of 0.3 and 0.5 by percentage of volume of concrete.

Casting of Specimen :

The cubes of 15x15 cm were cast for carrying compressive strength at 7, 28 and 90 days age.

The 15x30 cm cylinders were cast for split tensile at 7, 28 and 90 days age.

The 12 mm dia. M.S. bar was embedded in the cylinder to see the bond with steel at 28 days.

The additional 3 cylinders were cast for 'E' value at 28 days.

10x10x50 cm beam were cast for flexural strength at 7, 28 and 90 days.

15 cm dia. spheres were cast for abrasion resistance at 28 days.

TABLE NO. 1

Test result of cement sample

Lab. sample No.	Specific Gravity	Fineness cm^2/g	Normal consistency %	Setting time in Min.		Soundness %		Compressive Strength in kg/cm^2		
				Initial	Final	Autoclave	Leach	3 days	7 days	28 days
Laboratory Cement	3.048	3851	28.5	180	240	0.04	--	227	338	--
Limits		Min. 2250/ 3000		Min. 30	Max. 600	Max. 0.8		Min. 160	Min. 220	Min. 310

TABLE NO. 2

Test results of sand

Lab. No. & Mark	Grading individual % retained on sieves								Fineness modulus	Specific gravity	% Absorption	Silt content	Organic impurities
	10 mm	4.75 mm	2.36 mm	1.18 mm	600 Micron	300 Micron	150 Micron	Pan					
Local sand	--	4	8	20	32	25	10	1	3.0	2.66	3.5	1.5	Within limits

Max. limit as per I.S. 383-1970 : Silt % for Sand Max. 5%

TABLE NO. 3

Test results of aggregates

Lab. No. & Mark	Grading Individual % retained on Seives							Specific gravity	% Absorption	Unit weight Kg/ltr.		Remarks
	20 mm	12.5 mm	10 mm	4.75 mm	2.36 mm	1.18 mm	Pan			Loose	Rodded	
20 mm	5	60	30	5	-	-	-	2.87	2.0	1.4	1.56	-
10 mm	-	15	5	70	5	4	1	2.85	2.5	1.47	1.70	-

TABLE NO. 4

Physical properties of nylon and steel

Name of fibre	Diameter of fibre	Actual load in kg.	Tensile strength in kg/cm^2	Remarks
Nylon	0.20 mm	1.049	33.4	I.S. 280 - 1972 Limits (32-44) kg/cm^2
Steel	0.57 mm	12	46.0	

TABLE NO. 5

Statement showing the test results of fibre reinforce concrete with cement content 350 kg/m^3

Mix No.	Type of mix.	% Fibre by vol.	Aspect ratio	Cement kg/m^3	Water kg/m^3	Sand kg/m^3	Aggregate kg/m^3		Slump cm	Unit wt. kg/lit.	Compressive strength on 15 cm. cubes in kg/cm^2		
							20 mm	10 mm			7 days	28 days	90 days
1	2	3	4	5	6	7	8	9	10	11	12	13	14
FR-1	Control	-	-	350	190	806	552	552	7.0	2.29	104 (100)	182 (100)	290 (100)
FR-2	Nylon	0.3	100	350	190	806	552	552	3.5	2.31	127 (122)	228 (125)	329 (113)
FR-3	Nylon	0.5	100	350	190	806	552	552	2.0	2.33	127 (122)	223 (123)	303 (105)
FR-5	Nylon	0.3	300	350	190	806	552	552	1.0	2.32	138 (133)	215 (118)	311 (107)
FR-6*	Nylon	0.5	300	350	190	806	552	552	0.0	2.17	68 (65)	133 (73)	180 (95)
FR-4	Steel	0.3	35	350	190	806	552	552	4.0	2.37	111 (107)	198 (109)	295 (102)
FR-7	Steel	0.5	35	350	190	806	552	552	2.0	2.33	135 (130)	259 (142)	318 (110)
FR-8	Steel	0.3	105	350	190	806	552	552	3.0	2.35	155 (149)	260 (143)	303 (105)
FR-9	Steel	0.5	105	350	190	806	552	552	1.0	2.35	162 (156)	240 (132)	275 (95)

*Mix was very harsh and balling is observed.

TABLE NO. 5 (Contd.)

Mix No.	Split tensile on 15x30 cm cylinder in kg/cm^2			Flexural strength on 10x10x50 cm bars in kg/cm^2			Abrasion resistance on 15 cm spheres % loss in weight			'E' Value kg/cm^2 x 10^5	Bond with steel pullout test kg/cm^2
	7 days	28 days	90 days	7 days	28 days	90 days	1000	2500	5000		
1	15	16	17	18	19	20	21	22	23	24	25
FR-1	14 (100)	23 (100)	37 (100)	20 (100)	33 (100)	48 (100)	18	27.8	35.5	1.77 (100)	30 (100)
FR-2	16 (114)	30 (130)	41 (111)	20 (100)	40 (121)	60 (125)	9	14.5	19.2	2.0 (117)	38 (127)
FR-3	16 (114)	27 (117)	38 (103)	20 (100)	43 (130)	66 (138)	5.2	8.9	12.5	1.52 (86)	38 (127)
FR-5	17 (121)	28 (122)	40 (108)	28 (140)	40 (121)	52 (108)	3.8	7.5	10.9	3.2 (181)	39 (130)
FR-6	11 (79)	19 (83)	23 (62)	18 (90)	28 (85)	28 (58)	10.2	17	19.7	1.6 (90)	25 (83)
FR-4	13.3 (95)	26 (113)	28 (103)	24 (120)	38 (115)	60 (125)	7	12	16	1.45 (82)	37 (123)
FR-7	17 (121)	31 (135)	40 (108)	32 (160)	44 (133)	72 (150)	7.9	11.3	15.2	1.88 (106)	50 (167)
FR-8	22 (157)	33 (143)	42 (114)	28 (140)	38 (115)	54 (113)	7	12.6	16.9	2.28 (129)	45 (150)
FR-9	23 (164)	30 (130)	36 (97)	28 (140)	40 (121)	60 (125)	9.5	15.8	21.2	2.0 (113)	48 (160)

Figures in brackets shows percentage strength that of control.

TABLE NO. 6

Statement showing the test results of fibre reinforce concrete with cement content 450 kg/m^3

Mix. No.	Type of mix.	% Fibre by vol.	Aspect ratio	Cement kg/m^3	Water kg/m^3	Sand kg/m^3	Aggregate kg/m^3		Slump cm	Unit wt. kg/lit.	Compressive strength on 15 cm cubes in kg/cm^2		
							20 mm	10 mm			7 days	28 days	90 days
1	2	3	4	5	6	7	8	9	10	11	12	13	14
FR-11	Control	-	-	450	200	676	545	545	7.0	2.31	191 (100)	287 (100)	367 (100)
FR-12	Nylon	0.3	100	450	200	676	545	545	4.5	2.27	179 (94)	263 (92)	390 (136)
FR-13	Nylon	0.5	100	450	200	676	545	545	2.0	2.29	221 (116)	325 (113)	390 (136)
FR-14	Nylon	0.3	300	450	200	676	545	545	2.5	2.30	164 (86)	242 (84)	407 (142)
FR-15*	Nylon	0.5	300	450	200	676	545	545	1.5	2.26	151 (79)	283 (99)	317 (86)
FR-16	Steel	0.3	35	450	200	676	545	545	3.5	2.26	204 (107)	315 (110)	397 (138)
FR-17	Steel	0.5	35	450	200	676	545	545	2.0	2.29	229 (120)	385 (134)	484 (169)
FR-18	Steel	0.3	105	450	200	676	545	545	2.0	2.27	244 (128)	400 (139)	448 (156)
FR-19	Steel	0.5	105	450	200	676	545	545	1.0	2.33	257 (135)	416 (145)	466 (162)

* Mix was very harsh and balling is observed.

TABLE NO. 6 (Contd.)

Mix. No.	Split tensile on 15x30 cm cylinder in kg/cm^2			Flexural strength on 10x10x50 cm Bars in kg/cm^2			Abrasion resistance on 15 cm spheres % loss in weight			'E' Value kg/cm^2 x 10^5	Bond with steel pullout test kg/cm^2
	7 days	28 days	90 days	7 days	28 days	90 days	1000	2500	5000		
1	15	16	17	18	19	20	21	22	23	24	25
FR-11	26 (100)	38 (100)	55 (100)	34 (100)	64 (100)	70 (100)	10	20	30	2.1 (100)	42 (100)
FR-12	27 (104)	36 (95)	60 (105)	36 (106)	64 (100)	84 (120)	5.5	11	14	2.56 (122)	41 (98)
FR-13	27 (104)	36 (113)	60 (116)	36 (106)	76 (119)	95 (136)	5.5	11	14	3.2 (132)	46 (110)
FR-14	17 (65)	31 (82)	44 (80)	30 (88)	50 (78)	60 (86)	3	6	8	1.83 (87)	41 (98)
FR-15	19 (70)	27 (71)	41 (75)	30 (88)	60 (84)	62 (89)	8	15	20	2.0 (95)	32 (76)
FR-16	28 (108)	46 (121)	66 (120)	38 (112)	70 (109)	88 (126)	5	10	12	1.8 (86)	50 (119)
FR-17	30 (115)	50 (132)	74 (135)	41 (121)	76 (119)	92 (131)	6	10	12	2.0 (95)	60 (143)
FR-18	34 (131)	54 (142)	75 (136)	47 (138)	86 (134)	95 (136)	.5	8	13	2.1 (100)	55 (131)
FR-19	37 (142)	57 (150)	79 (144)	48 (141)	89 (139)	98 (140)	6	12	15	1.9 (90)	58 (138)

Figures in brackets shows percentage strength that of control.

COLLECTION OF DATA

Tables 1 to 4 present properties of various material used, as quoted in para 3.0 above. Tables 5 and 6 show various test results viz. compressive strength, flexural strength, split tensile strength, abrasion resistance, 'E' value, Bond with steel and test on fresh concrete i.e. slump and unit weight.

ANALYSIS OF DATA

Workability of Concrete:

It is observed that addition of fibres into the mix reduce the workability of concrete.

The nylon fibres when added in concrete reduce workability more than that of steel fibres. As the percentage of fibres is increased the reduction in workability is more which can be seen from Table 5.

As the aspect ratio is increased the workability of concrete is reduced abruptly.

If nylon fibres are considered with reinforcement of 0.5% and aspect ratio 300, the mix becomes unworkable and vibration is very difficult.

Compressive Strength:

The compressive strength is increased for steel fibres from 13% to 30% and for nylon fibres 10% to 45% at 28 days. But the addition of nylon fibres of 0.5% and having aspect ratio 300, the compressive strength is reduced. Due to random orientation of fibres, sometimes the strengths appear to be improved and vice versa.

Steel fibres reinforced concrete in general gives much better strengths than those obtained with the use of nylon fibres.

The maximum strength achieved with 0.5% addition of steel fibres having aspect ratio 105 and for nylon fibres it is at 0.5% addition having aspect ratio 100.

Split Tensile Strength :

The tensile strength of concrete is increased with addition of fibres. The increase in the split tensile strength is from 13 to 30% for nylon fibres and 13% to 50% for steel fibres at 28 days. The aspect ratio of fibres has no predominant effect on the tensile strength of concrete.

The maximum strength achieved is with steel fibres with 0.3% addition having aspect ratio 35 and for nylon fibres it is at 0.3% having aspect ratio 100.

It is observed from the study done so far that addition of nylon fibres should be limited to 0.5% with aspect ratio as 300. Increase in the addition of fibres beyond that the mixes are become very harsh and balling of fibre is observed.

Flexural Strength:

The flexural strength of concrete is increased by addition of fibres. The general increase in the flexural strength is in the order of 19% to 30% for nylon fibres and 10 to 40% for steel fibres. For the percentage of nylon fibres of the order of 0.5% and having aspect ratio 300, the flexural strength observed is less than the control concrete, due to the non-cohesive concrete and balling action of fibre. Steel fibres increase the flexural strength more than that of nylon fibres.

The maximum flexural strength with nylon fibres is achieved with 0.3% fibres and aspect ratio 300 and for steel it is with 0.5% fibres with aspect ratio 105.

Statis Modulus of Elasticity:

The static modulus of elasticity improved by addition of all types of fibres as compared to plain concrete.

The nylon fibres showed better modulus of elasticity as compared to steel. The general increase of 'E' value for nylon fibre is at the range of 17% to 81% and for steel fibres is at the range of 6% to 29%. The maximum 'E' value is observed at 0.3% nylon with aspect ratio 300. For steel fibres the maximum modulus of elasticity is observed for 0.3% having aspect ratio 105.

Abrasion Resistance:

The 15 cm dia. spheres were tested in the Los Angels Abrasion Testing Machine for abrasion resistance for different revolutions as shown in Table No.5 and 6.

It is observed that abrasion resistance to wear is improved by addition of fibres.

Bond with Steel:

The bond strength was determined by pull out tests as per reference in I.S. 2770 (Part-I)-1967.

The addition of fibres improved the bond strength with steel. The general improvement for bond with steel is 23 to 67%. The steel fibres give more bond than nylon fibres.

The maximum bond strength with steel fibres is achieved with 0.5% steel by volume of concrete having aspect ratio 35. The nylon fibres shows less bond with steel as compared to steel the maximum bond with addition of nylon fibres is with 0.3% and having aspect ratio 300.

CONCLUSIONS

In general, the compressive strength is increased by 5 to 57% with the addition of fibres. However, this increase is more predominant in case of steel fibres than nylon fibres.

The split tensile/flexural strengths are increased with the addition of fibres. The split tensile strength is increased by 15 to 45% and flexural strength from 20 to 60%.

Bond with steel is improved with the addition of fibres. Aspect ratio has very little effect on the bond strength of fibre reinforced concrete.

The addition of fibres reduce the workability of concrete. The workability is also reduced with the increase of aspect ratio.

It is generally observed that the properties of fibres reinforced concrete vary to a great extent mostly because of random orientation of fibres.

REFERENCES

1. IS:4031-1967 Hydraulic Cement, method of physical tests.
2. IS:10262-1982 Recommended guidelines for concrete mix design.
3. IS:2770-1967 Methods of testing bond in reinforced concrete.

Proceedings of the International Symposium on Fibre Reinforced Concrete, December 16-19, 1987, Madras, India

MOMENT REDISTRIBUTION CHARACTERISTICS OF A REINFORCED FIBROUS CONCRETE HINGING ZONE

Dr. M. Lakshmipathy
Assistant Professor in
Structural Engineering

Dr. A.R. Santhakumar
Professor of
Structural Engineering

College of Engineering
Anna University
MADRAS - 600 025

ABSTRACT

To provide information regarding moment redistribution characteristics of a reinforced fibrous concrete hinge, four numbers of two span continuous beams, two with fibrous concrete in the hinging zones and two beams with conventional concrete throughout were tested. This paper presents the details of this investigation. The results are compared with theoretical predictions with respect to deflections, stiffness, ultimate load and the extent of moment redistribution. The results conclusively indicate the better performance of beams with fibrous concrete in the hinging zones.

INTRODUCTION

Limit analysis of statically indeterminate structures is based on the fact that sufficient number of plastic hinges form to transform either the whole or part of the structure into a mechanism. For reinforced concrete structures, it is necessary to check that the requisite rotation capacity of the hinges are available for the desired overall ductility to be attained. The rotation capacity of the hinges must be sufficient to permit moment redistribution envisaged. This ductile performance of the plastic hinges is an essential requisite for the development of full ultimate strength as well as for ductile behaviour of the structure in disaster prone areas.

THEORETICAL CONSIDERATIONS

Moment Redistribution

Fig.1(a) shows a two span continuous beam having a uniform cross section. M and M' give elastic moments and Fig.1(c) shows the elastic moment variation. M'_u and M_u are the ultimate moment of resistances of the support section and mid span sections respectively. It is assumed that the section is adequately reinforced to suppress premature shear failure. It is also assumed that the moment curvature variation for the sections is the idealised bilinear relationship for a ductile section

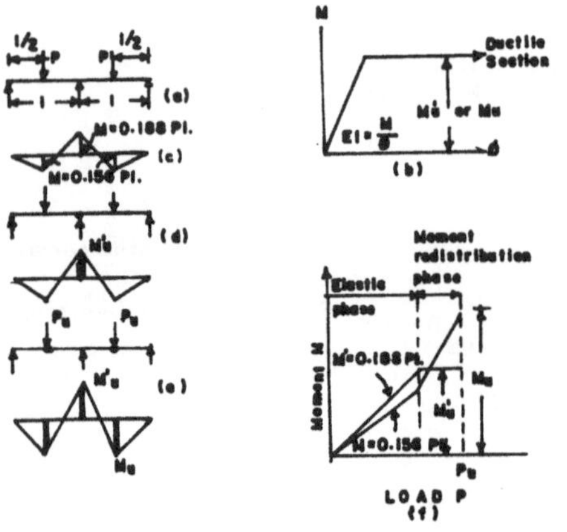

FIG.1 MOMENT REDISTRIBUTION AND FORMATION OF COLLAPSE MECHANISM FOR A CONTINUOUS BEAM

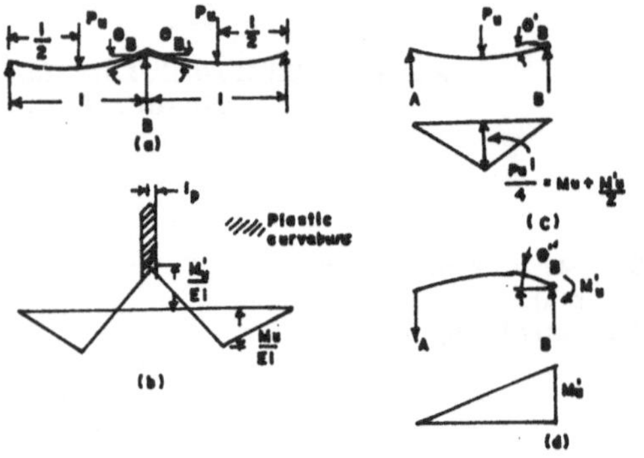

FIG.2 CALCULATION OF HINGE ROTATION FOR BEAM OF FIGURE - 1.

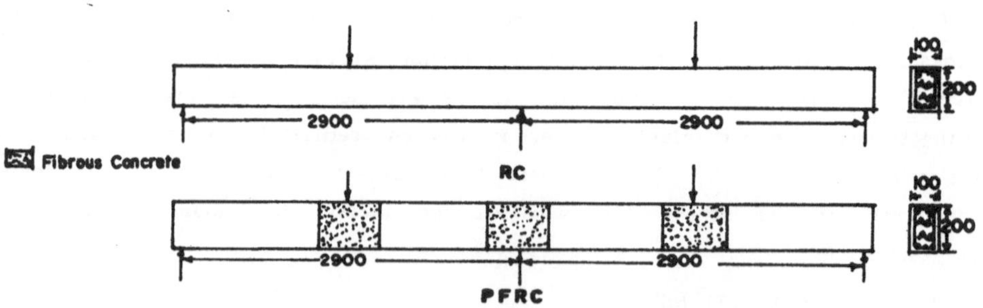

FIG.3 DETAILS OF TWO SPAN CONTINUOUS BEAMS

as shown in Fig.1(b). At low loads, the distribution of bending moment due to the two concentrated loads is in accordance with the elastic theory. See Fig.1(c). As the applied loads are increased the ultimate moment of resistance is reached at one critical section, say over the centre support, before it is reached at other sections. Then the moment at the centre support will be Mu' as in Fig.1(d). The extent to which further load can be carried by the beam depends on the ability of the hinge at centre support to undergo plastic rotation. If the support section is ductile, additional load can be carried because the plastic hinge at this point will rotate while maintaining its moment of resistance Mu'. Thus moment redistribution will occur until the maximum positive moment in mid span sections increase to Mu. This leads to the collapse mechanism in Fig.1(e). It traces the variation of bending moment at critical sections with loads on the beam as shown. (1)

In this example

$Mu'/Mu < M'/M = 1.2$

$M + M'/2 = Pl/4$

Therefore $P = 4/l\ (M + M'/2)$

In addition, if the moment at the centre support remains at Mu' until Mu develops at the mid span sections, we have

$P_u = 4/l\ (Mu + Mu'/2)$

Additional rotations after the formation of mechanism as shown in Fig.1(e) is considered favourable for earth quake resistant structures as at this stage energy dissipation occurs at constant ultimate load.

Requirement of Plastic Rotation for Redistribution

The beam and the curvature diagrams of Fig.2(a) and 2(b) shows the stage at which sufficient plastic rotations have occured at the centre support B to enable the ultimate moment Mu to be just developed at mid span. Hence Fig.2(a) and 2(b) indicates the stage when Pu is just reached. The plastic curvature is considered to occur over the equivalent plastic hinge length l_p either side of the critical section. The elastic curvature along with the length of the member is calculated using the elastic bending moment variation and the assumed constant flexural rigidty EI. The plastic rotation θ_p at the centre support B is the discontinuity of slope between the ends of the adjacent members and $\theta_p = 2\ \theta_B$ as indicated in Fig.2(a).

To find θ_p' consider the elastic deformations of the members supporting the load P_u. First replace the plastic hinge at B by a frictionless hinge as in Fig.2(c). Then from the moment area theorem the rotation at B due to the load P_u alone on one span is

P_u alone on one span is

$$\theta'_B = \frac{(Mu + 0.5\ Mu')\ l}{4\ EI}$$

considering the effect of the ultimate support moment Mu' acting at the frictionless hinge as in Fig.2(d), the rotation at B due to Mu' alone is

$$\theta''_B = Mu'\ l/3\ EI$$

Therefore $\theta_B = \theta'_B - \theta''_B = 1/4\ EI\ (Mu - 5/6\ Mu')$

$$\theta_p = 2\ \theta_B = l/2\ EI\ (Mu - 5/6\ Mu')$$

Equation 1 gives the rotation required at the plastic hinge at the centre support B (assuming Mu > 5/6 Mu')

The available plastic rotation required at the plastic hinge is calculated as

$$\theta_p \text{ available} = \left(\frac{\epsilon_c}{c} - \frac{\epsilon_{ce}}{Kd}\right) 2l_p$$

When Mu > 5/6 Mu' redistribution of moments can take place until the ultimate moment develops at each mid span section if

$$\frac{l}{2\ EI}(Mu - 5/6\ M'_u) \leq \left(\frac{\epsilon_c}{c} - \frac{\epsilon_{ce}}{Kd}\right) 2l_p$$

DETAILS OF TESTS

A two span beam with identical span lengths of 290 cm each as shown in Fig.3 was subjected to two concentrated loads each applied at centre of each span. The beams with reinforced concrete (RC) and partial Fibrous Reinforced Concrete (PFRC) are shown in Fig.3. The mid reaction point and centres of each spans are the possible locations of hinges and hence the fibrous concrete was provided at these locations only. Companion conventional reinforced concrete beams (RC) were also tested. Table-1, gives the details of 4 specimens tested.

Theoretical capacities of sections used

The detailed calculations based on theoretical consideration as per procedure by Lakshmipathy(2), showed the capacities and variables for the section as indicated in Table-1.

TEST SET UP AND TESTING

The entire set-up together with jack, test frames, supports etc., can be seen from the photographs reproduced in Fig.4. The test frames centred at 290 cm intervals were utilised as loading points for the continuous beam. Three supports were erected at 290 cm interval. All beams were tested to failure.

TABLE-1
OBSERVED AND THEORETICAL RESULTS

Designation	RC 1	RC 2	PFRC 1	PFRC 2
Percentage of Steel A'_s	0.5	0.5	0.5	0.5
A_s	1.2	1.2	1.2	1.2
Steel fibres by Volume	Nil	Nil	1%	1%
Cube compressive strength in N/mm^2	22.1	22.3	22.6	22.8
Yield strength of steel in N/mm^2	290.0	290.0	290.0	290.0
Modulus of Rupture N/mm^2	3.29	3.30	4.32	4.34
Cracking Moment KN.m	3.0331	3.0424	3.9827	4.0012
Load at cracking (T) KN	5.56	5.58	7.31	7.34
Load at cracking (O) KN	6.00	6.0	9.00	9.00
Ratio of load at cracking (O/T)	1.079	1.075	1.23	1.22
Ultimate Moment KN.m M_u	11.70	11.71	14.83	14.85
M'_u	4.98	4.98	9.58	9.59
Ultimate Load (T) KN	19.60	19.70	27.00	27.10
Ultimate Load (O) KN	25.00	25.50	30.00	31.00
Ratio of Ultimate Load(T/O)	o.784	0.772	0.90	0.874
Mid Span deflection at cracking (T) mm	11.00	11.00	14.40	14.40

T - Theoretical values O - Observed values

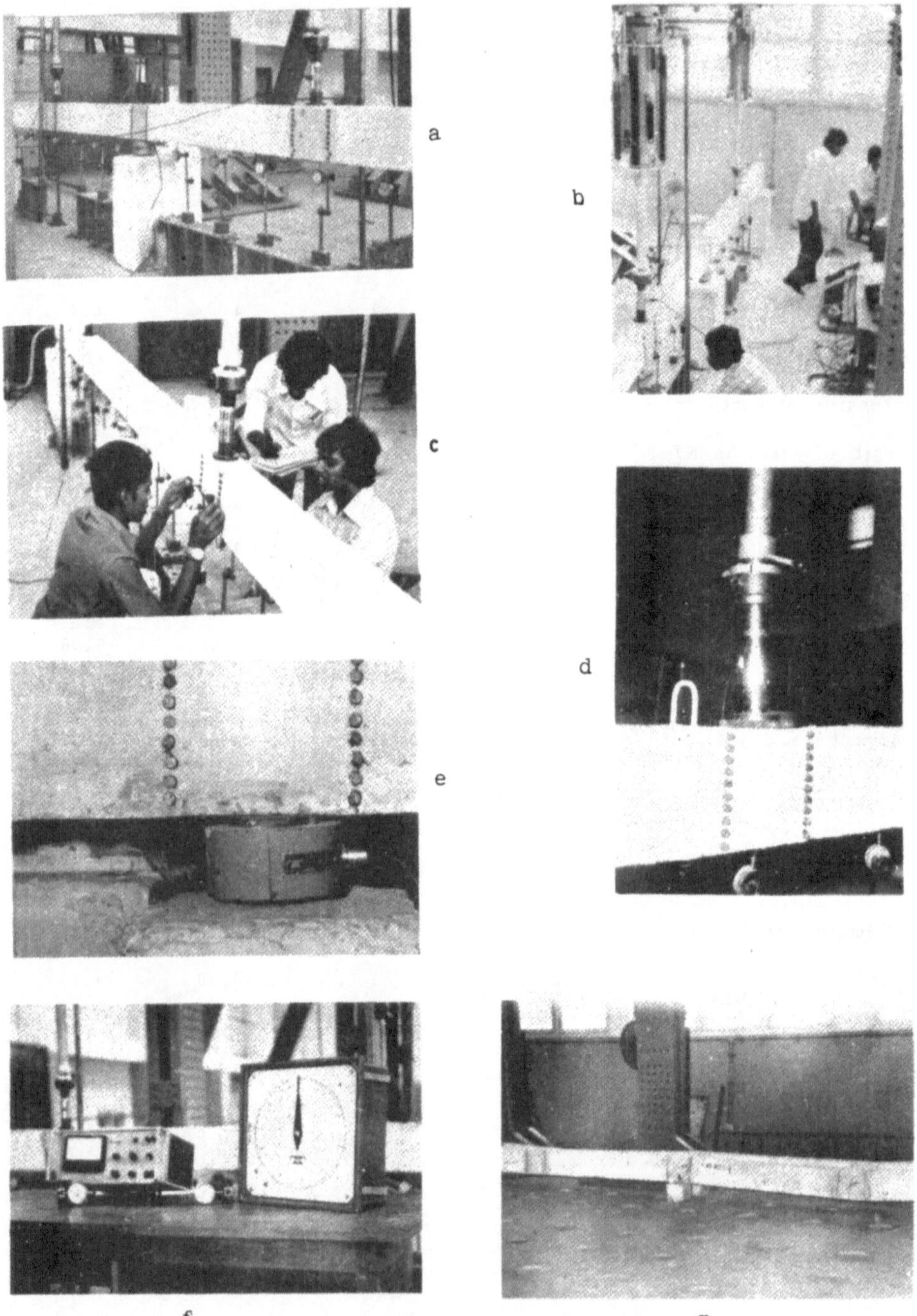

FIG. 4. TESTING OF TWO SPAN CONTINUOUS BEAMS

a,b,c. General set up together with jacks, test frames and supports. d. Load cell at loading point. e. Load cell at reaction point. f. Load indicators and demec gauges. g. Tested beams.

BEHAVIOUR WITNESSED DURING TEST

Loads at cracking and Ultimate

The observed and predicted loads for the four beams are shown in Table-1. The cracking load computed based on elastic moment variation and the observed first crack load corelate well. As expected the cracking load for PFRC beam was higher than that for RC beam by more than 30%.

The predicted ultimate load based on full moment redistribution and that observed have a ratio of less than 1 for both RC and PERC beams. This shows that full redistribution has occured before the mechanism has formed. Also, the predicted results as per the procedure used is in agreement with those observed.

Load deflection Behaviour

Fig.5(a) and (b) shows the load mid-span deflection curves for the RC and PFRC beams tested respectively. In the same figure theoretical stiffness are also shown. All the four beams show clear formation of the three hinges and full moment redistribution and have ductilities of the order of 4 or more. The stiffness at working load of the PFRC beams are 25% higher than RC Beam.

Moment Redistribution Characteristics

Fig.6(a) shows the predicted and observed load moment curves for RC beams and Fig.6(b) shows the similar curves for PFRC beams. In these figures theoretical load-moment variation has also been plotted. It is seen that the yield plateau for PFRC is larger than for RC beams. The moment in both hinges, pick up after plastic rotation, because of strain hardening effect. The increase in moment due to strain hardening is considerable for the first hinge formed. These curves clearly show the formation of the three hinges and full moment redistribution. The yield plateau for the first hinge is larger than that for the hinges formed subsequently.

Strain Variation Across the Depth

Fig.7(a) shows the strain variation across the depth for RC beam at mid-span and support sections and Fig.7(b) shows similar curves for PFRC beam. The compressive zones in RC beam is narrow compared to that of PFRC beam. Theoretical predictions agree with such a conclusion at the mid span section. Larger curvature variation are witnessed for the support section compared to the mid span section. The curvature obtained for the PFRC beams are two times larger than for RC beams. This can be considered to be a favourable behaviour for PFRC beams. A perusal of the figures shows that the shift in neutral axis is more in RC beams compared to PFRC beams for same load increment in the post cracking stage.

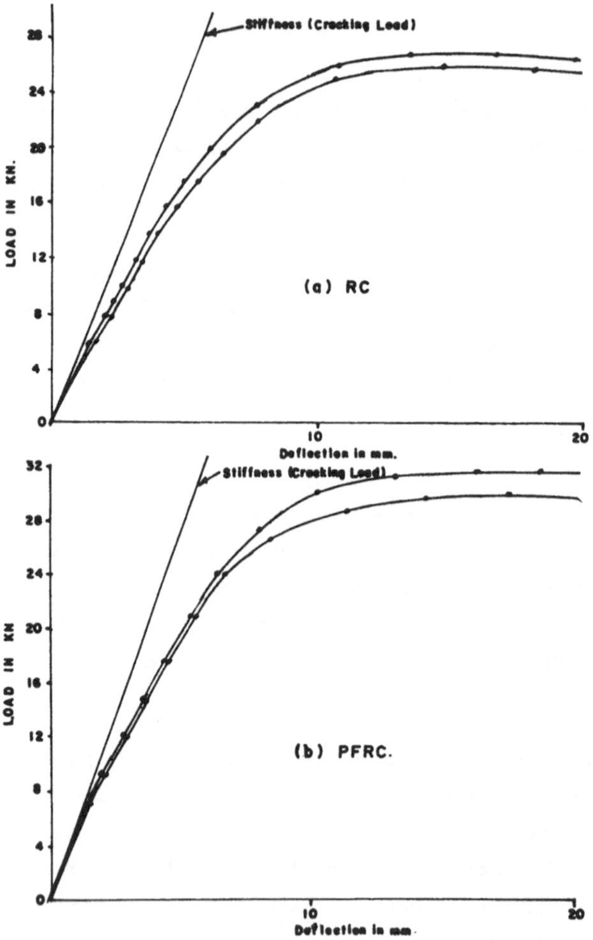

FIG.5 LOAD Vs MID SPAN DEFLECTION

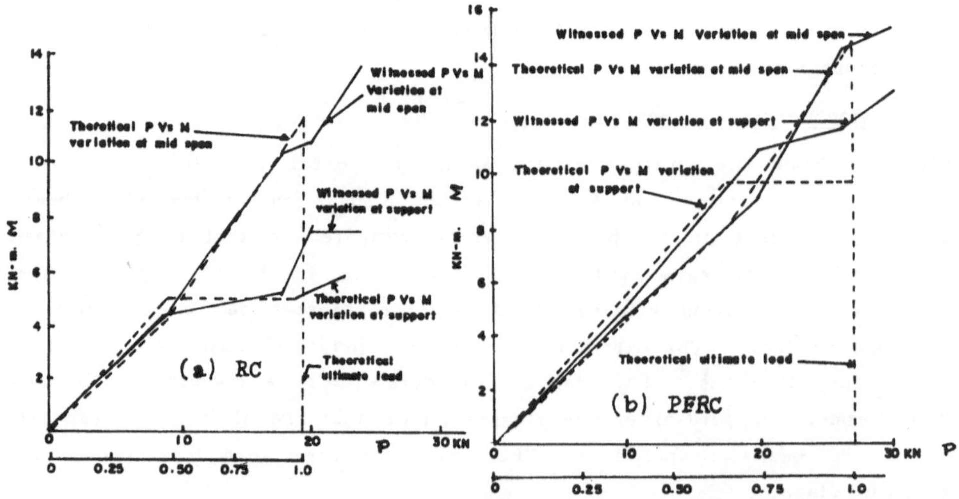

FIG.6 TYPICAL PREDICTED AND OBSERVESD LOAD-MOMENT CURVES

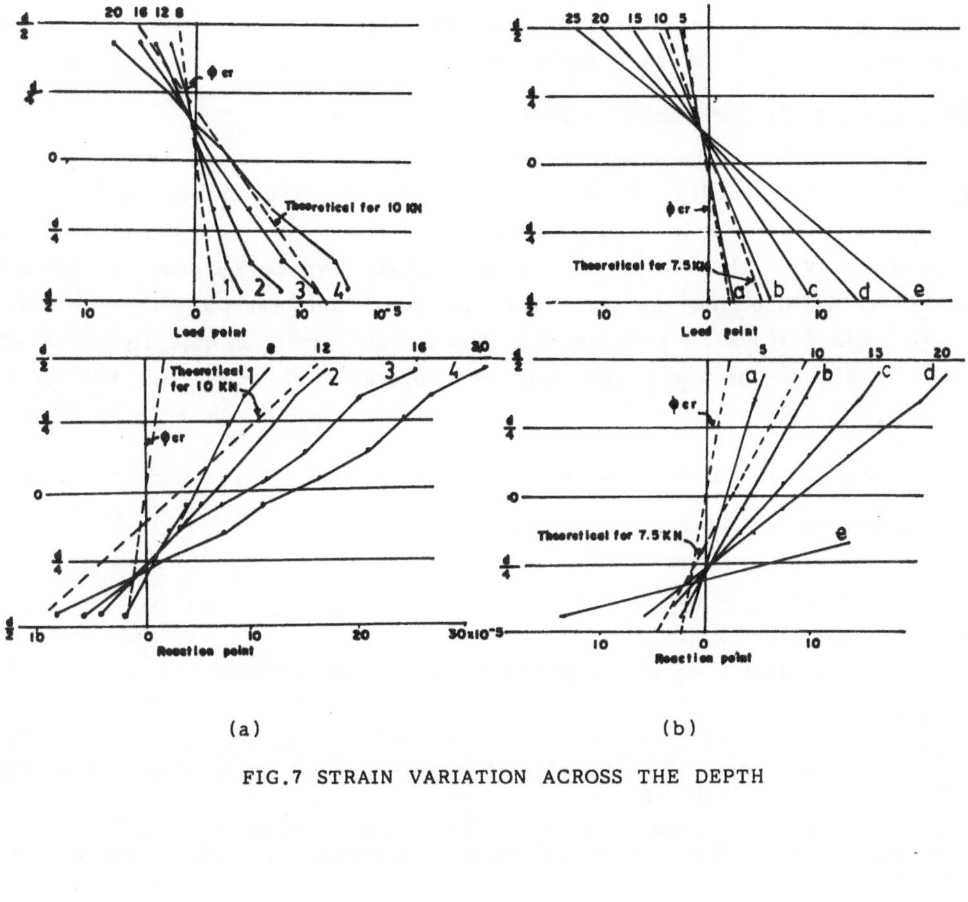

(a) (b)

FIG.7 STRAIN VARIATION ACROSS THE DEPTH

Cracking, Crack width and Disintegration

The cracking and disintegration of concrete in the damaged tension zone were quite significant for RC beams whereas it was considerably less for PFRC beams. Crack widths in PFRC beams in the hinging zones were found to be less by 30% compared to crackwidths in the same regions of the RC beams for comparable load level. At ultimate load stage, the cracks in tension zone of RC beams opened up considerably due to excessive deformation and spalling of concrete fragments were inevitable. On the other hand the PFRC beams showed shatter resistance and exhibited more even crack distribution.

Ductility and Energy Absorption Capacity

Section ductility as measured from distribution of strain across depth (Fig.7(a) and (b)) for PFRC is two times that for RC beam for the critical support section. The overall displacement ductility for both RC beams were of the order of 10. The energy absorption capacity was worked out for RC and PFRC beams by calculating the area under the load displacement diagrams obtained for both spans (each span separately). The energy absorption capacity for PFRC beam (54.2 KN. cm) is 25% larger than the comparable RC beam (42.8 KN. cm.).

CONCLUSIONS

The inclusion of fibrous concrete in the hinging zones has resulted in

i) 25% increase in stiffness at service load
ii) 15% increase in ultimate load
iii) 20% reduction in crack width
iv) 10% increase in ductility
v) 25% less stiffness degradation compared to conventional beams.

REFERENCES

1. Park.R., and Paulay.T., "Reinforced Concrete Structures" John Wiley and Sons, New York, 1975, pp. 769.
2. Lakshmipathy.M., "Ductility of Reinforced Fibrous Concrete Structural Members" Ph.D. Thesis presented to the University of Madras, April 1983, pp 1-221.

Proceedings of the International Symposium on Fibre Reinforced Concrete, December 16–19, 1987, Madras, India

A STUDY ON THE POST-CRACKING BEHAVIOUR OF SFRC WORKING TOGETHER WITH STEEL BAR IN UNIAXIAL TENSION

Liqiu Guan[1] and Guofan Zhao[2]
Dalian Institute of Technology, China

Synopsis: As the first step of the program being carried out in this laboratory, the results of an analytical and experimental study on the post-cracking behaviour of SFRC working together with steel bars in uniaxial tension are reported. Theoretical equations derived show a good agreement with the test data and therefore make it possible to facilitate computations in structural analysis of steel fibre and bar reinforced concrete structures.

INTRODUCTION

Since in engineering short steel fibres are often used in concrete members reinforced with conventional bars, it is necessary to study the behaviour of SFRC in this case, which may differ considerably from the behaviour of commonly tested concrete members reinforced with steel fibres alone. The research presented in this paper was carried out for this purpose.

TEST PROGRAM

In order to simulate concrete covers around steel bars in realistic engineering structures, the prisms with dimension of $60 \times 60 \times 180$ (mm) were prepared for tension, each with an indented steel bar, diameter 14 mm, placed through its central axis (Fig. 1).

Two types of steel fibres were used in the test with three different fibre volume fractions, V_f, varying from about 1 to 2 per cent. The water-cement ratio and the fibre-sand content were kept constant for all the specimens which were cast and cured under the same condition. The concrete mix and the details of the specimens are shown in Table 1 and 2 respectively.

Table 1. Concrete mix (kg/cm^3)

Specimen	Cement	Water	Sand	limestone $D_{max} = 12$mm	Steel fibres	W/C	V_f(%)
P-	564	239	544	891	0	0.42	0
75F-	575	240	469	891	75	0.42	0.96
110F-	587	244	434	891	110	0.42	1.41
150F-	599	248	394	891	150	0.41	1.92
110RF-	587	244	434	891	110	0.42	1.41

[1]Assistant
[2]Professor

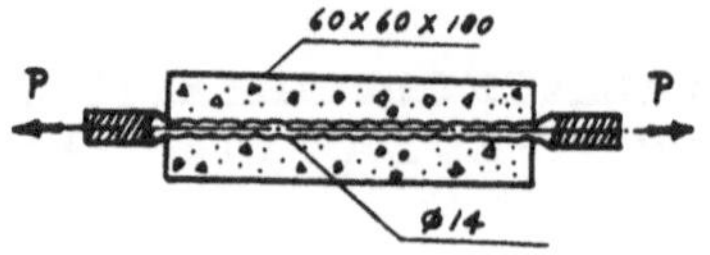

Figure 1: Tensile Specimen

Table 2. Details of the Specimens

Symbol	Fibre type (mm)	d_e (mm)	Fibre aspect ratio l/d	V_f(%)	Number of specimens
P-				0	2
75F-	0.4 × 0.6 × 25	0.48	52	0.96	2
110F-	0.4 × 0.6 × 25	0.48	52	1.41	2
150F-	0.4 × 0.6 × 25	0.48	52	1.92	2
110RF-	0.37 × 27	0.37	73	1.41	2

In Table 2 the equivalent diameters, d_e, are obtained by the following consideration.

because

$$\sigma = \frac{\tau}{4}\frac{P}{A_f}$$

for round fibres

$$P = \pi d, \; A_f = \pi d^2/4$$

thus

$$l/d = \sigma/\tau$$

for rectangular fibres

$$P = 2(a+b), \; A_f = ab$$

thus

$$\sigma/\tau = \frac{a+b}{2ab}l = l/d_e$$

hence the equivalent diameter is

$$d_e = \frac{2ab}{a+b}$$

Tests were carried out on standard material testing machine. The average longitudinal concrete strains on each of the four sides of the specimen were measured by four electrical resistance strain gauges, 8 × 80 mm gauge length, glued on the concrete surfaces. The average longitudinal strains of the central steel bar at each load intervals were measured by eight electrical resistance strain gauges, gauge length 3 × 5mm, placed in the channels on the two opposite sides of the bar (Fig. 2).

The channels were tightly and carefully sealed before casting. Tensile load was applied to the bar and transferred to the SFRC by bond stress.

TEST RESULTS

Typical results obtained are shown in Fig. 3[3]. And the curve shown in Fig. 4 is obtained based on the following calculations:

[3]this is the similar results obtained in repeated tensile loading test which will be reported elsewhere

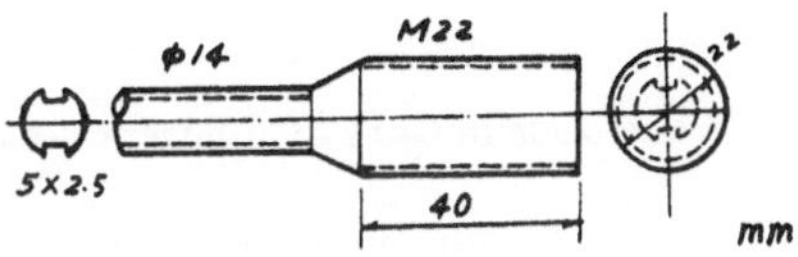

Figure 2: The details of the steel bar

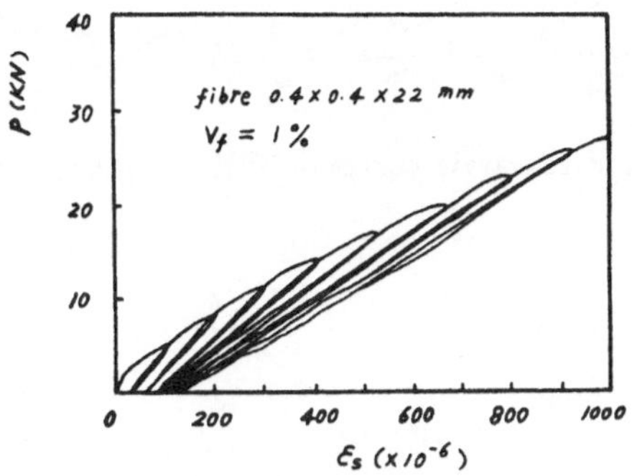

Figure 3: Load-steel stress curves of the specimen

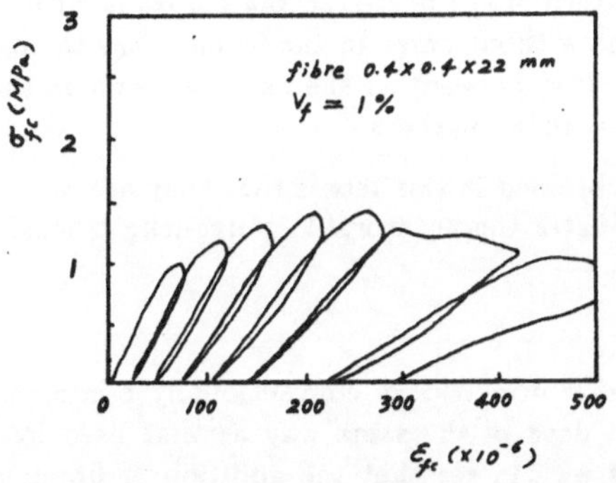

Figure 4: Tensile stress-strain curve of SFRC in the specimen

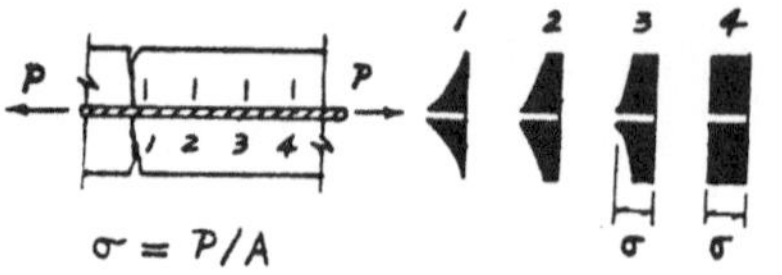

Figure 5: stress distributions at different sections

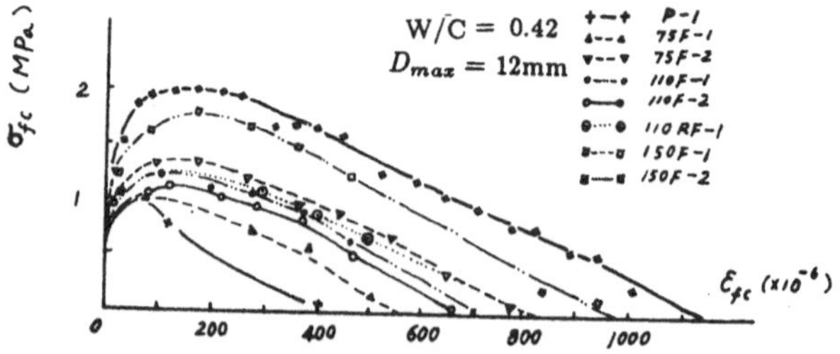

Figure 6: Average stress-strain curves of SFRC of the specimens

$$\sigma = \frac{P - \sigma_s A_s}{A - A_s} = \frac{P - E_s \varepsilon_s A_s}{A - A_s} \tag{1}$$

$$\varepsilon = \frac{1}{2}(\varepsilon_s + \varepsilon_c) \tag{2}$$

where σ, ε are average longitudinal stress and strain of SFRC respectively; P is the total load applied; A and A_s are the cross sectional areas of the specimen and that of the bar; σ_s, ε_s and E_s are respectively the stress, strain and Young's modulus of the steel bar; ε_c is the longitudinal strain of concrete on the surface of the specimen.

Equation (2) is based on the linear strain distribution assumption shown in Fig. 5.

The summary of the test results under monotonous tensile loading is illustrated in Fig. 6. It is shown from the figure that the pattern of the post-cracking curves of SFRC working together with bars differs considerably from those without bars. In the former case the curve descends gradually whereas in the latter case a drop often appears on the curve as soon as the peak load is reached, which was observed and reported by many authors.

Another feature of the curves obtained in this test is that they are not gauge length dependent. This is important because it eliminates the unfavorable gauge-length-sensitive feature observed in the case without bars [1].

ANALYSIS

The effect of fibres on the tensile properties of conventionally reinforced concrete becomes apparent when we interpret the test data in the same way as that used by Prof. F. Leonhardt [2] (Fig. 7 and 8). From the figures we can see that the addition of fibres into reinforced concrete affects mainly its post-cracking behaviour and increases both the strength and the rigidity of the

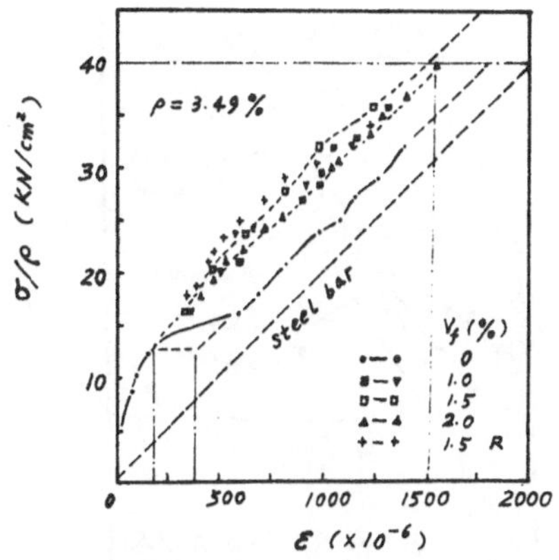

Figure 7: Stress factor-strain relationship of the specimens

member after the cracking of the matrix. The similar result in bending tests was also obtained by the authors [3].

According to the test data in Fig. 7, the stress-strain relationship can be simplified bilinearly for the case of SFBRC [4] and trilinearly for the case of CRC[5] as illustrated in Fig. 9.

By applying the law of mixtures and the mechanism of fibre composites described in [4] to conventionally reinforced concrete, the values at the critical points as those indicated in Fig. 9 can be obtained as follows:

$$\varepsilon_1 = \varepsilon_{mu}(1 + \alpha_c) \tag{3}$$

$$\varepsilon_2 = \varepsilon_y - \alpha_c \varepsilon_{mu} \tag{4}$$

$$\alpha_c = \frac{1}{2}\frac{E_c}{E_s}\frac{1-\rho}{\rho} \tag{5}$$

where E_c is Young's modulus of plain concrete; ρ is bar volume content and ε_{mu} is the ultimate tensile strain of the matrix, that is the ultimate tensile strain of concrete in this case. E_s and ε_y are young's modulus and the yielding strain of the bar respectively.

By extending the above theory to the idealized relationship of SFBRC, we can establish the following equations, i.e.

When $\varepsilon < \varepsilon_{mu}$,

$$\sigma = E\varepsilon \tag{6}$$

where σ, ε are the average stress and strain of the specimen and

$$E = E_m(1-\rho) + E_s\rho$$

In the case of CRC, E_m is the Young's modulus of plain concrete (i.e. $E_m = E_c$) and in SFBRC it is that of SFRC, i.e.

[4]SFBRC—Steel Fibre and Bar Reinforced Concrete
[5]CRC—Conventionally Reinforced Concrete

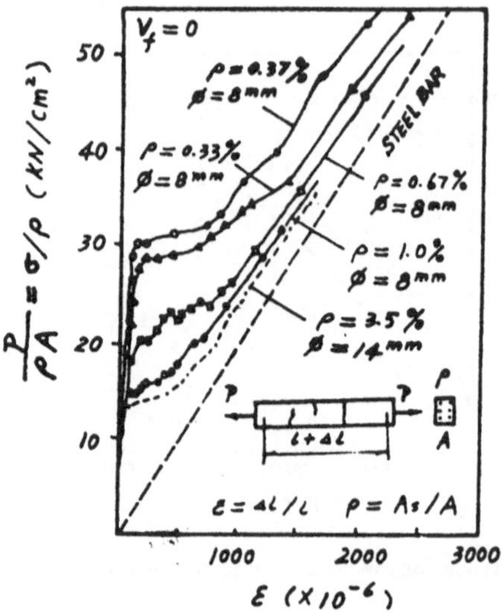

Figure 8: stress factor-strain curves reported in [2]

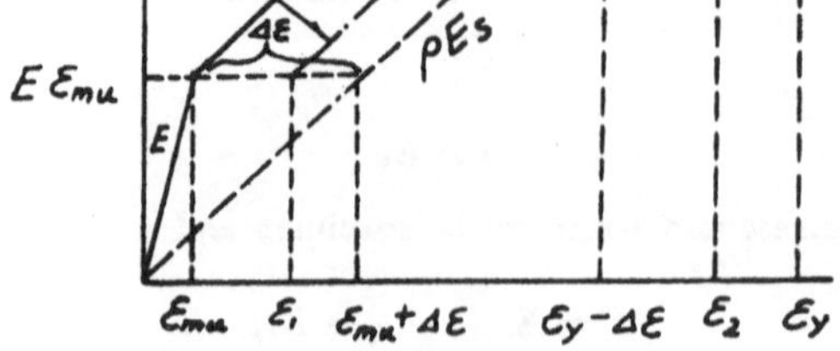

Figure 9: Idealized stress-strain relationship of SFBRC and CRC in uniaxial tension

$$E_m = E_c(1 - V_f) + \eta_o \eta_l E_f V_f$$

where η_o and η_l are fibre orientation and distribution efficiency factors respectively; E_f is the Young's modulus of the fibres.

When $\varepsilon > \varepsilon_{mu}$

$$\sigma = \rho E_s \varepsilon + \Delta\sigma \tag{7}$$

$$\Delta\sigma = (E - \rho E_s)\, \varepsilon_{mu} \tag{8}$$

from Fig. 9, it is obvious that

$$\Delta\varepsilon = (\frac{E}{\rho E_s} - 1)\varepsilon_{mu} \tag{9}$$

Hence the ultimate tensile strain of the SFBRC corresponding to the yielding strain of the bar (ε_y) is

$$\varepsilon_u = \varepsilon_y - \Delta\varepsilon = \varepsilon_y - (\frac{E}{\rho E_s} - 1)\varepsilon_{mu} \tag{10}$$

and

$$\sigma_u = \sigma_y \rho$$

Taking the specimens in this paper for example, $\varepsilon_y = 2000 \times 10^{-6}$, $\varepsilon_{mu} = 120 \times 10^{-6}$, $\rho = 3.49\%$, $E_s = 200GPa$. Let $E_m \approx E_c = 30GPa$, we have $E = 35.9GPa$, $\Delta\varepsilon = 497 \times 10^{-6}$. It therefore gives

$$\varepsilon_1 = 396 \times 10^{-6}$$

$$\varepsilon_2 = 1724 \times 10^{-6}$$

$$\sigma_{mu}/\rho = E\varepsilon_{mu}/\rho = 12.3KN/cm^2$$

and

$$\sigma_u/\rho = \sigma_y = 40KN/cm^2$$

$$\varepsilon_u = 1505 \times 10^{-6}$$

It can be seen from Fig. 7 that the calculated values above agree well with the test data, which verifies the assumption and the analysis.

The post-cracking stress-strain relationship of SFRC in SFBRC can also be derived based on the analysis above.

Since

$$\sigma_{fc} = \frac{P - \sigma_s A_s}{A - A_s} = \frac{\sigma - \rho\sigma_s}{1 - \rho} = \frac{\sigma - \rho E_s \varepsilon_s}{1 - \rho} \tag{11}$$

By substituting equation (7) and (8) into equation (11), we have

$$\sigma_{fc} = \frac{1}{1 - \rho}[\Delta\sigma - \rho E_s(\varepsilon_s - \varepsilon)] \tag{12}$$

where $\sigma = P/A$ is the average stress of the specimen, and σ_{fc} is the average stress of SFRC in the specimen.

In fact, ε_c reaches its minimum at the section of the crack. however it is still greater than zero because of the presence of fibres across the crack. If we suppose $\varepsilon_c = \alpha\varepsilon_s$, where α is a constant depending only on the bond related factors such as concrete mix, fibre type and volume fractions, bar type and its surface texture etc., we have from equation (2)

$$\varepsilon = \frac{1}{2}(\alpha + 1)\varepsilon_s = \beta\varepsilon_s \tag{13}$$

Thus equation (12) can be written as

$$\sigma_{fc} = \frac{1}{1-\rho}(\Delta\sigma - \rho C E_s \varepsilon) \tag{14}$$

where $\beta = \frac{1}{2}(\alpha + 1)$, $C = \frac{1}{\beta} - 1$.

If α is a constant, C is also a constant, which means that in the post-cracking range σ_{fc}, according to equation (14) should descend linearly as the average strain of the specimen, ε, increases. This is verified by the test results shown in Fig.6.

Let $\sigma_{fc} = 0$, the maximum tensile strain of SFRC working together with bars can be obtained from equation (14), i.e.

$$\varepsilon_{max} = \frac{\Delta\sigma}{\rho C E_s} = \frac{\Delta\varepsilon}{C} \tag{15}$$

According to the test data in this paper, C for each specimen is about the same and when fibre volume fraction varies from 1 to 2 per cent, C varies from 0.39 to 0.92. and C decreases with the increase of V_f. Since $\Delta\varepsilon = 497 \times 10^{-6}$ as calculated above, the maximum tensile strain of the specimens in this paper according to equation (15) is $\varepsilon_{max} = (540 \sim 1277) \times 10^{-6}$ (when$V_f = 1 \sim 2\%$). This also agree well with the test data shown in Fig. 6.

CONCLUSIONS

By the analysis above following conclusions can be drawn :

1. The post-cracking stress-strain relationship of SFRC in SFBRC subjected to uniaxial tension appears to be linear which facilitates computations in the structural analysis.

2. The linear descending pattern of the curve indicates that the cracking of the matrix in SFBRC does not result in the abrupt load increment in steel bars as what happens in conventionally reinforced concretes. This gradual load transfer gives rise to a more ductile failure, i.e. a greater energy dissipation capacity which is expected in seismic design.

3.The method used in this paper for analyzing the properties of steel fibre and bar reinforced concrete by regarding it as a whole and applying the theories of the composite proved to be worth trying.

4. the theoretical stress-strain equations for the ascending portion of SFRC in uniaxial tension discussed in [5] is recommended here for use in structural analysis.

ACKNOWLEDGEMENT

The authors are grateful to Mr. Zhao Yibing, engineer of the laboratory, for his great assistance in carrying out the tests. Thanks are also given to Mr. Huang Chengkuei, head of the lab., Mr. Jin Guili, Mr. Liuyi and Shen jina for their kind advice and help.

REFERENCES

1. Gopalaratnam, V.S. and Shah, S.P., Micromechanical model for the tensile fracture of steel fibre reinforced concrete, Vol.1, paper 1.3, *Proceedings of RILEM Symposium on Developments in Fibre Reinforced Cement and Concrete*, Ed. by R.N.Swamy et al, University of Sheffield, 1986.

2. Leonhardt, F., *Nachweis der Gebrauchsfähigkeit Rissebeschränkung, Formänderungen Momentenumlagerung und Bruchlinientheorie rin Stahlbetonbau*, Springer-Verlag, 1978

3. Guan Liqiu and Zhao Guofan, Analysis of the complete moment-çurvature curve of concrete beam reinforced with steel bar and steel fibres, accepted for presentation at the first International Congress of RILEM from Material Science to Material Engineering, Paris, Sept.1987.

4. Hannant, D.J., *Fiber Cements and Fiber Concretes*, John Wiley and Sons., 1978, pp14-20.

5. Cancho, A.V. and Sanchez-Galvez, V.A., Micromechanical model for the tensile stress-strain curve of the fiber reinforced cements, *Proceedings of RILEM Symposium on Developments in Fiber Reinforced Cement and Concrete*, Ed.by R.N.Swamy et al, Vol.1, paper 1.1, 1986.

Proceedings of the International Symposium on Fibre Reinforced Concrete, December 16–19, 1987, Madras, India

TESTING OF PARTIALLY STEEL - FIBRE REINFORCED MORTAR

W. Radomski

Warsaw University of Technology, Poland

SYNOPSIS

The test results of the fully and partially fibre-reinforced mortar specimens are presented. The round straight steel fibres 0.4 x 40 mm have been used. The specimens of 30 x 60 x 300 cut out from the plates have been tested in bending by means of Instron 1251 equipment. The central deflections and the strains on the compression and tension surfaces of the specimens have been measured. Three different thicknesses of the reinforced layer casted in tension and eight different nominal contents of the fibres have been used. The test results have indicated that the distribution of the fibres in the tension layer of specimens is much more effective for the flexural strength than the same amount of the fibres used as a full reinforcement of the member. Recommendations for the further investigations of this problem are formulated in the paper.

INTRODUCTION

During the last decade a dramatic development in various applications of steel fibre reinforced concrete or mortar in the constructions has been observed. However, in nearly all applications of these materials, it is important that only surfaces of the structural members be reinforced for strength and durability. Thus, the same performance may be achieved in a partially fibre-reinforced member as in a fully fibre-reinforced one and with a significant reduction in total fibre content. This concept of partially steel fibre reinforced members is very important for both technology and economy but it has been studied so far in relatively few publications, e.g. [1], [2] and recently [3]. The all published test results indicate that behaviour of the partially reinforced members in bending may be the same or better than behaviour of the fully reinforced members if only the thicknesses of the reinforced layers casted in the tension sides are sufficient. The authors of publication [2] have concluded that e.g. the beams of dimensions 76 x 76 x 380 mm with fibre reinforcement of 1.5% in which the thickness of the reinforced section is over one-sixth the depth of the beam are comparable in strength to a beam reinforced throughout and exhibit ductile failures. It should be also emphasized that for beams failing in a ductile manner and having superior strength in bending the most economical ones are those having a reinforced thickness approximately one-sixth the depth of the member because the cost of materials for such a partially reinforced member is only about one-half that for a fully reinforced one [2]. However, the concept of the partially steel-fibre reinforced members may be considered also from the other point of view. Namely, the determined amount of steel fibres may be distributed as a full reinforcement of the member or the same amount of fibres may be used as a more strong reinforcement of a layer of the member only. It leads to extent the fibre's effectiveness. This is a new problem requiring a fundamental research based on the experiments. The experimental results concerning this problem are reported herein.

The study presented in the paper is limited due to the number of specimens, the type of the steel fibres used for reinforcement and the scope of measurements. However, the main aim of the performed experiments has been preliminary checking a new concept of the partially steel-fibre reinforced members in respect to their behaviour in bending in comparison with the most often tested fully reinforced members. The obtained test results seem to be very interesting and indicate the necessity of the further investigations.

2. FABRICATION AND TESTING OF SPECIMENS

The information about the tested specimens is given in Table 1.

All specimens were fabricated from the mortar matrix with the following mix proportions per m^3 :

portland cement 250	550 kg
sand/grains up to diameter of 2 mm/	1650 kg
water	275 kg
w/c ratio	0.5

Round straight fibres 0.4 x 40mm /l/d=100/ were used.

All specimens of nominal dimensions 30 x 60 x 300 mm were cut out from the plates of 590 x 150 x 60 mm or 320 x 180 x 60 mm.

The plates were fabricated in two general stages.

Firstly, the mould was casted of steel fibre reinforced mortar layer with assumed thickness /h_f/. The thickness of the fibre reinforced layers and the volume contents of fibres /β/ were assumed as follows :

series A1 - h_f = 60 mm, $h_f : h$ = 1.00, β = 0.45%
A2 - h_f = 30 mm, $h_f : h$ = 0.50, β = 0.90%
A3 - h_f = 20 mm, $h_f : h$ = 0.33, β = 1.35%
A4 - h_f = 15 mm, $h_f : h$ = 0.25, β = 1.80%
Series B1 - h_f = 60 mm, $h_f : h$ = 1.00, β = 0.75%
B2 - h_f = 30 mm, $h_f : h$ = 0.50, β = 1.50%
B3 - h_f = 20 mm, $h_f : h$ = 0.33, β = 2.25%
B4 - h_f = 15 mm, $h_f : h$ = 0.25, β = 3.00% h - the depth of plates.

/Plates of series A1 and B1 were casted in one stage only/

The fibre reinforced layers A1 & B1 were consolidated by vibration. Secondly, immediately after casting and consolidation of fibre reinforced layers, the moulds were casted of mortar matrix without any fibres up to the full depth of plates, i.e. 60 mm. The full moulds were also vibrated for consolidation of the two different layers of the materials. After casting, the moulds were placed in a curing chamber. After 3 days the plates were removed from the moulds and stored under water until age 28 days.

Moreover, six cubes of 150 x 150 x 150 mm of mortar matrix without any fibres were fabricated for determining the compressive strength of material.

Next, the test specimens were cut out from the plates. The real dimensions of all specimens were measured with the accuracy of 0.1 mm. The X-ray photographs of all specimens were made. On the basis of the X-ray photographs a uniform distribution of fibres in the reinforced layers were found. Moreover, on the basis of these photographs the real thicknesses of fibre reinforced layers were measured. The average values of the real ratios of thicknesses of reinforced layers to depth of the specimens $/h_f:h/$ are listed in Table 1.

The specimens were tested in flexure with the constant bending moment and with the reinforced layers in tension. The Instron 1251 equipment for loading was used. The load to rupture was applied continuously and at the same constant rate of 0.2mm/min for all specimens. The central strains on the bottom and upper surfaces of specimens as well as the central deflections of specimens were measured. The schematic diagram of the measuring set is shown in Fig. 1. The load - strains and load - deflection curves up to the failure of specimens were obtained.

3. TEST RESULTS AND DISCUSSION

The test results obtained from the analysis of the load - deflection curves are mostly presented herein. The test results concerning the strains and the final analysis of the behaviour of partially steel-fibre reinforced members in bending will be published elsewhere. Two examples of the typical experimental load-deflection curves are shown in Fig. 2.

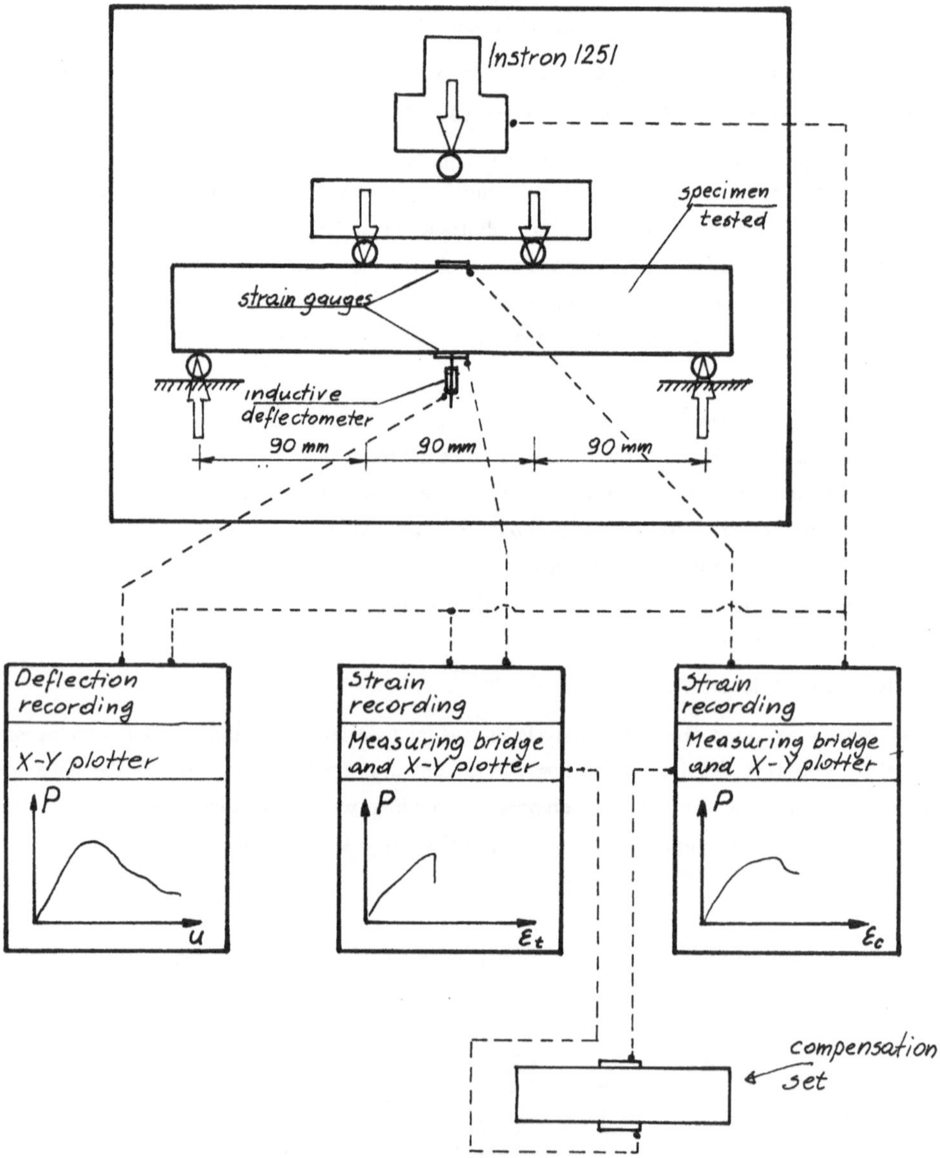

Fig. 1 Schemaatic diagram of measuring set

Tested specimens and test results

Table 1

Tested specimens						Test results					
Series	Assumed h_f:h	β [%]	Number of specimen	Measured h_f:h	β [%]	P_{max} [kN]	U_z [kNmm]	U_l [kNmm]	U_z / U_l	σ_R [MPa]	D/J 10^4[1/ m^4]
A1	1.00	0.45	A1/1	1.00	0.54	1.63	0.66	1.33	0.50	4.70	194
			A1/2	1.00	0.34	1.37	0.21	1.05	0.20	3.20	201
			A1/3	1.00	0.31	1.32	0.17	0.90	0.19	3.80	188
			A1/4	1.00	0.42	1.84	0.14	1.31	0.11	5.10	380
	Average values			**1.00**	**0.40**	**1.54**	**0.30**	**1.15**	**0.25**	**4.20**	**240**
A2	0.50	0.90	A2/1	0.49	0.38	1.32	0.16	0.47	0.34	2.50	111
			A2/2	0.45	0.76	1.47	0.27	0.94	0.29	2.80	211
			A2/3	0.45	0.46	1.89	0.44	1.30	0.34	3.70	137
			A2/4	0.51	0.74	2.60	0.47	1.65	0.28	5.20	154
	Average values			**0.48**	**0.59**	**1.82**	**0.34**	**1.09**	**0.31**	**3.55**	**154**
A3	0.33	1.35	A3/1	0.26	0.56	2.05	1.10	1.52	0.72	4.10	85
			A3/2	0.29	0.42	2.11	0.60	1.55	0.39	4.40	173
			A3/3	0.31	0.87	3.16	0.86	2.27	0.38	6.50	249
			A3/4	0.31	0.73	2.47	0.29	1.68	0.17	5.30	262
	Average values			**0.29**	**0.65**	**2.45**	**0.71**	**1.76**	**0.41**	**5.08**	**192**
A4	0.25	1.80	A4/1	0.24	1.51	3.32	0.95	1.94	0.49	6.60	141
			A4/2	0.24	0.87	1.74	0.35	1.14	0.31	3.40	196
			A4/3	0.23	0.66	2.00	0.52	1.14	0.46	3.90	121
			A4/4	0.23	2.28	3.00	0.62	2.07	0.30	7.60	275
			A4/5	0.29	1.73	2.84	0.65	2.01	0.32	6.50	241
			A4/6	0.25	1.60	2.74	1.33	2.60	0.51	6.40	276
	Average values			**0.25**	**1.28**	**2.61**	**0.74**	**1.82**	**0.39**	**5.73**	**208**
B1	1.00	0.75	B1/1	1.00	0.38	1.16	0.83	1.18	0.70	3.00	389
			B1/2	1.00	0.54	1.55	0.75	1.84	0.40	4.00	343
			B1/3	1.00	0.45	1.79	0.88	1.83	0.48	4.20	307
			B1/4	1.00	0.66	1.84	0.74	1.69	0.44	4.90	342
	Average values			**1.00**	**0.51**	**1.58**	**0.79**	**1.64**	**0.50**	**4.03**	**345**
B2	0.50	1.50	B2/1	0.39	0.77	3.08	1.02	2.98	0.34	5.80	277
			B2/2	0.45	0.89	2.39	0.69	2.11	0.33	4.50	160
			B2/3	0.41	0.98	2.53	0.88	2.14	0.41	5.10	90
			B2/4	0.39	1.16	2.32	0.68	2.15	0.32	4.80	154
			B2/5	0.44	0.95	2.84	0.75	2.25	0.33	5.30	158
	Average values			**0.42**	**0.95**	**2.63**	**0.80**	**2.33**	**0.35**	**5.10**	**168**
B3	0.33	2.25	B3/1	0.24	1.97	3.53	2.21	3.20	0.69	7.10	109
			B3/2	0.27	0.99	3.05	1.14	2.52	0.45	6.30	136
			B3/3	0.29	0.79	3.13	0.45	2.07	0.22	5.00	121
			B3/4	0.29	1.43	2.47	0.68	2.41	0.28	5.80	276
			B3/5	0.25	1.39	3.05	1.26	2.76	0.46	7.00	215
			B3/6	0.27	1.09	2.84	1.12	2.65	0.42	5.30	251
			B3/7	0.27	1.66	2.95	1.00	2.40	0.42	6.50	272
	Average values			**0.27**	**1.33**	**2.86**	**1.12**	**2.57**	**0.42**	**6.14**	**197**
B4	0.25	3.00	B4/1	0.24	2.14	2.53	0.89	2.19	0.41	6.20	285
			B4/2	0.26	1.66	2.95	1.70	2.80	0.61	7.20	198
			B4/3	0.25	2.59	3.76	2.06	3.48	0.59	8.30	182
			B4/4	0.21	1.60	3.00	1.51	3.02	0.50	7.10	201
			B4/5	0.25	2.56	3.68	1.44	3.43	0.42	7.90	197
	Average values			**0.24**	**2.11**	**3.18**	**1.52**	**2.98**	**0.51**	**7.34**	**212**

β = total volume of fibres in specimen as a percentage; β = determined real content of fibres
ı broken cross section of specimen as a percentage; any other symbols are described in the text
paper. The cube compression strength of mortar matrix - about 20 MPa. The tensile strength
steel fibres - about 680 MPa

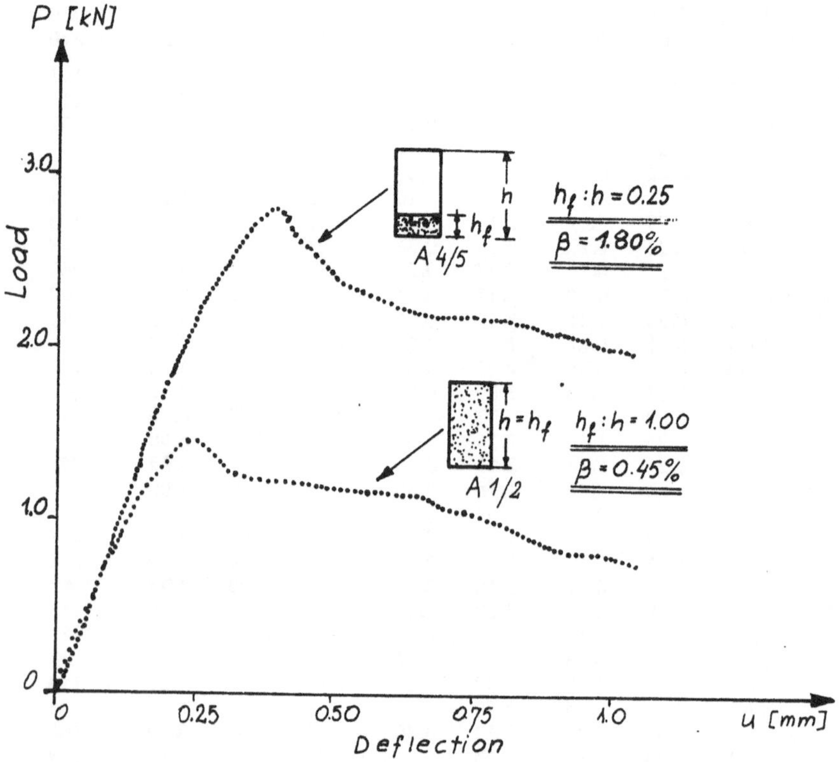

Fig. 2 Typical load-deflection curves

the basis of the all obtained curves the following characteristics were analysed:

P_{max} - the maximum load, i.e. the fracture load,

U_z - the fracture energy determined as an area under the load deflection curve up to the point corresponding to the fracture load P_{max},

U_1 - the fracture energy determined as an area under the load deflection curve up to the point corresponding to the deflection of, 1 mm,

D - so called experimental stiffness ' measured as a tangent of inclination angle of secant between the origin of coordinates P, u and the point corresponding to the maximum load P_{max},

σ_R - modulus of rupture determined as a quotient of the fracture bending moment and the sectional modulus of specimen.

The analyzed characteristics P_{max}, U_z, U_1 and D are shown in Fig. 3. For determining U_z and U_1, the areas under load-deflection curves were measured by means of a planimeter. Dividing the experimental stiffness /D/ by moment of intertia /J/ of the cross section, the new characteristics /D/J/ of the behaviour of the specimens was obtained. It should be emphasized that characteristics σ_R and D/J should be considered as the first approximation only due to the sectional modulus and the moment of inertia were calculated taking into account the real dimensions of the cross section but assuming the homogeneity of the material.

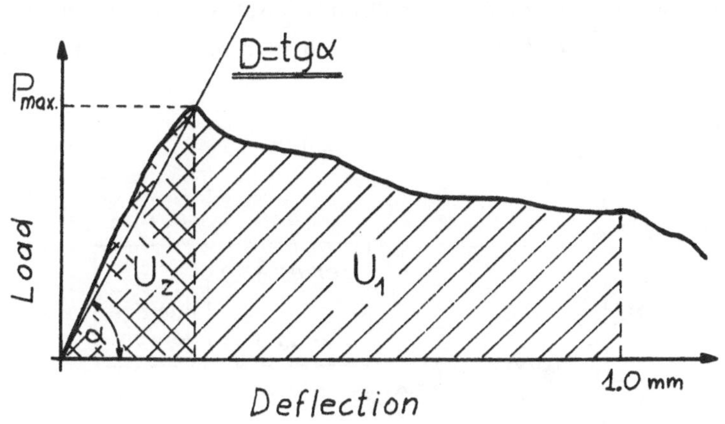

Fig. 3 Analyzed characteristics of specimens.

The test values of the all mentioned above characteristics are listed in Table 1.

It is known that in the fibre-reinforced concrete or mortar the distribution of the fibres in the material is not identical in the all cross sections of a member. When the bending moment is constant in the certain length of the specimen, the fracture occures in the cross section having the weakest reinforcement. For this reason the real content of fibres in the broken cross section of specimen should be determined. In the tests presented herein the real content of steel fibres /β_r/ was determined by counting the number of fibres on the visible faces of the broken pieces of specimens, as proposed and described by Kasperkiewicz [4 - 5]. The values of β_r are also listed in Table 1.

The test results are presented in Figures 4 - 8. Because of the limited number of tests carried out and the scatter of the test results due to the hetergeneity of the material, the results presented herein should be considered as quantative only. In spite of this the results indicate interesting and new information on the behaviour of partially steel fibre reinforced members in bending. This is discussed below. The nature of failure of all tested specimens, i.e. partially and fully reinforced, was ductile. No separations between the fibre reinforced layer and the layer without any fibres were occured inthe specimens. It is evident from Fig. 4 that distribution of the fibres in the bottom /tension/ layer of

specimens is more effective for the flexural strength than the same amount of the fibres use as a full reinforcement of the member. The test results indicate that the fracture load /P_{max}/ and the modulus of rupture / σ_R/ increase with decreasing the thicknesses of the reinforced layers. The maximum increase of the fracture load was 69% and 101% in the test series A and series B respectively. The maximum increase of the modulus of rupture was36% and 82%. It seems to be very interesting that the fracture load and the modulus of rupture increase in the nearly linear from with increasing the content of fibres in the broken cross section of the specimen as it is shown in Fig. 5. The directions of the straight lines shown in the figure were found by the least squares method.
The nearly linear form of the relationship between the fracture load and the fibre content in the broken cross section of the specimens confirms the validity of the approximate formula for determinationof the force /S_f/ acting in the fibres in reinforced layer proposed

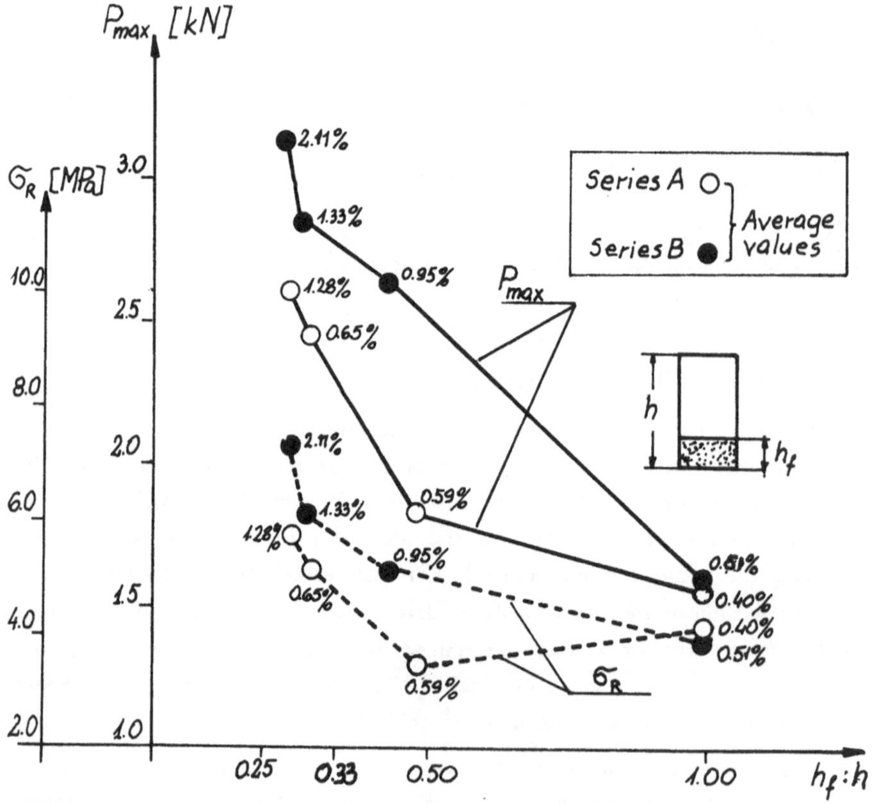

Fig. 4 Relationships between thickness of reinforced layer and fracture load as well as modulus of ruputure

by Kajfasz[1] and expressed as follows:

$$S_f = \frac{M_T - M_C}{h\left[1 - \frac{\lambda}{2} - \frac{1}{3}\,\frac{\xi_c}{\xi_c + \xi_t}\right]}$$

in which M_T = total bending moment; M_C = bending moment transfered by the material without any fibres; λ = ratio of thickness of fibre reinforced layer to depth of specimen; h = depth of specimen; ξ_t, ξ_c = strains on the tension and compression surfaces of a specimen respectively. The cited above formula has been formulated on the ground of the linear distribution of the strains in the cross section of a member. The determination of the strains and stresses distribution in the partially fibre reinforced members requires the further theoretical and experimental studies first of all due to the composed structure of the member and the heterogeneity of the reinforced layer. This problem is now analysed on the basis of the strain measurements. The results of the study will be published elsewhere. The concentration of the fibres in the tension layer of specimens tested in bending causes also increase in the fracture energy as it is shown in Fig. 6. The maximum increase

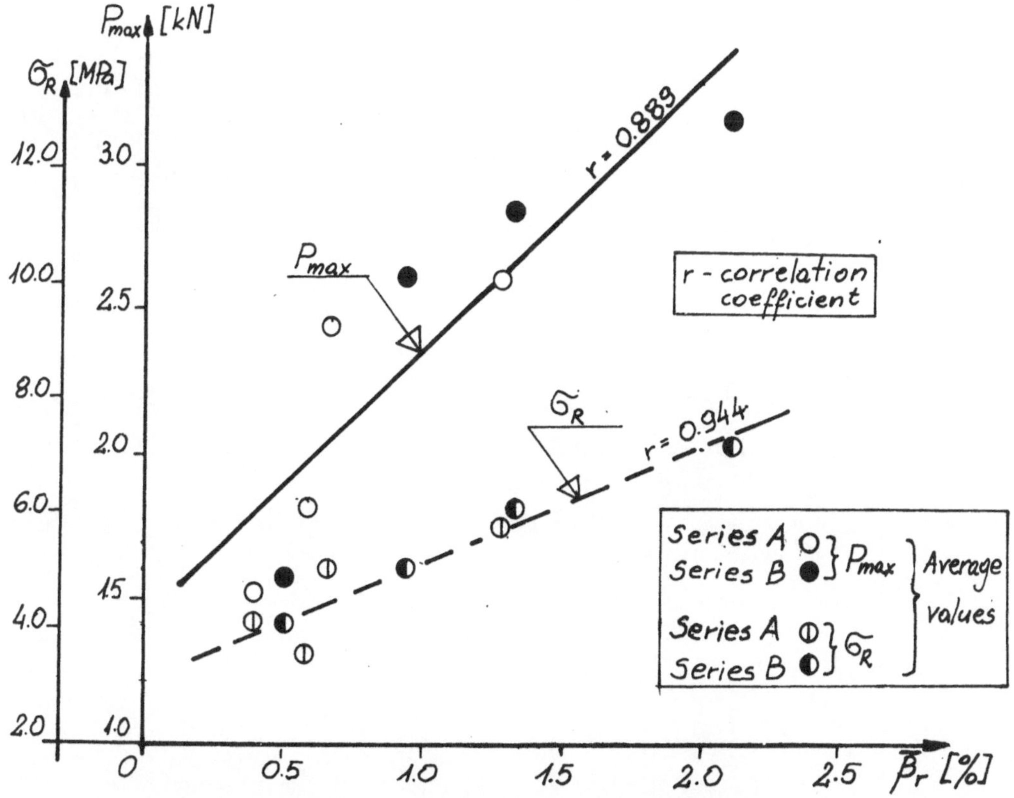

Fig. 5 Relationship between fibre content in broken cross section of specimen and fracture load as well as modulus of ruputure

of the fracture energy U_z was 146% and 92% in the test series A and series B respectively. The maximum increase of the fracture energy U_1 was 58% and 82% The relationship between the fracture energies U_z, U_1 and the content of fibres in the broken cross sections of the specimens was found also nearly linear as it is shown in Fig. 7. Clearly, as expected, the fracture energy U_z was less than the fracture energy U_l. The energy U_z was from 0.25 to 0.51 of energy U_l for the all specimens. The relationship between the thickness of fibre-reinforced layer and so called experimental stiffness'/characteristic D/J/ was similar in the both tested series of specimens as it is shown in Fig. 8. However, the effect of the thickness of the reinforced layer on the characteristic D/J is not so clear as in the cases concerning the fracture load and the fracture energy discussed previously. /The experimental stiffness/of the specimens reinforced throughout with fibres was evidently more than the stiffness of partially reinforced specimens. The minimum of the 'experimental stiffness' occured in the specimens with the ratio h_f:h varied from 0.33 to 0.50. The stiffness of the partially reinforced members is a new and very important problem for applications of this type of members as the structural elements. For this reason it requires the further studies.

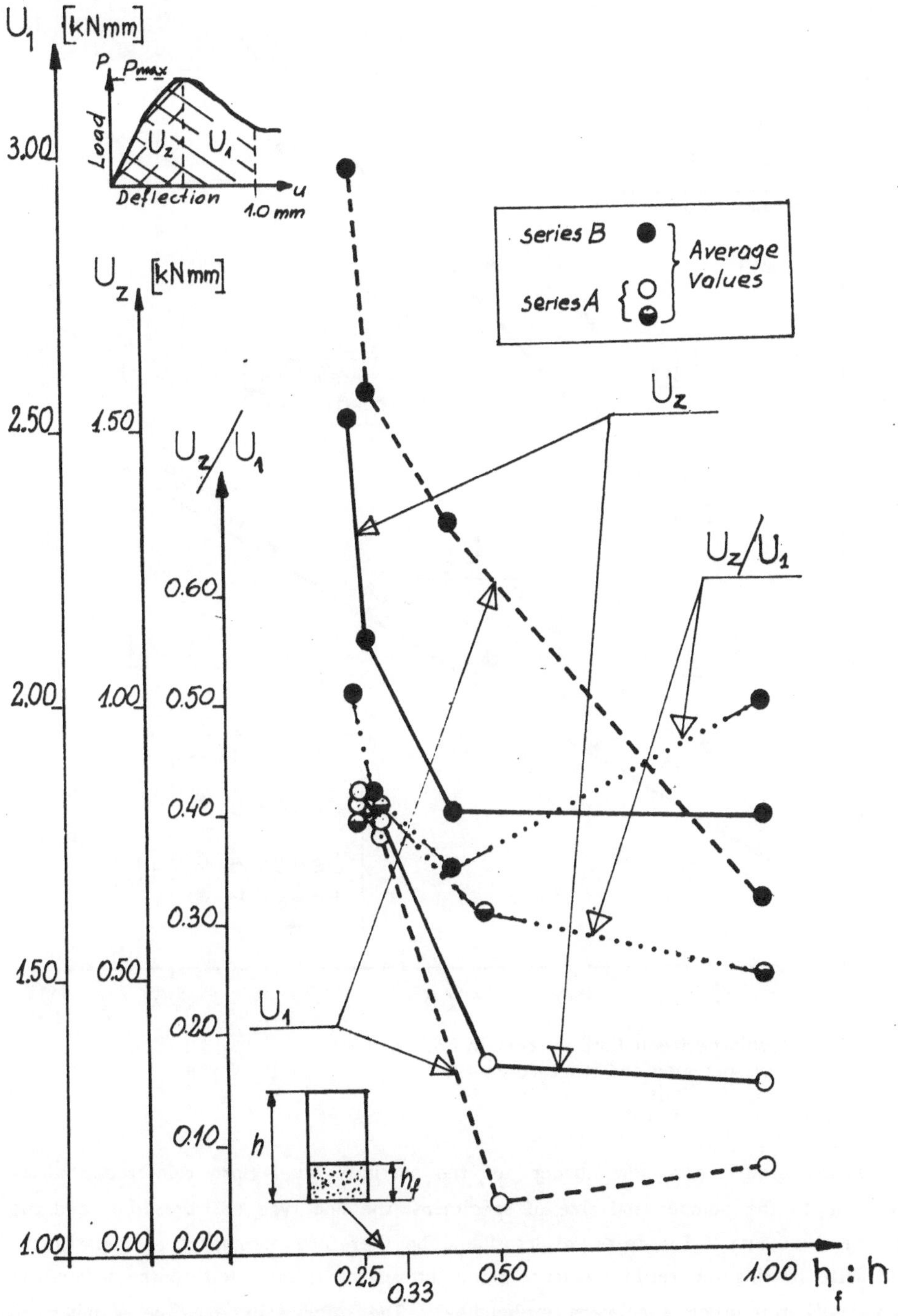

Fig. 6 Relationship between thickness of reinforced layer and fracture energy

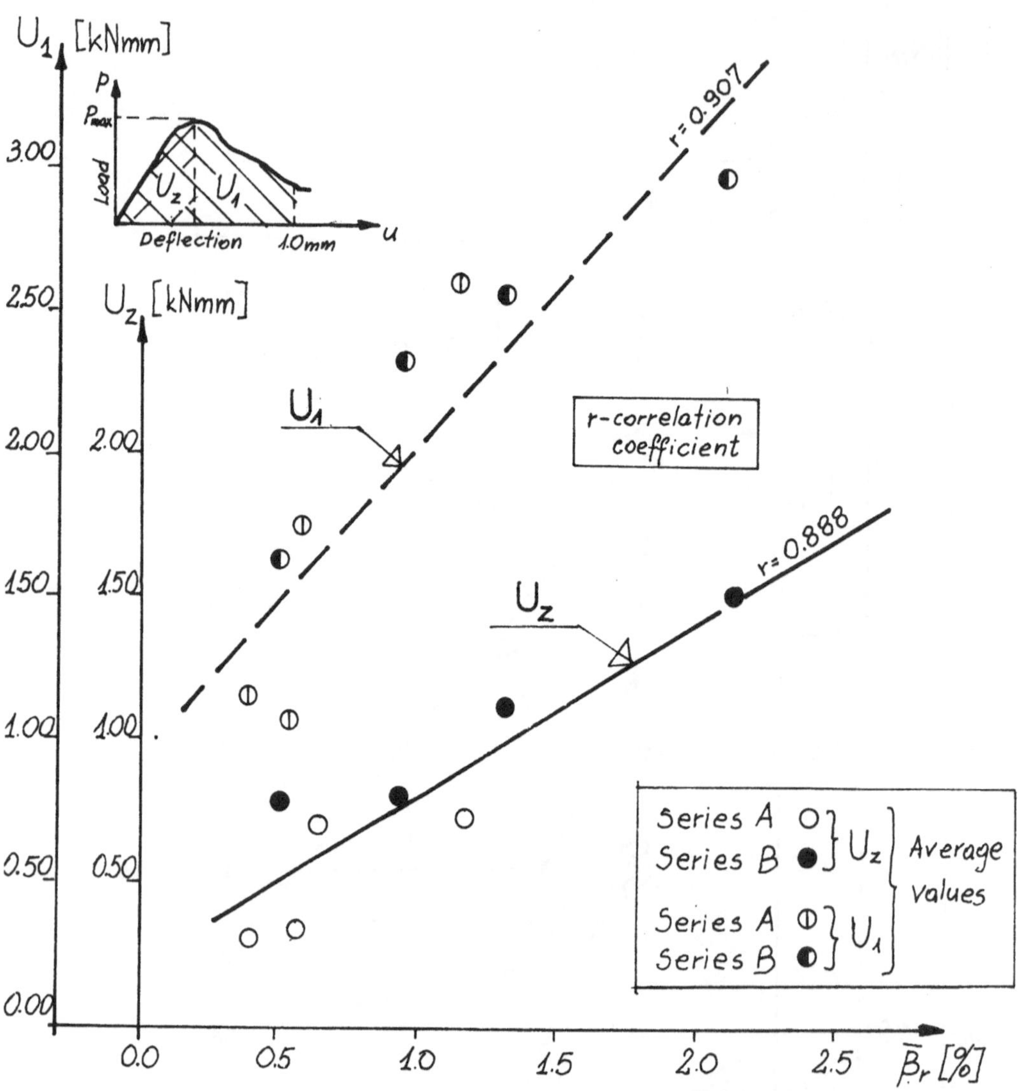

Fig. 7 Relationship between fibre content in broken cross section of specimen and fracture energy

4. CONCLUSIONS

This preliminary study on a new concept of the partially steel fibre reinforced mortar is limited due to the number and size of specimens, the one type of fibres only and the scope of measurements. The tests results should be therefore considered as quantitative only. In spite of this the results indicate some interesting information on the behaviours of the partially reinforced specimens in bending. The information may be summarized as follws.

1. The nature of failure of the all tested specimens was ductile and no separations between the fibre reinforced layer and the unreinforced layer were observed. The fracture of the

all specimens occured in the cross section situated in the central part of the members, between the located points. i.e. in the length of specimens with the constant bending moment.

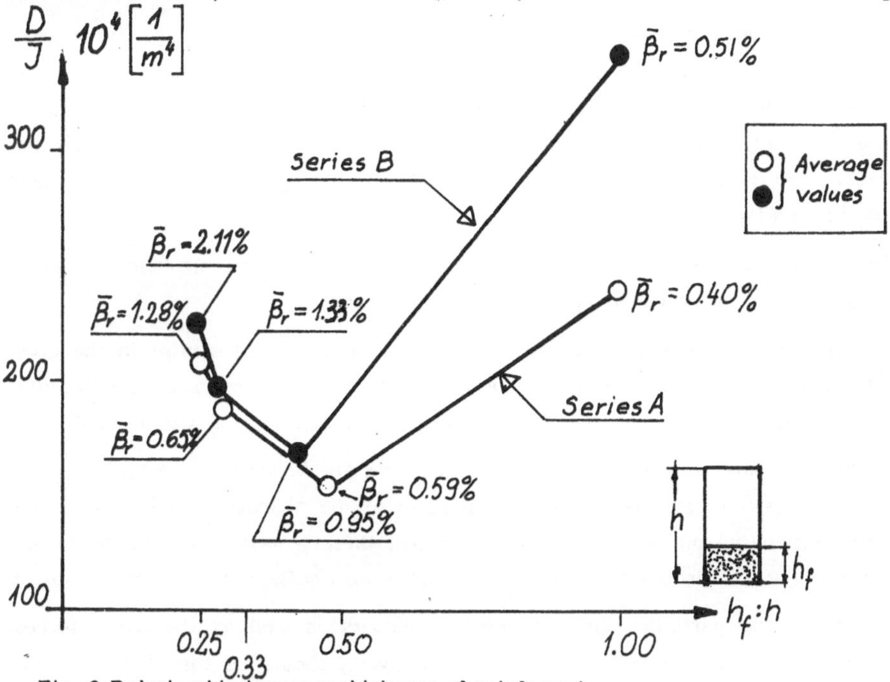

Fig. 8 Relationship between thickness of reinforced layer and experimental stiffness

2. The distribution of the fibres in the tension layer of specimens tested in bending is more effective for the strength than the same amount of fibres used as a full reinforcement of a member

3. The flexural strength of the tested specimens depended on both the thickness of the reinforced layer and the content of fibres in this layer. In the tests reported herein, the highest flexural strength exhibited the specimens with the smallest thickness of the reinforced layer, i.e. with h_f:h about 0.25, and having the most strong reinforcement, i.e. with β = 3.0%. In this case the fracture load varied from 69% to 101% greater than those of the fully reinforced specimens with the total volume contents of fibres 0.45% and 0.75%.

4. The relationship between the fracture load and the content of fibres in the broken cross section of specimens was found nearly linear. The same form exhibited the relationship between the fracture energy and the content of fibres in these sections. These both observations confirm the validity of the approximate formula for determination of the force acting in the fibres in the reinforced layer proposed by Kajfasz[1].

5. The 'experimental stiffness' of the specimens reinforced throughout with fibres was found more than the stiffness of the partially reinforced specimens. The minimum of the 'experimental stiffness' occured in the specimens with the ratio h_f :h varied from 0.33 to 0.50. The stiffness of specimens with h_f:h about 0.25 was more than those but evidently less than observed in the fully reinforced specimens.

5. RECOMMENDATIONS

On the basis of the performed tests, the following recommendations may be formulated.
1. The concept of the partially steel-fibre reinforced members seems to be very interesting of both technological and economical reasons. The further investigations of this subject are therefore necessary. These investigations should involve larger aggregate size, other specimen sizes, other types and contents of fibres and reinforcing on the compressive face as well as the tension face.

2. The effect of the partial fibre reinforcing on the strength of the members in which it is used should be determined experimentally and theoretically.

3. The theretical approach should be based on the distribution of the strains in the cross sections of the partially and fully reinfoced members determined by means of the measurements. The propositions concerning this problem are now preparing and will be published.

4. When the members are tested in flexure with the constant bending moment, the deflections should be measured not only in the central point of the member span but also inthe points on the tension bottom surface of the member, directly under the loaded points. In this way the more information on the relationship load-deflection may be obtained.

5. The stiffness of the partially fibre reinforced members as well as the shear forces, particularly between the reinforced and unreinforced layers, should be tested as the very important problems for various structural applications of the elements of this type.

ACKNOWLEDGEMENTS

The author acknowledges his graduate students Mr. P. Kazubski and Mr. A. Sadowski for preparation the specimens and the performance of the tests. The tests have been performed in the Laboratory of the Institute of the Fundamental Technological Research, Polish Academy of Sciences. Valuable talks with Professor A.M. Brandt and Dr. A. Burakiewicz from this Institute are deeply appreciated.

REFERENCES

1. kajfasz, S., The efficiency of a steel fibre external layer in modification of concrete properties in flexure, Proc. of the XXIV Konf. Nauk. PAN i PZITB, Krynica, 1978, pp. 133-142 /in Polish/.

2. Rahimi, M.M., and Kesler, C.E., Partially steel-fibre reinforced mortar, Jour. of the Struct. Div., ASCE, Vol. 105, No. ST1, Jan. 1979, pp. 101-109.

3. Fatma, E.El-Refai, and Monir M.K., Behaviour of polymer fibre reinforced concrete beams, Third International Symposium on Developments in Fibre Reinforced Cement and Concrete, RILEM Symposium, Sheffield 1986, Symposium Papers, Vol.2, pp. 185-194.

4. Kasperkiewicz, J., Apparent, spacing in fibre reinforced composites, Bull. Acad. Polon Sci., Ser. Sci, Techn., 26/1978, 1, pp. 55-63.

5. kasperkiewicz, J., Reinforcement parameter for fibre concrete, Bull. Acad. Polon. Sci., Ser. Sci. Techn., 26/1978, 1, pp. 65-72.

Proceedings of the International Symposium on Fibre Reinforced Concrete, December 16–19, 1987, Madras, India

FLEXURAL BEHAVIOUR OF PRESTRESSED FIBROUS CONCRETE BEAMS

M. Lakshmipathy
Assistant Professor in
Structural Engineering
College of Engineering
Anna University
MADRAS - 600 025

P. Hariraj
Assistant Executive
Engineer
Highways and Rural
Works
TAMILNADU

A.R. Santhakumar
Professor of
Structural Engineering
College of Engineering
Anna University
MADRAS - 600 025

ABSTRACT

The experimental investigation reported in this paper is based on the flexural tests on sixteen beams of prestressed fibrous concrete and conventional prestressed concrete. The tests have been carried out under monotonic and reversed cyclic loading. The results of the beams tested under monotonic loading exhibited superior performance with respect to load carrying capacity, stiffness, ductility and energy absorption characteristics for prestressed fibrous concrete members, when compared with conventional prestressed members.

The results of the beams tested under static reversed cyclic loading indicate superior performance with respect to load carrying capacity and cracking characteristics for prestressed fibrous concrete members. The behaviour of these beams are superior with respect to ductility and energy absorption before yield load level only. Also the experiments clearly reveal that no advantage is derived if the member is subjected to reversed cyclic loading beyond yield load level.

INTRODUCTION

The use of randomly distributed discontinuous steel fibres improve the tensile strength of concrete by inhibiting propagation of cracks and thereby imparts qualities of crack control, toughness, impact resistance and shatter resistance. As a result of this ability to arrest cracks, fibrous concrete possesses increased extensibility and tensile strength at first crack and at ultimate, particularly under flexural loading. Fibres in tension zone contribute to improve flexural stiffness. Fibre in compression zone delay the process of disintegration and preserve the integrity of concrete thus enabling the beams to develop greater overall ductility at failure. This leads to pronounced post cracking ductility, which substantially increases its energy absorption characteristics. O' Leary et al (1) have studied the behaviour of partially prestressed fibre reinforced concrete beams and found increase in flexural strength, stiffness. This paper describes the behaviour of prestressed fibrous concrete in flexure under monotonic and reversed cyclic loading (2).

EXPERIMENTAL PROGRAMME

The present investigation is confined to rectangular beams of size 10 x 20 x 200 cm (Fig.1) and the scheme is presented in Fig.2. For prestressing, mild steel anchor plates are fixed at the ends of the beams together with prestressing wires inserted through the duct holes, for post tensioning.

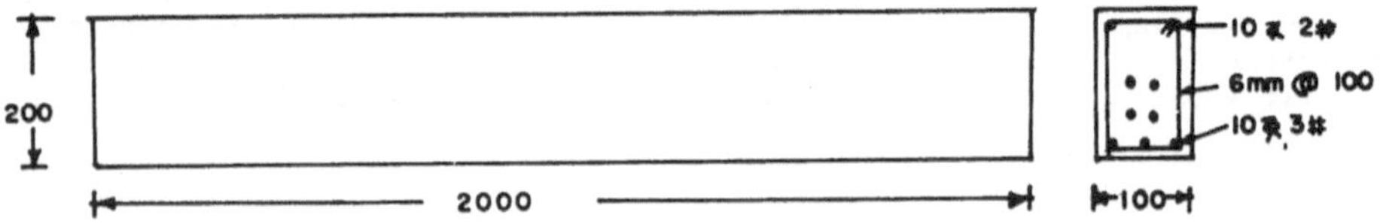

FIG.1 DETAILS OF SPECIMENS

TESTING

Test Set up and Testing Procedure

The testing arrangement is shown in Fig.3. It may be noted that with the arrangement shown, reversed cyclic loads can be effectively applied on the specimen. The specimen was subjected to loading as per the predetermined loading stages. The observation for deflections, strain across depth were taken. The loads were either monotonic or reversed cyclic depending on the specimen under test.

Single Point Loading

Two Point Loading

FIG.3 TESTING ARRANGEMENT

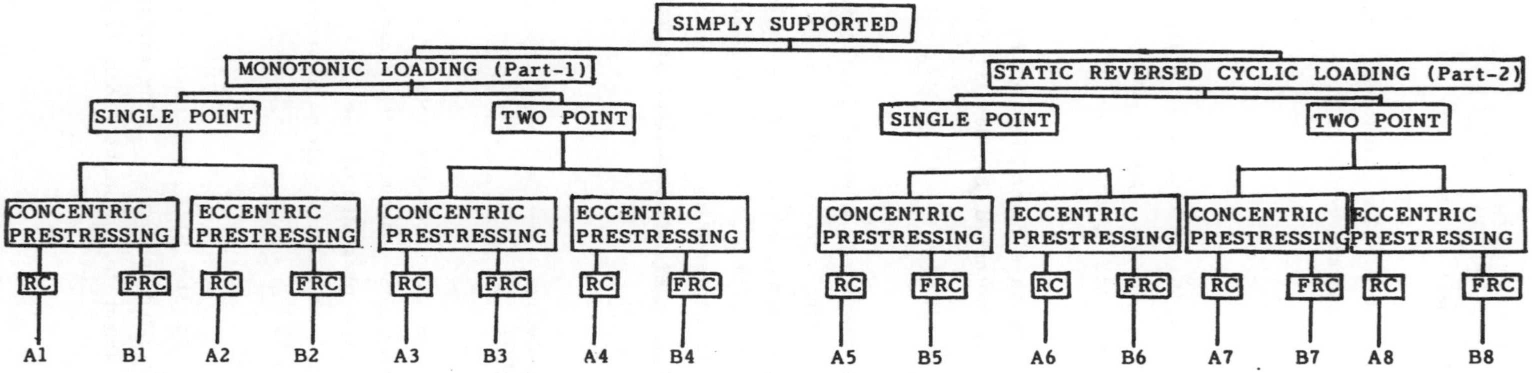

FIG.2 SCHEMES FOR EXPERIMENTAL INVESTIGATION

BEHAVIOUR WITNESSED DURING TEST

The results are presesnted in two parts. The part-1 deals with the two types of beams subjected to monotonic loading only. Part-2 deals with the two types of beams subjected to static reversed cyclic loading. The reinforced concrete beams are designated as type 'A' and the fibrous concrete beams as type 'B'.

MONOTONIC LOADING

a) Load Carrying Capacity

The beams are prestressed either at the centre of the section or at a fixed eccentricity of 40 mm. A net prestressing force of 50 KN was applied using two numbers of 5 mm cables. The ratio of theoretical load and to the observed load is presented in Table-1.

In general the fibrous concrete beams are found to have higher load carrying capacity than R.C. beams. The same trend has been observed by Henegar (3), Lakshmipathy (4) etc.,

b) Load Displacement Characteristics

The load midspan deflection characteristics are presented in Fig.4, for beams A1 to A4 and B1 to B4. In the same figure the stiffness at initial stages and after cracking have been shown. In general all the beams exhibited three stage behaviour. The first stage is upto elastic limit. The second stage, is non-linear and extends upto yielding of steel. The third stage representing the post yield behaviour and exposes the ductility characteristics. They are indicated in the figures presented.

C) Ductility

The ductility for the purposes of comparison is defined as:-

$$\text{The ductility } (\mu) = \frac{\text{Deflection at any load level}}{\text{Yield deflection}}$$

All the beams have been made to undergo large displacements upto failure to expose their ductility characteristics. Considering the observed yield loads, the ductilities reached for these beams are calculated and presented in the Table-1. The type 'B' beams exhibited 10% higher ductilities at ultimate stages compared to type 'A' beams.

d) Energy Absorption Capacity

The energy absorption capacity is calculated as the area under the load displacement diagram. The respective values are presented in Table-1. The energy absorption capacity for type 'B' beams have been found to be higher than type 'A' beams.

Table-1

PARTICULARS OF BEAMS SUBJECTED TO MONOTONIC LOADING (PART-1)

DESIGNATION OF BEAMS	A1	B1	A2	B2	A3	B3	A4	B4
Percentage of Steel A'_s	0.785	0.785	0.785	0.785	0.785	0.785	0.785	0.785
A_s	1.18	1.18	1.18	1.18	1.18	1.18	1.18	1.18
Steel fibres by volume	-	1%	-	1%	-	1%	-	1%
Cube compressive strength in N/mm^2	31.0	29.3	28.6	26.7	31.0	29.3	28.6	26.7
Ultimate Load (T) in KN	39.45	39.01	42.8	42.24	59.16	58.51	64.2	63.36
Ultimate Load (O) in KN	40.19	50.83	48.46	50.33	57.92	60.29	60.28	62.65
Ratio of ultimate load (T/O)	0.98	0.77	0.88	0.83	1.02	0.97	1.06	1.01
Ductility	3.38	3.77	2.65	2.77	3.96	-	2.15	2.27
Stiffness before cracking Θ_1 in KN/mm	30.00	32.50	20.00	27.00	23.50	24.00	29.00	34.00
Stiffness after cracking Θ_2 in KN/mm	2.60	2.70	2.81	3.00	3.13	4.54	3.18	3.25
Energy absorption in KN mm	128.66	177.69	142.04	142.62	-	-	143.19	162.38

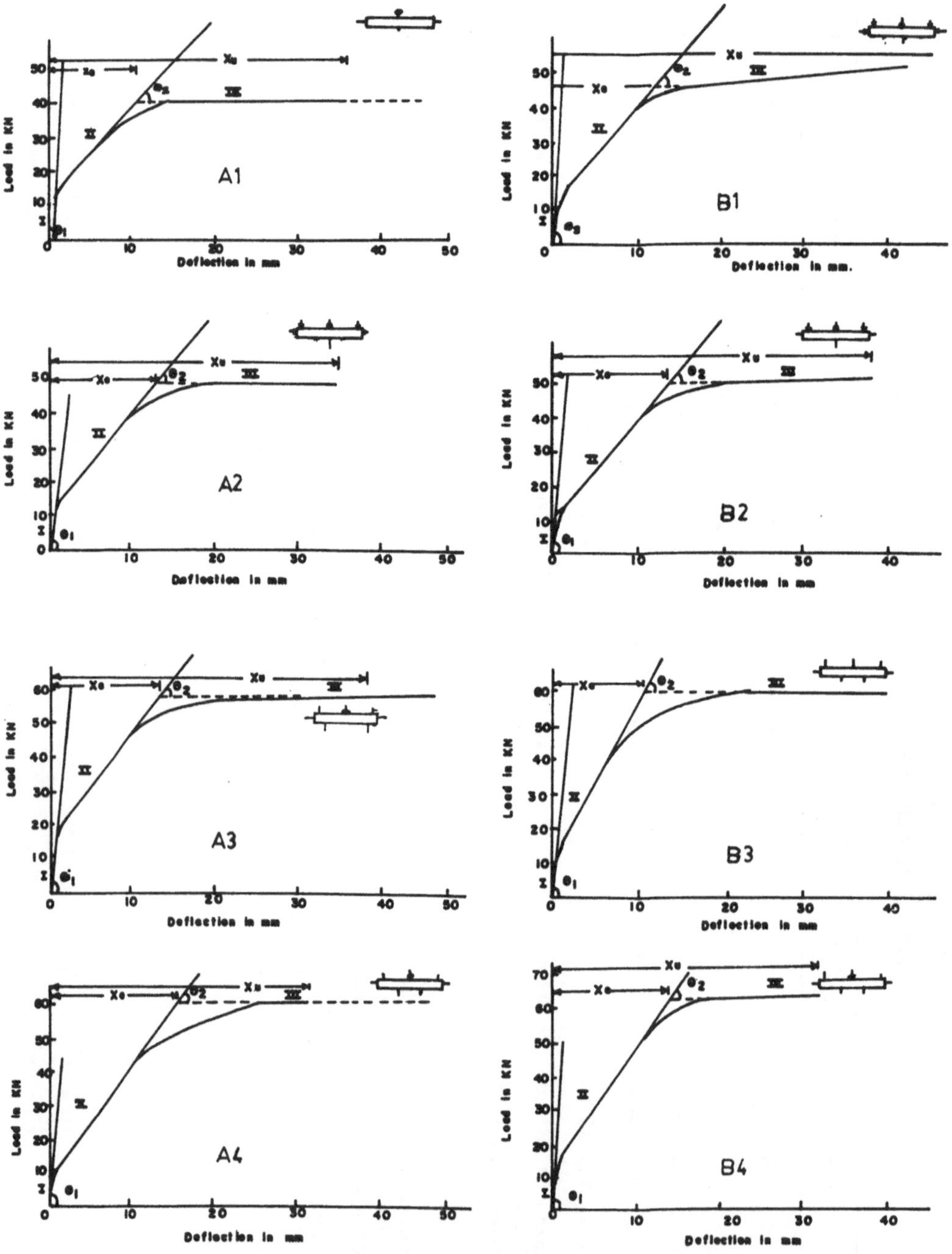

FIG.4 LOAD Vs MIDSPAN DEFLECTION

e) Cracking and Disintegration

The propagation of cracks, has been extensive in the case of type 'A' beams compared to type 'B' beams. In the case of type 'B' beams the widening of crack occured for one of the cracks at mid region of the beam and there was no significant spalling of concrete. The fibres have been found to yield before failing. Some of the fibres which were in the inclined direction to the beam axis were found to fail by pulling out rather than by fracture. The spalling of concrete was extensive in the case of type 'A' beams.

REVERSED CYCLIC LOADING

a) Load Carrying Capacity

The theoretical and observed load carrying capacity as well as the ratio of theoretical load to the observed load for fibrous concrete beams and their companion R.C. beams have been given in Table-2.

In general the fibrous concrete beams were found to have higher load carrying capacity than R.C. beams by about 10 percent.

b) Load Deflection Characteristics

The variation of load Vs midspan deflections have been presented in Fig.5, for beams A5 to A8 (R.C.) and B5 to B8 (F.R.C.).

From the figure, it can be seen that the load deflection curves showed stable hysterisis loops. Three to four load cycles have been performed before reaching the yield load level. After yielding, two to three cycles in the post yield stage have been performed till the beams reached their collapse load. It is also seen that the load carrying capacities in the forward cycles is higher than in the reversed cycles. The reasons for the above behaviour may be attributed to the increase in percentage of steel on the tension side and the existence of the eccentric prestressing force on the tension side of the beam. Further, before the yield load is reached, all the B type beams exhibited smaller deflection when compared with A type beams(i.e.) the spread of the residual deflection on the x-axis, when the load is released, is small for type 'B' (FRC) beams compared to type 'A' (R.C.) beams. This is due to the effectiveness of fibres to close the cracks on removal of the load.

From the load deflection curves the ductility and energy absorption capacity have been worked out for every load cycle.

c) Ductility

The yield deflection for both forward and reversed load cycles are calculated separately as the load carrying capacities in forward and reversed directions are different. Generally all the type 'B' beams exhibited higher ductilities in the last two or three cycles, compared to type 'A' beams. The cumulative ducti-

Table-2

PARTICULARS OF BEAMS SUBJECTED TO STATIC REVERSED CYCLIC LOADING (Part-2)

DESIGNATION OF BEAMS	A5	B5	A6	B6	A7	B7	A8	B8
Percentage of Steel A'_s.	0.785	0.785	0.785	0.785	0.785	0.785	0.785	0.785
A_s	1.18	1.18	1.18	1.18	1.18	1.18	1.18	1.18
Steel fibres by volume	-	1%	-	1%	-	1%	-	1%
Cube compressive strength in N/mm^2	31.0	29.3	28.6	26.7	31.0	29.3	28.6	26.7
Ultimate Load (T) in KN	39.45	39.01	42.8	42.24	59.16	58.51	64.2	63.36
Ultimate Load (O) in KN	54.95	42.55	47.1	51.81	54.75	69.00	70.65	62.80
Ratio of ultimate load (T/O)	0.71	0.92	0.91	0.82	1.08	0.84	0.91	1.01
Cumulative ductility	22.07	16.19	13.1714	04	20.10	19.96	19.47	20.46
Cumulative Energy Absorption in KN m	10.41	10.22	4.86	4.77	7.95	9.98	9.43	8.79

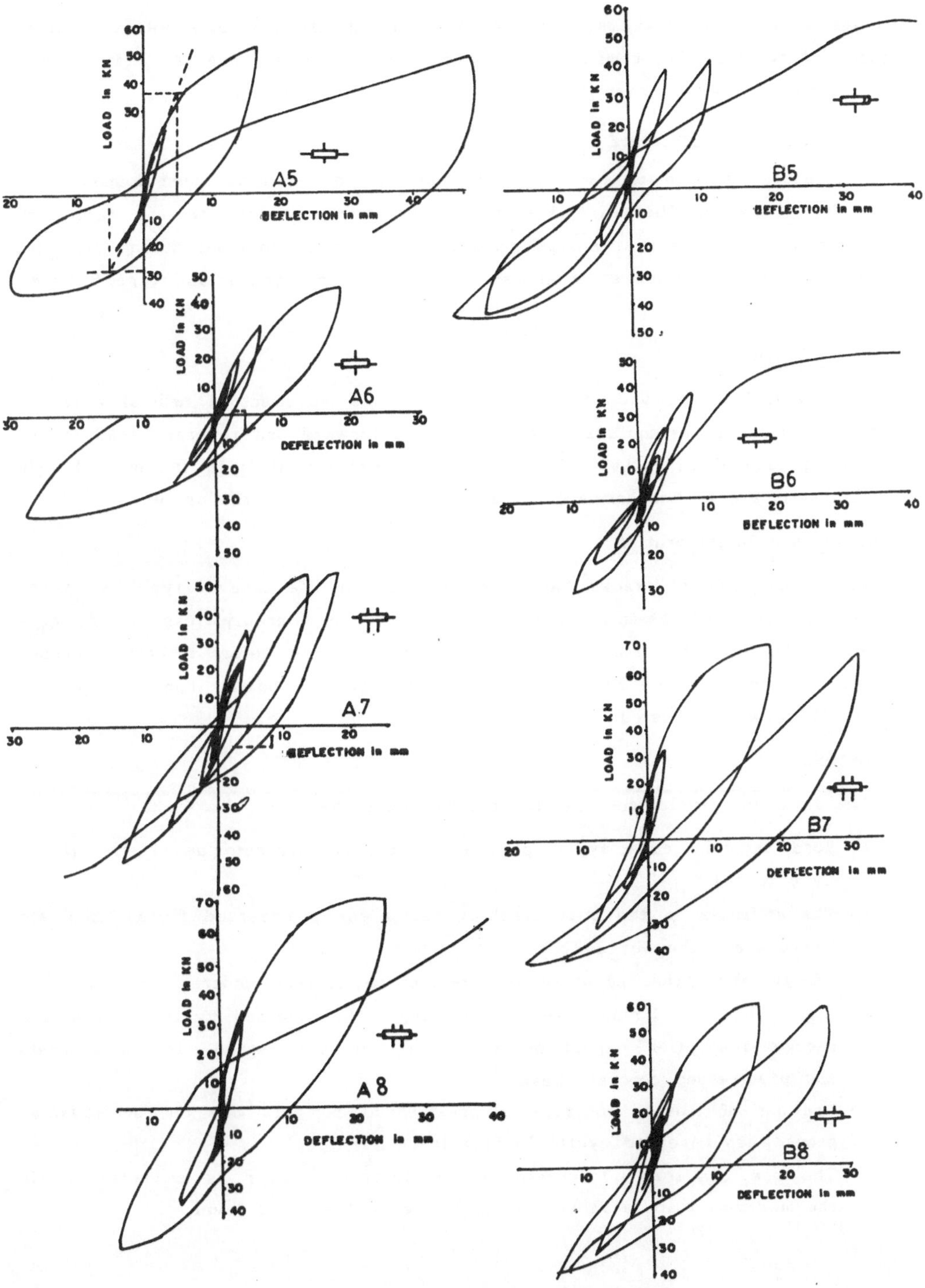

FIG.5 LOAD Vs MIDSPAN DEFLECTION

lities are worked out as the sum of the individual ductility factors obtained during various load cycles. The comparison of these values between type 'A' and type 'B' beams showed higher values for type B beams (Table-2).

d) Energy Absorption Capacity

The energy absorption capacity for type 'B' beams have been found to be higher than type 'A' beams. The cumulative energy absorption are calculated as the sum of the individual energy absorption capacities obtained during different load cycles. These values show almost same values for type A and type B beams (Table-2).

e) Strain Variation

The typical strain variation, which shows the movement of neutral axis for the cross section at various load stages for both forward and reversed load cycles have been presented in Fig.6. It is observed that the shift in the neutral axis for R.C. beams are higher compared to fibrous beams in both the cycles.

f) Cracking and Disintegration

The propogation of cracks has been extensive in the case of type 'A' beams compared to type 'B' beams. Spalling has been considerably less in the case of type 'B' beams. The fibres have been found to yield before failure. Number of cracks are found to be less in the case of type 'B' beams than in type 'A' beams. Fig.7 shows the tested specimens.

CONCLUSIONS

The following conclusions are drawn from the study

a) Load carrying capacity of prestressed fibrous concrete beams are higher by 6 to 10%.
b) The stiffness in the post cracking region for prestressed fibrous concrete beams are higher by 10%.
c) Energy absorption capacity of prestressed fibrous concrete members are higher by 18% during monotonic loading. In reversed cyclic loading the energy absorption capacities are equal for prestressed fibrous concrete and prestressed concrete beams.
d) The use of fibrous concrete in prestressed members may not be advantageous when reversed cyclic loading is anticipated.
e) The use of fibrous concrete in prestressed member is advantageous if the maximum load is restricted to be below the yield level.

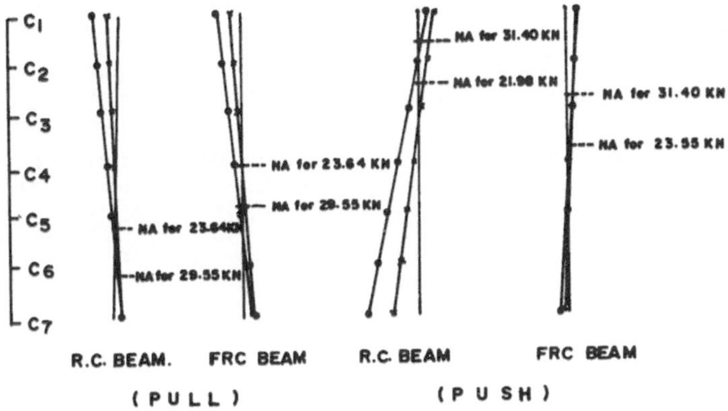

FIG.6 TYPICAL STRAIN VARIATION

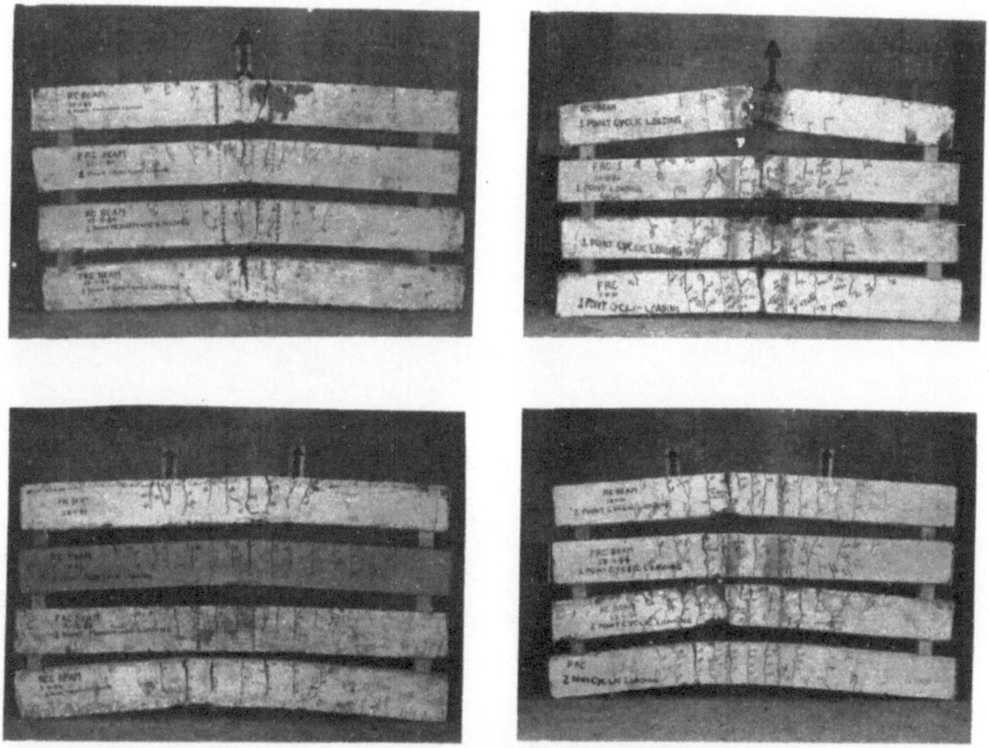

FIG.7 TESTED SPECIMENS

REFERENCES

1. O' Leary, D.C., Dave, N.J., and Saunders,J., "Steel fibres in partially prestressed composite Beams" Fibre Reinforced Concrete, ACI Publication SP-44, 1974 Detroit, pp 477-495.

2. Hariraj,P., "Use of fibrous concrete in prestressed concrete members", M.E. Thesis presented to the Anna University, Madras Jan. 1985, pp 1-159.

3. Henager, C.H., and Doherty, T.J., "Analysis of Reinforced Fibrous Concrete Beams", ASCE, Journal of Structural Division Vol. 102, No.ST1, Jan 1976, pp 177-188.

4. Lakshmipathy, M., "Ductility of Reinforced fibrous concrete Structural members" Ph.D. Thesis, presented to the University of Madras, April 1983, pp 1-221.

Proceedings of the International Symposium on Fibre Reinforced Concrete, December 16–19, 1987, Madras, India

BEHAVIOUR OF FIBRE REINFORCED CONCRETE ARCH PANELS

Agrawal T.P., Sinha, A.N., Kumar, V. and Singh, R.P.
Department of Civil Engineering
Institute of Technology
Banaras Hindu University
Varanasi - 221 005.
INDIA

SYNOPSIS

In this paper authors have presented the results of model testing of concrete arch panels. The testing of fibre reinforced concrete arch panels has been carried out to investigate the effect of the rise on their behaviour. A comparison of the behaviour of fibre reinforced concrete arch panels with that of concrete arch panels reinforced with wires and hexagonal chicken mesh has been done. In each case three cases of rise 40 mm, 70 mm and 100 mm have been considered. For this purpose concrete arch panels 0.5 m wide, 1 m span and of 35 mm thickness have been tested. It has been found that the fibre reinforced concrete arch panels with 100 mm rise give highest ultimate load carrying capacity.

INTRODUCTION

Fibre reinforcement of cement and concrete is not a new idea. In last two decades scientists have started using synthetic fibres to improve the performance of conventional mouldable construction materials such as concrete. Since concrete possesses limited ductility and resistance to cracking so for using concrete as a load bearing element, it is necessary to impart tensile resistance properties to concrete structural member. The fibre reinforcement can be used for this purpose. The fibre reinforced concrete can be used for the construction of prefabricated roofs for low income group houses. This will result in reducing construction time by 40 percent and cost of roof by 75 percent.

Arch construction has been used since ancient times. For this type of construction highly skilled labour is required even than because of economy it is popularly used. Brick tiles [1, 2] or fibre reinforced concrete [3] can be used for the construction of arch panels. In fibre reinforced concrete portland cement, coarse sand, 6 mm size aggregate and 0.4 mm dia fibres of aspect ratio (length/dia. of fibres) 100 are used at the rate of 32 Kg/m^3.

TESTING OF PANELS

Fibre reinforced concrete arch panels 0.5 m wide, 1 m span and thickness 35 mm are fabricated with 40 mm, 70 mm and 100 mm rise. These are tested by applying uniformly distributed load over the whole surface. The support conditions are approximately same as in the case of a two hinged arch. Fig.1 shows the variation of vertical deflection at the crown with load for the arch panels of rise of 40, 70 and 100 mm. This graph shows that the variation of deflection with load is approximately linear for all the above three cases of rise. This fig. shows that the vertical deflection is largest in case of arch panel with 40 mm rise and minimum for arch panel with 70 mm rise. A comparison of fibre reinforced concrete arch panels with wire reinforced, and chicken mesh reinforced concrete arch panels shows that the rate of increase of deflection with load is minimum in case of fibre reinforced concrete arch panels and largest in case of wire reinforced concrete arch panels for all the cases of rise. At the load of 7.5 KN/m^2 the deflections in an arch panel with 40 mm rise and reinforced with wire mesh and chicken mesh, are larger by 36% and 28.7% as compared to the deflection in an fibre reinforced arch panel. Similar behaviour is observed in case of panels with 70 mm and 100 mm rise. In case of 100 mm rise at 7.5 KN/m^2 load the deflection in the case of wire mesh and chicken mesh is 47% and 20% larger as compared to that of fibre reinforcement At the load of 7.5 KN/m^2 in the case of fibre reinforced arch panels the deflections are larger in case of 40 mm rise and 100 mm rise approximately by 55% and 12.6% respectively as compared to the deflection in an arch panel with 70 mm rise. A similar behaviour is observed in the case of panels reinforced with chicken mesh and wire mesh. The panels reinforced with fibre inforcement with 40 mm, 70 mm and 100 mm rise show the first crack at loads 9.14 KN/m^2, 10.9 KN/m^2 and 15.34 KN/m^2 respectively. For above rises in case of chicken mesh the first crack appeared at the load of 10.5, 11.96 and 13.6 KN/m^2 and for wire reinforced panels 8.30, 9.50 and 10.54 KN/m^2 respectively.

THEORETICAL ANALYSIS

The theoretical analysis is carried out on an ICL 1904 main frame computer at B.H.U. computer Centre. The arches are descretized into seventeen straight line elements with eighteen nodal points as shown in Fig. 2. The value of young's modulus and poisson's ratio are taken as 22.08 KN/mm^2 and 0.0 respectively. The elements are loaded with a uniformly distributed load of 10 KN/m^2. The stress resultants namely axial compressive force, radial shear and bending moment for the arch panels with 40, 70 and 100 mm rise are computed and shown in Figs. 3, and 5 respectively. A detailed discussion of these stress resultants is given below.

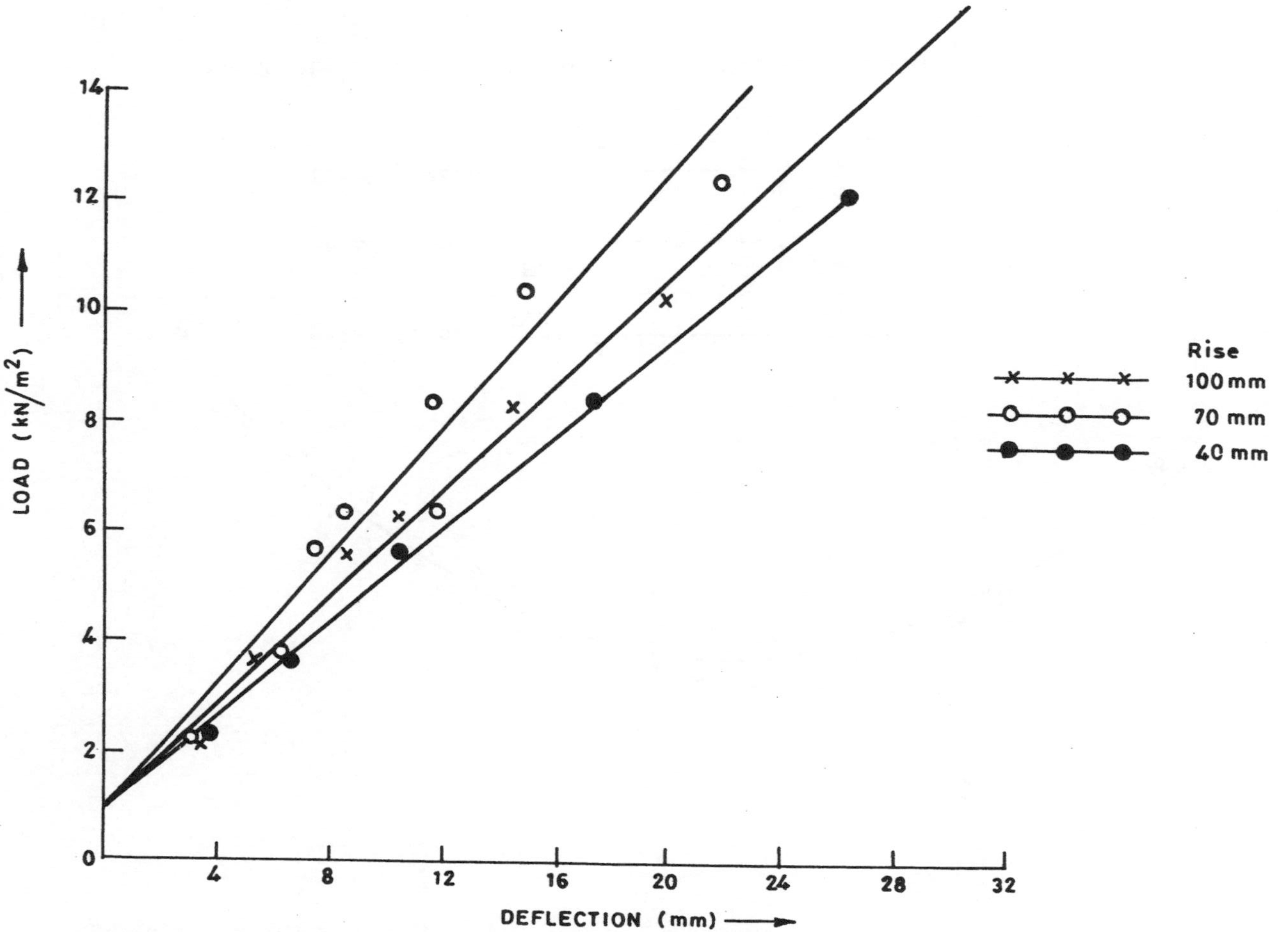

Fig 1. Load-deflection relationship for fibre reinforced concrete arch panels with 40 mm, 70 mm and 100 mm rise.

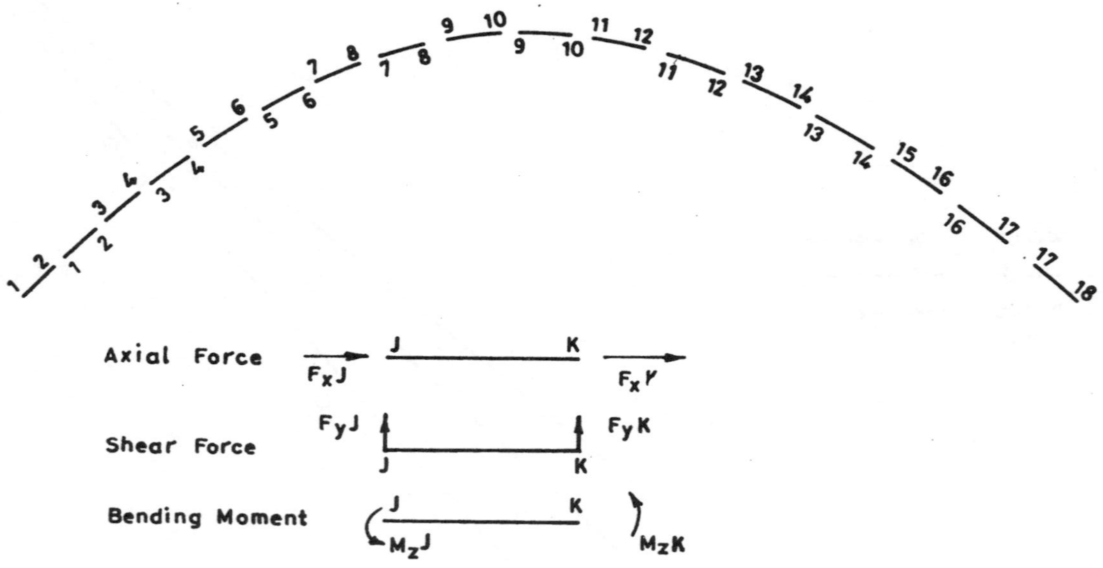

Fig. 2. Mathematical model of arch panel for theoretical analysis.

Axial Compressive Force

The axial compressive force is almost uniform throughout the span of the arch panel. The axial force is slightly smaller in the central element and is the largest in the end elements. The difference between the axial force in the end and central element increases with the increase in rise.

Radial Shear Force

The variation of radial shear force across the span is curvilinear and convex in nature for 40 mm rise. The shape of the curve changes from convex to concave as the rise increases from 40 mm to 100 mm. The values of radial shears at the end of end elements are given in Figs. 3, 4 and 5.

Bending Moment

The shape of the bending moment diagram is approximately parabolic. In the case of arch panels with 40 mm and 70 mm rise is maximum at the centre of middle element. In case of arch panels with 70 mm rise the bending moment is almost constant in the three central elements. In case of arch panels with 100 mm rise the maximum bending moment is obtained approximately a distance 'L/3' from the supports where L is the span of the arch panel. The bending moment diagram in the central zone changes from convex to concave as the rise increases 40 mm to 100 mm.

CONCLUSIONS

Based on the experimental investigations discussed in the above article, following conclusions are obtained:

(i) Load-deflection curve is approximately linear for all the cases of reinforcements and rise of panels.

(ii) Load carrying capacity is maximum for the arch panel with rise 100 mm reinforced with fibre reinforcement, and is minimum for the arch panel with rise 40 mm reinforced with wire mesh.

(iii) From theoretical analysis it is seen that the axial force in all the cases of arch panels is almost uniform throughout the span of the panel. The axial force is maximum at the ends and minimum at the centre.

(iv) The nature of the shear force diagram changes from convex to concave as the rise increases from 40 mm to 100 mm

(v) The bending moment curve changes its shape in the central L/3 portion from convex to concave as the rise increases from 40 mm to 100 mm.

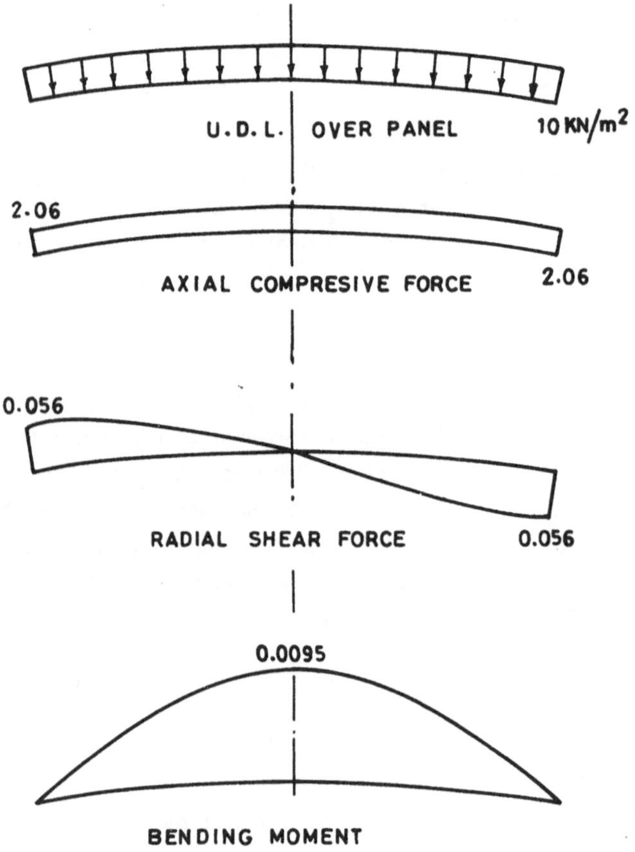

Fig. 3. Stress resultants for arch panel with 40 mm rise.

(vi) The maximum bending moment in the case of 40 mm and 70 mm rise is at the crown while in the case of 100 mm rise it is approximately at a distance of L/6 from the centre.

(vii) During the testing of arch panels it was observed that, in general, the panels with rise of 70 mm and 100 mm did not crack at the crown but slightly away from the crown which is confirmed by the theoretical analysis also.

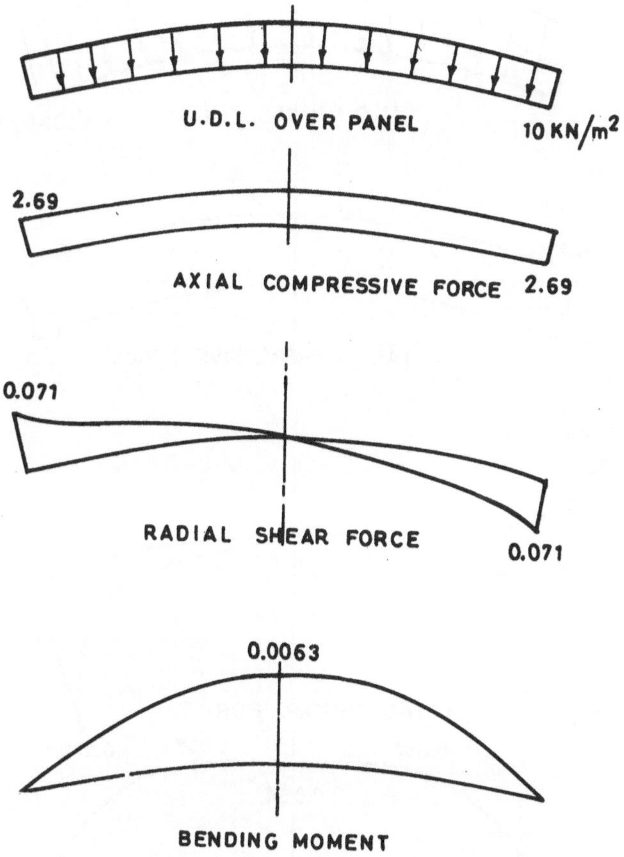

Fig. 4. Stress resultants for arch panel with 70 mm rise

ACKNOWLEDGEMENT

Authors sincerely acknowledge the help given by Shri M.M. Shukla, Technical Assistant, the project on "Analysis of Cable Stayed Bridges", for the preparation of manuscript of the paper.

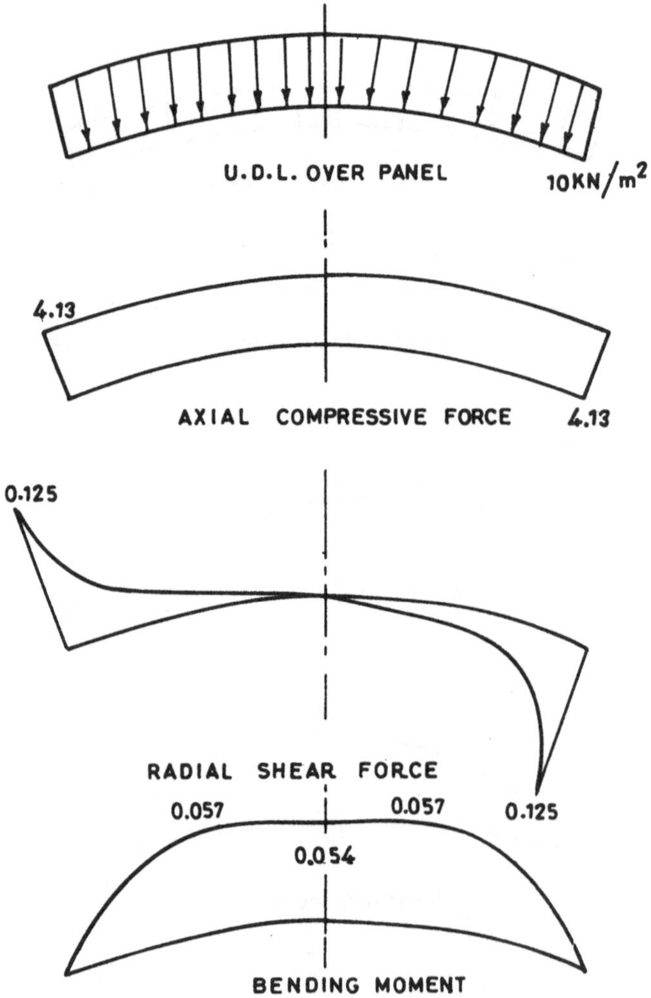

Fig. 5. Stress resultants for arch panel with 100 mm rise.

REFERENCES

1. Yadav, A., An Experimental Study of Prefabricated Arch Brick Panel for Low Cost Houses, M.Tech. Dissertation, Civil Engineering (Structural Engg.), I.T., B.H.U., Varanasi 1979.

2. Ranjan, A., An Experimental Investigation of Prefabricated Arched Brick-Tile panels for a Low - Cost Roofing System, M.Tech. Dissertation, Civil Engg. (Structural Engg.),

I.T., B.H.U., Varanasi 1982.

3. Singh, R.P., Internal Force Distribution at Cracking in Prefabricated Arched Concrete Panels, A Dissertation Submitted in Partial Fulfilment of the Requirements for the Award of the Degree of Master of Technology in Civil Engg. (Structural Engg.), Dept. of Civil Engg., I.T., B.H.U., Varanasi - 221 005, India 1985.

Proceedings of the International Symposium on Fibre Reinforced Concrete, December 16-19, 1987, Madras, India

INVESTIGATION OF THE PULL-OUT CHARACTERISTICS OF STEEL FIBERS FROM MORTAR MATRICES

Vellore S. Gopalaratnam and Halim J. Abu-Mathkour

Department of Civil Engineering
University of Missouri-Columbia
Columbia, Missouri 65211, U.S.A.

ABSTRACT

The article describes a servo-controlled pull-out test developed for an ongoing study on the behavior of nonlinear interfaces in ductile fiber-brittle matrix composites. Results from the experimental component of the study are also discussed in detail and compared wherever relevant with those from some of the earlier pull-out studies. Primary parameters taken up for the study include fiber length, fiber diameter and matrix quality. Experimental observations from the study reinforce the hypothesis that inelastic frictional bond may more significantly influence the composite failure characteristics than earlier acknowledged.

INTRODUCTION

It has been recognized that knowledge of the nonlinear fiber-matrix interfacial response is essential to analytically characterize the fracture behavior of short ductile fiber (steel)-brittle matrix (mortar) composites [1]. Even the limited amount of systematic information presently available on the subject [2-6] are from pull-out tests that were not conducted under stable servo-controlled test conditions. The stability of the pull-out test is vital to proper monitoring of the fracture behavior of such brittle composites exhibiting softening. In addition, a direct slip control ensures better pre- and post-peak stability, unlike crosshead or overall displacement controlled tests. Of particular concern is the very low initial stiffness [7] recorded in the latter types of tests.

Also, there now is considerable interest in the study of progressive fracture of interfaces [8-13]. Since local study of the steel-fiber-mortar-matrix interface is not possible without associated instrumentation adversely affecting such test results, indirect macroscopic studies of progressive interfacial failures are needed. These indirect studies can be accomplished through cyclic (static) tests. Unloading/reloading tests enable observations on progressive stiffness degradation and accumulation of residual slip [10]. Such data is not presently available from the earlier fiber pull-out studies.

EXPERIMENTAL INVESTIGATION

In addition to being concerned about the stability of the pull-out test, one needs to minimize the parasitic influence of adverse gripping stresses in the vicinity of the interface being studied and also minimize eccentricities in the loading set-up, in order to obtain reliable pull-out load-slip characteristics [3]. The manner in which all of these concerns were addressed in the present investigation is described in this section. Fig. 1 shows a schematic of the slip-controlled closed-loop pull-out test set-up.

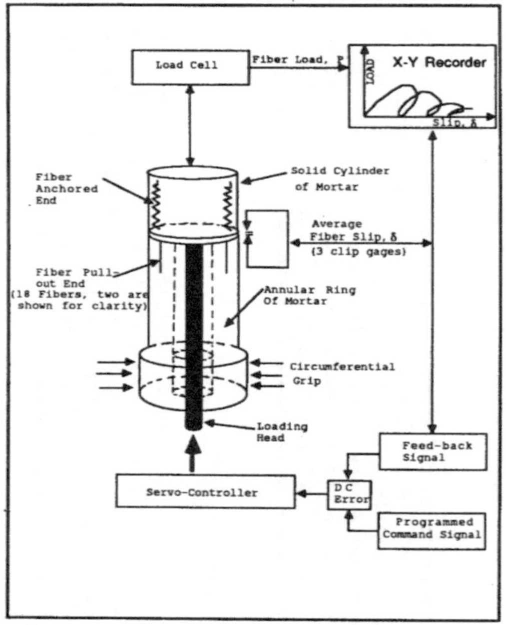

Fig. 1. Schematic diagram of the servo-controlled pull-out test.

(a)

(b)

(c)

(d)

Fig. 2. Stages in the pull-out specimen preparation process. (a) Template holding the fibers in place for the first stage of casting. (b) Specimen after the first stage of casting. Fibers protruding conform to a 1" embedment length configuration. (c) Specimen after the second stage of casting. Note the disjointed interface and the inner mold left in place. (d) Specimen ready for testing showing the clip gage seating devices.

Pull-Out Specimen

Specimens chosen for the study were similar (except for the dimensions and other minor details) to those used by Gokoz and Naaman [14]. Eighteen smooth straight steel fibers that were uniformly spaced circumferentially in an annular cylinder of mortar matrix were simultaneously pulled out. All these fibers were anchored at their other ends in a solid cylinder of the mortar matrix, Fig. 2. The specimen was cast in two stages. Fig. 2 shows details of the specimen preparation. The fibers were held in the template to achieve the desired embedment length and spacing (Fig. 2a). An external mold was placed over the fiber-template assembly and the first stage of the specimen was cast (Fig. 2b, solid cylinder holding the embedded portion of the fibers). After 24 hours the external mold and template were removed. A thin polyethylene sheet was placed between the first and second stage of casting so that there was no bonding between the solid mortar cylinder of the first stage and the annular mortar cylinder of the second stage. Outer and inner molds were securely placed on top of the solid cylinder so that the fiber-ends exposed from the first stage were positioned in the annular region between the molds of the second stage. Mortar was poured in the annular region embedding the pull-out end of the fiber (Fig. 2c). The annular cylinder of mortar had an inside diameter of 2.375 in. and an outside diameter of 4 in. (60.3, 101.6 mm). The fiber to matrix area fraction at the pull-out section varied between 0.025%-0.0841% depending upon fiber radius used. Interaction between the fibers as a result can be neglected. The outer mold was removed while the inner mold was left inside the specimen. After sealing the junction with waterproofing tape to prevent rusting of fibers at this location, the specimens were cured at 100% R.H. until they were tested. Age at testing typically varied between 28-35 days. A day prior to testing the specimens were left to dry so that the clip-gage seating devices could be glued (Fig. 2d).

Test Fixtures

A cylindrical frictional device, Fig. 3, designed to align and grip the specimen was mounted to the fixed cross-head of the MTS machine. The amount of gripping stresses on the annular cylinder was controlled based on preliminary tests to ensure that no slip occurred between the specimen and the grips even at the maximum pull-out loads. Care was taken to ensure no tensile splitting cracks were introduced due to the gripping stresses.

The loading head used to push out the solid cylinder of mortar was mounted on the hydraulic actuator. Fibers embedded in the annular cylinder of mortar were pulled out as the actuator moved up, Fig. 1. A 2,000 lb (8.9 kN) full-bridge tension-compression load cell was used in series, sandwiched between the actuator and the loading head.

Three full bridge clip-gages (0.0544 in. range, 1.382 mm) were used 120° apart on the specimen to measure the average slip at the junction between the first and the second stages of casting. Specially designed clip-gage seating devices glued on to the specimen prior to the testing enabled zeroing of the clip-gages individually, Fig. 3.

Experimental Program

As stated earlier, the experimental investigation was conducted with a view to study the effect of the following three parameters on the fiber pull-out characteristics; fiber embedment length, fiber diameter and matrix quality. Fiber embedment lengths of 0.5, 1.0 and 2.0 in. (12.7, 25.4 and 50.8 mm) were used. Fiber diameters of 0.012, 0.015 and 0.022 in. (0.31, 0.38 and 0.56 mm) were used. Two different matrix mixes were used. Mix M1 was proportioned 1:1.5:0.35 (C:S:W by weight) and mix M2 was proportioned 1:3.0:0.5. Table 1 gives details of the experimental program and series designations of the specimens. Totally 10 series of specimens were tested during the study. Between 3 and 6 specimens were tested (at least one cyclic and two static tests were performed) for each series. Brass coated high tensile strength carbon steel fibers donated by National Standard Company were used for the study, Table 2.

Fig. 3. Close-up of the gripping device. Photograph also shows the clip gages used to measure average interfacial slip.

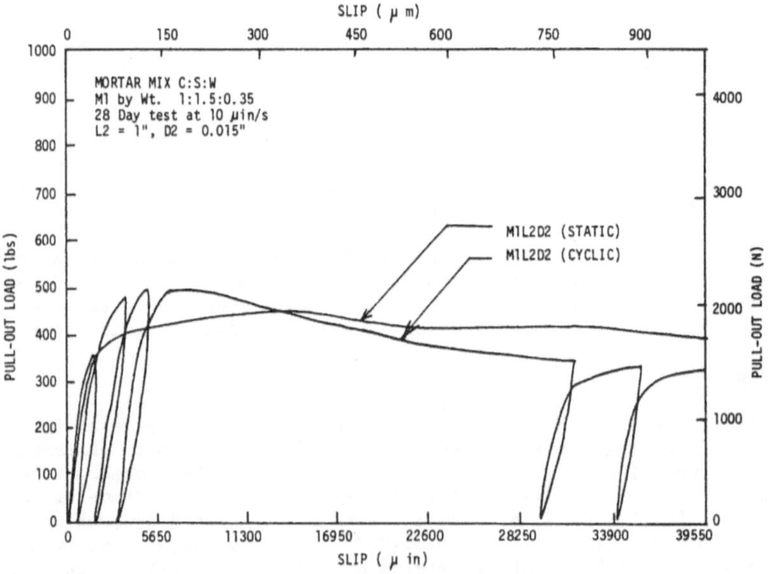

Fig. 4. Typical plot of the fiber pull-out load versus fiber slip showing a static and a cyclic test.

Table 1. Details of the Experimental Program

Mortar Matrix Designation, Mix Proportion C:S:W(by weight)		Fiber Embedment Length in., (mm)			Fiber Diameter in.,(mm)
		L1=0.5(12.7)	L2=1.0(25.4)	L3=2.0(50.8)	
		Series Designation			
M1	1.0:1.5:0.35	---	M1L2D1	---	D1=0.012(0.31)
		M1L1D2	M1L2D2	M1L3D2	D2=0.015(0.38)
		---	M1L2D3	---	D3=0.022(0.56)
M2	1.0:3.0:0.50	---	M2L2D1	---	D1=0.012(0.31)
		M2L1D2	M2L2D2	M2L3D2	D2=0.015(0.38)
		---	M2L2D3	---	D3=0.022(0.56)

Table 2. Properties of the Constituent Materials

Mortar Mix	Compressive Strength psi (MPa)	Peak Strain	Modulus of Elasticity psi (GPa)
M1	7500 (51.72)	0.018	5.3×10^6 (36.55)
M2	5500 (37.93)	0.002	4.5×10^6 (31.03)
Steel Fiber Property	Tensile Strength psi (GPa)	Peak Strain	Modulus of Elasticity psi(GPa)
	$3\text{-}3.5 \times 10^5$ (2-2.4)	0.02	29×10^6 (200.00)

Table 3. Average§ Peak Pull-Out Loads and Average§ Bond Strength

Series Designation*	Average Peak** Pull-Out Load lbs (N)	Average Bond Strength psi(MPa)
M1L2D1	158+ (703)	233+ (1.61)
M1L1D2	258 (1148)	608 (4.19)
M1L2D2	492 (2189)	580 (4.00)
M1L3D2	585++ (2603)	345++ (2.38)
M1L2D3	640 (2848)	514 (3.55)
M2L2D1	281 (1250)	414 (2.86)
M2L1D2	292 (1299)	689 (4.75)
M2L2D2	514 (2287)	606 (4.18)
M2L3D2	526++ (2341)	310++ (2.14)
M2L2D3	701 (3119)	564 (3.89)

§Average values from at least 3 specimens per series have been reported.
*See Table 1 for Series Designations
**Load is reported for 18 Fibers
+Tested at 14 days. All other series tested between 28-35 days.
++Most of the L3 series specimens failed by fiber fracture.

Test Procedure

The experiment involves the pulling-out of 18 equally spaced (along a constant radius) aligned fibers from a mortar matrix using a "fiber withdrawal rate" controlled test. The servo-controlled test was performed using an MTS closed-loop testing machine, Fig. 1. Details of the test set-up are provided in reference 10.

Specimens removed from the curing room after requisite curing were left to dry under laboratory conditions (50-60% RH, $75^{o}F$). Clip-gage seating devices were glued on to the specimen. These devices were designed to hold the clip-gages (Fig. 3) 120° apart from each other along the circumference of the specimen. The compressed gage length of the gages were approximately 0.5 in. (12.7 mm). These gages had a range of 0.0544 in. (1.382 mm). The seating devices were glued in aligned pairs across the junction where the fiber pulled out (Fig. 2d). Screws provided on the devices enabled proper zeroing of the individual gages (Fig. 3). Summed output signal from the three clip-gages was used for the feed-back control. Monotonic static tests were conducted at a fiber slip rate of 10 in./s. A ramp generator was used for the command signal. Tests were conducted beyond the clip-gage range using the actuator LVDT signal for control. It should be noted that most specimens had already been loaded beyond slips corresponding to the peak pull-out load much prior to reaching the limiting range of the clip-gages.

Cyclic unloading/reloading tests were conducted by manually unloading/reloading at approximately the rate used for the monotonic static tests. Load-slip plots were recorded using a HP 7046A XY recorder. Load and slip values were also monitored at regular intervals using voltmeters.

RESULTS AND DISCUSSION

Test Results

Properties of the constituent materials are presented in Table 2. Average peak pull-out loads recorded in the tests and corresponding average bond strength (assuming uniform shear stress distribution) are listed in Table 3. Typical pull-out test records of load-slip characteristics for a monotonic and cyclic test are shown in Fig. 4. Displacements measured by the clip-gages directly yield slip at the interface resulting from the end of the fiber that is pulled out. Matrix deformations within the gage length and slip of the embedded end are considered negligible (rigidly anchored, Fig. 2a).

Typical trend observed for the load-slip behavior from monotonic test is as follows. Linear elastic behavior is observed up to slip values of the order of 0.001 in. (0.025 mm). At this stage load levels are already in the 50-90% of the peak-pull-out loads depending upon the fiber embedment lengths (higher percentage for smaller embedment lengths). This also implies a much larger initial stiffness of the pull-out specimens (an order of magnitude or more higher) unlike those reported from the earlier tests [2-7, 15]. Load-slip behavior is significantly inelastic beyond this stage with the extent directly governed by the fiber embedment length. Due to the relatively flat nature of the load-slip characteristics in the vicinity of the peak-load, considerable scatter in the slip corresponding to the peak pull-out load was observed. For specimens that failed due to fiber pull-out (embedment lengths less than 2 inches) slip at peak load varied in the 0.01-0.04 in. range (0.25-1 mm). Post-peak load-slip characteristic was considerably flatter than those obtained under crosshead/overall displacement controlled conditions [2-6, 15], Fig. 4. A slip-controlled test ensures greater stability during specimen softening. This explains the rapid post-peak unloading reported in some earlier displacement controlled pull-out studies. This is much like the trends observed in tension softening tests of unreinforced concrete [16, 17]. For specimens that failed by fiber fracture (L3 series), slip at peak load of around 0.2 in. (2.54 mm) was recorded. This provides evidence that prior to fiber breakage (at fiber strain of 0.02), stick-slip phenomenon has occured with significant slip before failure. It is hence also possible that for the matrix quality and other parameters used, the critical embedment length is very close to 2 inches. All L3 series specimens had fiber breakage of greater than 83% (15 fibers or more out of 18 fractured). All other specimens failed by fiber pull-out.

Cyclic tests were conducted by manually unloading/reloading. The unloading/reloading rates approximately correspond to those used for the monotonic tests. However, since manual unloading/reloading involves a relatively jerky operation, momentarily the rates may have been significantly higher. This is evidenced by the fact that the peak cyclic loads were consistently higher (even though marginal) than the corresponding monotonic envelopes, Fig. 4. In addition to the relatively flat load-slip response near the peak-load, the faster rate of unloading/reloading particularly near the envelope, may have been responsible for the lower slip values at the peak-load (local peaking) for cyclic tests, Fig. 4. To confirm whether the initial stiffnesses were as high as recorded in monotonic tests, load cycling was performed in the linear elastic and the inelastic stages. This cycling confirmed that the high inital stiffnesses recorded were indeed true pull-out specimen characteristics and not due to other forms of artificial stiffness (for example due to bonding between first and second stage of mortar) Fig. 4. Detailed discussion of the cyclic test results is presented later in this section.

Effect of Fiber Embedment Length

Fig. 5 shows a plot of the average interfacial bond strength versus fiber embedment length for a constant fiber diameter (0.015 in., 0.38 mm). While the scatter for smaller embedment length is large, it still does not mask the trend (dotted line) that the average bond strength is inversely related to the fiber embedment length. For the largest embedment length used (2 in., 50.8 mm), at least 15 out of the 18 fibers fractured in each test. As noted earlier, fiber breakage for the L3 series was also preceded by some slip as evidenced by the slip values being larger than the fiber failure strains (Table 2). Based on values given in Table 2, an approximate value of 2 inches (50.8 mm) for the critical embedment length can be obtained. For fiber embedment lengths larger than 2 inches, it is expected that the "equivalent average bond strength" will be independent of fiber embedment length (failure by fiber fracture). One important consequence of the observations in this section is that since average bond strength computed based on the assumption of uniform shear strength is embedment length dependent, strength values reported in the literature should not be used for embedment length other than that reported. Also since efficiency factors derived to relate mean embedment length to random "shorter" embedment length across a crack [18] ignore this strength dependence on embedment length, they may need some corrections. Slip and energy absorbed corresponding to the peak-pull-out load greatly increased with larger embedment length although there was a considerable scatter in these values due to the reason stated earlier.

Effect of Fiber Diameter

Fig. 6 shows a plot of the average interfacial bond strength versus fiber diameter for a constant fiber embedment length of 1.0 in. (25.4 mm). Note that series M1L2D1 was tested at 14 days due to experimental scheduling problems, unlike all other series which were tested between 28-35 days. As a result significantly lower bond strength values were obtained for that series, Fig. 6. Diameter unlike embedment length does not significantly influence the average bond strength. However increases in fiber diameter does result in small increases in the average bond strength (trend indicated by the dotted line). Slip corresponding to the peak-load and energy absorbed per unit bonded area up to peak loads also increases with larger fiber diameters [11], although the trend was not always consistent. This may in part be due to experimental scatter for the small range of diameters investigated (0.012, 0.015 and 0.022 in., 0.31, 0.38 and 0.56 mm).

Effect of Matrix Quality

The peak pull-out loads for matrix mix M2 was almost consistently higher than that obtained for M1, Figs. 5-6 and Table 3. This may seem inconsistent with the popular notion that a higher matrix compressive strength necessarily improves the bond strength and hence would yield higher peak pull-out loads. Based on analytical [1, 9,

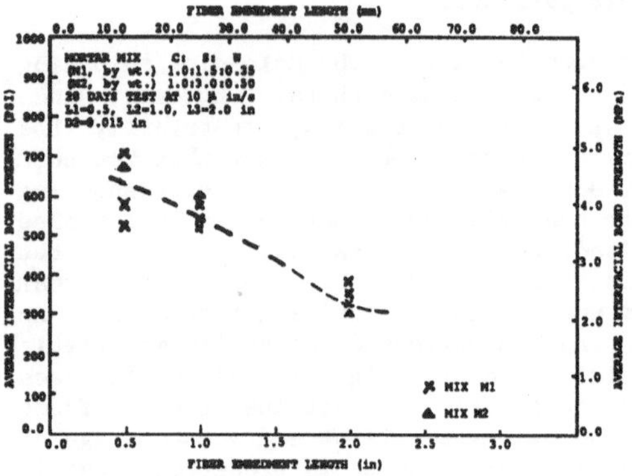

Fig. 5. Average interfacial shear strength versus fiber embedment length. Many of the 2" fibers fractured instead of pulling out.

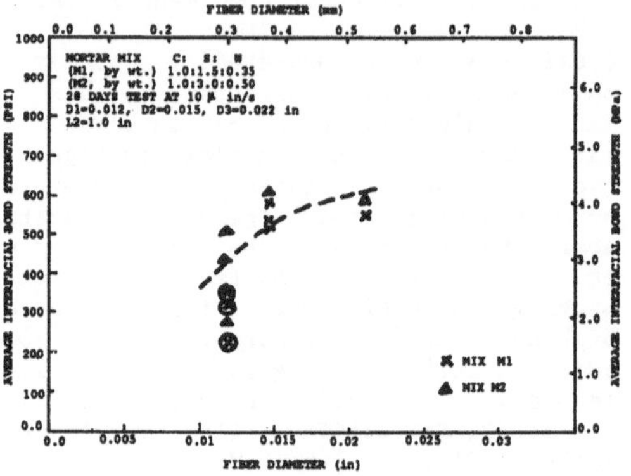

Fig. 6. Average interfacial shear strength versus fiber diameter. ⊗-14 day test.

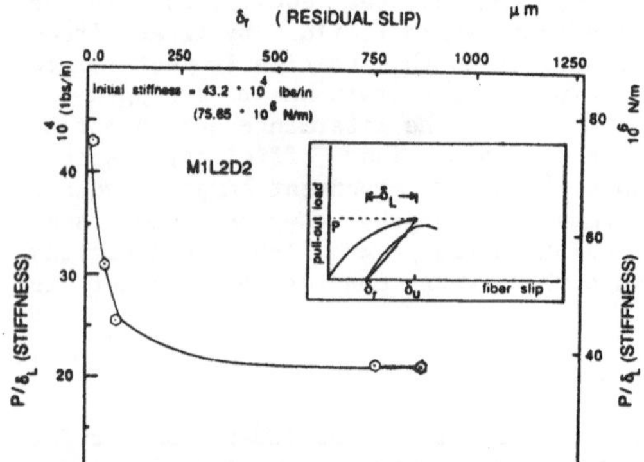

Fig. 7. Stiffness versus residual slip from a cyclic test.

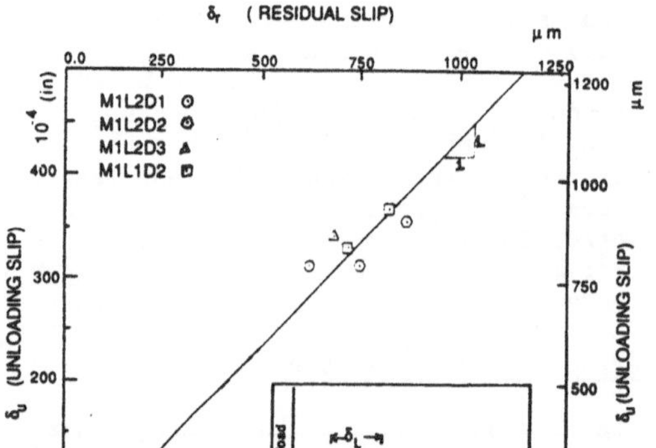

Fig. 8. Slip at unloading versus residual slip from a cyclic test.

13, 19-21], experimental [22] and numerical [12] models that recognize the importance of the contribution of the frictional softening characteristics of the interface, it has been shown that [13,14], weaker mixes with superior frictional softening characteristics may be capable of providing larger resistance to pull-out (load, corresponding slip as well as energy absorption). It should be noted that mix M2, while being weaker than mix M1, has twice the sand content (Tables 1 and 2) of mix M1. Results from this experimental study clearly show that frictional softening bond may at least play as important a role as the elastic bond particularly for steel fiber reinforced cement composites. The extent of pre-peak inelastic behavior may be governed by a stable frictional softening of the fiber-matrix interface following the failure of elastic bond.

Stiffness Degradation and Residual Slip

Cyclic tests were conducted to study stiffness degradation and the accumulation of residual slip. These parameters yield some useful insights to the progressive failure of the elastic bond and the subsequent frictional softening of the interface, since direct local instrumentation of the interface is not practical. Fig. 7 shows a plot of the unloading/reloading stiffness ($P/\delta L$, where δL is defined in the inset) versus the residual slip δr. The values of stiffnesses reported are based on cyclic loading tests on specimens containing 18 fibers. As noted earlier the initial stiffness value observed in this study are more than a magnitude higher than those reported earlier [7]. Better slip measurements in the present study may have contributed to more representative values of the initial stiffness. Based on average shear stress and strains (assuming a uniform shear stress and strain distribution along the fiber embedment length) a shear modulus for the interface of 0.51×10^6 psi was obtained ($(P/\delta L)_{initial} = 43.2 \times 10^4$ lb/in for 18 fibers). Accounting for the actual nonlinear elastic shear stress distribution along the interface and a weaker "interface transitional ring" [15] compared to the matrix may improve the shear modulus value further. The shear modulus value obtained in the present study seems more reasonable (matrix shear modulus = 2.5×10^6 psi) than those reported in any of the earlier studies.

From the cyclic tests, prior to the peak pull-out loads some degradation of the stiffness was observed (Figs. 4 and 7). At large unloading displacements δu, little or no degradation was observed. Fig. 8 includes a plot of the unloading slip, δn versus the residual slip δr. From an initial slope of larger than 1, this plot tends to a slope of 1, indicating pre-peak debonding and subsequent frictional pull-out (all post-peak displacements applied result in nonrecoverable slip).

CONCLUSIONS

1. A slip-controlled closed loop pull-out test yields a stable load-slip characteristic and hence may be more reliable for fracture studies of the interface.

2. Fiber embedment length plays an important role in the stress transfer mechanism in the pull-out test. Average bond strength is inversely related to the embedment length. Critical fiber length for the matrix mix and fibers used in the study is approximately 2 inches. Slip corresponding to the peak pull-out load and energy absorbed are directly related to the fiber embedment length. Larger extent of pre-peak inelastic behavior was observed for the pull-out of longer fibers.

3. Although fiber diameter does not play as significant role as the fiber embedment length, the average bond strength of an interface increases with an increase in fiber diameter.

4. From the cyclic tests, degradation of specimen stiffness was observed prior to the peak pull-out load. Little additional stiffness degradation was observed beyond slips corresponding to the peak pull-out load.

5. The peak pull-out load for identical fibers being pulled out from matrices of higher compressive strength is not necessarily higher. Higher frictional bond

strength which might be unrelated to matrix compressive strength is perhaps the reason for the higher pull-out loads observed for the weaker mix.

ACKNOWLEDGEMENTS

The research reported here was made possible by a Weldon Springs Endowment Program research grant awarded to the first author. Acknowledgements are due to Mr. Stan Baker of National Standard Company, Corbin, Kentucky, for donating the fibers used in the study. Thanks are also due to Mr. Delbert Morton who helped fabricate the fixtures used in the test. The authors would like to thank Mrs. Mary McCush for help with the preparation of the manuscript.

REFERENCES

1. Gopalaratnam, V.S., and Shah, S.P., "Tensile Failure of Steel Fiber-Reinforced Mortar," Journal of Engineering Mechanics, ASCE, Vol. 113, No. 5, May 1987, pp. 635-652.

2. Burakiewicz, A., "Testing of Fibre Bond Strength in Cement Matrix," RILEM Symposium on Testing and Test Methods of Fibre Cement Composites, Construction Press, England, April 1978, pp. 355-365.

3. Gray, R.J., and Johnston, C.D., "The Measurement of Fibre-Matrix Interfacial Bond Strength in Steel Fibre Reinforced Cementious Composites," RILEM Symposium on Testing and Test Methods of Fibre Cement Composites, Construction Press, England, April 1987, pp. 317-328.

4. Maage, M., "Fibre Bond and Friction in Cement and Concrete," RILEM Symposium on Testing and Test Methods of Fibre Cement Composites, Construction Press, England, April 1978, pp. 329-336.

5. Naaman, A.E., and Shah, S.P., "Pull-Out Mechanism in Steel Fiber Reinforced Concrete," Journal of Structural Division, ASCE, Vol. 102, No. ST8, August 1976, pp. 1537-1548.

6. Stroeven, P., "Micro- and Macromechanical Behavior of Steel Fibre Reinforced Mortar in Tension," Heron, Vol. 24, No. 4, 1979, pp. 7-40.

7. Soroushian, P., and Bayasi, Z., "Prediction of the Tensile Strength of Fiber Reinforced Concrete: A Critique of the Composite Material Concept," ACI Special Publication, November 1986, 18 pp.

8. Stang, H., and Shah, S.P., "Failure of Fiber Reinforced Composites by Pull-Out Fracture," Journal of Materials Science, Vol. 21, 1986, pp. 953-957.

9. Gopalaratnam, V.S., and Shah, S.P., "Failure Mechanisms and Fracture of Fiber Reinforced Concrete," ACI Special Publication, November 1986, 25 pp.

10. Abu-Mathkour, H., "Experimental Investigation of the Pull-Out Characteristics of Short Steel Fibers from Mortar Matrix," CE 400 Report, University of Missouri-Columbia, Columbia, Missouri, December 1986, 76 pp.

11. Morrison, J.K., Shah, S.P., and Jenq, Y.S., "Analysis of the Debonding and Pull-Out Process in Fiber Composites," Journal of Engineering Mechanics, ASCE, Submitted for Publication, 1987, 23 pp.

12. Sahudin, A.H., "Nonlinear Finite Element Study of Axisymmetric Fiber Pull-Out," M.S. Thesis, University of Missouri-Columbia, Columbia, Missouri, July 1987, 110 pp.

13. Jin, C., "An Analytical Study of Inelastic Interface in Fibrous Composites," M.S. Thesis, University of Missouri-Columbia, Columbia, Missouri, in progress under the supervision of Dr. Gopalaratnam, 1987.

14. Gokoz, U.N., and Naaman, A.E., "Effect of Strain Rate on the Pull-Out Behavior of Fibers in Mortar," The International Journal of Cement Composites, Vol. 3, No. 3, Septmeber 1981, pp. 187-202.

15. Wei, S., Mandel, J.A., and Said, S., "Study of the Interface Strength in Steel Fiber Reinforced Cement-Based Composites," ACI Journal, Vol. 83, No. 4, July-August 1986, pp. 597-605.

16. Gopalaratnam, V.S., and Shah, S.P., "Softening Response of Plain Concrete in Direct Tension," ACI Journal, Vol. 82, No. 3, May-June 1985, pp. 310-323.

17. Reinhardt, H.W., "Fracture Mechanics of an Elastic Softening Material Like Concrete," Heron, Vol. 29, No. 2, 1984, 42 pp.

18. Laws, V., "The Efficiency of Fibrous Reinforcement of Brittle Materials, Journal of Physics D: Applied Physics, Vol. 4, 1971, pp. 1737-1746.

19. Lawrence, P., "Some Theoretical Considerations of Fiber Pull-Out From an Elastic Matrix," Journal of Materials Science, Vol. 7, 1972, pp. 1-6.

20. Bartos, P., "Analysis of Pull-Out Tests in Fibres Embedded in Brittle Matrices," Journal of Materials Science, Vol. 15, 1980, pp. 3122-3128.

21. Gray, R.J., "Analysis of the Effect of Embedded Fibre Length on Fibre Debonding and Pull-Out from an Elastic Matrix, Part 1: Review of Theories," Journal of Materials Science, Vol. 19, 1984, pp. 861-870.

22. Pinchin, D.J., and Tabor, D., "Interfacial Contact Pressure and Frictional Stress Transfer in Steel Fibre Cement," RILEM Symposium on Testing and Test Methods of Fibre Cement Composites, Construction Press, England, April 1978, pp. 337-344.

Proceedings of the International Symposium on Fibre Reinforced Concrete, December 16–19, 1987, Madras, India

STUDIES ON FIBROUS REINFORCED CONCRETE BEAMS SUBJECTED TO BENDING AND TORSION

M.F. Sharief, Jawaharlal Nehru Technological University, Hyderabad, India.

T. Abdul Hafeez Khan, University of Constantine, Algeria.

V.V. Govinda Rao, Consulting Engineer, New Delhi.

SYNOPSIS

The behaviour of steel fibre reinforced concrete beams subjected to combined bending and torsion is investigated by testing sixteen beams of size 150 x 200 x 3000 mm with varying percentage of volume of fibres (0, 0.5, 0.75, 1.0%) and varying torque to moment ratios (2.5, 1.67, 1.25, 1.0). All the beams were tested to failure. The effect of fibre content on first crack strength, ultimate strength and crack width of beams is studied. The fibre used was of plain steel with tensile strength of 190 N/mm^2 with aspect ratio of 62.5. It is found that an increase of 8%-67% in the ultimate strength is obtained by the inclusion of 0.75% fibre for all the torque to moment ratios studied.

INTRODUCTION

The superior performance characteristics of fibre reinforced concrete have been well established in tension, flexure (1,2). Also studies have been carried out on fibre reinforced concrete members in pure torsion and combined bending and torsion (3,4,5,6). Studies on fibre reinforced concrete full scale beams have also been carried out (2).

The investigation reported in this paper was aimed at studying the behaviour of full size reinforced concrete beams with steel fibres and comparing it with the behaviour of reinforced concrete beams without fibres. Sixteen beams have been tested under combined bending and torsion with varying percentages of steel fibres and varying the torque to moment ratios.

EXPERIMENTAL PROGRAMME :

All the beams tested were 150 mm wide 200 mm deep and 3000 mm over all length. Torsion brackets extending in opposite directions and at suitable distance apart to give the desired T/M ratios were provided. All the beams had identical conventional reinforcement. The schematic diagram of loading arrangement and the details of conventional reinforcement are shown in Figure 1 and 2 respectively.

Plain straight steel fibres of 19 N/mm^2 yield strength were used. A concrete mix of (1:2.4:4) by weight with coarse aggregate of 20 mm max. nominal size and water cement ratio of 0.55 was used. Plain mild steel bars of 12 mm diameter were used for longitudinal reinforcement while 6 mm diameter plain bars were used as shear reinforcement.

For preparation of the mix manual mixing was adopted by adding fibres in small quantities continuously with necessary precautions to avoid balling effect. The specimens were cast in timber moulds and were compacted using a needle vibrator. Control cube and cylinder specimens were cast and tested to determine the compressive strength.

The beams and control specimen were demoulded after 24 hours and cured under identical conditions for 28 days. The beams were tested in a 20 tonne loading frame using hydraulic jack and load was measured by proving ring. Deflections were measured under the loads using dial gauges with least count of 0.02 mm and there from angles of twist were calculated. The crack widths were measured using a microscope.

The loading was applied in increments recording the dial gauge readings and noting the propogation of crank after each increment. The beams were tested upto failure. The physical characteristics of the beams tested are given in Table 1.

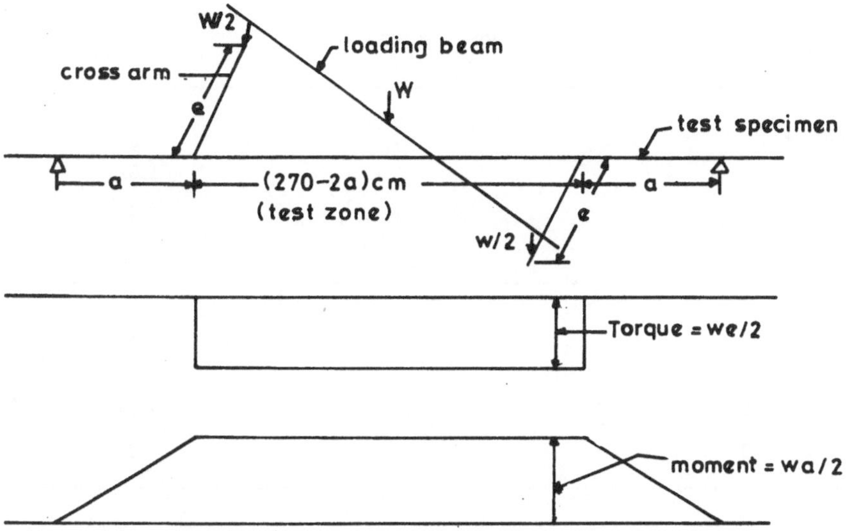

FIG:I. GENERAL LOADING ARRANGEMENT.

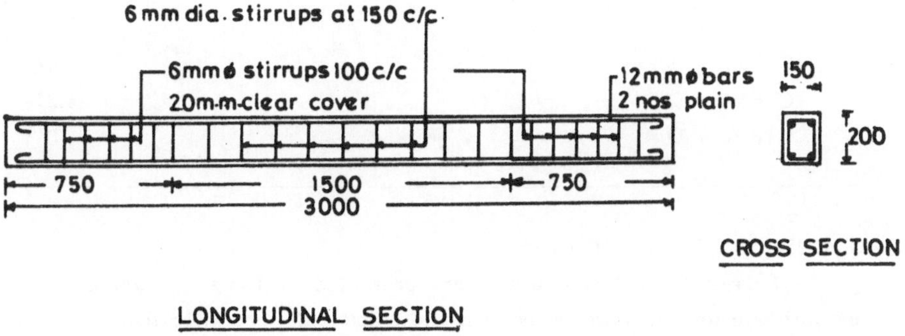

FIG :2. REINFORCEMENT DETAILS OF TEST BEAMS.

note: all dimensions are in m.m

TABLE (1) DETAILS OF BEAM SPECIMEN

Specimen	Fibre content V_f (%)	T/M ratio
CTRC - 2	0	2.5
CTRC - 15		1.67
CTRC - 16		1.25
CTRC - 1		1.0
CTFRC - 5	0.5	2.5
CTFRC - 6		1.67
CTFRC - 4		1.25
CTFRC - 3		1.0
CTFRC - 8	0.75	2.5
CTFRC - 7		1.67
CTFRC - 9		1.25
CTFRC - 10		1.0
CTFRC - 13	1.0	2.5
CTFRC - 14		1.67
CTFRC - 12		1.25
CTFRC - 11		1.0

DISCUSSION OF TEST RESULTS :

The results of tested beams are presented in Table 2. Where the cracking and ultimate torque, ultimate moment and crack width are indicated.

TABLE (2) SUMMARY OF TEST BEAM RESULTS :

Specimen	Fibre content (V_f%)	$\frac{T_{cr}}{\sigma_{cu} \, bd^2}$	$\frac{T_u}{\sigma_{cu} \, bd^2}$	$\frac{M_u}{\sigma_{cu} \, bd^2}$	Measured crack width (mm)
a) For T/M 2.5					
CTRC 2	0	13.47	27.45	10.98	2.00
CTFRC 5	0.5	22.97	29.73	11.89	1.60
CTFRC 8	0.75	30.82	42.68	17.07	2.00
CTFRC 13	1.00	23.95	53.11	21.25	0.90
b) For T/M 1.67					
CTRC 15	0	25.35	44.79	26.88	3.00
CTFRC 6	0.50	27.19	33.11	19.87	1.50
CTFRC 7	0.75	33.78	51.58	30.95	2.40
CTFRC 14	1.00	22.69	49.29	29.58	1.00
c) For T/M 1.25					
CTRC 16	0	25.35	41.37	33.09	2.50
CTFRC 4	0.5	20.15	30.48	24.39	2.50
CTFRC 9	0.75	32.27	54.07	43.26	0.80
CTFRC 12	1.00	19.62	43.39	34.71	0.60
d) For T/M 1.0					
CTRC 1	0	21.78	33.26	33.26	2.60
CTFRC 3	0.5	21.37	31.69	31.69	1.80
CTFRC 10	0.75	30.98	54.06	54.06	0.80
CTFRC 11	1.00	19.62	29.35	29.35	2.00

EFFECT OF FIBRE CONTENT ON TORQUE TWIST BEHAVIOUR :

The volume percent of fibres has a marked effect on the torque twist relationship for the various T/M ratios for which beams were tested as seen from Figures 3 and 4. Though increased ductility was expected with the inclusion of fibres, the

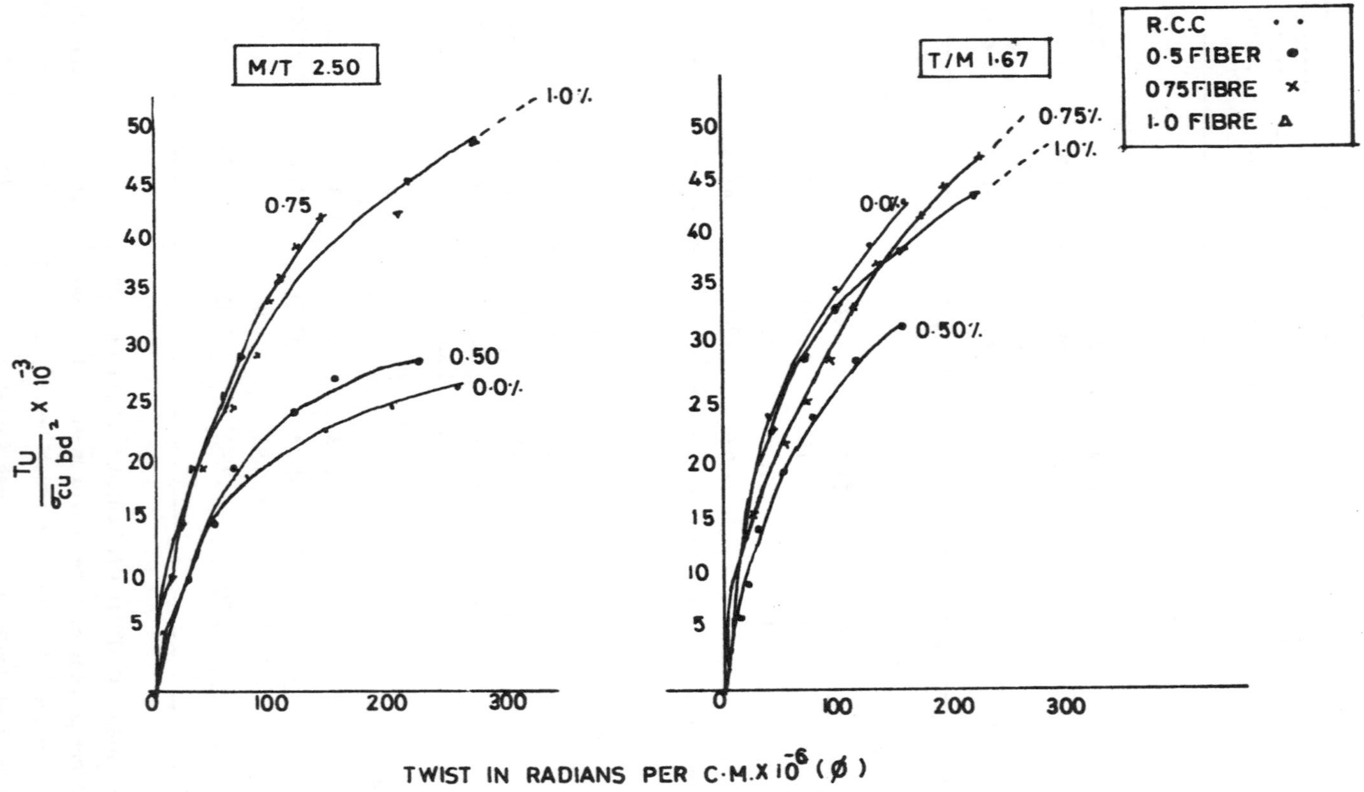

FIG: 3 EFFECT OF FIBRE CONTENT ON TORQUE TWIST RELATIONSHIP

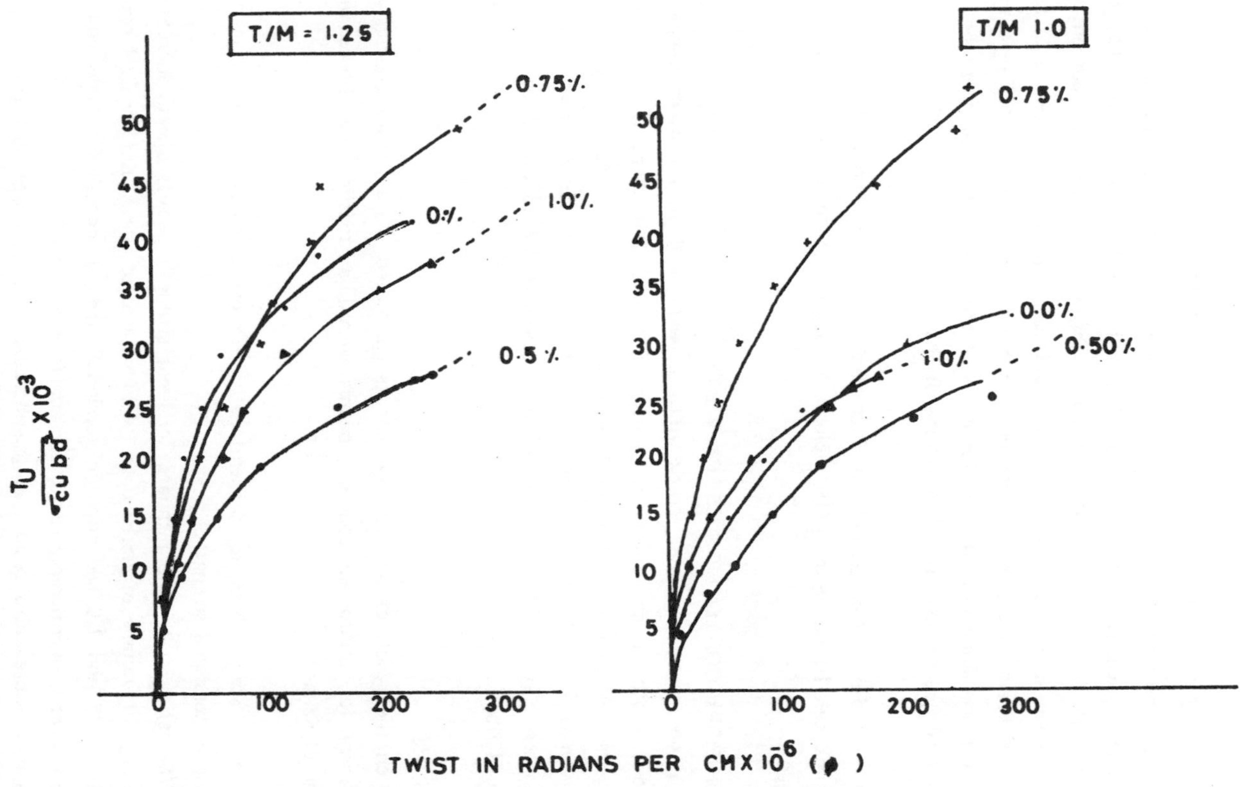

FIG: 4. EFFECT OF FIBRE CONTENT ON TORQUE TWIST RELATIONSHIP.

test results do not show any marked increase in ductility for beams under combined bending and torsion. The torque-twist relationship was obtained only upto ultimate load due to constraint of instrumentation.

EFFECT OF FIBRE CONTENT ON THE STRENGTH OF BEAMS:

The volume percent has a marked effect both on the torque at first crack and at ultimate stage, as seen from Table 2 and Figure 5. It is observed that for T/M ratio of 2.5 there is increase in ultimate torque with increasing percentage of fibres. Exactly similar behaviour is not found for T/M ratios of 1.67, 1.25 and 1.0. For these beams there was decrease in the ultimate torque at 0.5% volume of fibres, over the conventionally reinforced concrete beams. But for these beams the maximum torque occured at 0.75% volume of fibres.

Inclusion of fibres has resulted in increased first crack torque over conventionally reinforced beams upto 0.75% volume of fibre and then there is decrease in the torque.

EFFECT OF FIBRE CONTENT ON CRACK WIDTH:

There is decrease in the max. crack width of beams with fibre content compared to the crack width of conventionally reinforced beams. But there is no definite pattern in the decrease of crack width with increasing percentage of volume of fibres. This may be attributed to random orientation of the fibres. However, the decrease in the crack width in the presence of fibres is due to the crack arresting mechanism of fibres.

CONCLUSIONS

Based on the result of limited experiments conducted the following conclusions may be drawn for reinforced concrete beams with high tensile steel fibres with aspect ratio of 62.5.

1. Addition of steel fibres to conventionally reinforced concrete results in the increase of torsional strength.
2. For T/M ratio of 2.5 the ultimate torsional strength increases with increasing percentage of volume of fibres upto 1% of volume of fibres. But for T/M ratios of 1.67, 1.25 and 1.0 reduction in torsional strength occurs for 0.5% volume of fibres over conventionally reinforced concrete beams.
3. Inclusion of fibres also results in the increase of first crack strength of fibre reinforced concrete beams upto 0.75% volume of fibres.
4. The increase in ultimate torsional strength is of the order of 8% - 67% for 0.75% fibre content.

5. The increase in first crack strength of the beams is of the order of 27% - 128% for 0.75% fibre content.
6. More experimental work is needed to study the effect of volume percent on the torsional strength using wide range of volume percent of fibres.

ACKNOWLEDGEMENTS :

The paper is based on the M.Tech dissertation of the third author. The experimental work was conducted at the Structural Engineering Laboratory of the Civil Engineering Department of Jawaharlal Nehru Technological University College of Engineering, Anantapur. The authors are thankful to the University and College authorities for providing the facilities for the investigation.

REFERENCES :

1. Swamy, R.N. "Fibre reinforcement of cement and concrete", Materials and Structures, Research and testing V 8 No.45, May/June 1975 pp 235-254.
2. Abdul Wahab, H.M.S, Al-Ausi, M.A, Tawfiq, S.H. "Steel fibre reinforced concrete members under combined bending, shear and torsion" Developments in fibre reinforced cement and concrete, Rilem symposium (FRC86) Volume 2, July 1986.
3. Narayanan, R and Toorani, G.Z. "Fibre reinforced concrete in pure torsion and combined bending and torsion" proceedings of Institution of Civil Engg. London Part 2 Vol 67 Dec, 1979 pp 987-1001.
4. Mansur, M.A. and Paramasivam, P "Steel fibre reinforced concrete beams in pure torsion". The Int. Journal of cement composites and light weight concrete V=-4 No.1 Feb, 1982 pp 39-45.
5. Hafeez Khan, T.A. etal "An experimental study of fibre reinforced concrete beams under pure torsion, Indian Concrete Journal, Vol. No. 50 October, 1976 pp 314-317.
6. Sadananda Murthy, P "An experimental study of fibre reinforced concrete beams under pure torsion - M.Tech Thesis submitted to Jawaharlal Nehru Technological University, Hyderabad, India 1975.
7. Abdul Hafeez Khan, T, Shariff, M.F. and Yella Reddy, Y "An experimental study of Sisal fibre reinforced concrete beams under pure bending and torsion" Developments in fibre reinforced concrete, Rilem symposium (FRC86) Vol 2, July 1986.
8. Govinda Rao, V.V. "Fibrous reinforced concrete beams under combined bending and torsion - An experimental Investigation" M.Tech Thesis submitted to Jawaharlal Nehru Technological University, Hyderabad, India 1983.

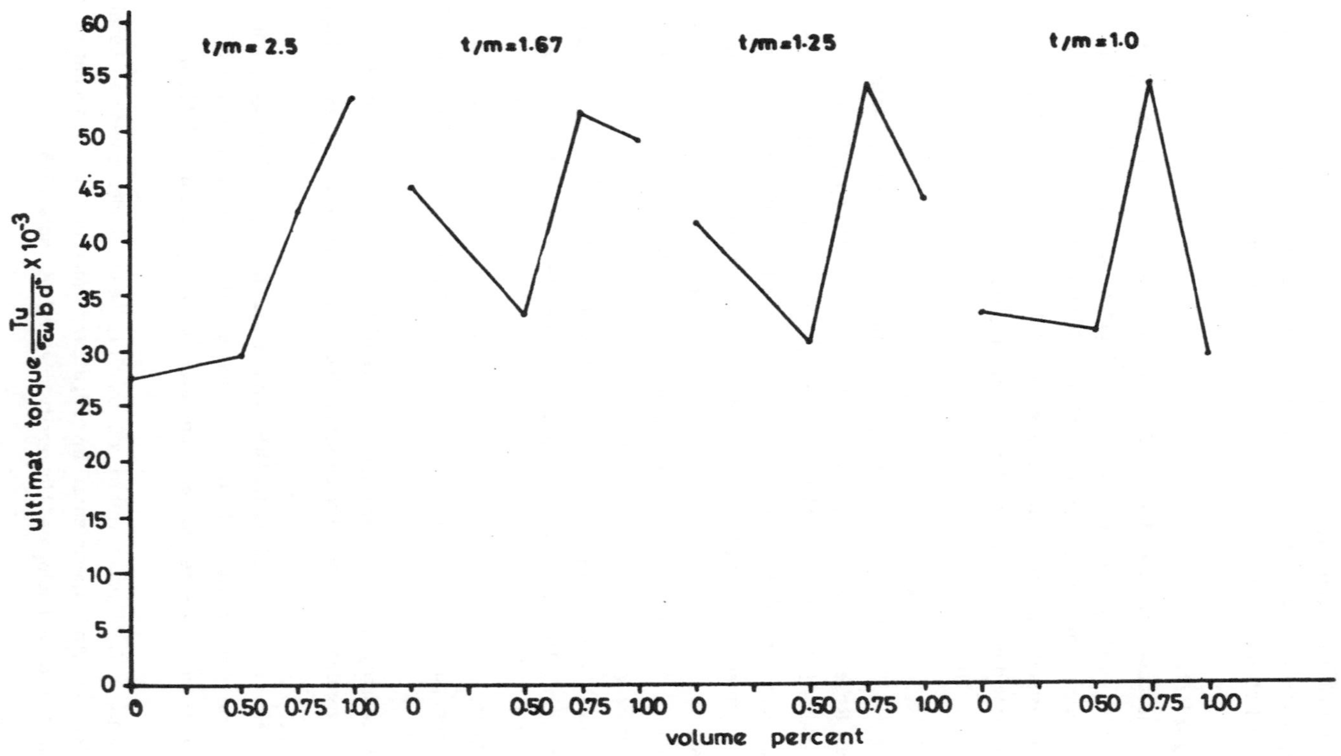

FIG 5. EFFECT OF VOLUME PERCENT ON ULTIMATE TORQUE.

NOTATIONS AND ABBREVIATIONS :

The following symbols are used in this paper

T	=	Torsional moment
M	=	Bending moment
V_f	=	Volume percent of fibres
T_{cr}	=	Torsional moment at first crack
T_u	=	Ultimate torsional moment
M_u	=	Ultimate bending moment
σ_{cu}	=	Cube compressive strength of concrete
b	=	Breadth of beam
d	=	Depth of beam
CTRC	=	Conventionally reinforced beams
CTFRC	=	Conventionally reinforced beams with fibres.

Proceedings of the International Symposium on Fibre Reinforced Concrete, December 16-19, 1987, Madras, India

STRAIN RATE EFFECTS ON THE COMPRESSIVE PROPERTIES OF FIBER REINFORCED CONCRETE

D. E. Otter and A. E. Naaman
University of Michigan, Ann Arbor, Michigan, USA

SYNOPSIS:

Compression test results on plain and fiber reinforced concrete cylinders subjected to three different strain rates are presented. High loading rates increased the strength and toughness of the composite but did not have a pronounced effect on the shape of the stress-strain curve. Differences in the fiber material (steel, glass, or polypropylene) or the type of fiber (straight or hooked) also had little effect within the range of strain rates used. On the average about a 10% increase in strength and 15% increase in toughness were obtained when the strain rate was increased from 100 μ/s to 30000 μ/s. The influence of fiber addition to plain concrete is found to be very comparable to that of confinement.

INTRODUCTION

There is increased need for information on the properties of fiber reinforced at various rates of loading; however, a wide range of loading rates can be considered depending on the type of application. For each range of strain rate (from creep to impact) a different testing technique may be needed. Strain rate (low impact) effects on the tensile, flexural and pull-out properties of fiber reinforced concrete have been described in previous publications [1-5].

Little information exists on the effect of strain rate on the compressive properties of fiber reinforced concrete. This study is aimed at developing useful information on the compressive stress-strain response of fiber reinforced concrete at rates of loading simulating seismic loads. Indeed one of the potential applications of fiber reinforced concrete is its use in the joints of reinforced concrete frames to improve their ductility, energy absorption, and hysteretic response. The overall study includes tests under monotonic loading as well as three different cyclic loading regimes. Some results on one of the cyclic loading regimes can be found in [6]. Additional information related to the cyclic tests will be available in future publications.

EXPERIMENTAL PROGRAM

The mix proportions of the various matrices used in this investigation are shown in Table 1. The physical properties of the fibers used are described in Table 2. The variables used in the experimental program are summarized in Table 3. It can be observed that for each series of tests, three strain rates of loading were considered. The loading rate of 100 μ/s (microstrain per second) was used to determine the compressive strength of concrete cylinders simulating static loads. Two additional loading rates used were 10000 μ/s and 30000 μ/s. This last strain rate was found to be well within the range of strain rates induced in concrete structures by earthquake forces. Additional variables included the type of concrete matrix, the volume fraction and aspect ratio of the fibers, the type of fiber (hooked or straight), and the fiber material (steel, glass, or polypropylene). In all 42 series of tests were conducted and for each series four specimens were tested. In this paper average results will be discussed.

Specimens consisted of four by eight inch (10 by 20 cm) cylinders. After casting, they were allowed to set for one day, then cured in water for seven days, removed, and kept in a laboratory environment for about one year before testing. They were capped with Cylcap, a high strength sulfur compound, to insure smooth loading surfaces at testing.

Testing was undertaken using a computer controlled servohydraulic universal testing machine (Instron System 8000) connected to a data acquisition system capable of recording up to 800 readings per second. The load was read directly from the load cell of the machine. To obtain a reliable deformation reading, the axial deformation of the specimen was measured by three linear variable differential transformers (LVDTs). They were placed in a horizontal circular arrangement at 120 degree angular spacing, and were positioned vertically between the two plattens of the testing machine (Fig. 1). Hence the deformations and the corresponding values of the strains recorded are the overall values along the length of the specimen. The signals from the three LVDTs were averaged through a signal conditioner before transfer to and storage in the data acquisition system. During the tests, load and deformation data were recorded and stored digitally. Later the data were sorted, analysed, and plotted as shown in the figures described below.

RESULTS

Typical stress-strain curves of plain and fiber reinforced concrete at the three strain rates considered are shown in Figs. 2a and 2b. Two observations can be made: First, the presence of 2% steel fibers by volume of mortar leads to a significant increase in the compressive strength at any strain rate. This effect is more pronounced for hooked than for straight fibers. Second, an increase in strain rate from 100 to 30000 μ/s leads to only about a 10% increase in compressive strength for either plain concrete or fiber reinforced concrete. The first observation differs slightly from previous findings which stated that the compressive strength of concrete is not influenced much by the presence of fibers. However, it is believed that the efficiency of fiber reinforcement may be higher for low strength concrete and also increase with the age of the concrete. Furthermore, fiber reinforcement may also be more effective for a concrete with a much larger maximum aggregate size, as was used in these tests, because the larger aggregates result in increased roughness of fracture surfaces and increased mechanical interlock. The second observation agrees well with previous work on plain and confined concrete.

In Figs. 3a and 3b the influence of aspect ratio on the stress-strain curve in compression is illustrated for two different strain rates. No correlation could be established between aspect ratio and strain rate. This is perhaps because steel fibers pulled out at all strain rates and the pull-out process, which is a friction process, is not rate sensitive.

Figure 4a describes the variation of normalized compressive strength, termed dynamic strength ratio, with the strain rate of loading; the ratio shown is the ratio of average compressive strength at a given strain rate to that obtained at the rate of 100 μ/s for a given series of tests. All combinations of steel fibers used in Mix 1 (see Table 3) are shown in Fig. 4. A large scatter is observed but the trend is somewhat consistent with previous findings on fiber reinforced and confined concrete [1, 7-12]. A similar conclusion can be derived from plotting the dynamic toughness ratio versus strain rate (Fig. 4b). The dynamic toughness ratio was computed from the areas under the stress-strain curve up to a strain of 0.020. Although the scatter is substantial, it can be said that an average 15% increase in compressive toughness can be expected when the strain rate increases from 100 to 30000 μ/s.

The effect of strain rate on the compressive strength of fiber reinforced concrete reinforced with different fiber materials is illustrated in Fig. 5. No consistent trend could be derived except that strain rate seems to affect fiber reinforced concrete less than plain concrete. Differences between fiber types could not be clearly isolated since large scatters were observed.

Three prediction equations proposed in [8, 10, 12] were used to predict the increase in compressive strength with strain rate. They are summarized in Table 4. Comparison of predicted values with the experimental results obtained in this study (Fig. 5) suggests that the equation developed by Tadros [8] presents a lower bound to the data while the equation proposed by Dilger [12] offers an upper bound. It is noted that the equation proposed by Tadros was developed for strain rates below 100 μ/s. A modified equation that takes into consideration the influence of the compressive strength of concrete was developed by Ahmad [10]. In order to plot it, a value of compressive strength equal to 41.4 MPa (6 ksi) was selected as it represents the range of strengths observed in the specimens tested. The corresponding curve fits somewhat in between the two above mentioned bounds and seems to be closer to the data (Fig.5a). It was not clear from the data, however, if the compressive strength was also a significant factor in strain rate effects.

Using a similar approach, average normalized toughness results illustrating the influence of type of fiber are shown in Fig. 5b.

CONCLUSIONS

1. Considerable scatter in the data can be expected when studying the properties of fiber reinforced concrete at different strain rates. The scatter in the data makes it extremely difficult to isolate the separate effects due to fiber type, volume fraction, or aspect ratio. The scatter obtained in the toughness data was larger than that obtained in strength data.

2. Increasing the strain rate of loading affects the compressive properties of fiber reinforced concrete in a similar manner but possibly to a lesser extent than it does plain concrete. A similar conclusion was reached in [12] when comparing plain and confined concrete. In this study, increases of about 10% in strength and 15% in toughness were observed when the strain rate of loading was increased from 100 μ/s to 30000 μ/s.

3. The strain rate of loading does not seem to affect the shape of the stress-strain curve of fiber reinforced concrete in compression.

4. On the average, specimens reinforced with glass or polypropylene fibers were more sensitive to the rate of loading than specimens reinforced with steel fibers; however, the differences between fiber types are hard to delimit due to large variations.

REFERENCES

1. Naaman, A. E., "Fiber Reinforced Concrete Under Dynamic Loading," ACI Special Publication, SP-81, American Concrete Institute, Detroit, 1984, pp. 169-186.

2. Naaman, A. E., and Gopalaratnum, V. S., "Impact Properties of Steel Fiber Reinforced Concrete," *The International Journal of Cement Composites and Lightweight Concrete*, Vol. 5, No. 4, Nov. 1983, pp. 225-233.

3. Gokoz, U., and Naaman, A. E., "Effect of Strain Rate on the Pull-Out Behavior of Fibers in Mortar," *The International Journal of Cement Composites*, Vol. 3, No. 3, Aug. 1981, pp. 187-202.

4. Nammur, G. G., and Naaman, A. E., "Strain Rate Effects on Tensile Properties of Fiber Reinforced Mortar," Proceedings of MRS Symposium on Cement Based Composites: Strain Rate Effects on Fracture," S. Mindess, Editor, Materials Research Society, Vol. 64, Pittsburgh, 1986, pp. 97-118.

5. Gopalaratnum, V. S., and Shah, S. P., "Properties of Steel Fiber Reinforced Concrete Subjected to Impact Loading," *ACI Journal*, Vol. 83, No. 1, 1986, pp. 117-126.

6. Otter, D., and Naaman, A. E., "Steel Fiber Reinforced Concrete Under Static and Cyclic Compressive Loading, Proceedings of Third International Conference on Fiber Reinforced Cement and Concrete, Edited by R. N. Swamy, R. L. Wagstaffe, and D. R. Oakley, Scheffield, England, July 1986.

7. Watstein, D., "Effect of Straining Rate on the Compressive Strength and Elastic Properties of Concrete," *ACI Journal*, Proceedings, Vol. 49, No. 8, April 1953, pp. 729-744.

8. Tadros, G. S., "Plastic Rotation of Reinforced Concrete Members Subjected to Bending, Axial Load and Shear," Ph.D. Thesis, University of Calgary, 1970, 292pp.

9. Bresler, B., and Bertero, V. V., "Influence of High Strain Rate and Cyclic Loading on the Behavior of Unconfined and Confined Concrete in Compression," Proceedings, Second Canadian Conference on Earthquake Engineering, MacMaster University, Hamilton, Ontario, Canada, June 1975.

10. Ahmad, S., "Properties of Confined Concrete Subjected to Static and Dynamic Loads," Ph.D. Thesis, University of Illinois at Chicago Circle, 1981, 342 pp.

11. Hughes, B. P., and Gregory, R., "Concrete Subjected to High Rates of Loading in Compression," *Magazine of Concrete Research*, Vol. 24, No. 78, March 1982.

12. Dilger, W. H., Koch, R., and Kowalczyk, "Ductility of Plain and Confined Concrete Under Different Strain Rates," *ACI Journal*, Proceedings, Vol. 81, No.1, January-February 1984, pp. 73-81.

Table 1: Concrete Mix Proportions

	Water	Cement	Fine Aggregate	Coarse Aggregate
Mortar	0.46	1.00	2.04	-----
Mix 1	0.47	1.00	2.56	2.14
Mix 2	0.38	1.00	2.04	2.41

Cement: ASTM Type I

Coarse Aggregate: 19 mm maximum size crushed limestone.

Proportions by weight are based on saturated surface dry aggregates.

Fiber content is based on the volume of mortar in a mix.

A superplasticizer, Melment, was used as necessary for workability.

Table 2: Fibers Used

Straight Steel, $l/\phi = 45$:	Mitchell Fibercon, 0.25mm x 0.56mm x 19mm
Straight Steel, $l/\phi = 60$:	Mitchell Fibercon, 0.25mm x 0.56mm x 25mm
Straight Steel, $l/\phi = 80$:	Baekert, 0.25mm diam x 20mm
Hooked Steel, $l/\phi = 60$:	Baekert Dramix, 0.50mm diam x 30mm
Polypropylene, $l/\phi = 200$:	Fibermesh, 19mm monofilament
Glass, $l/\phi = 68^*$:	Nippon Electric Glass, Alkali Resistant, 13mm, 200 filaments

*Equivalent aspect ratio

Table 3: Test Parameters Used for Each of Three Strain Rates (100, 10000, 30000 microstrain per second).

Concrete Mix	Fiber Type	Volume Fraction	Aspect Ratio
Normal Strength (Mix 1)	Control	0%	---
	Hooked Steel	1%	60
		2%	60
		3%	60
	Straight Steel	2%	45
		2%	60
		2%	80
	Polypropylene	1%	200
		2%	200
	Glass	1%	68*
High Strength (Mix 2)	Control	0%	---
	Hooked Steel	2%	60
Mortar	Control	0%	---
	Hooked Steel	2%	60

Note: For each combination of test parameters and each strain rate four specimens were tested.

*Equivalent aspect ratio

Table 4: Equations Predicting Strain Rate Effect.

For all equations, $f_{c\,16}'$ refers to the compressive strength at a strain rate of 16 microstrain per second, and ε is in units of 1/second.

Tadros: $f_c'/f_{c\,16}' = 1.14 + 0.03 \log \varepsilon$

Dilger: $f_c'/f_{c\,16}' = 1.38 + 0.08 \log \varepsilon$

Ahmad: $f_c'/f_{c\,16}' = [0.95 + 0.27(\log \varepsilon + 6)/f_{c\,16}']^{\alpha}$

where

$f_{c\,16}'$ is in units of ksi (1ksi = 6.9 MPa)

$\alpha = 0.85 + 0.9\,d - 0.02\,h$

h = specimen height in inches (1 inch = 25.4 mm)

d = specimen diameter in inches

Fig. 1 : Testing Setup

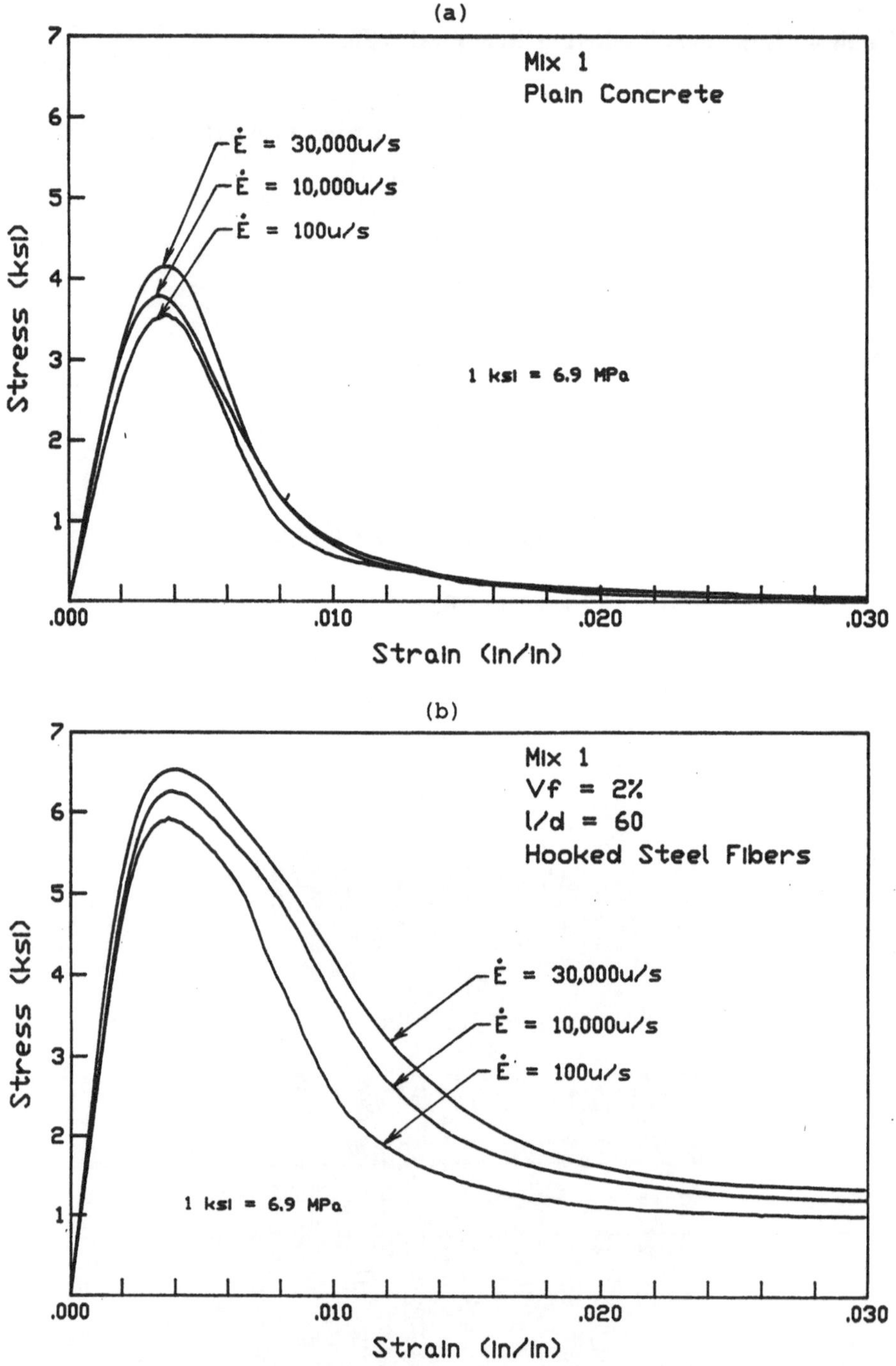

Fig. 2 : Typical Effect of Strain Rate on the Stress-Strain Response in Compression for Plain Concrete and Fiber Reinforced Concrete.

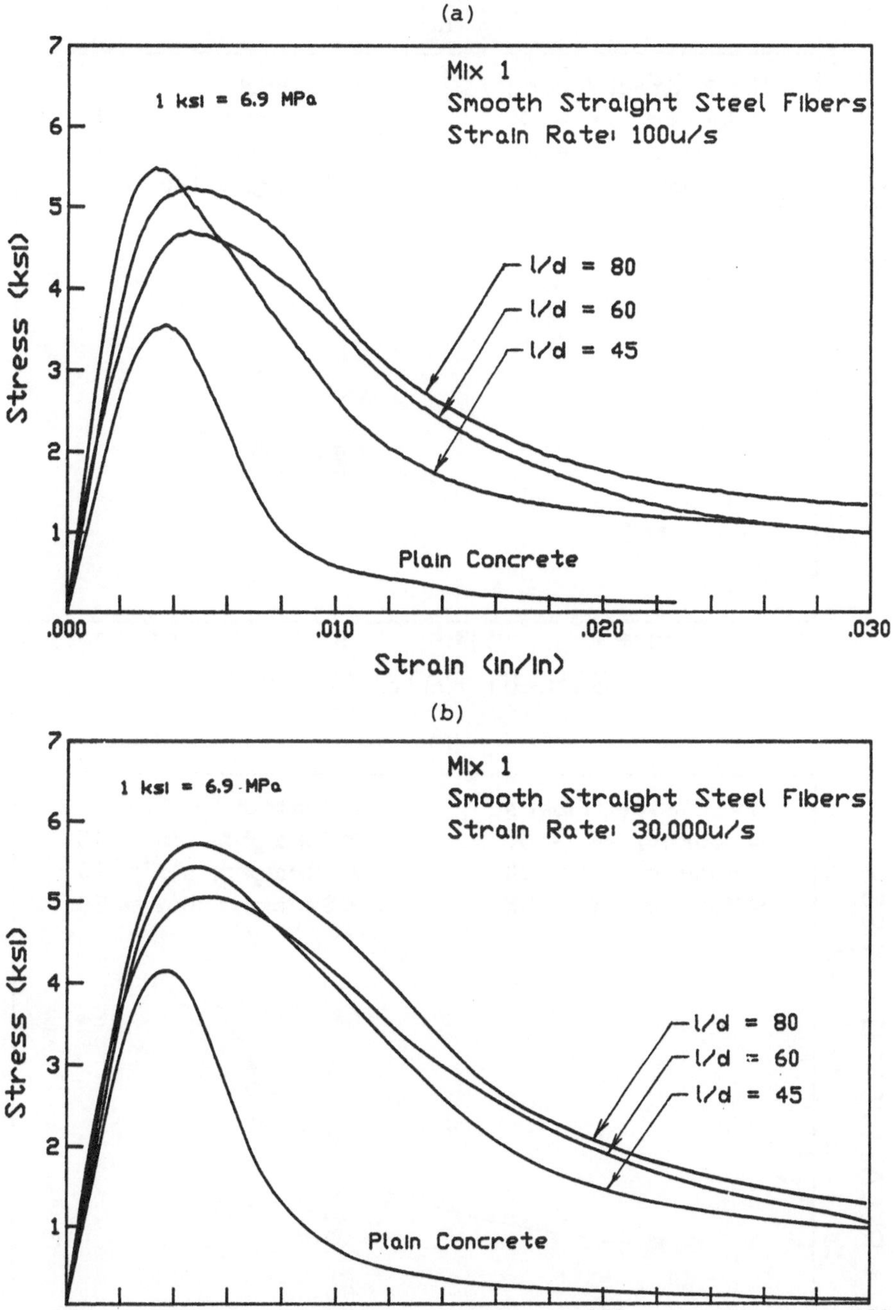

Fig. 3 : Typical Effect of the Aspect Ratio of Fibers on the Stress-Strain Response in Compression at Two Different Strain Rates.

(a)

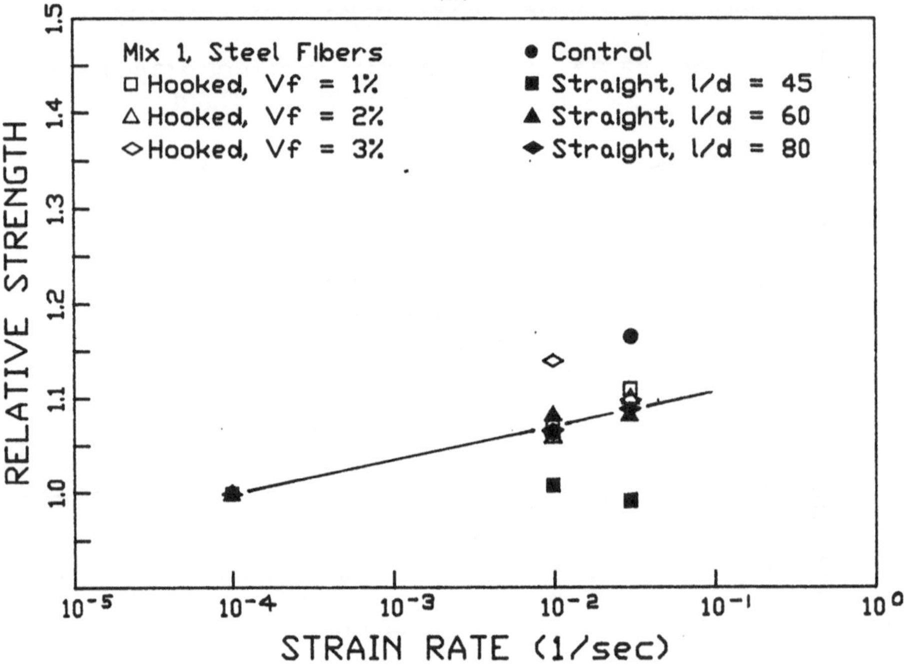

(b)

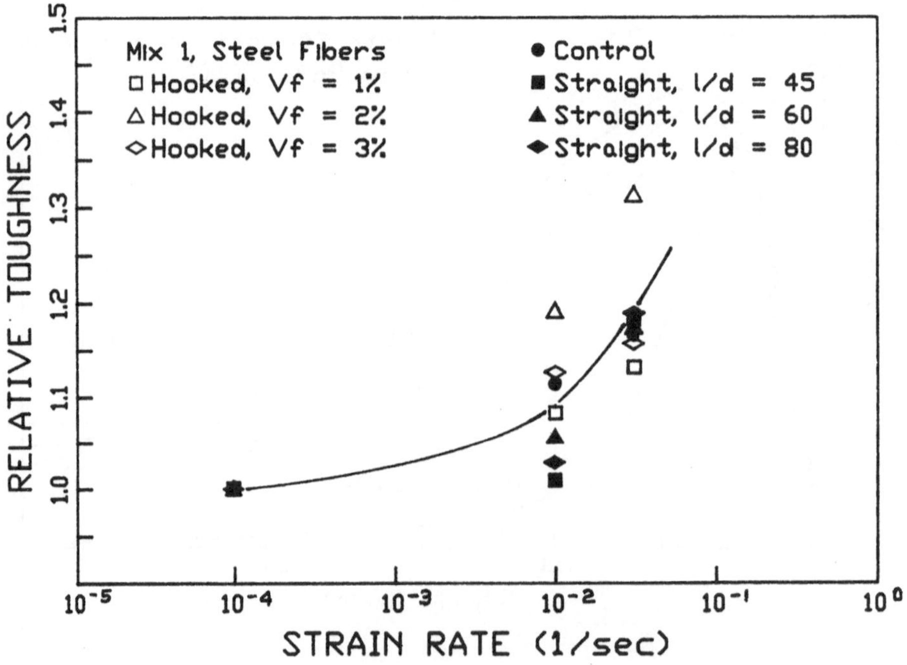

Fig. 4 : Relative Increase in Strength and Toughness With Increasing Strain Rate for All Series of Tests Involving Steel Fibers.

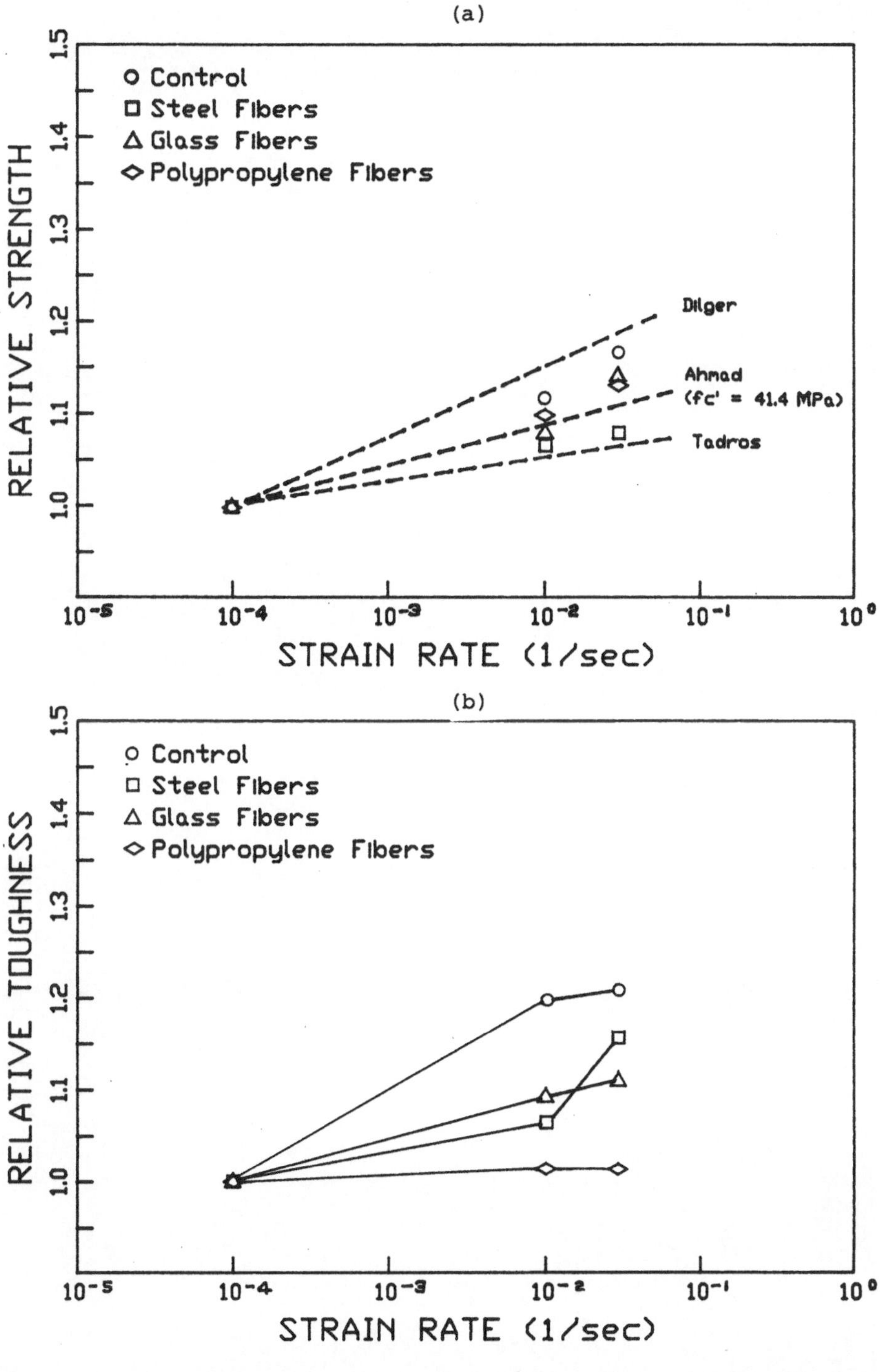

Fig. 5 : Comparison of the Relative Increase a) in Strength, and b) in Toughness with Strain Rate for Tests Series Containing Steel, Glass, and Polypropylene Fibers.

Proceedings of the International Symposium on Fibre Reinforced Concrete, December 16-19, 1987, Madras, India

ULTIMATE STRENGTH OF FIBRE REINFORCED CONCRETE SLABS

Dr. C.B. Kukreja,
Professor in Civil Engineering,
Thapar Institute of Engg.&Tech;
(Deemed University), Patiala.

Dr. M.B. Kanchi,
Bechtel Energy Corporation,
Housten, Texas, U.S.A.

Dr. S.K. Kaushik,
Professor in Civil Engineering,
University of Roorkee, Roorkee.

Dr. O.P. Jain,
Consulting Structural Engineer,
Munirka Enclave, New Delhi.

SYNOPSIS

Superior crack resistance, greater ductility and distinct post cracking behaviour are some of the important static properties of fibre reinforced concrete. Fibrous concrete will be advantageous in designing thinner sections, particularly in water tanks, pipes and concrete pavements where tensile strength and cracking are the criteria for design.

In the present study, two way reinforced fibre concrete slabs have been tested for various volume percentages of fibres and with an optimum aspect ratio of 80. It has been observed that the fibrous slabs exhibit increased first cracking and ultimate strengths. The deflictions and crack widths at serviceability limit state get reduced. The deflections at ultimate state increase while the crack widths decrease.

INTRODUCTION:

Plain concrete has a very little tensile strength and very small resistance to cracking. Internal microcracks are inherently present in the concrete. These cracks propagate on the application of load, which leads to the brittle fracture of the concrete matrix.

To improve the tensile strength of concrete, conventional flexural steel and prestressing have been used. But these methods do not increase the inherent tensile strength of the concrete. It has been recognised that the addition of small fibres to the concrete matrix improves its basic properties. The fibres inhibit crack formation exerting pinching forces at the tips of the cracks, thus acting as crack arresters.

Fibrous concrete is specially useful in hot climates where concrete shrinkage causes such cracks. Another application of crack arresting is adding steel fibres to the conventionally reinforced liquids storage structures, which are designed to sustain the hoop tension in the concrete shell. Fibres have also been used for shear reinforcement in frames, beams and flat slabs.

HISTORICAL BACKGROUND:

There is very little literature available on the study of reinforced fibre concrete slabs. An experimental investigation on the car park at London (Heathrow) Airport was made by Swamy and Kent (1). They conducted tests on 1.1m x 1.1m square slabs of 50 mm and 63.5mm thickness with fibre contents varying from 1 to 4 percent by weight. Control slabs without fibres were also tested for comparison purposes. All the slabs were subjected to point loads transmitted through an area 304.8mm square. The slabs were tested at the age of 28 days. Their results show that 63.5mm thick slabs with 3 percent fibre content by weight used in the car park have a reserve of strength in all cases and adequate safety factor for the design load. In fact the data showed that a 50mm thick slab with three percent fibre content might well have sufficed to carry the design load with a load factor of 2.0. In spite of the inherent variability involved in the mannual dispersion and mixing of fibres, the results showed a satisfactory degree of consistency although large variations occur between slabs of low fibre content.

In December,1977, Dinesh Mohan and Rehsi (2) reported that the strength and structural performance of cellular concrete units and doubly curved concrete tiles developed by central Building Research Institute, Roorkee (India) is improved by incorporating steel fibres having an aspect ratio of 75 and to an amount of 0.5 per cent by volume of concrete. The results showed that the steel fibre reinforced units developed approximately 20% higher strength at 28 days as compared to corresponding unreinforced units.

Rehsi (3) concluded that steel fibre reinforced cement flat sheets of size 1800x900 x6mm suitable for use as wall claddings, partitions etc. can be produced using 5% steel fibre (by weight of cement) of diameter 0.5mm and aspect ratio 74. A cement sand mix of 1.05:1 or 1:1 proportion by weight can be used in place of neat cement paste.

Ghalib (4) presented an analytical approach based on experimental data to determine the ultimate moment capacity and the cracking moment of fibre reinforced concrete slabs. His analytical prediction matched the experimental data for both the one way and two day slabs.

Pawan Kumar (6) carried out an experimental investigation on one way and two day fibre reinforced concrete slabs. The author concluded that the first crack and ultimate strength increases with the increase of fibre content. A higher ultimate load ratio of 1.87 was achieved with addition of 1.0% of fibres by volume and having aspect ratio of 75.

EXPERIMENTAL PROGRAMME:

Black annealed round mild steel fibres having average diameter equal to 0.46 mm were used. The fibres were obtained by cutting the wire on a hand operated shearing machine in length 36.8mm and having aspect ratio 80. The fibre wire was tested for its yielo and ultimate limits and these values and other properties are listed in table(1).

The details of the reinforcement provided in the two was slabs is shown in fig.(1), while the physical properties of the slabs are shown Table (2). The concrete was compacted in the slabs manually and by using needle vibrator. Three numbers control cube specimens for each series were cast along with slabs.

TESTING PROCEDURE:

The slabs were simply supported on all the four edges and corners were not held down. At the centre of an I section frame a 20mm rod was welded and the slab was placed on this frame for testing. To simulate a uniformly distributed load, the load was applied at 16 points with the help of a 20 tons capacity hydraulic jack operated by a test cylinder plant. The slabs were tested in a 300 tons test frame. The loading arrangement is shown in figure (2). The effective span of the slab was 900 x 1190mm with bearing of 37.5mm on each edge. The load was applied in 15 increments at intervals of about 10 minutes. The tensile side of the slabs was white washed for better crack observations. Deflections were measured with dial gauges located at the centre of the slab and at quarter points on both the centre lines of the slab. The crack widths were measured with a crack measuring instrument and the cracks were marked after each load increment. The control cubes were also tested on the corresponding day and their average values are listed in column 9 of table 2.

TEST RESULTS AND DISCUSSION:

Four two way slabs were cast, one without fibres and three with fibres containing fibres, in 0.5, 1.0 and 1.5 volume percentage, and having an aspect ratio 80. Table 3 and 4 contain the summary of test loads, deflections and crack widths.

(a) LOAD CARRYING CAPACITY:

Columns 4 and 5 of table 3 indicate that the first cracking strength of reinforced fibre slabs increase as the volume percentage of fibres increases. The maximum increase Of 35 per cent over the slab without fibres was obtained by incorporating fibres in 1.5 volumetric per-centage.

The ultimate strength of the reinforced fibre slabs also increased by adding fibres. The maximum increase of 80.92 percent was obtained by using 1.0 percent fibres. The load at which the slab without fibres failed is about 44 per cent more as compared to the theoretical load of 77200 N for the same slab. The theoretical load has been calculated from yield line theory which does not take into account the development of the compressive membrane forces(7). These membrane forces increase the loading capacity of the slabs beyond the johensen's yield load (9). First crack and ultimate load ratio versus volume percentage of fibres has been plotted in fig.(7).

(b) DEFLECTION:

Columns 4 and 5 of table 4 show that the deflections at serviceability state i.e. at design loads in fibrous slabs are lesser as compared to the slab (SA) without fibres. This deflection goes on decreasing with the increase in the fibre content.

The deflections at failure in reinforced fibre concrete slabs vary from 1 1/2 to 2 times of the deflection in slab without fibres. This shows that the slabs with fibres will take sufficient deflection before complete failure.

The load deflection curves have been plotted in figures 3 to 6. Three locations were selected for measurement of deflections i.e. central point, quarter point along short span & quarter point along long span. All the curves are almost linear upto the first cracking load and thereafter take the shape of a regular curve. The plots of limit state and maximum central deflections against volume percentage of fibres are shown in fig. (8).

(c) CRACK WIDTH:

The crack with at design loads in all the slabs containing fibre is lesser as compared to slab without fibres The reduction in crack width is directly related to increase in fibre content. The maximum decrease of 36 percent at limit state of cracking was obtained in slab SB, having fibres with volumetric content of 1.5 percent.

The maximum crack width at failure also decreased with the increase in fibre content. The maximum reduction of 30 percent was again obtained by incorporating 1.5 percent of fibres. The limit state and maximum crack widths versus volume percentage of fibres are plotted in fig. (9).

(d) MODE OF FAILURE AND CRACK PATTERN:

The crack pattern developed in all the four slabs was of the orthogonal type and agreed very well with the predicted crack pattern. Yield lines connecting the corners and the centre of the slab started developing at about 70% of the ultimate load. The corner effects were prominently seen and marked on the slabs. The fracture surfaces after failure for the four slabs are shown in photographs 1 and 2. These photographs emphasize several important properties of reinforced fibre concrete. Firstly the presence of fibres reduces the crack width at failure compared to slab without fibres. Secondly slabs with fibres showed an increasing degree of ductility at failure with increasing fibre content, while the slab without fibre showed a brittle fracture. The presence of fibres thus ensures a distinct post cracking stage and ductile behaviour at failure. Thirdly the failure modes show a more well defined failure pattern similar to the formation of yield lines. The number of cracks increased with increase in fibre content as is evident from photographs 2 and 3. The cracks were well distributed over the surface of the slab.

CONCLUSIONS:

From the experimental investigations, the following conclusions may be drawn which are valid within the range of parameters i.e. fibres having aspect ratio 80 and volume percentage 1.5.

(1) The first cracking strength of reinforced fibre concrete slabs increases with the increase in fibre content. The maximum increase of 35 percent is obtained. The ultimate strength also increases and the maximum increase of 80.92 percent is obtained.

(2) The deflections at serviceability limit state in reinforced fibre concrete slabs are about 21 percent lesser with fibres having aspect ratio 80, and volumetric percentage of 1.5. The deflections at failure loads in reinforced fibre concrete slabs are nearly 2 times of deflections in slabs without fibres.

(3) The crack widths at limit state of cracking in reinforced fibre concrete slabs get reduced. The maximum reduction of 36 percent is obtained. The crack widths at failure also get reduced. Hence lesser cover to conventional reinforcement can be recommended for reinforced fibre concrete slabs.

(4) Much thinner sections can be used in exposed conditions than would otherwise be possible, thus greatly reducing the self weight of the slab. Improved post-cracking behaviour and greater ductility will enable higher load carrying capacity for reinforced fibre concrete slabs.

ACKNOWLEDGEMENTS:

The present report is a part of doctoral thesis of the first author, submitted to the University of Roorkee, Roorkee (India). The authors are indebted to the University authorities for providing the laboratory facilities. The authors are grateful to the Ministry of Human Resources and Development, Government of India, New Delhi for the financial support extended to the authors for the above investigation.

REFERENCES:

1. Swamy R.N. and Brian Kent, "Some Practical Structural Applications of Steel Fibre Reinforced Concrete". American Concrete Institute Special Publication 44-18, 1974, pp. 319-336.

2. Dinesh Mohan and Rehsi, S.S., "Steel Fibre Reinforced Concrete Roofing Units". Indian Concrete Journal, December,1977, pp.370.

3. Rehsi S.S., "Steel Fibre Reinforced Cement Flat Sheets Indian Concrete Journal. V.53, No.9, September, 1979, pp. 247-248.

4. Mudhafar A.Ghalib, "Moment Capacity of Steel Fibre Reinforced Concrete Small Concrete Slabs". A C I Journal, July-August, 1980.

5. Kukreja, C.B., "Structural Characteristics of steel Fibre Reinforced Concrete", Ph.D.Thesis, University of Roorkee, Roorkee, 1981.

6. Pawan Kumar, "Limit Strength of Fibre, Reinforced Concrete Slabs". M.E.Thesis, Punjabi University, Patiala, December, 1981.

7. Kemp, K.O., "Yield of a Square Reinforced Concrete Slab on Simple Supports, Allowing for Membrane Forces". The Structural Engineer (London), Vol.45, No.7, June, 1967. pp. 235-240.

8. I.S: 456-1978, "Indian Standard Code of Practice for Plain & reinforced concrete". Indian Standard Institution, New Delhi.

9. Johensen, K.W., "Yield Line Theory", English Translation. Cement and Concrete Association, London, 1962.

10. A C I : 318-1983 - Building Code Requirements for Reinforced Concrete. American Concrete Institute U.S.A.

11. CP: 110-1972, "The Structural Use of Concrete. British Standard Institution, London.

TABLE - 1

Salient Properties of Fibre Steel

Type	Standard Wire Gauge	Average diameter (mm)	Young's Modulus of Elasticity (N/mm^2)	Yield Strength (N/mm^2)	Ultimate strength (N/mm^2)	Ultimate Bond Stress between fibre wire and concrete (N/mm^2)
1	2	3	4	5	6	7
Mild Steel Black Annealed Wire	26	0.46	$2.1x10^6$	313.5	410	1.525

TABLE - 2

Physical properties of Two Way Slabs

Sr.No.	Slab Series	Slab Type	Total Thickness (mm)	Clear Cover (mm)	Effective Depth or thickness (mm)	Reinforcement ($\phi_1 x \phi_2$)	Reinforcement Spacing ($S_1 x S_2$)	Average Control Cube Strength (N/mm^2)
1	2	3	4	5	6	7	8	9
1.	SA	Reinforced Concrete	50	7.5	39.5	6mmx6mm	10cmx15cm	20.66
2.	SB	RFC-80-1.5*	50	7.5	39.5	6mmx6mm	10cmx15cm	26.04
3.	SC	RFC-80-1.0	50	7.5	39.5	6mmx6mm	10cmx15cm	25.06
4.	SD	RFC-80-0.5	50	7.5	39.5	6mmx6mm	10cmx15cm	22.95

* RFC-80-1.5 stands for Reinforced Fibre Concrete Slab with fibres having aspect ratio equal to 80 and Volume percentage 1.5.

TABLE - 3

CRACKING AND ULTIMATE LOADS

Sr.No	Slab Series	Slab Type	Observed first crack load (Newtons)	Observed first crack load ratio	Observed Ultimate load (Newtons)	Observed Ultimate Load ratio	Predicted failure mode	Observed failure mode
1	2	3	4	5	6	7	8	9
1.	SA	Reinforced Concrete	34220	1.00	111220*	1.0	Orthogonal Crack Pattern	Orthogonal** Crack Pattern
2.	SB	RFC-80-1.5	44220	1.29	201220	1.81	-do-	-do-
3.	SC	RFC-80-1.0	44220	1.29	201220	1.81	-do-	-do-
4.	SD	RFC-80-0.5	36220	1.06	149220	1.34	-do-	-do-

* Against a theoretical ultimate load of 77200 N based on yield line theory

**This slab failed near one of the edges along long span.

TABLE - 4

DEFLECTIONS AND CRACK WIDTHS

Permissible deflection for reinforced concrete slab = 4.76 mm
Maximum theoretical crack width for reinforced concrete slab = 0.32 mm
Maximum permissible crack width for reinforced concrete slab = 0.30 mm

Sr.No.	Slab Series	Slab Type	Central Deflection at design load (mm)	Central deflection ratio at design load	Maximum Central deflection (mm)	Maximum Central deflection ratio	Crack Width at Design Load (mm)	Design Load Crack Width Ratio	Maximum Crack Width (mm)	Maximum Crack Width ratio
1	2	3	4	5	6	7	8	9	10	11
1.	SA	Reinforced Concrete	7.33	1.00	9.640	1.0	0.14	1.00	0.60	1.00
2.	SB	RFC-80-1.5	5.81	0.79	18.32	1.90	0.09	0.64	0.42	0.70
3.	SC	RFC-80-1.0	6.33	0.86	19.47	2.02	0.10	0.71	0.46	0.76
4.	SD	RFC-80-0.5	6.51	0.89	13.08	1.35	0.12	0.86	0.52	0.86

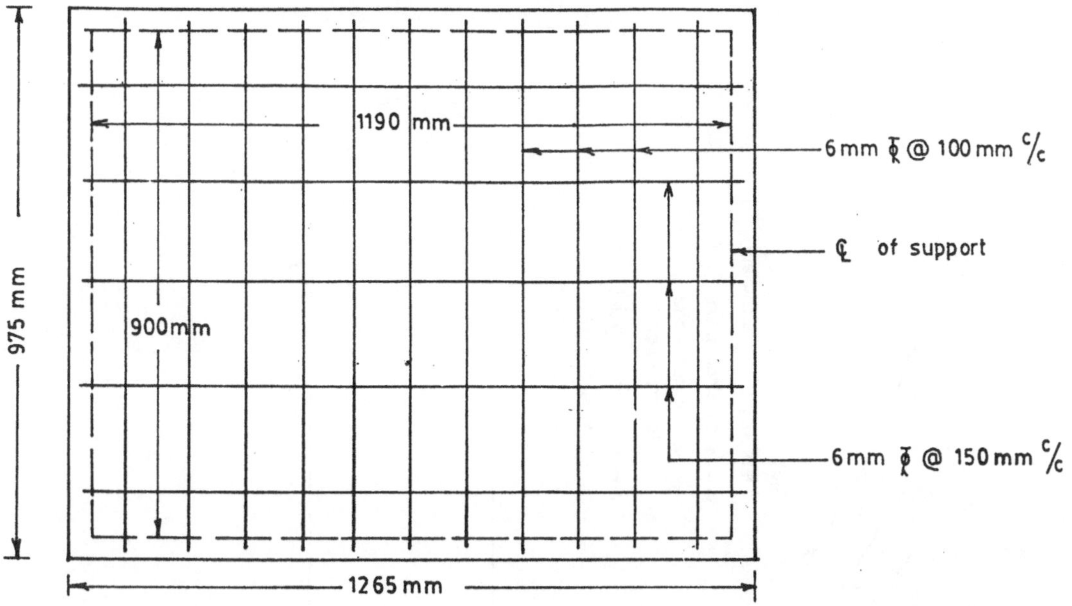

Fig. 1. Plan of bottom reinforcement in slabs.(Ref. table - 2)

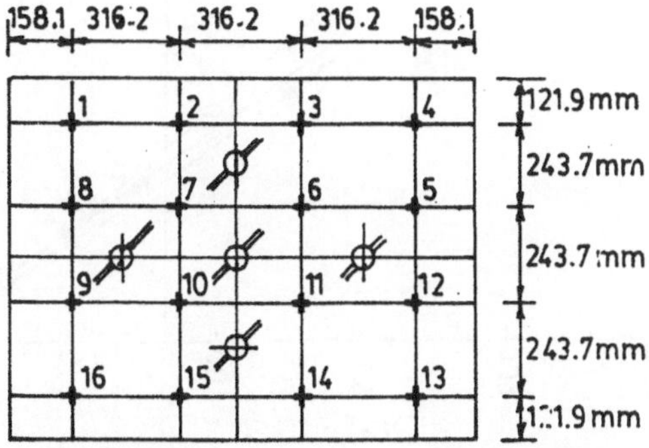

Fig. 2. Location of loading points and dial gauges.

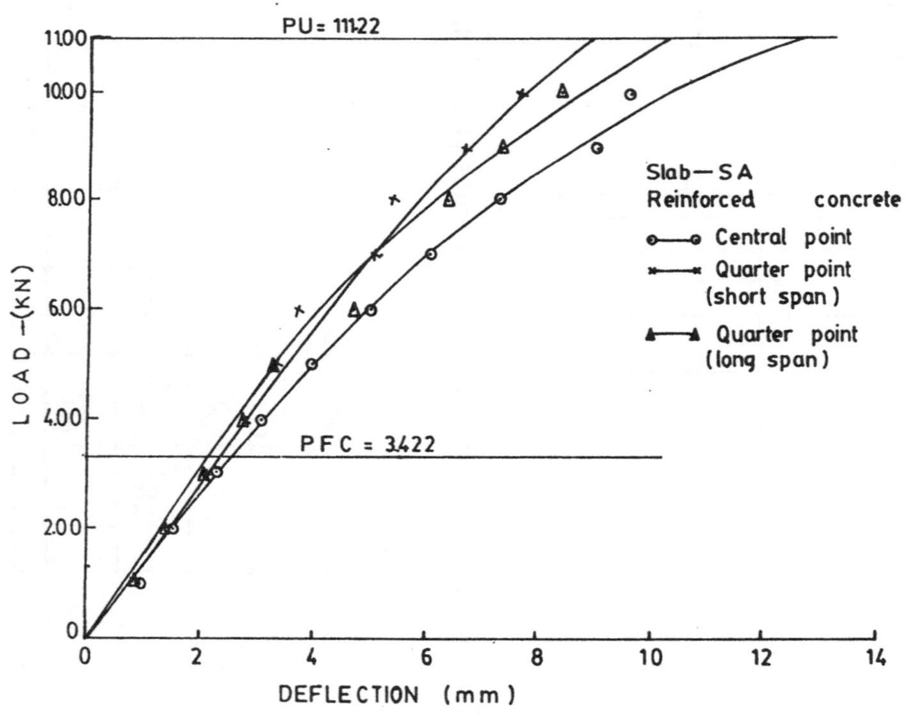

Fig. 3. Load – deflection curves.

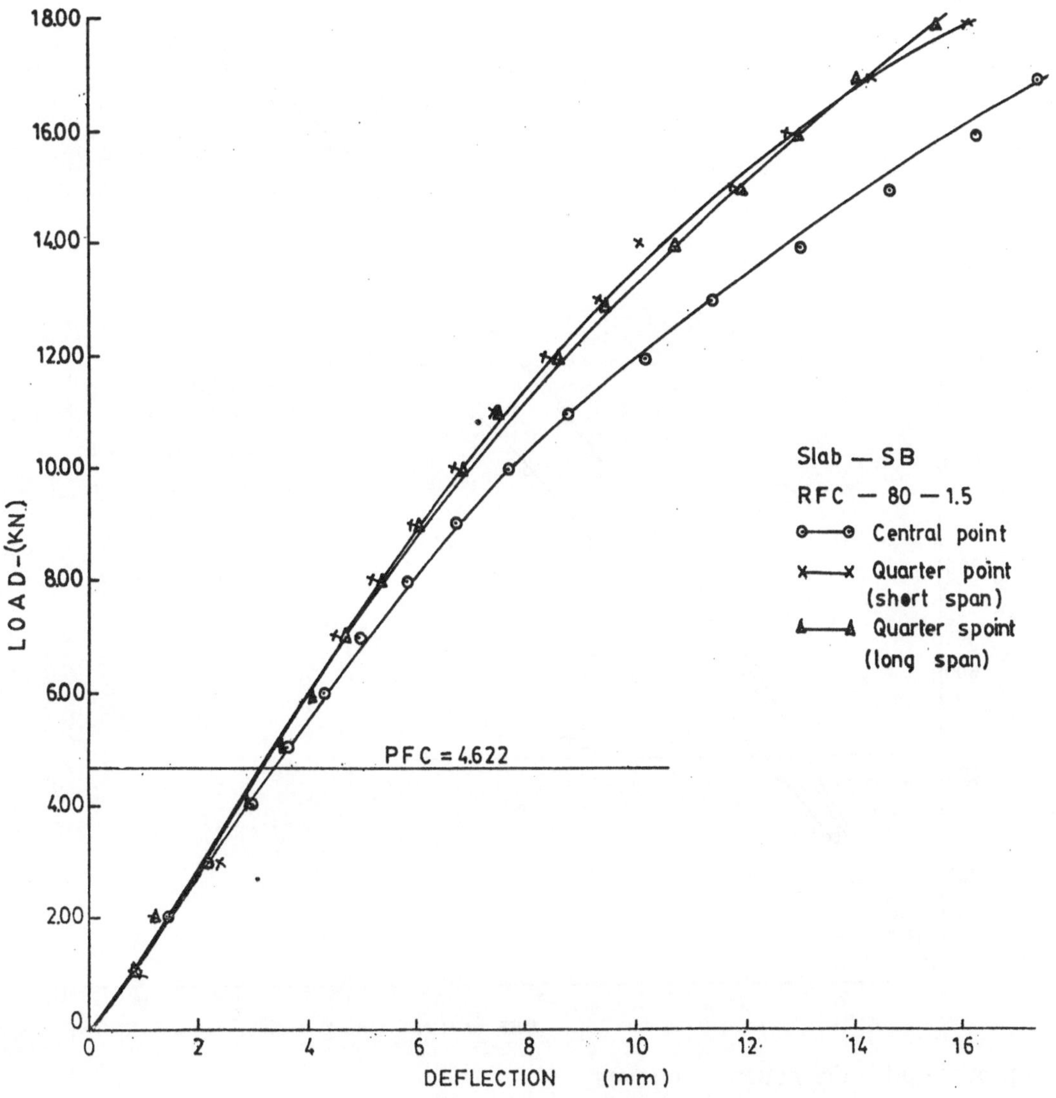

Fig. 4. Load — deflection curves.

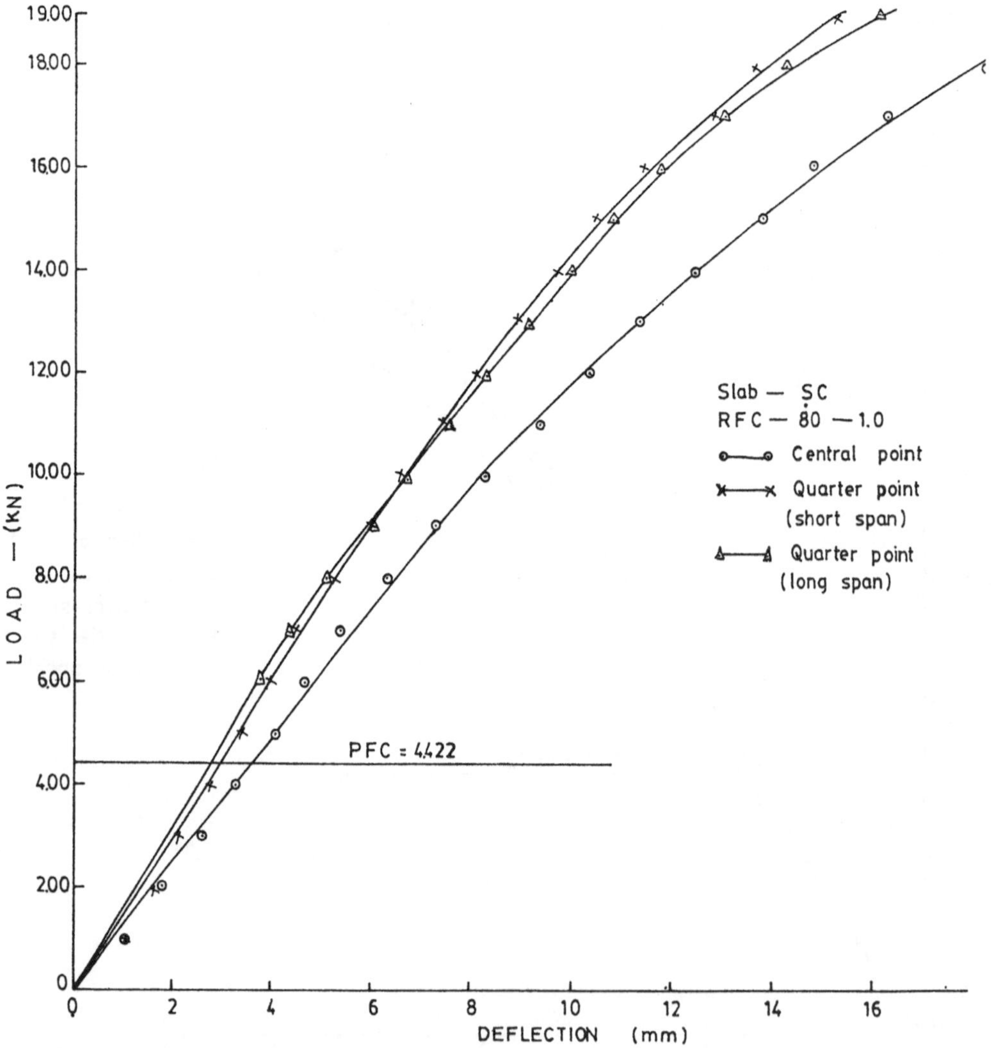

Fig. 5. Load – deflection curves.

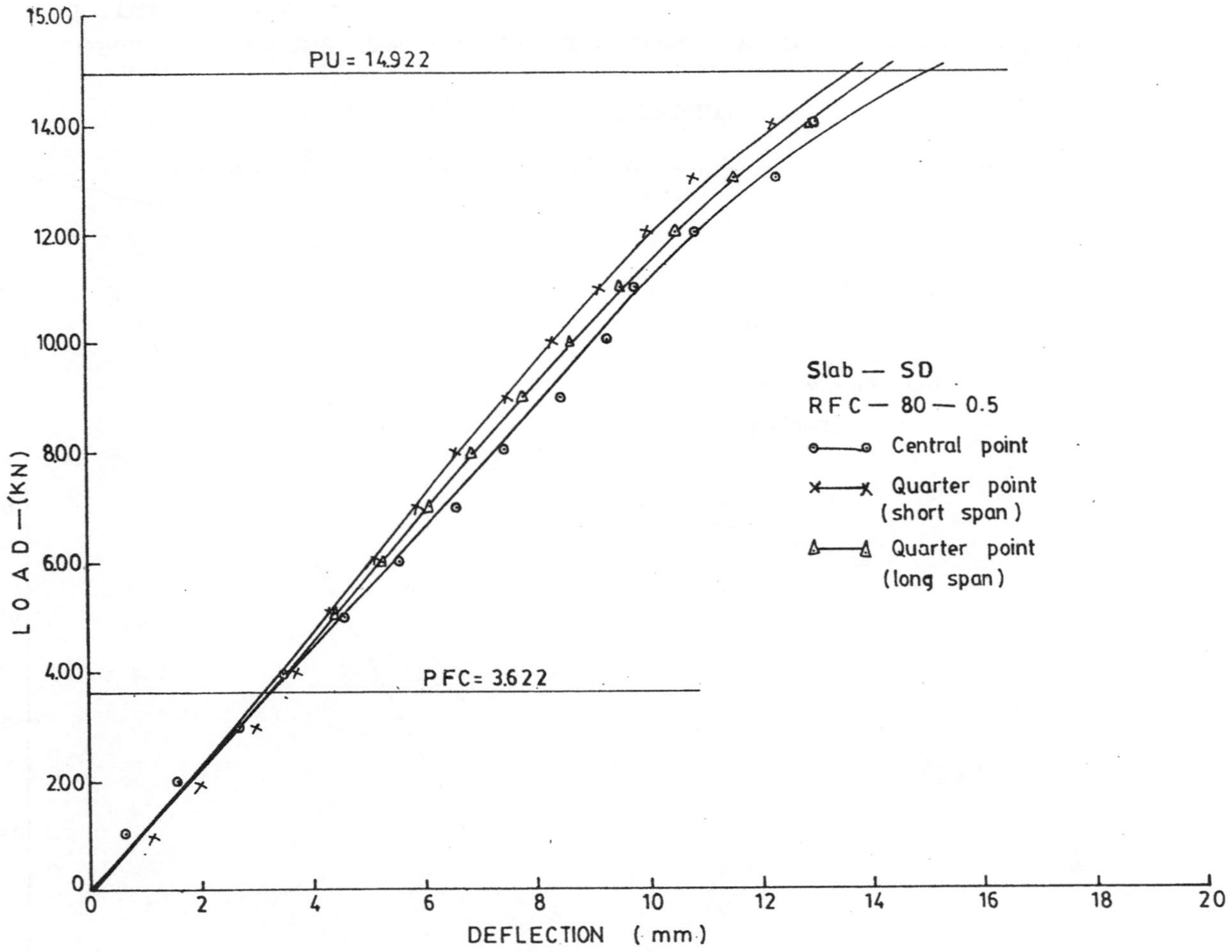

Fig. 6. Load — deflection curves.

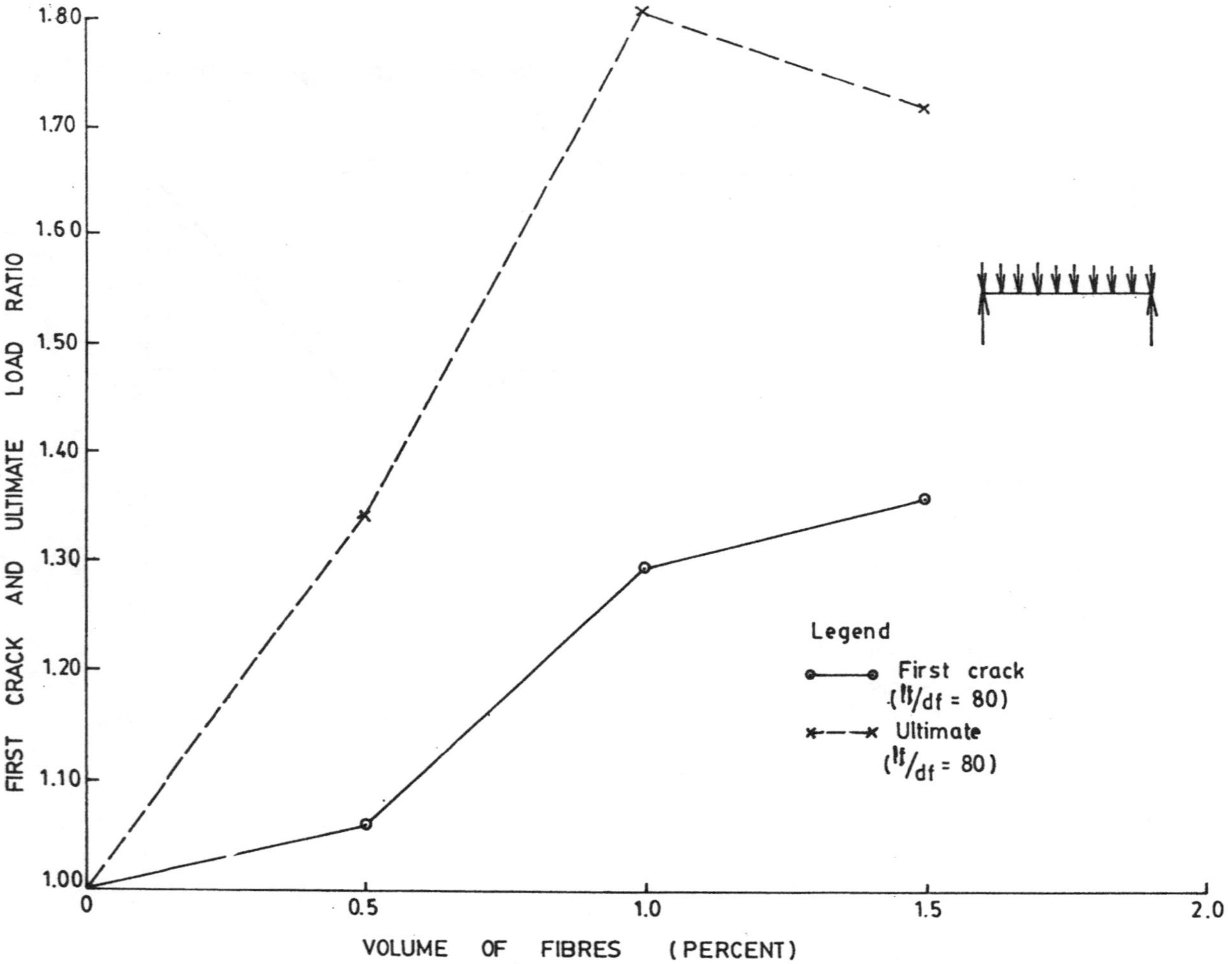

Fig. 7. Influence of volume percentage of fibres on first crack and ulimate load ratio. (Ref. Table 2)

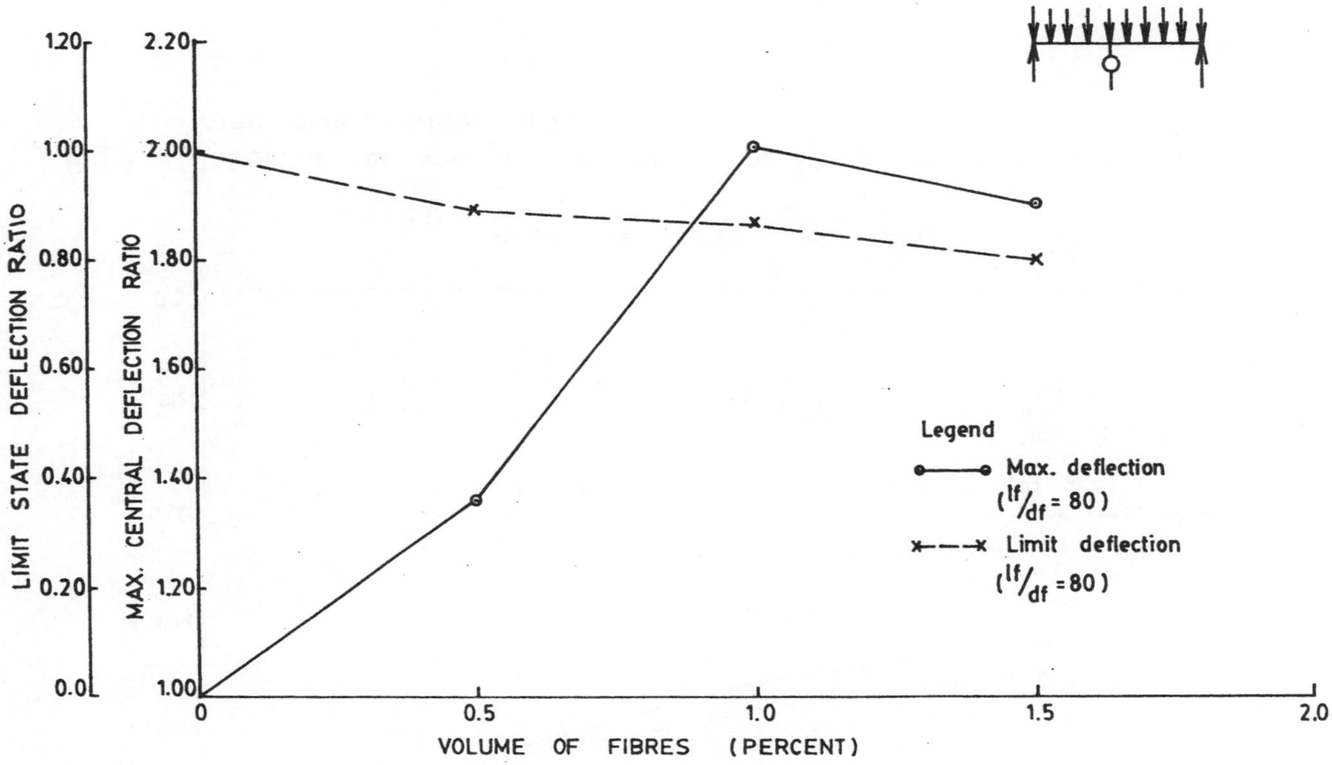

Fig. 8. Limit state and max. central deflection ratio vs. volume percentage of fibres. (Ref. Table 3)

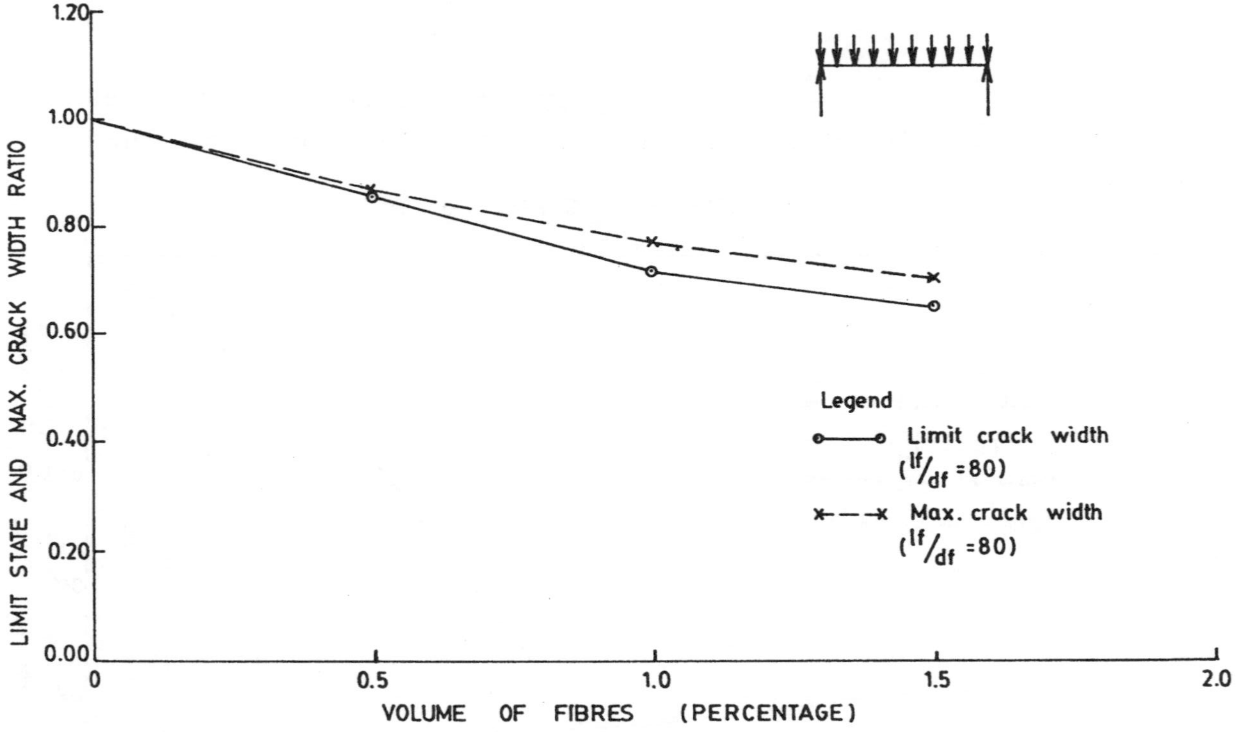

Fig.9. Influence of volume percentage of fibres on limit state and max. crack width ratio.

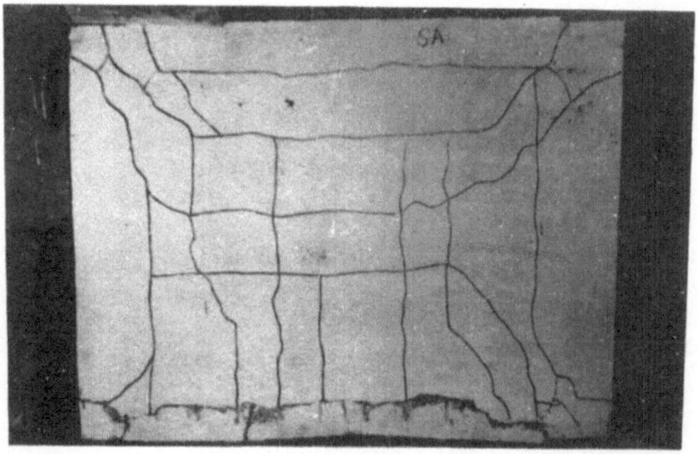

PHOTO - 1

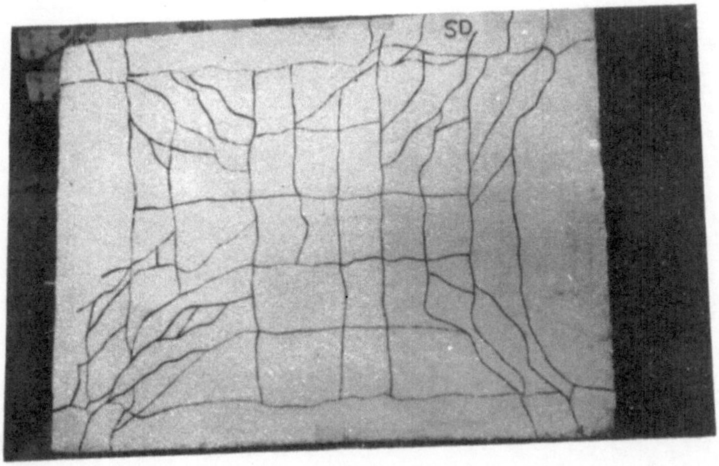

PHOTO - 2

Proceedings of the International Symposium on Fibre Reinforced Concrete, December 16–19, 1987, Madras, India

DYNAMIC LOADING RESISTANCE OF FIBRE REINFORCED CONCRETE CONTAINING THIN AND SHORT FIBRES

Jakob Šušteršič, Stanislav Urbančič, Viljem Kuhar

ZRMK - Institute for Material and Structure Research
Dimičeva 12, Ljubljana, Yugoslavia

ABSTRACT

The properties of concretes containing thin and short fibres were investigated. The use of these fibres leads to considerable increases in fatigue resistance and resistance to blows, while the improvement in static strength is moderate. The workability and compactibility of fresh concrete are not significantly affected by the addition of these fibres. The corrosion resistance of the fibres is within the expected limits.

1.0 INTRODUCTION

Fibre reinforced concretes, composed of hydraulic cements, aggregate and non-continuous, unconnected fibres, and made according to definition [1] or used as shotcrete [2], are discussed in this paper.

There are two fundamental problems in engineering practice: feasibility and fatigue [3]. The second problem is much more troublesome than the first, especially when dealing with composite materials like concrete. Fatigue resistance can be considerably improved by adding fibres, which represent micro-reinforcement in the concrete. Dynamic strength and resistance to blows are affected more than static strength (compressive, flexural).

Patents for different kinds of fibres have been known for a long time [4]. The improvement of hardened concrete properties, taking into account the requirements for fresh concrete (appropriate mixability and placeability) is the general purpose. The intention of our approach to the investigations of fibre reinforced concretes containing short and thin steel fibres [5] was the same.

The fibres used are oval in cross-section with a long-axis average length of about 105 μm and a short-axis average length of about 30 μm. The corresponding round cross-sectional diameter is 64 μm. The average length of the fibres is about 3.0 mm, while the l/d ratio is approximately 30.

Containing these fibres, concretes are easier to mix and place as compared to concretes with longer fibres. When used for shotcrete, a lower level of rebound is achieved. On the other hand, the properties of hardening and hardened concrete (expecially toughness and crack resistance), are not as good as in the case when longer fibres are used.

In the first place, we wanted to find out the properties of concretes containing thin and short steel fibres. We concentrated on dynamic loading resistance tests (fatigue resistance) and toughness (resistance to blows) tests. Some initial results are given and discussed in this paper.

2.0 PROGRAM CONCEPT

2.1 Materials and Concrete Mix Proportions

Concrete mixes were prepared with the following contents of fibres: 0.1 and 0.5 vol.% for shotcrete, and 1.3, 2.6 and 5.1 vol.%. for normal concretes. In addition, one mix without fibres was made for each kind of concrete.

Portland cement with 30% of slag was used for shotcrete. The aggregate used was crushed gravel with a maximum particle size of 4 mm. An accelerator was added as a chemical admixture.

Normal concretes were prepared with the same cement. The aggregate maximum particle size was 8 mm. The concretes were superplasticized.

Both groups of concretes were designed at a w/c ratio of 0.50.

2.2 Preparation, Mixing, Placing and Curing

The dry method was used for shotcreting. All the components except water were pre-mixed in a 100 l laboratory mixer. The average humidity rate of aggregate was 5.7%. The "dry" mix was put into the shotcrete equipment through a net. The shotcrete was jetted in the horizontal direction into 50x50x12 cm wooden moulds. Test samples were sawn from the shotcrete blocks.

Normal concretes were mixed in a laboratory mixer. First the dry components and the fibres were mixed up thoroughly, and then water and the superplasticizer were added. Concrete was cast into moulds and compacted on the vibration table.

All the samples were cured in a climate chamber at a temperature of 20°C and at 95% relative humidity.

2.3 Program of Tests

The results presented and discussed in this paper were obtained from the following tests:

shotcrete: compressive and flexural strength, static modulus of elasticity, watertightness and dynamic loading (fatigue) resistance

normal concrete: compressive and flexural strength, static modulus of elasticity, corrosion of fibres and resistance to blows (toughness) at flexure

The results of resistance to static loadings, as well as watertightness results and corrosion measurements, are presented in the following items. To complete the information about the toughness and fatigue resistance of fibre reinforced concretes, the results of static parameters are given relative to concretes without fibres.

3.0 STATIC LOADING RESISTANCE, WATERTIGHTNESS AND FIBRE CORROSION

3.1 Static Loading Resistance

The increase in compressive strength and modulus of elasticity due to the addition of short and thin fibres is moderate. The effect of these fibres concerning flexural strength is greater. The results relative to the concretes without fibres are shown in Fig. 1 for shotcrete and in Fig. 2 for normal concretes.

Fig. 1: Fibre reinforced shotcrete - static loading resistance results relative to shotcrete without fibres

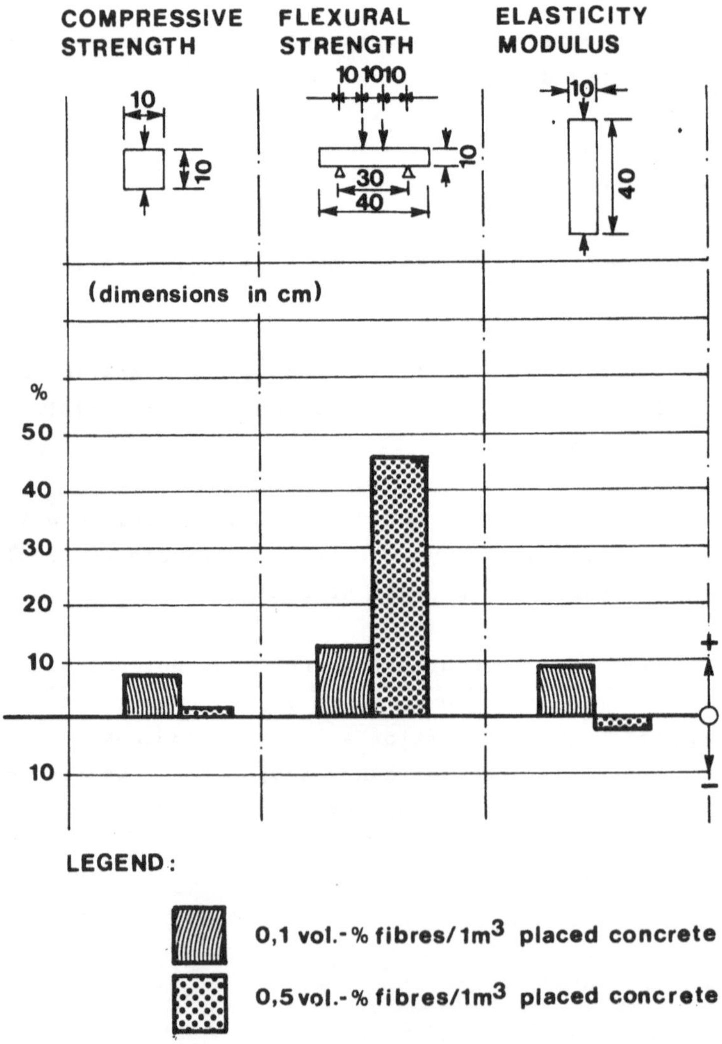

Fig. 2: Fibre reinforced concrete - static loading resistance results relative to concrete without fibres

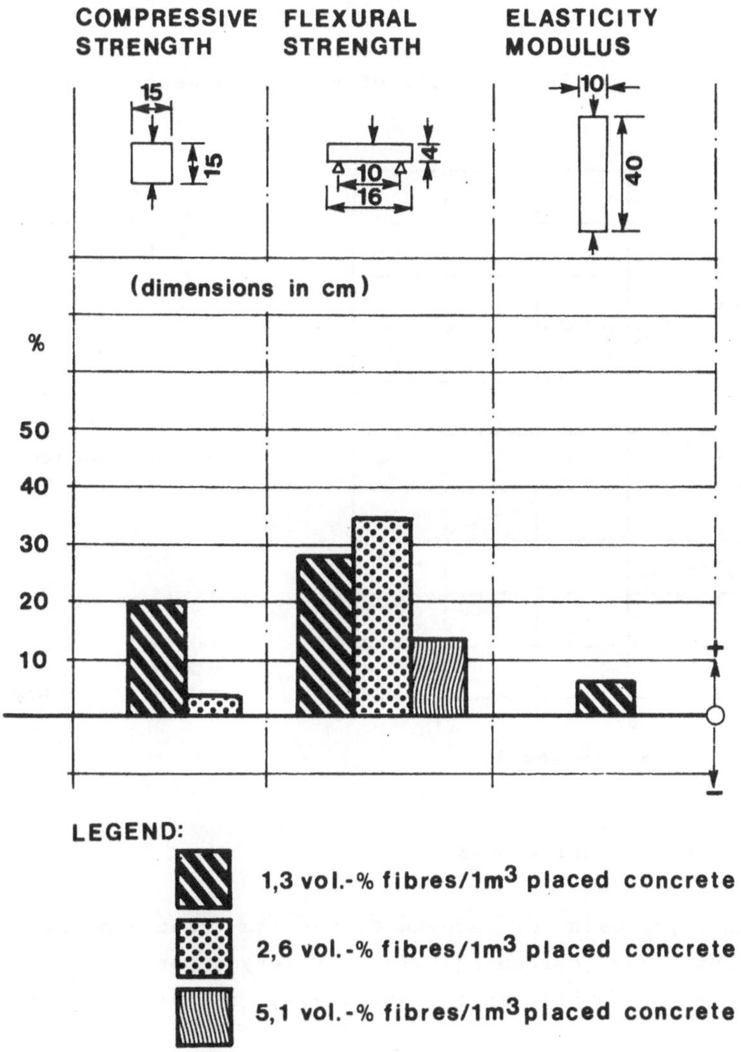

3.2 Watertightness

Watertightness is related to the appearence of microcracks in concrete. The maximum depth of water penetration (e max.) was measured on two series of samples according to the method given in DIN 1048. 10x10x20 cm specimens were sawn from shotcrete blocks. Half of them were cured in a climate chamber at conditions specified in the standard. The others were dried at 105°C and kept in a vacuum after having achieved constant weight.

A modest increase in the depth of water penetration was noticed at a low quantity of added fibres in comparison to concretes without fibres. Relatively, the increase in water penetration (and appearence of microcracks) is smaller in the case of concretes which were dried.

Fig. 3: Influence of the quantity of fibres on watertightness

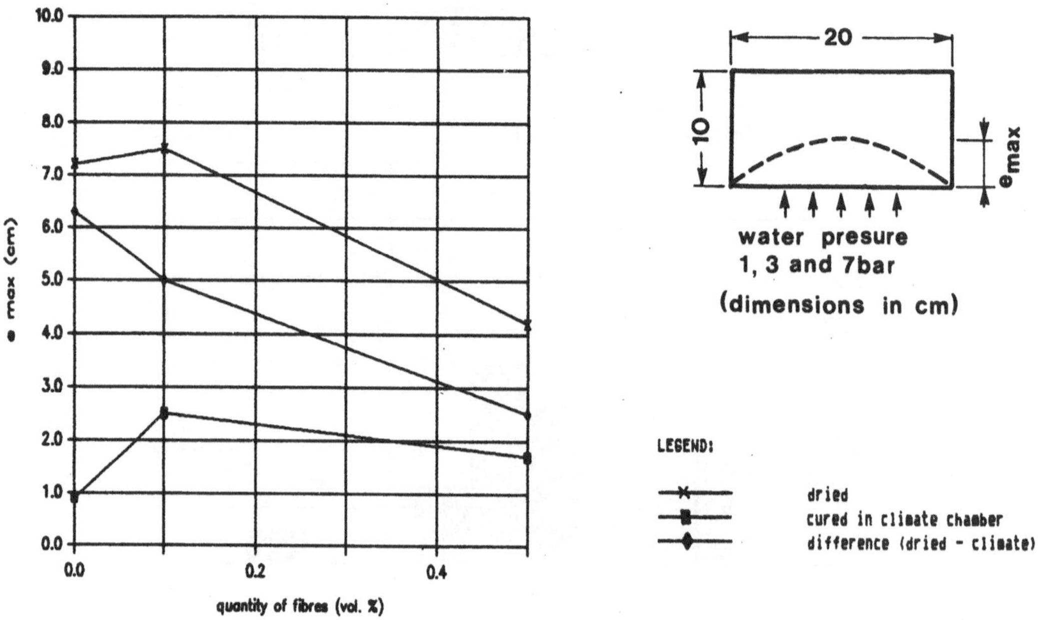

3.3 Corrosion of Steel Fibres

One of the purposes of adding fibres to concrete is to prevent reinforcement from corroding. This is very important in the case of concrete screeds, as they have to be compact and without cracks.

Due to steel fibres used as micro-reinforcement, the surface corrosion of concrete may appear. This surface corrosion can penetrate inwards in certain corrosion conditions.

The aim of our tests was to determine the corrosion influence of fibres in a wet climate. The folowing tests were performed: visual evaluation, analysis of Cl^- and SO_4^{2-} contents, determination of pH value and electrochemical tests.

4x4x16 cm prisms with different quantities of fibres (0.0, 1.3, 2.6 and 5.1 vol.%) were cured at 20°C and 95% relative humidity for approximately 460 days.

The results show that all the tested concretes have normal corrosion resistance. Steel fibres are easily compacted in the concrete mass. This was evident also from the results of concrete compactness.

Surface corrosion appeared in the case of the samples with a high fibre content. However, it influenced only the external appearence of the concrete.

4.0 RESISTANCE TO BLOWS (TOUGHNESS)

Resistance to blows was investigated using the method of Charpy's striking hammer. The hammer hit prismatic specimens (4x4x16cm) on two supports (Fig. 4). Toughness is defined as the work per unit of cross-sectional area which is needed to break the prism.

The results are shown in Fig. 4. The effect of added fibres is evident.

Fig. 4: Dependance of toughness on quantity of fibres

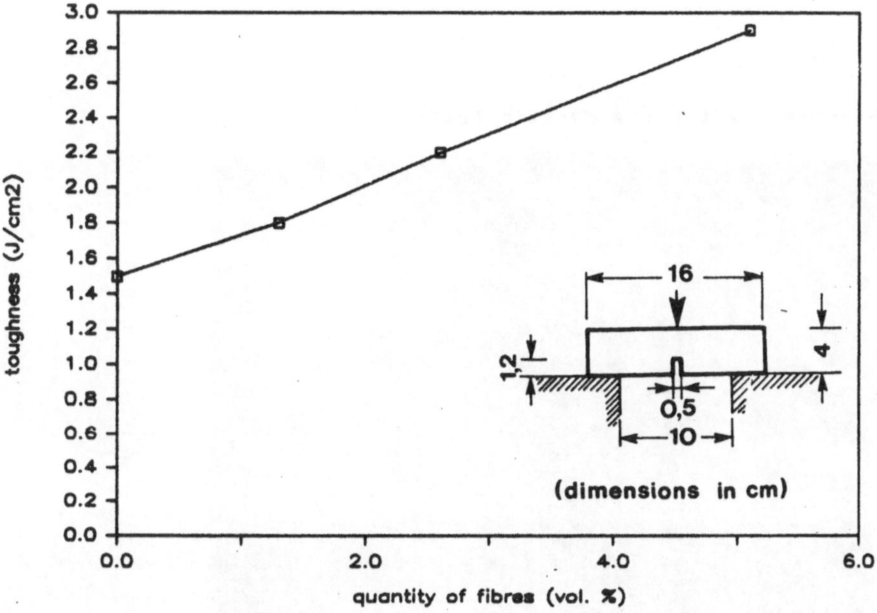

5.0 DYNAMIC LOADING RESISTANCE (FATIGUE RESISTANCE)

Prismatic concrete specimens (10x10x40cm) were vibrated on a pulsator with a working range of 0 to 250 kN, at a frequency of 8Hz. Prisms were loaded longitudinally. The loaded cross-section area was thererore 100 cm2.

The average static failure load (Pbs) of the prisms of all concretes was 210 kN. All the dynamic tests were performed within the same loading range, from Pld=21kN (10% Pbs) to Phd=189kN (90% Pbs). This range represents the largest amplitude of dynamic tests performed on fibre reinforced concretes [6].

The average results of dynamic load resistance (fatigue resistance) are presented in table 1.

Table 1: Dynamic tests results

vol.% of fibres	number of cycles
0.0	2,100
0.1	1,000,000
0.5	245,000

Fig. 5: Specimens after the dynamic loading test

The small number of tests does not permit definite conclusions. The tests so far carried out indicate that the best results were achieved with concrete containing 0.1% of fibres. The test was stopped after 1,000,000 cycles. The specimen did not fail. Concretes with 0.5% of fibres resisted on average about 250,000 cycles. Characteristic damage consisted of vertical cracks (Fig. 5, 6). These cracks occured long before the collapse at approx. 170,000 cycles. The concretes without fibres collapsed at cca. 2,000 cycles.

The surface stresses were measured on the specimen that resisted 1,000,000 cycles. A strain gauge was attached to one of the specimen's sides in the longitudinal direction (Fig. 6). The average compression (force per unit of cross-sectional area) was oscillating between 2.1 and 18.9 MPa. The values of compressions on the surface show that compressions were not distributed uniformly inside the specimen. In fact, compressions on the surface were oscillating in a narrow range of 1.0 Mpa (5.8% of the range of 'total' compressions) between 13.8 and 14.8 MPa with the middle value of 14.3 MPa which lies at 73% of the 'total' range. (Fig. 7). It

Fig 6: Specimens after the dynamic loading test

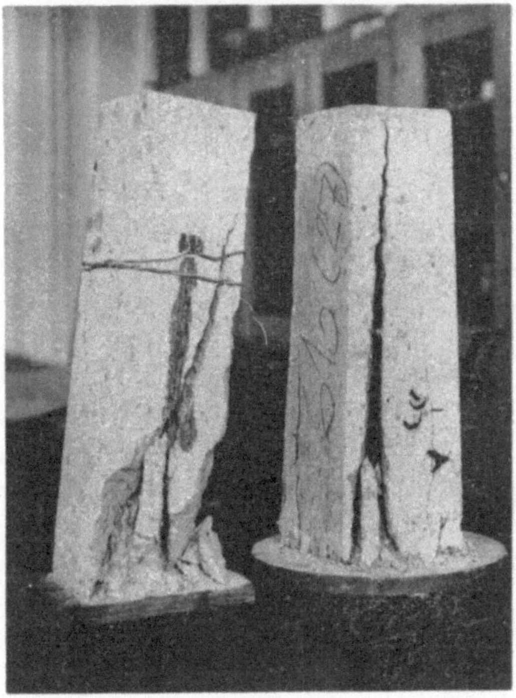

Fig. 7: Surface stresses

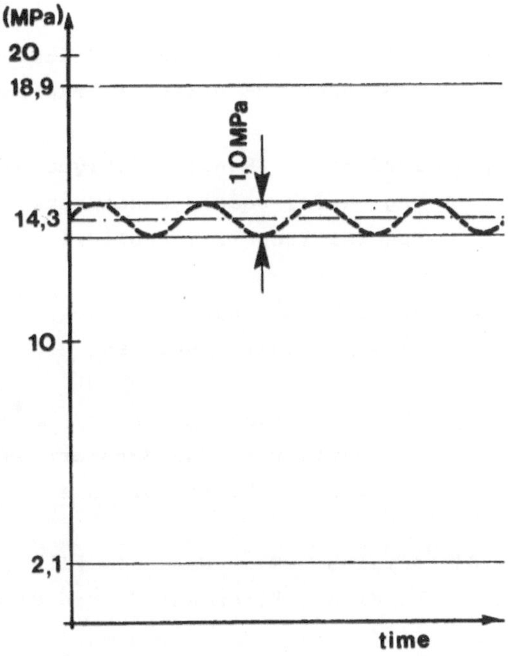

indicates that the material cannot follow quick changes of stresses when exposed to such a frequency (8 Hz) and responds with internal redistribution of stresses.

6.0 CONCLUSION

First investigations of concretes containing thin and short steel fibres show that considerable increases in resistance to blows and fatigue resistance can be expected. Flexural strength and crack resistance increase with the use of these fibres too.

A more detailed evaluation of the stated improvements will be given after further investigations. It will be necessary to study the fatigue mechanism, which represents a complex problem in composite materials because it depends on several parameters which must all be determined or known.

REFERENCES

[1] ACI Committee 544: State-of-the-Art Report on Fiber Reinforced Concrete, ACI Journal, Nov. 1973

[2] ACI Committee 506: State-of-the-Art Report on Fiber Reinforced Shotcrete, Concrete International, Dec.1984

[3] Jeffrey T. Fong: Fatigue Mechanism - Key to the Solution of the Engineer's Second Fundamental Problem, Fatigue Mechanisms, ASTM Special Technical Publication 675, Oct. 1979

[4] Antoine E. Naaman: Fiber Reinforcement for Concrete, Concrete International, Mar. 1985

[5] J. Šušteršič et. al.: Development and Use of Fibre Reinforced Concrete, Part One: Properties of Fibre Reinforced Shotcrete, ZRMK, Ljubljana 1986 (in Slovene)

[6] Antoine E. Naaman: Fiber Reinforced Concrete under Dynamic Loading, Fiber Reinforced Concrete, International Symposium ACI, 1984

CHAPTER III

PROPERTIES OF FIBRE REINFORCED CONCRETE USING FIBRES OTHER THAN STEEL

Proceedings of the International Symposium on Fibre Reinforced Concrete, December 16-19, 1987, Madras, India

THE DEVELOPMENT AND PERFORMANCE OF ALKALI RESISTANT GLASS FIBRES FOR CEMENT REINFORCEMENT

B. A. Proctor
Pilkington Brothers PLC, R&D Laboratories, Lathom,
Ormskirk, Lancashire, UK.

SYNOPSIS

The great potential for a glass fibre reinforced cement material (GRC) was recognised in the early days of reinforced plastics (GRP). Unfortunately, conventional borosilicate glass fibres, made from E-glass, were rapidly and severaly attacked in the highly alkaline conditions of hydraulic cements. The need for a suitable alkali resistant glass fibre composition led to the development of high zirconia glasses whose compositions and main features are described. The effect of zirconia content on fibre strength retention is discussed and the performance of zirconia glasses is compared with that of other suggested alkali resistant compositions. Recent and future developments are discussed.

The successful acceptance and use of GRC products required evidence of satisfactory long term strength retention. The development of accelerated ageing procedures is described and the results related to long term exposure in real weather in several climates.

INTRODUCTION

Virtually all commercial glass compositions, and certainly all commercial glass fibre compositions, are based on fused silica. Other oxides, of alkali metals, alkaline earths, boron, etc, are added to reduce the melting temperature; modify viscosity, crystallisation and durability; and to provide other specific characteristics: but the chemical, physical and mechanical integrity of the glass are provided by a continuous 3-dimensional network of silicon-oxygen-silicon bonds (Ref.1). It so happens that this bond is particularly susceptible to hydroxyl attack according to

$$\text{-Si-O-Si-} + OH^- \longrightarrow \text{-Si-OH} + SiO^- \text{ (in solution)}$$

This results in dissolution of some commercial glasses in strong alkali and drastic strength loss of all glass fibres in solutions of high pH (Fig.1, Ref.2). It held back the development of glass fibre reinforced cement materials for more than three decades because all the commercially significant hydraulic cements provide highly alkaline environments during the initial setting period, and whenever they are damp and moist.

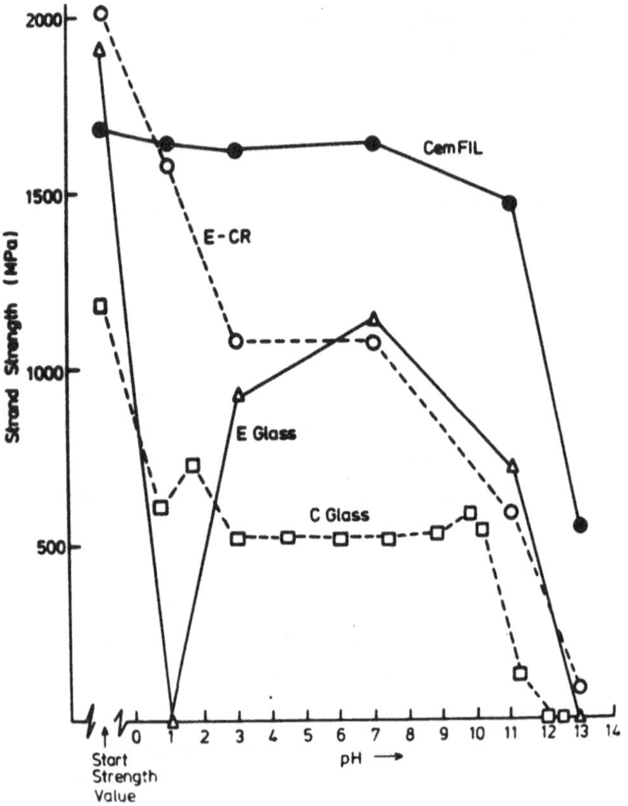

Figure 1. Strengths of Glassfibre Strands after immersion for 2 days at 80°C in Solutions of Different pH.

The first attempts to avoid these problems appear to have been made in Russia in the 1950's when Biryukovitch and Biryukovitch (Ref.3) tried to make composites with E-glass fibres in the rather less alkaline high alumina cement (HAC). However these attempts failed and we now know that even this cement attacks E-glass fibres quite drastically (Ref.4). The real development of practical glassfibre reinforced cement (GRC or GFRC) materials began with the work of Nurse, Majumdar and colleagues (Ref.5) who realised that a new glass composition would be required in order to achieve a high degree of alkali resistance - and hence provide adequate long term strength in cement composites. The real target was to be able to provide that property in the most alkaline, most useful and most widely used cement - Portland cement.

In 1926 Dimbleby and Turner (Ref.6) had shown that the addition of significant amounts of zirconia to sodium silicate compositions conferred a high degree of alkali resistance. However these compositions were used for bulk glasses and it was widely believed that they were not suitable for fibre drawing. Majumdar and his colleagues (Refs.5, 7, 8) produced relatively small quantities of fibre from a single tip bushing and succeeded in drawing a high zirconia content glass (G.20, Table 1) which had very much better strength retention in aqueous cement extract than conventional glasses such as Pyrex, E-glass or A-glass. Further evaluation and development of composites technology showed that G.20 had sufficient durability to provide useful reinforcement in Portland cement. An indication of the very marked improvement in alkali resistance achieved with high zirconia alkali resistant glasses is shown in Fig.1 for Cem-FIL.

TABLE 1 TYPICAL GLASS COMPOSITIONS (WT%)

	A-GLASS	E-GLASS	PYREX	G.20	CEM-FIL	AR FIBRE SUPER
SiO_2	73	54	81	71	62	56.4
Na_2O	13	-	4	11	14.8	15.3
CaO	8	22	-	-	5.6	-
MgO	4	0.5	-	-	-	
K_2O	0.5	0.8	-	-	-	0.9
Li_2O	-	-	-	1	-	-
Al_2O_3	1	15	2	1	0.8	-
Fe_2O_3	0.1	0.3	-	-	-	-
B_2O_3	-	7	13	-	-	-
ZrO_2	-	-	-	16	16.7	16.9
TiO_2	-	-	-	-	0.1	-
Rare Earth Oxides	-	-	-	-	-	10.3

DEVELOPMENT OF COMMERCIAL A.R. FIBRES

The requirements of large scale, economic fibre production impose severe additional limitations and restrictions on the choice of glass composition. To meet the requirements of low cost and high volume production several hundred individual filaments are drawn in parallel from specially shaped and closely spaced "tips" or holes in the base of an electrically heated platinum-rhodium alloy bushing. Breakage of only one filament will cause a drip of molten glass to fall across the remaining filaments - leading to the breakdown of the drawing operation for the whole bushing, with severe effects on operating cost.

Fibre breakage can be due to mechanical effects in the fibre winding such as abrasion or snagging; to hydrodynamic instability in the viscous flow from the tip; or to particles or inhomogeneity in the glass due to poor melting or devitrification.

To achieve stable hydrodynamic conditions the glass viscosity must be in a fairly narrow range close to 100 Nsm^{-2} which imposes one limitation on the fiberising temperature/viscosity characteristic of the glass. To avoid devitrification problems the liquidus temperature (T_L) must be at least 40°C (and preferably nearer 80°C) below the fiberising temperature (T_F). Compositions of high viscosity also require higher melting temperatures (with consequent fuel and furnace costs) to dissolve the raw materials and avoid breakages at the drawing tips. In addition to these factors, with current materials, bushing life decreases rapidly as the fiberising temperature exceeds 1300°C -another important influence on cost.

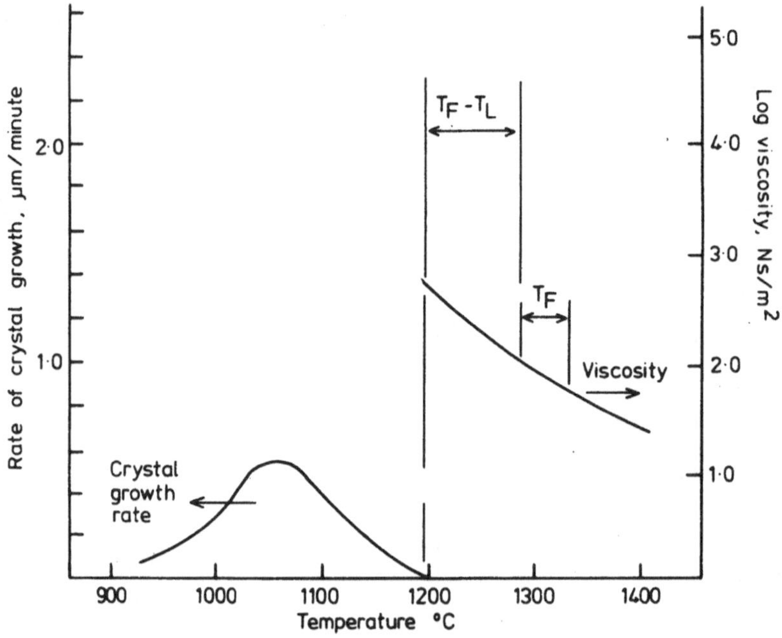

Figure 2 Relationship between important fibre drawing parameters

The interaction of these factors is illustrated in Fig.2. (Ref.9). The matching of the required fiberising characteristics with a high degree of alkali resistance, whilst using materials of reasonable batch cost in order to produce an economically viable fibre for use in an inexpensive basic construction material, represented a severe challenge to the glassmaker.

The initial G.20 alkali resistant glass was basically a 3 component Na_2O-ZrO_2-SiO_2 composition (Table 1). Development of practical, large scale fiberising characteristics, with no detriment to the alkali resistance (Ref.9) was achieved by the addition of some calcia for silica and an increase in soda content to give the four component Cem-FIL glass listed in Table 1. This was the first commercially available A.R. Glassfibre for cement reinforcement. (Ref.4)

Other commercially available alkali resistant glasses have followed the same route of achieving alkali resistance by zirconia additions. In this family of glasses, strength retention in cement is essentially controlled by zirconia content up to about 17% by weight as indicated by direct measurements of (fibre) strength retention in a cement environment by means of the Strand-in-Cement (SIC) test, Fig.3, (Ref.9). Above this level there is less improvement and the consequent increase in fiberising temperature and liquidus temperature leads to significantly higher cost of production.

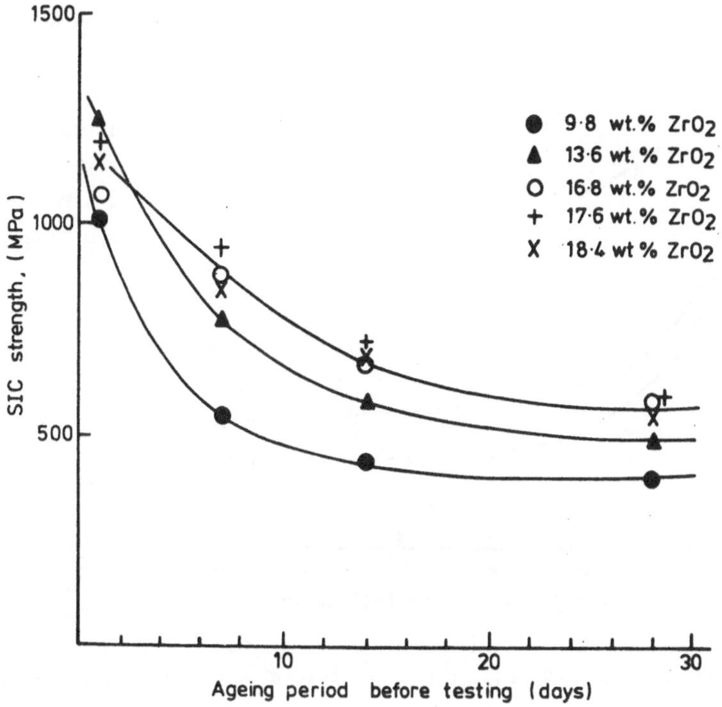

Figure 3. Effect of Zirconia content on SIC strengths in water at 50°C

For a short period some manufacturers produced lower zirconia content glasses, at some sacrifice in durability. Also a number of research workers have been tempted to investigate glasses in other compositional fields to avoid the difficulty and expense of fiberising high zirconia glass. In particular, there have been suggestions that glasses of naturally occurring slate or rock compositions, and glasses containing manganese and iron, have alkali resistance equivalent to the zirconia compositions (Refs.10, 11, 12). However SIC Test (Ref.13) results show such glasses to be much inferior to the high zirconia type of composition, Fig.4, (Ref.9) and the latter appear to be the only practicable types of alkali resistant fibres which are likely to be commercially available for some time.

We have seen that the alkali resistance of the high zirconia A.R. glassfibres currently available is of a different order to that of the borosilicate E-glass which they replace (Fig.1). Nonetheless the conditions within a cement matrix are so aggressive that some fibre/cement reaction still occurs, the fibre surface is roughened, fibre strength is reduced and composite properties reduce with time (Refs.4, 9, 14, 15).

As discussed below, the development of accelerated ageing procedures combined with extensive laboratory and field testing (Refs. 16, 17) showed that, despite this

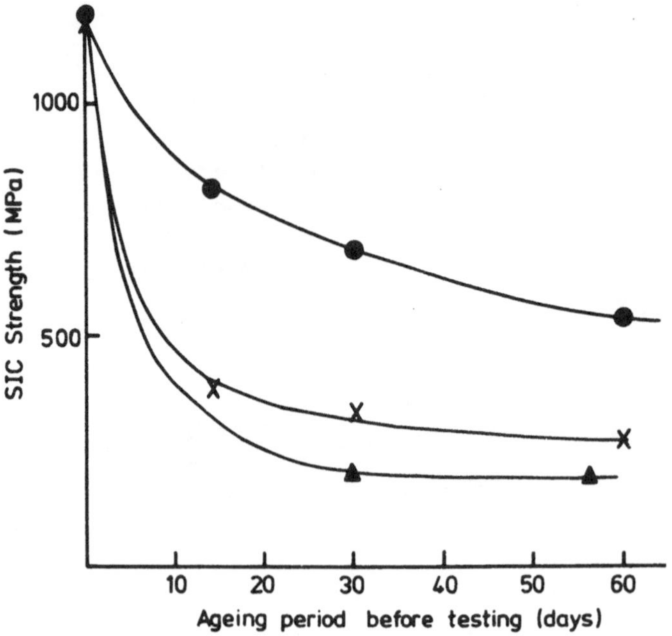

Figure 4.
SIC Comparison of Experimental glasses in water at 50°C
• = Cem-FIL
x = Rockwool
▲ = Iron/Manganese

loss of initial property, GRC composites reinforced with a good quality A.R. glassfibre have useful and stable long term strength properties. However fibre producers recognised the desirability of improving fibre strength retention and retaining higher strength, impact strength and toughness in GRC composites for considerably longer periods. The first commercial development in this direction was taken by Pilkington in 1979 with the launch of Cem-FIL 2 fibre (Ref.18). A surface treatment, involving the incorporation of a chemical inhibitor in the size coating which is slowly released around the fibre in the cement, led to significant improvements in fibre strength retention (Fig.5, Ref.9) which were reflected in improved retention of composite strength, failure strain and toughness - see for example Fig.6 and Refs.19, 20.

Glass fibre manufacturers have also continued to investigate durability improvements based on further glass composition development. Over the last 10 to 15 years over 100 patents have been filed. An evaluation of these reveals that all practically useful developments are still being based on high zirconia glasses with additions of such other oxide combinations as rare earths; rare earths plus titania; chromium and tin oxides; and tri-valent chromium plus titania plus rare earths. Recently Asahi Glass Company announced the commercial availability of an improved durability fibre "A.R. Fibre-Super" (Ref.21). In this some 10% of rare earth oxides have been added to the basic glass composition instead of calcia and silica (Table 1) Further extensive studies of the effect of more durable glass compositions on fibre strength retention and composite properties have also been reported by Fyles et al (Refs.22, 23) who have also drawn attention to the possibility of combining the

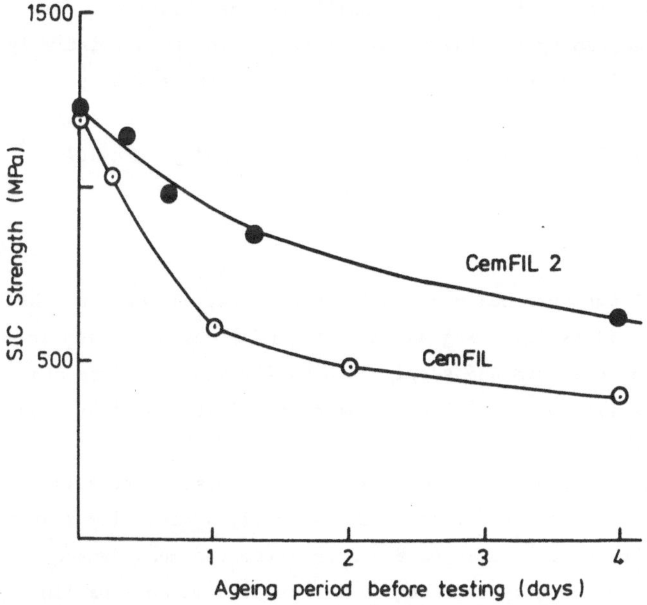

Figure 5. Strand in Cement (SIC) strengths at 80°C.

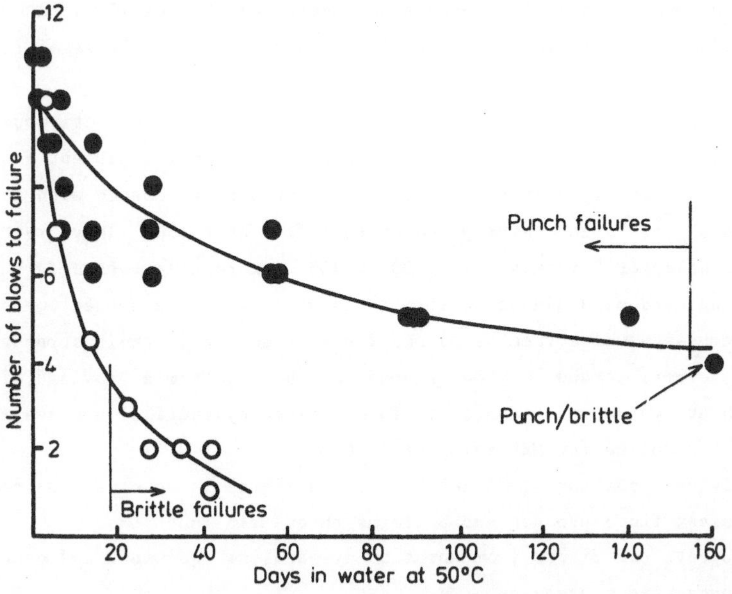

Figure 6. Dropping ball impact tests on artificially aged GRC panels
O = Cem-FIL fibre: • = Cem-FIL 2 fibre

effects of inhibitor/surface treatments of the Cem-FIL 2 type with glass composition changes.

Whether or not these developments become commercially available and widely applied over the next few years depends very much on a balance between the perceived need for more durable fibres, leading to a "market pull", and the additional cost. The surface treatment approach adopted by Pilkington for Cem-FIL 2 led to a relatively small increase in cost, but this has still held back the use of these fibres in some market areas. Improvements based on composition lead to more significant cost increases due to more expensive raw materials and greater difficulty in melting and fiberising.

FIBRE HANDLING TECHNOLOGY

Much of the present GRC business has been built up on materials made by GRP techniques (the spray-up method) as first suggested by Majumdar and Ryder as early as 1968 (Ref.5). For this reason the form and supply of alkali resistant fibre has followed the traditions and styles of continuous filament reinforcement fibre used in GRP (or FRP).

Glass fibres are extremely strong, but their strength is also critically dependant on the perfection of the fibre surface. It is easily drastically reduced by abrasion (Ref.24) and the fibre surface must be protected from too much damage.

To incorporate significant amounts of fibre in a composite without balling, fluffing and excessive voidage the fibres need to be essentially straight: often it is easier to incorporate fibres if they are in bundles of 10's or 100's rather than as individual filaments, certainly it is easier to handle large volumes of fibre in this way.

To meet these requirements reinforcement fibres are drawn continuously from a multitip bushing and before they make contact with any other equipment - or with each other - they pass lightly over an applicator roller and are coated with an aqueous polymer emulsion, or "size" which protects them from abrasion. They are then gathered together in bundles or "strands" of ~ 100 to 200 or more filaments before being wound onto a rotating cardboard sleeve to form a "cake". Cakes are heated to dry or cure the polymer emulsion which then binds the fibres together in their strands. Subsequently several strands may be grouped together to form a "roving", i.e. parallel strands which are wound on a mandrel to form a dense cylindrical package which is the normal product supplied for GRC spray production.

The polymeric coating or "size" applied to fibres is an essential component of the reinforcement fibre product and performs three main functions.

(i) It protects the fibres from abrasion by handling equipment and hence prevents or reduces fibre breakage.

(ii) It binds the individual filaments into convenient bundles for handling and incorporation in the composite.

(iii) It may improve the final composite properties in some way. For example by improving the interfacial bond strength in reinforced plastics or the fibre strength retention in Cem-FIL 2 GRC.

In order to carry out these functions reinforcement fibre sizes normally contain three major components:

A. a film forming polymer which, on drying/curing, binds the individual filaments together and provides a protective coating.

B. a lubricant which helps to help to protect both the glass fibre itself and the polymer film from abrasion.

C. an active agent.

In reinforced plastics this is normally an organosilane coupling agent which provides a degree of chemical bonding between the inorganic glass fibre and the organic resin matrix.

In the first generation of commercial fibres for cement reinforcement the size had virtually no effect on fibre/cement bonding nor on fibre strength, once the fibres had been incorporated in cement. That is, the size contained no ingredient which provided chemical protection against attack. In the second generation Cem-FIL 2 type of fibre however the size contains an active agent which is a chemical inhibitor. This reduces the rate of the alkali/glass chemical reaction and hence improves the strength retention of the fibres in a cement environment.

Thus the geometric form and surface coating treatments of cement reinforcement fibres have been derived from the practices of fibre reinforced plastics technology and are particularly suited to the manufacture of GRC composites based on GRP (or FRP) methods. Indeed they are essential if reasonable strength fibres are to be incorporated in adequate quantities, and the required orientations, in cement matrices.

Similar fibre geometries and coatings have also proved suitable for premix production although this method is more akin to concrete technology than GRP (FRP) processing. However there are similar requirements of protection of the fibre against abrasion (both in handling and in the cement mixer) and prevention of balling in the mixing process which favour the use of fibres in size coated bundle format.

PRACTICAL GRC MATERIALS

In principle all fibre reinforced composite materials can be produced in an almost infinite variety of forms depending on fabrication method and end use requirement. This is true for GRC and important specialised materials include continuous filament wound concrete pipes and short chopped fibres dispersed as individual filaments in asbestos substitution products. In the main however, GRC falls into two simplifying categories - "premix" materials, based on concrete mixing technology, and "spray-up" materials based on glass reinforced plastics technology as predicted by Majumdar and Ryder (Ref.5). The latter materials give the highest mechanical properties and provide the most extensive use of GRC in thin sheet form of one type or another.

"Premix" GRC involves the blending together of cement, sand, water and admixtures in a conventional concrete mixer. Chopped glass fibre strands are added at a late stage and the mixing time carefully controlled to avoid excessive fibre damage. Typically about 3.5% (wt) of 12 mm long chopped strands are used but both fibre content and length may be varied to suit the product and process.

"Premix" mixes may be vibration cast in moulds, press-moulded or even sprayed onto formers. Typical products include complex patterned sun-screens tiles, small junction boxes, litter bins etc. Mechanical properties (see Table 2) are lower than sprayed materials because of lower fibre contents, poorer fibre/matrix bonding and more random fibre orientation, which will however be process dependent.

TABLE 2 TYPICAL INITIAL MEAN PROPERTY VALUES OF CEM-FIL GRC

PROPERTY \ PROCESS		SPRAY DEWATERED(1)	HAND SPRAY (2)	VIBRATION CAST PREMIX (3)
Dry bulk density	(t/m^3)	2.0 - 2.05	1.9 - 2.1	1.9 - 2.0
Compressive strength	MPa	60 - 100	50 - 80	40 - 60
Modulus	GPa	13 - 25	10 - 20	13 - 18
Impact strength	Nmm/mm^2	11 - 25	10 - 25	8 - 14
Poisson's Ratio		0.24	0.24	0.24
Bending:				
LOP	MPa	9 - 13	7 - 11	5 - 8
MOR	MPa	28 - 42	21 - 31	10 - 14
Direct Tension:				
BOP	MPa	7 - 9	5 - 7	4 - 6
UTS	MPa	10 - 17	8 - 11	4 - 7
Strain to failure %		0.7 - 1.3	0.6 - 1.2	0.1 - 0.2
Shear:				
In-plane	MPa	10 - 17	8 - 11	4 - 7
Interlaminar	MPa	3 - 5	3 - 5	N.A.

Notes:

1. Sand:cement ratios of 0:1 to 0.5:1 and water:cement ratio 0.28.
2. Sand:cement ratios of 0.33:1 to 1:1 and water:cement ratio 0.33.
3. Sand:cement ratio of 0.5:1 and water:cement ratio of 0.33.

In the manufacture of "spray-up" materials a continuous fibre roving is fed to a glass-fibre chopping gun which cuts the fibre strands into lengths in the range 25-40 mm, typically 38 mm. Simultaneous sprays of chopped fibre and cement/sand mortar slurry are then blown down onto a mould surface and compacted by rolling or de-watering to give a material containing about 5% (wt) of glass fibre arranged in a random 2-dimensional array parallel to the mould surface. The spray gun may be hand held, or part of a mechanical spray rig. In their initial work the UK Building Research Establishment concentrated on de-watered material (Refs.5, 14) which gives high compaction and the highest initial strengths (Table 2). This material has been extensively used by Pilkington in its property researches (e.g. Refs.16, 25). In subsequent commercial developments, non-dewatered roller compacted material has been more extensively used. This has somewhat lower, but satisfactory, initial strength (Table 2).

STRENGTH PROPERTIES AND AGEING BEHAVIOUR

In wet conditions and normal weathering exposure the highly alkaline conditions in moist cement lead to some attack on the surface of even alkali resistant fibres. Some GRC composites may become weaker and there are changes in the fracture characteristics of all GRC composites over a period of years which lead to embrittlement and loss of impact strength. In order to give confidence in the use of GRC, and to gain acceptance of it as a construction material, it was obviously essential to be able to predict the full extent of these property changes - and to be sure that the material reached a stable long term state.

To achieve this extensive studies were made of the long term strength behaviour of spray dewatered GRC, which has the highest initial strength and is thus the most sensitive to fibre controlled property changes. The results and understanding obtained are directly applicable to the more widely used handspray and premix materials.

The main object of adding glass fibres to cementitious materials is to provide useful tensile and flexural strength properties. They also enable large sheets of thin section material to be made without cracking. The tensile and bending characteristics of good quality sprayed GRC sheet after an adequate cure are illustrated in Fig.7a. There is an initial linear region, followed by a pseudo yielding behaviour achieved by fine multiple cracking in the cement matrix (Refs.26, 27). Strain to failure is ~ 1% and the material is tough and impact resistant. Present durability data, now extending to 10 years, indicate very little change from this initial behaviour pattern in dry indoor storage conditions, with long term predictions indicating a relatively small reduction in strength with toughness and strain to failure little reduced (Refs.14, 15). However in wet and natural weathering exposure there is a loss of strength, together with a more marked loss of strain to failure as indicated in Fig.7b. It is the extent of these changes which needed to be predictable if GRC were to be extensively and safely used.

The oldest GRC samples were 'laid down' for long term weathering and test programmes by the Building Research Establishment (BRE) in 1968/69, they were tested at intervals up to 5 years and then at 10 years in 1978/79. The test results were analysed in detail by a joint BRE/Pilkington Working Party to give guidance on the expected long term strength of GRC over 20 years (Refs.14, 15).

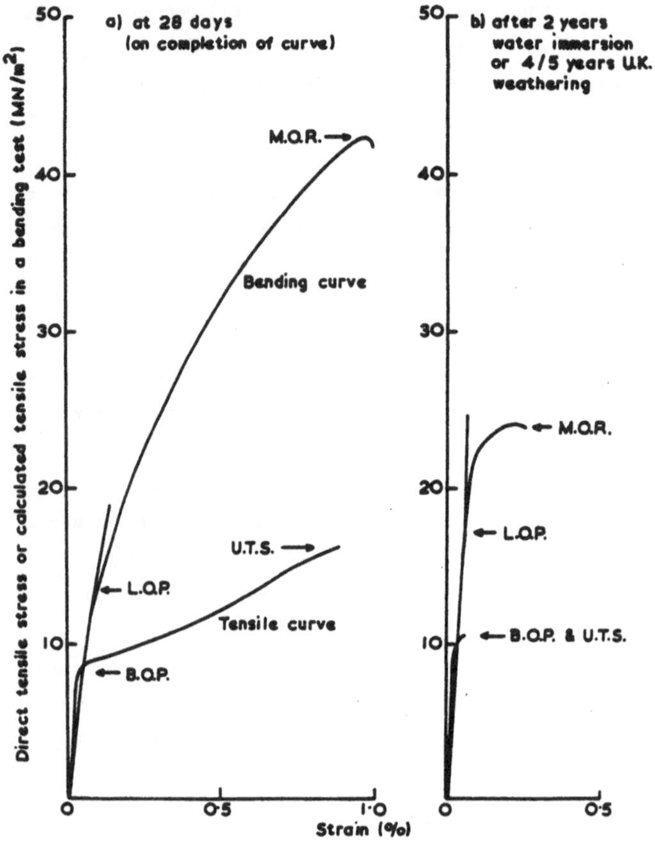

Figure 7.
Tensile Stress-Strain Curves for spray dewatered GRC.

From 1972 Pilkington initiated weathering programmes at various sites about the world and since then very considerable experience of accelerated testing has been obtained and has led to an ability to predict glass fibre strength values in GRC out to much longer periods. In fact a quantitative relationship has been established between accelerated ageing in hot water in the laboratory and real weathering exposure in different climates (Refs. 16, 19). This has been based on three main areas of work:

(i) studies of the extent and kinetics of glass fibre strength reduction in a cement environment in hot/wet accelerated conditions.

(ii) studies of the strength loss in GRC composites in similar accelerated conditions.

(iii) comparison of the above results with strength loss in GRC composites after weathering over several years in different climates.

The procedure and results have been described in detail in Reference 16.

The glass fibre strength changes were studied by means of a specially developed test (The Strand-in-Cement or SIC test) in which a short length of typical glassfibre reinforcement strand is embedded in a small cement block. The direct tensile strength of the glassfibre strand, after ageing in a water bath, was measured by pulling on the ends of the strand (Refs.13, 16). The time taken for the strand strength to fall to a given value (e.g. 1000, 900, 800 MN/m^2 etc.) was taken as a measure of the rate of the strength controlling cement glass interaction and plotting the logarithm of this time against the inverse of the absolute temperature gave a family of parallel straight lines (Ref.16). This indicated that one reaction mechanism, or certainly only one activation energy, was applicable across the whole temperature and strength ranges - suggesting the validity of extrapolating long term behaviour at low temperatures from data obtained from hot, short term, tests.

The strength changes for GRC composites in both hot wet accelerated tests and real weathering exposure is summarised in Fig.8. The accelerated ageing results from tests after exposure in water at 50°C, 60°C and 80°C, on the left hand side of Fig.8, show a consistent pattern of falling strength, where the composite strength is directly proportional to the SIC strength (Ref.19) followed by long term stability. The weathering results on the right hand side of Fig.8 which extend over 10 years in the UK and from 2 to 5 years in other climates show a parallel initial fall in strength, with no unexpected changes in pattern due to variations of drying out, wetting, temperature, freeze-thaw, etc, experienced in real climates.

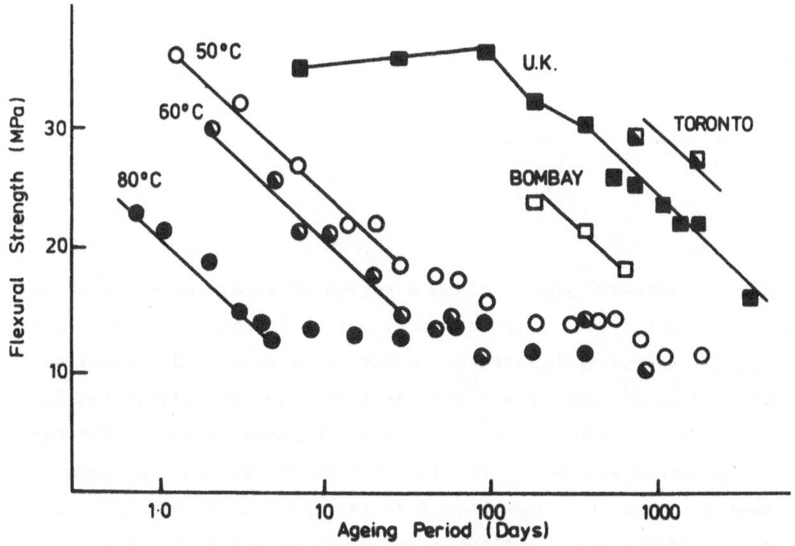

Figure 8. Flexural Strengths of GRC Composites in Accelerated Ageing and Real Weathering

Further reaction rate analysis of these results indicated that the same activation energy applied to strand strength changes in hot wet immersion, composite strength changes in similar hot wet accelerated conditions and, if the mean annual temperature of the climate was used to define the weathering condition, to composite strength changes in different climates. This is illustrated in Fig.9 by means of a "normalised" type of Arrhenius plot in which the relative rate of strength loss, across a wide range of strength levels for both strands and composites, is plotted against 1/T°K (Ref.16).

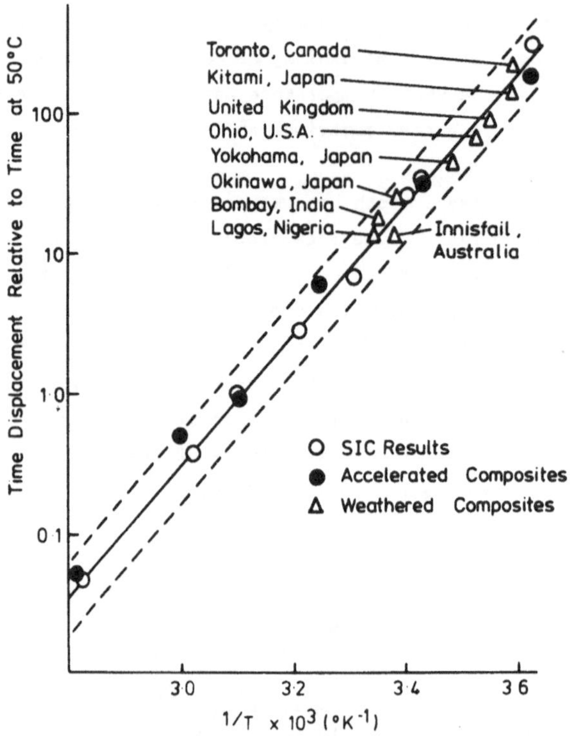

Figure 9. Normalised rates of strength loss in accelerated ageing and real weathering conditions. SIC's and GRC composites.

"Acceleration factors" which relate a period of accelerated ageing at some temperature to a lifetime in a given climate, may be deduced from the relative time ordinate of Fig.9. Use of such factors enables the accelerated composite strengths of Fig.8 to be plotted on an appropriate time axis in order to predict the long term strength behaviour of GRC composites. This has been done in Fig.10 for the UK climate and compared with actual weathering results over 10 years. The agreement is very close. The same procedure has been used more recently to average and summarise weathering results from a number of hot climates to show (Ref.17) that the levelling out of composite strength values to a long term stable value is also beginning to be observed in real weathering conditions. The most recent weathering results from several climates are summarised in Fig.11. Average flexural strength results from the UK up to 12 years are plotted as solid points. Results from hot climates have been transposed to the appropriate age in UK weather using transposition factors based on mean annual temperatures for the different sites (Ref.16, 17) in order to provide a better overall comparison. The intersite agreement is seen to be very good and the stabilising of composite strength in the long term is clearly seen.

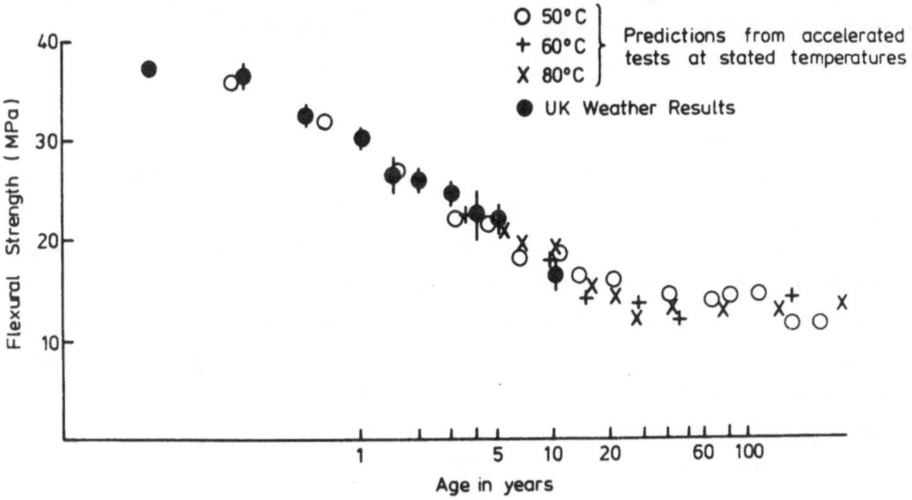

Figure 10. Comparison of predicted and observed strengths for GRC Composites in UK Weather

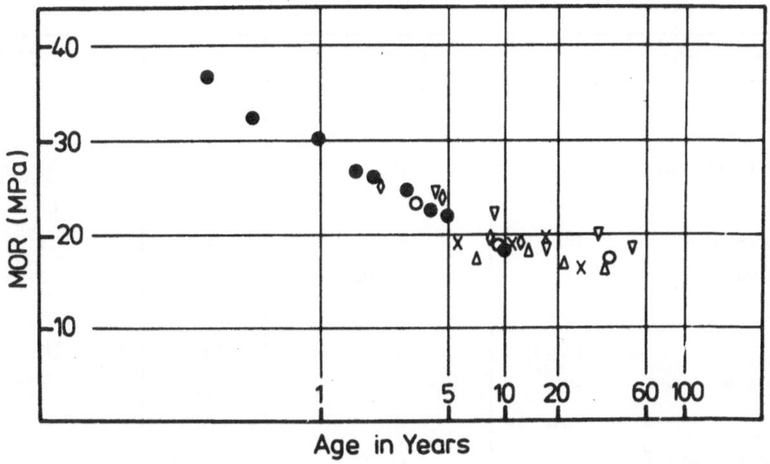

Figure 11. Strengths of GRC composites in different climates transposed to UK mean temperatures

• = UK (actual age), Δ = Cloncurry, x = Innisfail, ◇ = Arizona, O = Lagos, ∇ = Bombay; all transposed to UK.

The results summarised in Figures 9, 10 and 11 establish the validity of a methodology of predicting the long term strength of GRC from the results of short term accelerated tests. This has been used to provide quantitative long term strength predictions, for design purposes, for good quality sprayed GRC - the methodology has also been used to provide a type test on the suitability of components made from non-standard GRC formulations by performing load tests after a period of accelerated ageing of the component.

The results shown in Figures 7 to 11 above were obtained with the original, first generation, Cem-FIL A.R. Glassfibre. Subsequent work showed that the same accelerated ageing techniques could be applied to the improved durability Cem-FIL 2 fibre, enabling reasonable predictions to be made of the improvement in long term composite performance (Refs.19, 20, 28, 29).

Overall then we see that sprayed GRC materials lose strength and toughness over a number of years weathering exposure. The rate at which that happens depends on the climate of exposure-being faster for hot climates. However a stable strength value is reached independent of climate and this gives useful properties for long term application. Significant reduction in the rate of strength loss and extension of the period of "tough" behaviour can be obtained by the use of more durable fibres such as Cem-FIL 2. The long term end property values remain the same however. For spray dewatered material the stable long term bending strength is estimated to be about 14 MPa, for direct spray/handspray about 13 MPa. Premix materials have lower initial strengths and long term strengths of about 10 MPa depending on fibre content and process (Table 3).

TABLE 3 ESTIMATED LONG TERM STABLE PROPERTY VALUES FOR CEM-FIL GRC

PROPERTY \ PROCESS		SPRAY DEWATERED	HAND SPRAY	VIBRATION CAST PREMIX
Compressive strength	MPa	80	75	50
Modulus	GPa	30	25	20
Impact strength	Nmm/mm^2	4	4	3
Poissons'Ratio		0.24	0.24	0.24
Bending: LOP	MPa	12	10	8
MOR	MPa	14	13	10
Tension: BOP	MPa	6	5.5	4.5
UTS	MPa	6	5.5	4.5
Strain-to-failure %		0.04	0.04	0.03
In-plane shear strength	MPa	6	5.5	4.5
Interlaminar shear	MPa	4	4	N.A.

Pilkington recommendations for design stress levels take account, and have always taken account, of these strength changes on ageing (Table 4). For sprayed materials the stress level of 6 MPa in bending, for instance, is less than half the estimated long term strength, well within the linear portion of the stress-strain curve and below the matrix cracking stress level.

TABLE 4

STRESS TYPE	LOADING EXAMPLE	DESIGN STRESSES MPa	
		5% SPRAY GRC	3.5% PREMIX GRC
Compressive	Compressive	12	12
Bending	Bending solid beams or plates	6	4
Tensile	Cylindrical hoop stress	3	2
Tensile	Bending sandwich panels	3	2
Web Shear	In-plane shear of webs in box sections	2	1
Bearing Shear	Shear loading at bearing positions	1	1

CONCLUSIONS

Additions of relatively large proportions of zirconia to silicate glasses have led to the development of a range of commercially available alkali resistant glass fibres with sufficient strength retention to provide useful reinforcement in cement product.

Fibre strength loss and composite property loss still occurs on ageing. Some progress has been made in significantly reducing the rate at which this occurs by using surface treated fibres - further progress via composition development has been shown to be possible - whether it is commercialised or not depends on the perceived balance of cost/performance advantage factors in the market place.

Extensive weathering programmes and accelerated test work in the laboratory show that currently available GRC materials do provide stable and useable long term strength, albeit with a brittle failure characteristic. Design must take account of this, which has always been done by responsible GRC users.

ACKNOWLEDGEMENT

This paper is published with the permission of the Directors of Pilkington Brothers P.L.C. and of Mr. A. S. Robinson, Director of Group Research and Development.

REFERENCES

1. Warren, B. E., Krutter, H. and Morningstar, O., Jnl. American Ceramic Soc., Vol.19, 1936, pp.202-206.

2. Proctor B. A., The Long Term Behaviour of Glassfibre Reinforced Composites, Glass - Current Issues, Proceedings NATO Advanced Study Institute April 1984, edited A. F. Wright and J. Dupuy, Martinus Nijhoff, 1985, pp.524-550.
3. Biryukovitch K. L. and Biryukovitch Yu. L., Stroit. Mater. (II) Vol.1, 1961, pp.18-20.
4. Proctor B. A. and Yale B., Phil. Trans. Roy. Soc. Lond., Vol.A294, 1980, pp.427-436.
5. Majumdar A. J. and Ryder J. F., Glass Technology Vol.9, No.3, June 1968, pp.78-84.
6. Dimbleby V. and Turner W. E. S., Jnl. Soc. Glass Technology, Vol.10 1926, pp.304-358.
7. Majumdar A. J., Proc. Roy. Soc. Lond. Vol.A319, 1970, pp.69-78.
8. Larner L. J., Speakman K. and Majumdar A. J., Jnl. Non-crystalline Solids, Vol.20, 1976, pp.43-74.
9. Proctor, B. A., Alkali Resistant Glass Fibres for Reinforcement of Cement, Glass - Current Issues, Proceedings NATO Advanced Study Institute April 1984, edited A. F. Wright and J. Dupuy, Martinus Nijhoff, 1984, pp.555-573.
10. MacKenzie J. D., US Bureau of Mines, Materials Research Contract Report J0199056, 1981, University of California.
11. Ramachandron B. E., Velpari V. and Balasubramanian N., Jnl. Materials Science, Vol.16 1981, pp.3393-3397.
12. Paul A. and Youssefi A., Jnl. Materials Science, Vol.13, 1978, pp.97-107.
13. Litherland K. L., Maguire P. and Proctor B. A., Int. Jnl. Cement Composites and Lightweight Concrete, Vol.6 No.1, 1984, pp.39-45.
14. Building Research Establishment Current Paper CP38/76, June 1976, The Department of the Environment, UK.
15. Building Research Establishment Information Paper IP36/79, November 1979, The Department of the Environment, UK.
16. Litherland K. L., Oakley D. R. and Proctor B. A., Cement and Concrete Research, Vol.11, 1981, pp.455-466.
17. Aindow A. J., Oakley D. R. and Proctor B. A., Cement and Concrete Research, Vol.14, 1984, pp.271-274.
18. Gair H., Proceedings Int. Congress on Glassfibre Reinforced Cement, London 1979, edited Hayden Jeffery, The Glassfibre Reinforced Cement Association, p.121.
19. Proctor B. A., Proceedings Int. Congress of GRCA, November 1981, edited Vincent Blake, The Glassfibre Reinforced Cement Association, pp.50-67.
20. Proctor B. A., Oakley D. R. and Litherland K. L., Composites, April 1972, pp.173-179.
21. "ARFIBRE: Alkali-Resistant Glass Fiber for GRC", Asahi Glass Co. Ltd., GRC Division Literature.

22. Fyles K., Litherland K. L. and Proctor B. A., "The Effect of Glass Fibre Compositions on the Strength Retention of GRC", RILEM Symposium FRC 86, Developments in Fibre Reinforced Cement and Concrete, Vol.2, Paper 7.5.

23. Litherland K. L. and Proctor B. A., "The Effect of Matrix Formulation, Fibre Content and Fibre Composition on the Durability of Glassfibre Reinforced Cement", Proceedings, Durability of Glass Fiber Reinforced Concrete Symposium, edited S. Diamond, Prestressed Concrete Institute 1986, pp.124-135.

24. Proctor B. A., Composites, June 1971, pp.85-92.

25. Proctor B. A., "Properties and Performance of GRC", Proceedings Fibrous Concrete Symposium, Concrete International 1980, The Concrete Society, pp.69-86.

26. Proctor B. A., Oakley D. R. and Wiechers W. 'Tensile stress/strain characteristics of glass-fibre reinforced cement', 'Composites - Standard, Testing and Design' IPC Science and Technology Press, 1974, pp.106-107.

27. Aveston J., Cooper G. A. and Kelly A., 'Single and multiple fracture' in The Properties of Fibre Composites, IPC Science and Technology Press, 1971, pp.15-26.

28. Oakley, D. R., Litherland, K. L. and Proctor, B. A., Development of a Method of Predicting the Weathering Behaviour of Glass Reinforced Cement Composites, Proceedings Second International Conference on the Durability of Building Materials and Components, National Bureau of Standards, Gaithersburg, Maryland, USA, 14th-16th September 1981, pp.225-231.

29. Litherland, K. L. and Proctor, B. A., Predicting the Long Term Strength of Glass Fibre Cement Composites, RILEM Symposium FRC 86, Developments in Fibre Reinforced Cement & Concrete, Vol.2, Paper 7.6.

Proceedings of the International Symposium on Fibre Reinforced Concrete, December 16–19, 1987, Madras, India

DEVELOPMENT OF LIGHTWEIGHT CARBON FIBER REINFORCED FLY ASH-CEMENT COMPOSITES

Y.Ohama, K. Demura and Y.Sato
College of Engineering , Nihon University
Koriyama, Fukushima-ken, 963 Japan

Synopsis: For the purpose of an effective application of coal ash which is discharged in large quantities at thermal power stations, the process technology of lightweight carbon fiber reinforced fly ash-cement composites is developed and their properties are examined in the present paper. The lightweight carbon fiber reinforced fly ash-cement composites using perlite and a small amount of styrene-butadiene rubber latex are prepared with various carbon fiber and foaming agent contents, and autoclaved under proper conditions. After autoclaving, their properties such as strengths, deformation characteristics, water absorption, drying shrinkage and impact resistance are tested. As a result, their optimum mix proportions are successfully proposed. The polymer-modified lightweight carbon fiber reinforced fly ash-cement composites obtained by applying the mix proportions have a bulk density of 1.2 to 1.3kg/dm^3 and a flexural strength of 7.5 to 11MPa, and can be nailed, sawed and planed like wood.

INTRODUCTION

In recent years, a need for the development of lightweight and tough building materials has evidently been highlighted with the advancement in building technology which is represented in the construction of highrise buildings. Lightweight carbon fiber reinforced concrete as an example of such building materials has already been developed by Akihama et al., and used in practical applications [1]. On the other hand, a large quantity of coal ash has been discharged from thermal power plants, and creates a serious problem necessitating the finding of its useful applications in Japan. Recently the research and development of useful cementitious composites using the coal ash as its new applications have actively been conducted [2],[3]. From these investigations, it seems that substitution of fly ash for cement content of about 50% is possible in the production of such composites. The purpose of this study is to develop the lightweight and tough building composites using a large amount of fly ash, and to make clear the properties of the composites.

In this paper, lightweight carbon fiber reinforced fly ash-cement composites (lightweight CFR-ACCs) are prepared with various carbon fiber and foaming agent contents by applying autoclave cure, and tested for flexural and compressive strengths. The effect of polymer modification on the properties such as strengths, deformation characteristics, water absorption, drying shrinkage and impact resistance of the lightweight CFR-ACCs is also examined using styrene-butadiene latex. From the test results, the optimum mix proportions of the lightweight CFR-ACCs are discussed.

MATERIALS

Materials for Matrixes

Commercial fly ash (A), ordinatry portland cement(C) and perlite (bulk specific gravity, 0.34; particle size, 1.2-2.5mm) specified in JIS (Japanese Industrial Standard) A 6201 (Fly Ash), JIS R 5210 (Portland Cement) and JIS A 5007 (Perlite) respectively were used for matrixes. The chemical and physical properties of the fly ash is listed in Table 1.

Admixtures

Commercially available higher alcohol-sulfate type foaming agent (FA) for incorporation of air bubbles and dispersion of carbon fiber and styrene-butadiene rubber (SBR) latex (total solids, 44.6%; specific gravity, 1.020 at 20°C; pH, 7.83 at 20°C; viscosity, 147mPa·s at 20°C)were employed.

Carbon Fibers for Reinforcements

Commercially available low modulus carbon fibers were used for reinforcements, whose physical properties are listed in Table 2.

Table 1. Chemical Compositions and Physical Properties of Fly Ash.

Chemical Compositions (%)									
ig. loss	SiO_2	Al_2O_3	Fe_2O_3	CaO	SO_3	Na_2O	MgO	K_2O	Total
3.5	59.4	23.6	3.6	3.2	0.2	0.8	1.8	1.9	98.0

Physical Properties			
Size (mm)	Specific Gravity (20 °C)	Water Content (%)	Organic Impurities
$<4.4 \times 10^{-2}$	2.10	< 0.1	Nil

Table 2. Physical Properties of Carbon Fibers.

Carbon Fiber Type	Average Diameter (μm)	Average Length (mm)	Specific Gravity (20 °C)	Young's Modulus (MPa)	Tensile Strength (MPa)
Low Modulus Carbon Fibers	14.5	3.0	1.63	386	8.3

EFFECTS OF CARBON FIBER AND FOAMING AGENT CONTENTS ON FLEXURAL AND COMPRESSIVE STRENGTHS OF LIGHTWEIGHT CFR-ACCS

Testing Procedures

Preparation of lightweight CFR-ACC specimens

According to JIS R 5201 (Physical Testing Methods for Cement), lightweight CFR-ACCs were mixed with the mix proportions indicated in Table 3 by using an ordinary mortar

Table 3. Mix Proportions of Lightweight CFR-ACCs.

A : C (By Weight)	Perlite Content (vol% of A+C)	FA Content (wt% of A+C)	CF Content (vol%)	Water-(A+C) Ratio (%)	Flow
1 : 1	30	0.0	0.0	35.5	183
			0.5	36.8	178
			1.0	40.3	181
			1.5	47.0	180
			2.0 *	—	—
		0.1	0.0	36.5	182
			0.5	38.0	183
			1.0	42.0	185
			1.5	45.5	178
			2.0	50.0	179
		0.2	0.0	37.5	183
			0.5	39.5	182
			1.0	43.0	183
			1.5	47.0	185
			2.0	52.5	183
		0.3	0.0	39.0	183
			0.5	41.0	177
			1.0	44.0	175
			1.5	48.0	185
			2.0	55.0	183

Note; * Unable to mix.

mixer. Lightweight CFR-ACC speciments 40x40x160mm were molded, and then precured at 20°C and 80% relative humidity for 24 hours. Immediately after precuring, the specimens were cured in autoclave in which the maximum temperature of 180°C was maintained under a pressure of 1MPa for 3 hours.

Strength tests

The cured lightweight CFR-ACC specimens were tested for flexural strength with center-point loading at a crosshead speed of 0.5mm/min by using the Instron universal testing machine. The portions of the lightweight CFR-ACC specimens broken in flexural test were tested for compressive strength according to JIS R 5201.

Test Results and Discussion

The effects of carbon fiber and foaming agent contents on the flexural and compressive strengths of lightweight CFR-ACCs are illustrated in Figs 1. and 2. Generally the flexural strength of the lightweight CFR-ACCs increases with raising carbon fiber content. Regardless of the foaming agent content, the flexural strength of the lightweight CFR-ACCs provides the maximum value at a carbon fiber content of 1.5vol%. In particular, the flexural strength of the lightweight CFR-ACCs with a foaming agent content of 0.1% attains to approximately 7.5MPa at a carbon fiber content of 1.5vol%, and this value is about 2.5 times that of unreinforced lightweight ACCs. Such a flexural strength reduction of the lightweight CFR-ACCs at a carbon fiber content of 2.0vol% is likely to be due to the formation of some fiber balls in mixing process. On the other hand, the flexural strength decreases at a foaming agent content of 0.2% or more irrespective of the carbon fiber content. This may be explained to be due to the fact that the addition of foaming agent causes an increase in entrained air in the CFR-

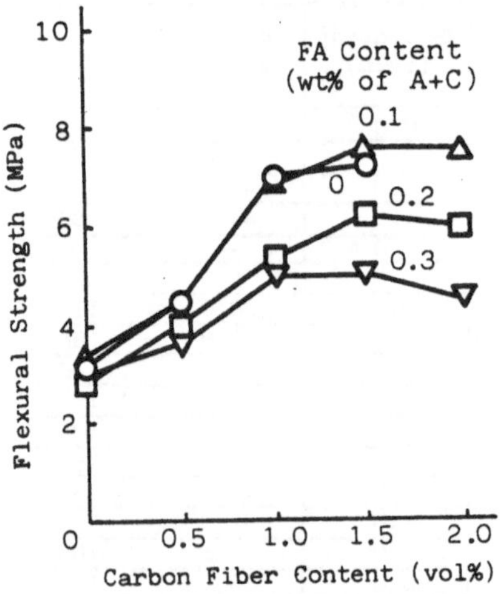

Fig.1. Effects of Carbon Fiber and Foaming Aagnt Contents on Compressive Strength of Lightweight CFR-ACCs.

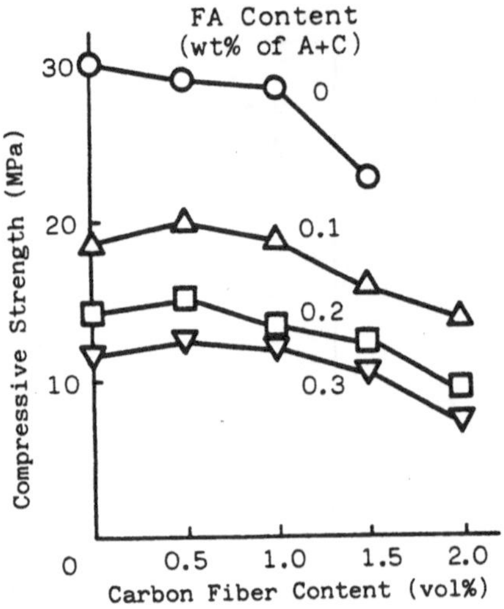

Fig.2. Effects of Carbon Fiber and Foaming Agent Contents on Flexural Strength of Lightweight CFR-ACCs.

ACCs, and a decrease in the bond between the carbon fibers and the matrix. In general, the compressive strength of the lightweight CFR-ACCs decreases with increasing carbon fiber and foaming agent contents. The effect of the foaming agent content on the compressive strenght decrease is more notable than that of the carbon fiber content. From the above test results, the optimum mix proportions of the lightweight CFR-ACCs are recommended as follows:

Fly ash : cement = 1 : 1 (by weight)
Perlite content (vol% of fly ash+cement) : 30
Foaming agent content (wt% of fly ash+cement) : 0.1
Carbon fiber content (vol%) : 1.5
Water-(fly ash+cement) ratio (%) : 45.5

PROPERTIES OF LIGHTWEIGHT CFR-ACCS

Testing Procedures

Preparation of lightweight CFR-ACC specimens

Lightweight CFR-ACC specimens were prepared with the mix proportions and geometry shown in Tables 4 and 5 respectively by the same methods as described above.

Tests for physical properties

Lightweight CFR-ACC beam specimens 40x40x160mm and cylindrical specimens 50x100mm were tested for flexural and compressive strengths by the same methods as stated above. The load-deflection curves were automatically recorded in the flexural strength test.

Lightweight CFR-ACC briquet specimens were tested for tensile strength according to ASTM C 190 (Test Method for Tensile Strength of Hydraulic Cement Mortars).

Lightweight CFR-ACC beam specimens 40x40x160mm were tested for water absorption by

Table 4. Mix Proportions of Lightweight CFR-ACCs.

Type of Composite	A : C (By Weight)	Perlite Content (vol% of A+C)	FA Content (wt% of A+C)	CF Content (vol%)	Polymer-(A+C) Ratio(%)	Water-(A+C) Ratio(%)	Flow
Un-modified	1 : 1	30	0.1	0.0	0.0	36.5	182
				0.5	0.0	38.0	183
				1.0	0.0	42.0	185
				1.5	0.0	45.5	178
				2.0	0.0	50.0	179
SBR-modified	1 : 1	30	0.0	0.0	5.0	31.0	180
				0.5	5.0	34.0	185
				1.0	5.0	38.4	185
				1.5	5.0	41.0	179
				2.0	5.0	55.0	180

Table 5. Geometry of Lightweight CFR-ACC Specimens.

Type of Test	Geometry of Specimen
Flexural Strength	Beam, 40x40x160 mm
Compressive Strength	Cylinder, 50x100 mm
Tensile Strength	Briquet Specified in ASTM C 190
Water Absorption	Beam, 40x40x160 mm
Impact Resistance	Plate, 200x200x10 mm
Drying Shrinkage	Beam, 40x40x160 mm

a water immersion method in accordance with JIS A 6203 (Polymer Dispersions for Cement Modifiers).

Lightweight CFR-ACC plate specimens 200x200x10mm were placed on sand layer with a thickness of 10cm, and tested for impact resistance with consecutive drop blows of a steel ball with a weight of 80g from a height of 50cm in conformity with JIS A 4121 (Method of Impact Test for Boards of Buildings). The number of blows required to cause the complete failure of the specimens was counted.

Drying shrinkage test was performed on lightweight CFR-ACC beam specimens 40x40x160mm. Immdiately after autoclaving, their original gage length was measured. Then they were cured at 20°C and 50% relative humidity for the desired period, and their drying shrinkage was determined by the comparator method specified in JIS A 1129 (Methods of Test for Length Change of Mortar and Concrete).

Test Results and Discussion

The effects of carbon fiber content and polymer modification on the flexural strength of lightweight CFR-ACCs are shown in Fig.3. The flexural strength of the unmodified and SBR-modified lightweight CFR-ACCs attains to the maximum at a carbon fiber content of 1.5vol%. Regardless of the carbon fiber content, the flexural strength of the SBR-modified lightweight CFR-ACCs is much higher than that of the unmodified lightweight CFR-ACCs. In particular, the highest flexural strength of the SBR-modified lightweight CFR-ACCs with a carbon fiber content of 1.5vol% is about 11MPa. This value is considerably high as compared to 7.5MPa of the data obtained on lightweight carbon fiber reinforced concrete with a carbon fiber content of 2.0vol% and a bulk density of about 1.0kg/dm^3 by Akihama et al. [1], although there are differences in the mix

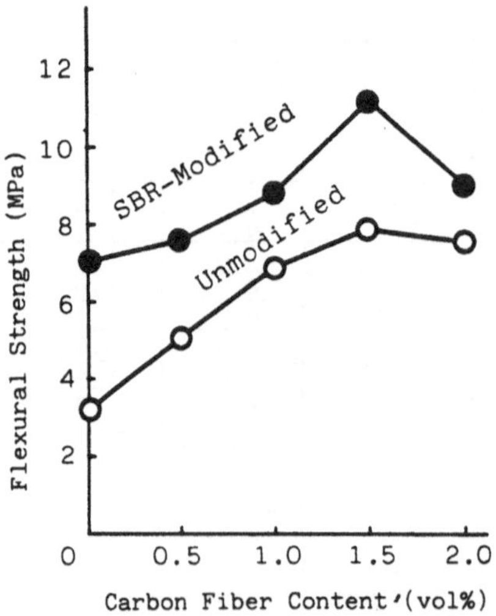

Fig.3. Carbon Fiber Content vs. Flexural Strength of Lightweight CFR-ACCs.

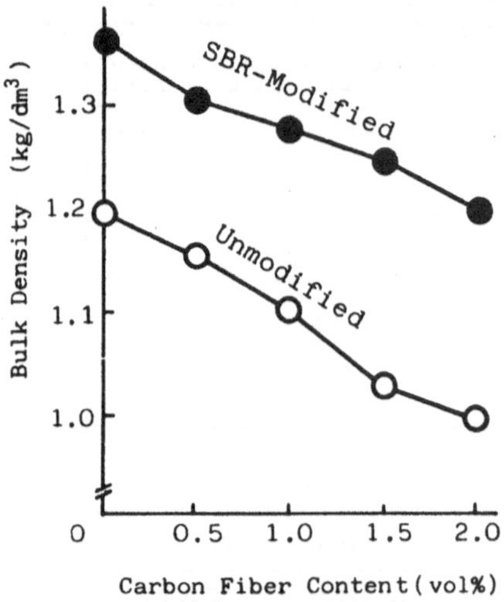

Fig.4. Carbon Fiber Content vs. Bulk Density of Lightweight CFR-ACCs.

proportions, specimen size, etc. between both. The relationship between the carbon fiber content and bulk density of the lightweight CFR-ACCs is presented in Fig.4. As seen in Fig.4, the bulk density decreases with an increase in the carbon fiber content. Regardless of the carbon fiber content, the bulk density of the unmodified CFR-ACCs is somewhat lower than that of the SBR-modified CFR-ACCs. Fig.5 demonstrates the specific flexural strength calculated as a ratio of the flexural strength, σ_f (MPa) to the bulk density, ρ (kg/dm^3). From Fig.5, the effective carbon fiber content is found to be 1.5vol%.

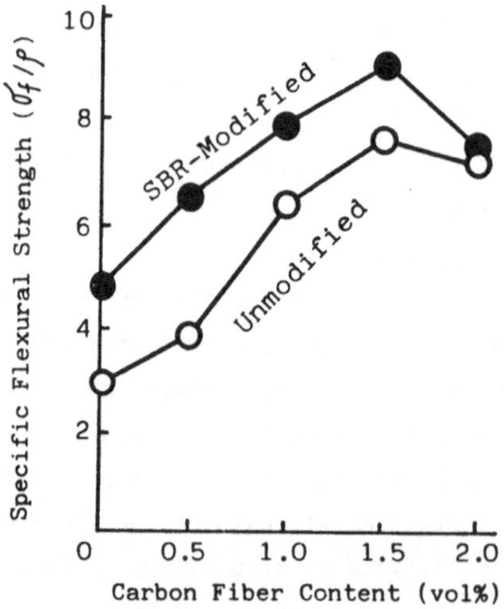

Fig.5. Carbon Fiber Content vs. Specific Flexural Strength of Lightweight CFR-ACCs.

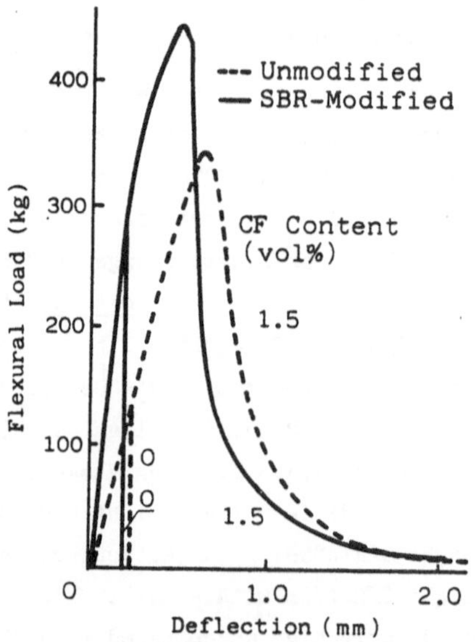

Fig.6. Flexural Load-Deflection Curves for Lightweight CFR-ACCs with Carbon Fiber Content of 1.5 vol%.

Fig.6 exhibits the flexural load-deflection curves for lightweight CFR-ACCs with carbon fiber contents of 0.0 and 1.5vol%. The flexural deformation behavior of the unmodified and SBR-modified lightweight CFR-ACCs is improved with carbon fiber reinforcement and polymer modification. A drop in the flexural load of the unmodified and SBR-modified lightweight CFR-ACCs with a carbon fiber content of 1.5vol% immediately after cracking is relatively gentle in comparison with unreinforced ones.

Fig.7 represents the relation between the carbon fiber content and compressive strength of lightweight CFR-ACCs. The compressive strength of the unmodified and SBR-modified lightweight CFR-ACCs decreases with raising carbon fiber content. This tendency well corresponds to a reduction in the bulk density with increasing carbon fiber content as seen in Fig.4.

Fig.8 indicates the relation between the carbon fiber content and tensile strength of lightweight CFR-ACCs. This relation is almost the same as that between the flexural strength and carbon fiber content as shown in Fig.3. The highest tensile strength values of the unmodified and SBR-modified lightweight CFR-ACCs at a carbon fiber content of 1.5vol% are approximately 2.1 and 3.5MPa respectively, and about 1.4 and 2.3 times those of the unreinforced unmodified and SBR-modified lightweight ACCs respectively.

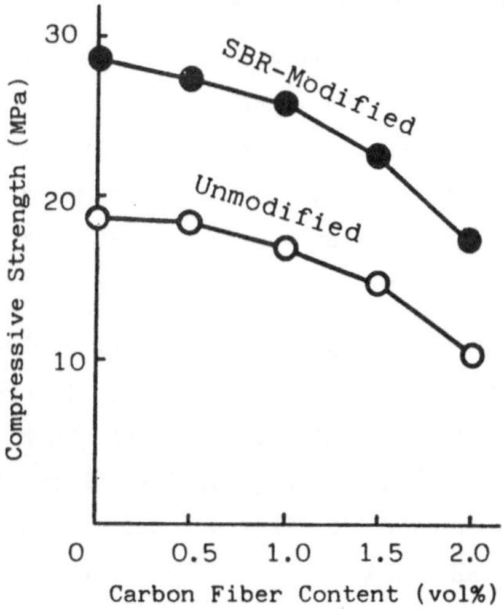

Fig.7. Carbon Fiber Content vs. Compressive Strength of Lightweight CFR-ACCs.

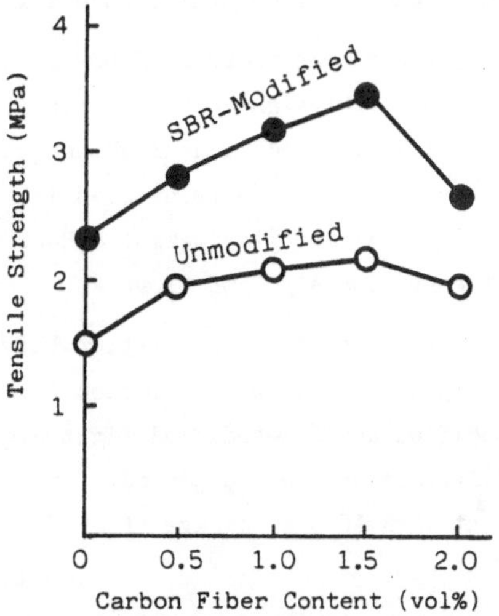

Fig.8. Carbon Fiber Content vs. Tensile Strength of Lightweight CFR-ACCs.

The relation between the carbon fiber content and water absorption of lightweight CFR-ACCs is shown in Fig.9. The water absorption of the unmodified and SBR-modified lightweight CFR-ACCs tends to increase with an increase in the carbon fiber content. This well corresponds to a trend in the bulk density. Regardless of the dry conditions of the lightweight CFR-ACCs, the water absorption of the SBR-modified lightweight CFR-ACCs is lower than that of the unmodified lightweight CFR-ACCs. This is attributed to the improved resistance to water absorption of the SBR-modified lightweight CFR-ACCs due

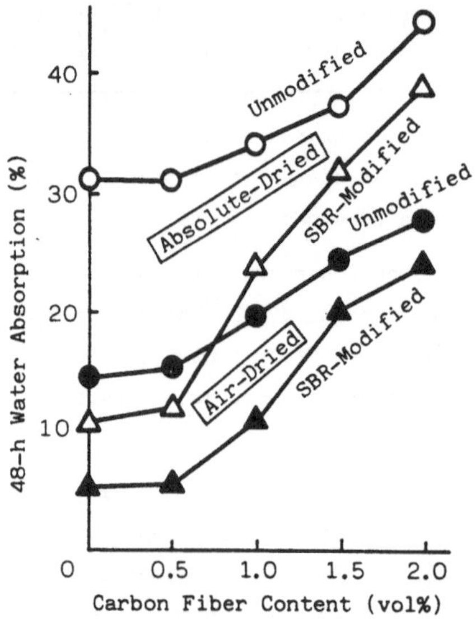

Fig.9. Carbon Fiber Content vs. Water Absorption of Lightweight CFR-ACCs.

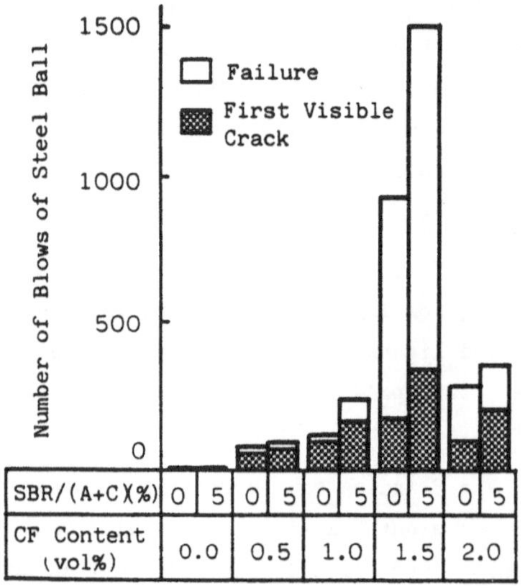

Fig.10. Effects of Carbon Fiber Content and Polymer Modification on Impact Resistance of CFR-ACCs.

to the impermeable polymer films formed in the matrixes by polymer modification.

The effects of carbon fiber content and polymer modification on the impact resistance of lightweight CFR-ACCs are given in Fig.10. The impact resistance of the unmodified and SBR-modified lightweight CFR-ACCs is markedly improved with raising carbon fiber content. Unreinforced lightweight ACCs are fractured at about ten blows, but the SBR-modified lightweight CFR-ACCs with a carbon fiber content of 1.5vol% are broken at approximately 1500 blows.

Fig.11 illustrates the relation between the dry curing period and drying shrinkage of lightweight CFR-ACCs with a carbon fiber content of 1.5vol%. The drying shrinkage of the unmodified and SBR-modified lightweight CFR-ACCs after autoclave curing increases with additional dry curing period. However, the drying shrinkage of the SBR-modified lightweight CFR-ACCs at an age of 28 days is much smaller than that of the unmodified

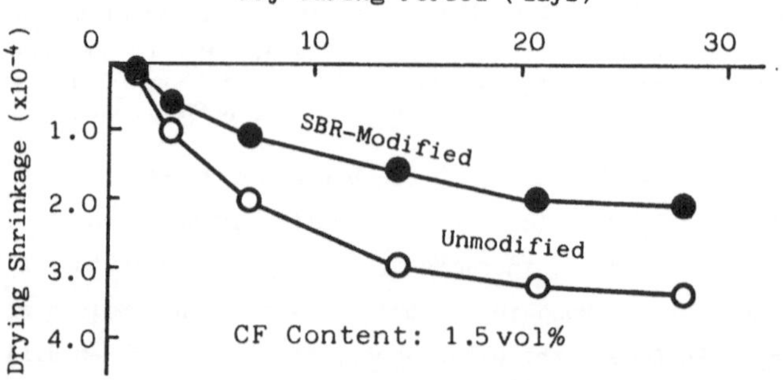

Fig.11. Dry Curing Period vs. Drying Shrinkage of Lightweight CFR-ACCs.

lightweight CFR-ACCs, and about 3/5 of that of unmodified lightweight CFR-ACCs.

Furthermore, lightweight CFR-ACCs can be nailed, sawed and planed in the same manner as wood.

CONCLUSIONS

(1) The properties such as flexural and tensile strengths, impact resistance and drying shrinkage of lightweight CFR-ACCs are greatly improved by the addition of a small amount of SBR latex. In paticular, the flexural strength of SBR-modified lightweight CFR-ACC with a carbon fiber content of 1.5vol% attains to about 11MPa with a low bulk density of $1.3 kg/dm^3$. The lightweight CFR-ACCs can be nailed, sawed and planed like wood.

(2) The optimum mix proportions of unmodified and SBR-modified lightweight CFR-ACCs are recommended as shown in Table 6. By applying such mix proportions and autoclave curing, it is possible to produce the lightweight CFR-ACCs having excellent properties in a short time, using a large amount of fly ash as a matrix component.

Table 6. Optimum Mix Proportions of Unmodified and SBR-Modified Lightweight CFR-ACCs.

Type of Composite		Lightweight CFR-ACC	SBR-Modified Lightweight CFR-ACC
Mix Proportion (kg/m^3)	Fly Ash	455.0	437.0
	Cement	455.0	437.0
	Perlite	34.8	35.3
	Foaming Agent	0.9	0.0
	SBR Latex	0.0	98.0
	Carbon Fibers	21.6	21.6
	Water	418.0	339.0

REFERENCES

1. Akihama, S. and Suenaga, T., *Cement & Concrete*, No 449, July 1984, pp 38-47.
2. Torii, K., Kawamura, M., Hasaba, S. and Taniguchi, K., *Transactions of the Japan Concrete Institute*, Vol 7, Dec 1985, pp 1-8.
3. Kasai, Y., Matsui, I. and Kasami, H., *ibid.*, pp 9-16.

Proceedings of the International Symposium on Fibre Reinforced Concrete, December 16-19, 1987, Madras, India

FIBRE REINFORCED CONCRETE MACHINE FOUNDATIONS

A.K. SACHAN[1], D.K. SHARMA[2], C.V.S. KAMESWARA RAO[1]

1. DEPARTMENT OF CIVIL ENGINEERING, HARCOURT BUTLER TECHNOLOGICAL INSTITUTE, KANPUR, INDIA.

2. SMALL INDUSTRIES SERVICE INSTITUTE, KALPI ROAD, KANPUR, INDIA.

SYNOPSIS

Conventionally machine foundation base plates are made of cast iron. However, the high cost and the difficulties in casting the plates with cast iron point out the need for development of alternative materials. The special conditions under which the foundation plates and blocks have to serve need high strength,toughness, ductility and durability under oil exposure conditions. Concrete improved with polymers and fibres is a promising alternative material. The paper presents the results of an experimental investigation in which different mechanical properties of an epoxy glass fibre concrete are investigated.

INTRODUCTION

The suitability and cost effectiveness of fibre reinforced concrete for a number of structural applications are well established [1], [2]. Among the newer and promosing areas of application of fibre concrete is the construction of machine foundation blocks [3]. Machine foundation blocks are generally made of grey cost iron. However the high cost, long time of manufacture and the brittleness of cast eron motivate search for alternative materials. Several mechanical properties are relevant to the functional and structural design of machine bases. The paper presents the results of an experimental investigation in which the compressive strength, flexural tensile strength, fracture toughness, oil absorption of epoxy glass fibre concrete are determined.

EXPERIMENTAL PROGRAMME

Materials

Test sample were prepared using a basic mix, consisting of fine sand, coarse sand, coarse aggregate, high strength portland cement, epoxy and hardner. The relative proportions by weight respectively of the components were 105:110:610:55: 57:6. The mix was arrived at after some preliminary trials and economic viability consideration. Sand used was ordinary river sand; sand passing through 600 micron sieve and retained on 150 micron sieve was used as fine sand, sand passing through 2.36 mm sieve and retained on 1.18 mm sieve was used as coarse sand. Crushed granite stones passing through 15 mm sieve and retained on 10 mm sieve was used as coarse aggregate. Epoxy used was Araldite CY-212 and hardner was HY-951 grade. Chopped glass fibres were used as fibre reinforcement and the fibre content was varied between 0 to 1 % by weight of dry materials.

Test Specimens

Compressive strength tests were conducted on cubes of side 70.7 mm. Flexural strength tests were conducted on beams of size 40 mm x 40 mm x 160 mm. Fracture toughness tests were conducted on notched beam specimens of size 40 mm x 40 mm x 200 mm with notch depths being 10 mm, 15 mm and 20 mm. The tests were conducted on a Instron machine. Oil and cutting fluid absorption tests were conducted on specimens of size 40 mm x 62 mm x 160 mm.

EXPERIMENTAL RESULTS AND DISCUSSIONS

The variation of flexural strength with fibre content is shown in Fig.1. The

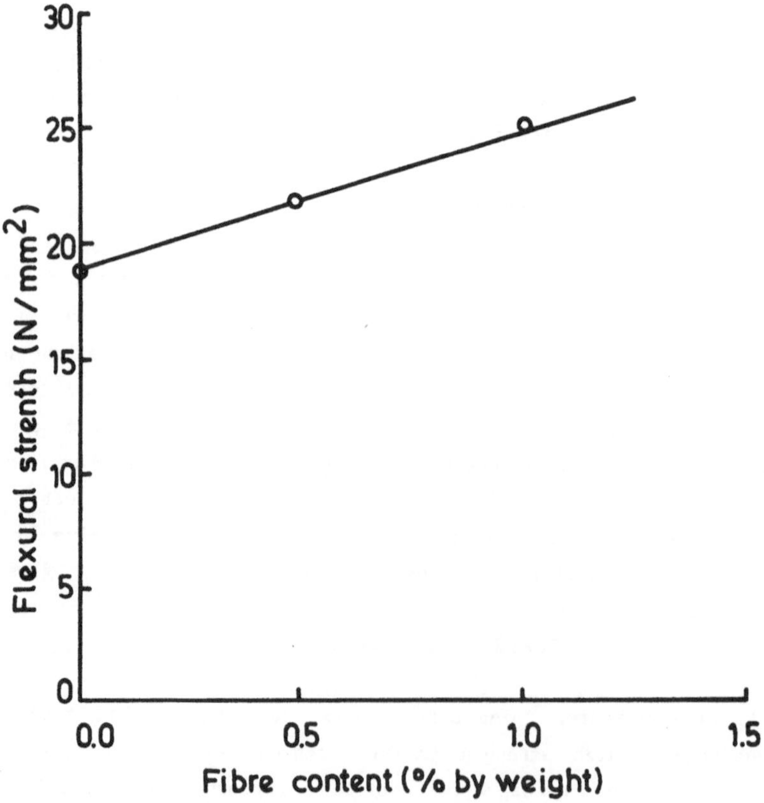

Fig. 1 Variation of flexural strength with fibre content.

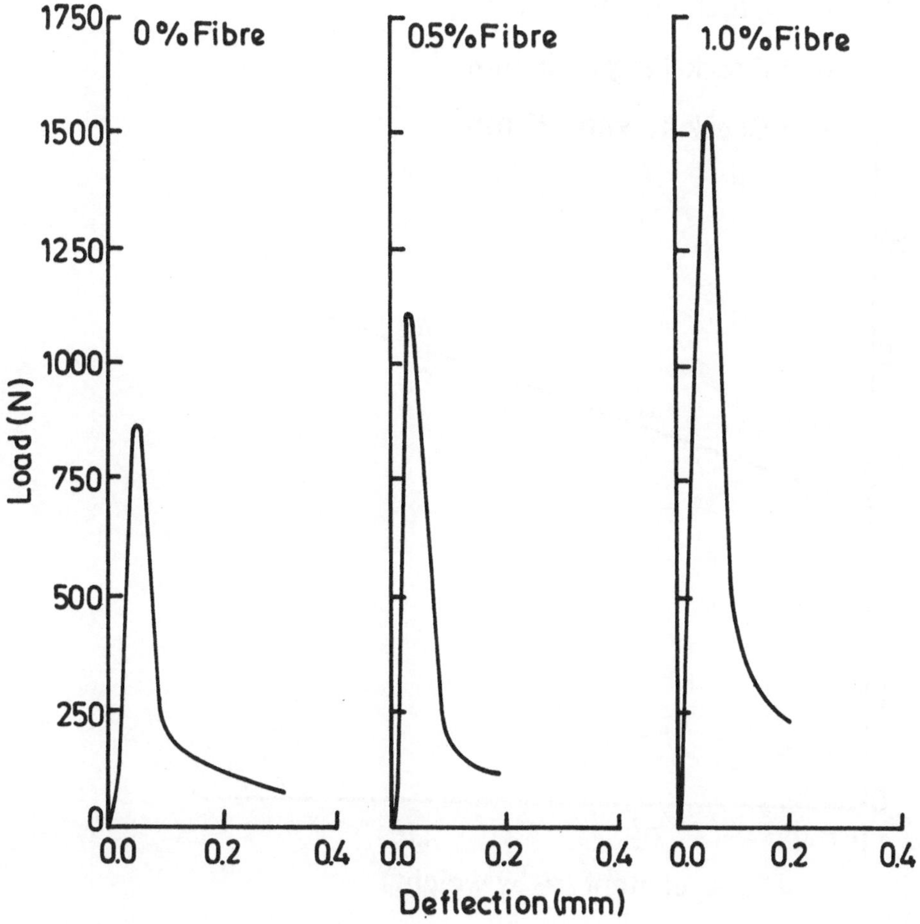

Fig. 2 Load deflection curves for different fibre content and notch depth of 10 mm.

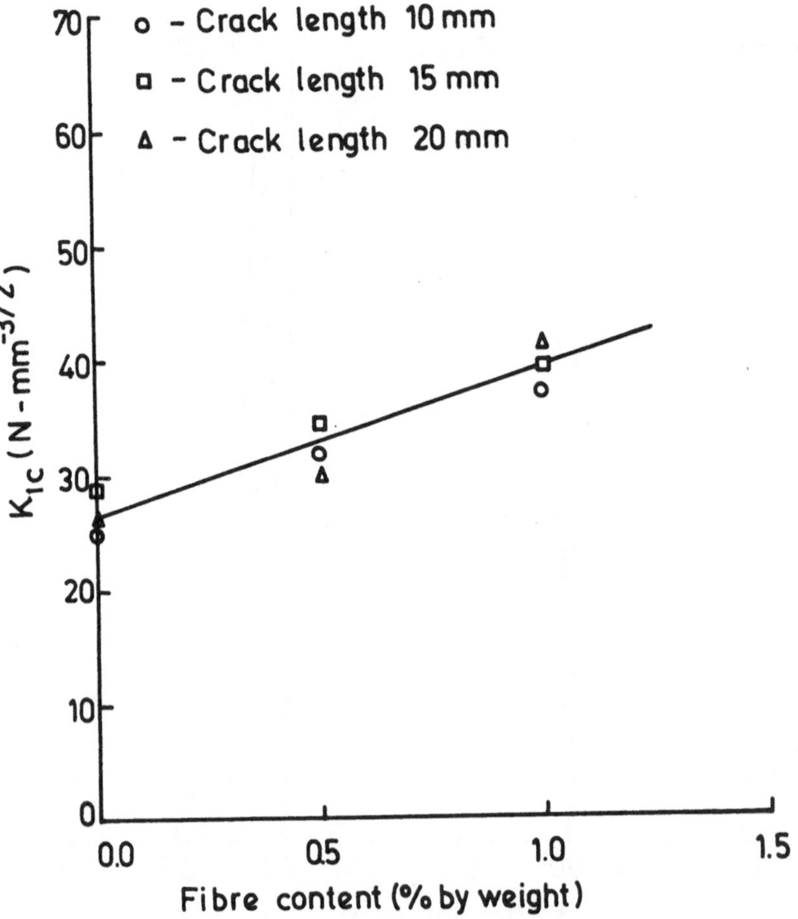

Fig. 3 Variation of fracture toughness with fibre content.

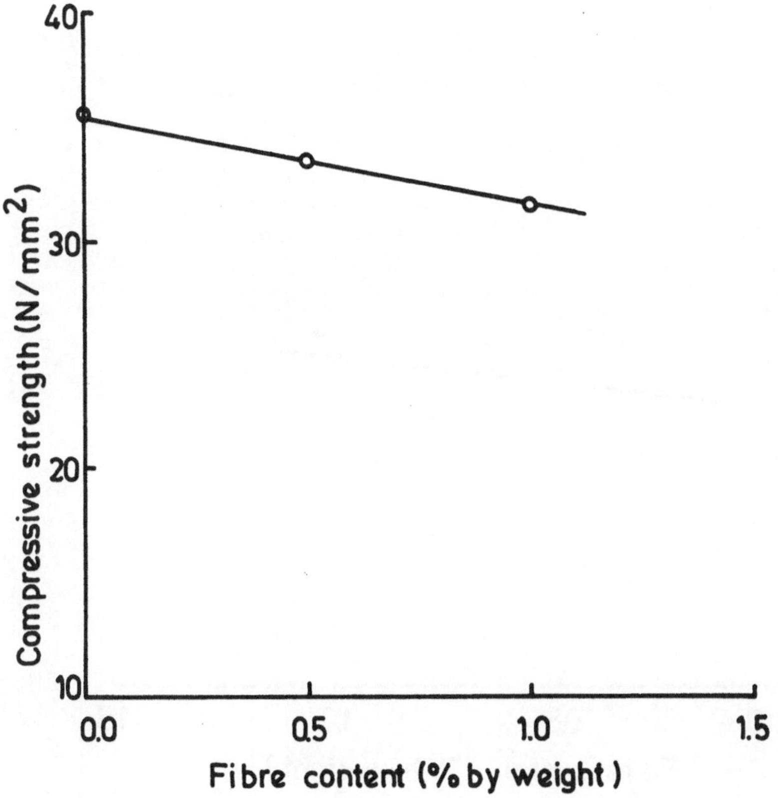

Fig. 4 Variation of compressive strength with fibre content.

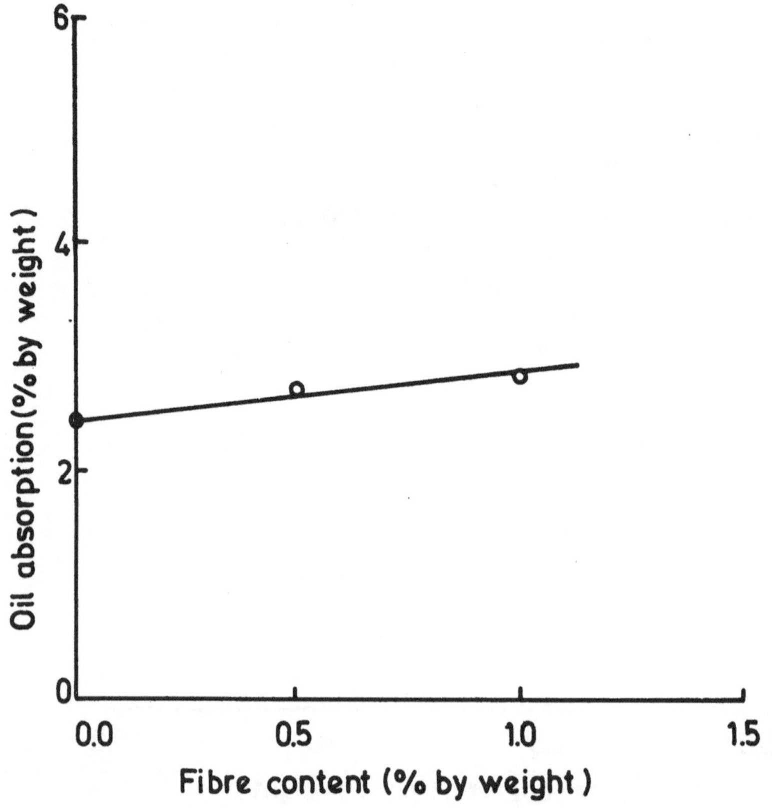

Fig. 5 Variation of oil absorption with fibre content.

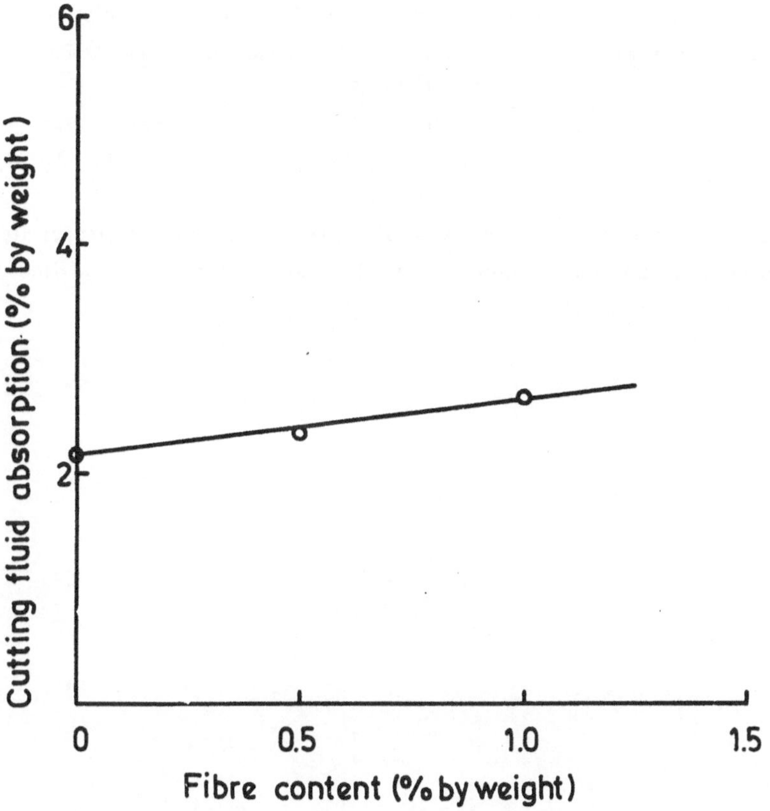

Fig. 6 Variation of cutting fluid absorption with fibre content.

load deflection plots as obtained from tests on Instron machine on notched beams are typical of those shown in Fig. 2. The variation of fracture toughness K_{1c} with fibre content is shown in Fig. 3. It can be seen that both the fracture toughness and flexrual strength are increased with fibre content. However the compressive strength was found to slightly decrease with fibre content as shown in Fig. 4. The variation in the oil and cutting fluid absorption tests with fibre contents are as shown in Figs. 5 and 6. Even though the oil absorption is found to slightly increase with fibre content it is within acceptable limits for practical applications.

The behaviours with reference to compressive strength and oil absorption tests reflects the porosity and adverse effect of having high fibre content mixes.

It was observed that the interface bond between the aggregate and matrix was so good that in all the load conditions cracks propagated through the granite aggregate resulting in its fracture.

Further test results from dynamic and field tests are under progress and will be reported in due course.

CONCLUSION

The test data indicate that the epoxy glass fibre concrete with 0.5 % fibre content by weight is a promising material for use in machine foundations.

REFERENCES

1. International Symposium on Fibre Reinforced concrete, ACI SP44, 1973.
2. Conference of Fibre Reinforced materials, Design and Engineering Applications, Institutions of Civil Engineers, London, March 1977.
3. Zygmunt Jamrozy and Swamy, R.N., Use of fibre reinforced for impact Resistance and machinery foundations , International Journal of cement composites Vol. 1, No. 1 May 1979, pp 65-77.

Proceedings of the International Symposium on Fibre Reinforced Concrete, December 16–19, 1987, Madras, India

STUDY OF FIBRE-REINFORCED CONCRETE WITH NATURAL FIBRES

MD. RAFIQUL ISLAM - RESEARCH OFFICER,
HOUSING & BUILDING
RESEARCH INSTITUTE,
MIRPUR,DHAKA,BANGLADESH.

A.K.M.KHORSHED ALAM- PRINCIPAL RESEARCH OFFICER,
HOUSING & BUILDING
RESEARCH INSTITUTE,
MIRPUR,DHAKA,BANGLADESH.

This research project, which forms a part of a continuing line of fibre-reinforced concrete and investigated the effect of introducing various percentages and length of different types of choped fibres into concrete. In this part of the study, the investigation has been made into the effect of several variables of jute and coconut coir fibres on the compressive and tensile strength of concrete which are most abundantly available in Bangladesh. It was also studied the fibre properties as reinforcement in concrete in fresh and hardened state, physical performance, factors effecting the properties of concrete. Presented data shows the effect of fibre length, percentage content on both ultimate compressive strength & tensile strength in concrete. The addition of jute and coconut coir fibres to concrete, decrease the compressive strength but slightly increase the tensile strength. Among this two fibres (Jute and coconut) jute fibre is preferable.

INTRODUCTION

Natural fibres are prospective reinforceing Materials and their use untill now has been more traditional than technical. From ancient times, it has been a common practice in our country to strengthen the mud walls with some kind of fibres like jute, straws etc. used to perform certain function satisfactory.

Cementitious materials in the form of mortars or concretes are altractive for use as constructional materials since they are readily moulded so that complex shaper may be fabricated. Their deficiencies lie in their brittle characteristics, poor tensile and impact strength and in their susceptibility to moisture movements. Reinforcement by fibres can offer a convenient practical and economical method of over coming these deficiencies. Their has been growing in recent years in utilizing natural fibres for reinforceing concrete. The investigations began on the possibility of manufacturing products of natural fibre reinforced concrete whose properties are comparable to those of asbestos-cement. Economics and other related factors in Bangladesh where natural fibres, jute and coconuts etc., orgin are abundantly available, demand that construction engineers and builders apply appropriate technology to utilize these natural fibres a effectively and economically as possible to produce good quality fibre-reinforced composite materials for housing & others needs.

EXPERIMENTAL INVESTIGATION

MATERIALS

a) Fibres

In this investigation two types of fibre were used.

1) Jute fibres
2) Coconut coir fibres.

The fibres were collected locally & cut into pieces in different sizes, such as 0.5 inch, 0.75 inch, 1.0 inch etc. and maximum size of the pieces were not more than 2 inches. The fibres contained in the concrete was 0.1, 0.2, 0.3, 0.4, 0.5, 0.6, 0.7, 0.8 and 0.9 percent by weight to cement weight. The physical properties of these two fibres are given Table - I.

b) Cement

Ordinary portland cement was used. A single batch of the cement was stored and used throughout the whole program. The physical & chemical properties are given in Table - 2.

c) Aggregates

Brick chips (khoia) used as coarse aggregate & river sand used as fine aggregate. The sands was reconsituted for 30% of a 1.5 - 0.6 Fm fraction and 70% of a less 0.6 Fm fraction to provide granding zone 3 of BS 882. The properties of both aggregates are given in Table - 3.

PREPARATION OF SPECIMENS

During the investigation, concrete specimens were tested. In this preliminary investigation we have transformed the widely used in Bangladesh proportion 1:2:4 by Vol. into proportion 1:2.43:3.17 by weight. This was done to related our test result with conventional concreting. Water cement ratio was 0.65.

Materials, except the fibre, were first mixed, then water was added. After the mixture uniformily, the fibres were added slowly to prevent bounding of the fibres and insure uniform distribution. The resulting mixure was than cast into molds.

The specimens were allowed to set during the following 24 hours and then the moulds were stripped and the specimens were stored under water at room temperature for curing until the test day.

Concrete specimens have cast in iron mold of 6 inch dia and 12 inch length cylinder for compressive and split cylinder test and also 6 inch cube for compressive strength.

All specimens were tested at 28 days using the universal testing machine. Three specimens were prepared to get one average point.

Mixing and preparation of sample and Testing has been in accordance with the standard procedure of ASTM.

TEST RESULT AND DISCUSSIONS

All test data are shown in Table - 4 to 12. The discussion was conducted based on the data.

Compressive Strength

The variation of strength with respect to plain concrete strength for different percentage of fibres used are shown in different figures. The test data of compressive strength of cube and cylinder for jute fibre are shown in Table - 11 and 7 and that for coconut fibre are shown in Table - 12. Tables 4 & 6, 7 & 8, and 11 & 12 show the variation of percentage deviation of strength with respect to plain concrete strength of cube and cylinder for jute and coconut fibres. Tab. 7 & 8 show the variation of strength for different percentage of jute and coconut fibres. For both fibres, compressive strength for cylinder and cube are always below plain concrete strength. Initially with the increase of percentage of jute and coconut fibres strength shown decreasing but with the increase the percentage of this fibre strength increases gradually with respect to previous percentage upto a certain percentage of fibre and again decreases for increasing amount of fibres.

Tensile Properties

a) Solit cylinder strength

The test of split cylinder test for jute and coconut fibres are shown in Table - 9 and Table - 10 respectively. Table 5 shows the variation of percentage deviation of strength with respect to plain concrete and Tables 9,10 show the strength variation among the percentage of fibres used. The strength varies gradually with increasing percentage of fibres upto a certain percentage and then decreases for both jute and coconut fibres. The strength exceeds the plain concrete strength for certain percentage of fibres.

ACKNOWLEDGEMENTS :

Helpful suggestions of Mr. Kazi Ataul Haque, Director of Housing & Building Research Institute, are gratefully acknowledged. The authors are also wishes to acknowledged to Mr. Abdul Mazid Khan, Research Assistant, CTRL - II and personnel in Physical Testing and Research Laboratory for their assistance with this study.

TABLE 1 : PHYSICAL PROPERTIES OF FIBRES

Properties	Jute Fibre	Coconut coir fibre
1. Specific gravity	1.02 - 1.04	1.12 - 1.15
2. Bulk density (kg/m^3)	-	145 - 280
3. Ultimate tensile strength PSI ($x10^3$)	36.25 - 50.75	17.40 - 29.00
4. Elongation at break, percent.	1.5 - 1.9	10 - 25
5. Fibre length inch.	2 - 14	72 - 110
6. Fibre diameter inch.	$4x10^{-3}$-0.016	$4x10^{-3}$- $8x10^{-3}$

TABLE 2 : PROPERTIES OF THE CEMENT USED

A. Chemical Properties

Constituents	Percent by weight
Calcium Oxide (CaO)	61.57
Silicon di-oxide (SiO_2)	21.53
Aluminium oxide (Al_2O_3)	5.64
Ferric oxide (Fe_2O_3)	3.14
Magnesium oxide (MgO)	3.55
Sulfur Tri-oxide (SO_3)	1.45
Loss on ignition	1.98
Insoluble residue	0.88

B. Physical Properties

Normal consistency, percent	23
Initial setting time, min.	120
Final setting time, min.	155
Specific gravity	3.14
Compressive Strength :	PSI
i) 3 days	1568.00
ii) 7 days	1960.00
iii) 28 days	2165.00

Tensile Strength :	
i) 3 days	186.00
ii) 7 days	210.06
iii) 28 days	262.97

TABLE 3 : **PROPERTIES OF AGGREGATE**

A. **Coarse Aggregate (Brick chips).**	
Bulk density (SSD), Kg/m^3	974.69
Specific gravity	
i) Bulk	1.81
ii) Apparent	2.49
Water absorption, percent	14.43
Particle shape,	Irregular
B. **Fine Aggregate**	
Bulk density (SSD) kg/m^3	1496.39
Specific gravity	2.6
Water absorption, percent	1.5

C. **Seive Analysis**

i) **Coarse Aggregate**

No. of Sieve	Percentage passing
¾	90
⅜	35
4	7
8	0

ii) **Fine Aggregate**

No. of Seive	Percentage passing
4	100
8	98.4
16	88.00
30	63.4
50	48.8
100	8.8
Pan	0

TABLE-4

COMPRESSION TEST OF CYLINDER

PLAIN CONCRETE STRENGTH (1:2.43:3.17)

Batch No.	Sample No.	Weight (Lbs)	Strength (Psi)	Average St.(Psi)
	1	25.60	3050	
1	2	25.75	3245	3150
	3	25.30	3155	

TABLE-5

SPLIT CYLINDER TEST

PLAIN CONCRETE STRENGTH (1:2.43:3.17)

Batch No.	Sample No.	Weight (Lbs)	Strength (Psi)	Average St.(Psi)
	1	24.80	208	
1	2	25.30	211	208
	3			

TABLE-6

COMPRESSION TEST OF CUBE

PLAIN CONCRETE STRENGTH (1:2.43:3.17)

Batch No.	Sample No.	Weight (Lbs)	Strength (Psi)	Average St. (Psi)
	1	16.80	3522	
	2	17.10	3351	3420
	3	16.65	3387	

TABLE-7

COMPRESSION TEST OF CYLINDER

FIBRE REINFORCED CONCRETE (JUTE FIBRE)

% of Jute fibre (W/W)	Sample No.	Weights (Lbs)	Strength (Psi)	Average strength	% deviation $\frac{F.S - P.S.}{P.S.} \times 100$
0.1	1	26.85	2060		
	2	26.88	2060	2060	-34.60
	3	26.75	2060		
0.2	1	25.66	2187		
	2	25.76	2180	2183	-30.69
	3	25.50	2182		
0.3	1	26.60	2290		
	2	26.52	2466	2378	-24.50
	3	2543	2378		
0.4	1	26.20	2415		
	2	26.00	2407	2411	-2346
0.5	1	24.70	2165		
	2	24.70	2075	2120	-3225
	3	24.30	2120		
0.6	1	25.45	2019		
	2	25.90	2015	2017	-35.96
	3	25.62	2017		
0.7	1	25.22	1991		
	2	26.00	1998	1995	-36.67
	3	25.80	1995		
0.8	1	25.30	1960		
	2	26.00	1972	1966	-37.58
	3	25.75	1966		
0.9	1	26.10	1845		
	2	26.00	1885	1865	-40.79
	3	26.25	1865		

TABLE-8

COMPRESSION TEST OF CYLINDER.

FIBRE REINFORCEMENT CONCRETE (COCONUT FIBRE)

% of fibre w/w	Sample No.	Weight (Lbs)	Strength (Psi)	Average Strength (Psi)	% deviation (F.S. - P.S.x100) / P.S.
0.1	1	26.51	1981		
	2	26.56	1995	1988	-36.88
	3	26.48	1988		
0.2	1	26.62	2119		
	2	26.31	2015	2067	-34.38
	3	26.66			
0.3	1	25.99	2197		
	2	25.93	2265	2231	-29.17
	3	25.82	2231		
0.4	1	26.65	2345		
	2	26.84	2427	2386	-24.25
	3	26.63	2486		
0.5	1	27.11	2018		
	2	27.16	2070	2044	-3511
	3	27.19	2044		
0.6	1	25.45	1926		
	2	25.19	1946	1936	-38.53
	3	25.17	1936		
0.7	1	25.22	1900	1825	
	2	26.00	1750	1825	-42.06
	3	26.30	1825		
0.8	1	25.30	1822		
	2	26.00	1588	1705	-45.87
	3	26.45	1705		
0.9	1	26.10	1632		
	2	26.00	1732	1682	-46.56
	3	26.35	1682		

TABLE-9

SPLIT CYLINDER TEST OF CYLINDER.

FIBRE REINFORCED CONCRETE (JUTE FIBRE)

% of fibre w/w	Sample No.	Weight (Lbs)	Strength(Psi) $\frac{2P}{DL}$	Average strength (Psi)	% deviation $\frac{F.S.-P.S.}{P.S.}$ x 100
0.1	1	26.82	222	219	+ 5.28
	2	27.08	216		
	3	26.85	219		
0.2	1	25.86	248		
	2	25.80	238	243	+ 16.82
	3	25.50	243		
0.3	1	25.52	257		
	2	26.64	258	257	+ 23.55
	3	26.75	257		
0.4	1	25.80	208		
	2	26.15	228	218	+ 4.80
	3	25.90	218		
0.5	1	24.75	198		
	2	24.40	206	202	-2.88
	3	24.80	202		
0.6	1	25.90	198		
	2	25.85	194	196	-5.76
0.7	1	25.55	186		
	2	25.70	182	184	-11.54
	3	25.45	184		
0.8	1	25.70	160		
	2	25.70	169	165	-20.67
	3	25.25	165		
0.9	1	26.10	150		
	2	25.95	162	156	-25.00
	3	26.20	156		

TABLE - 10

SPLIT CYLINDER TEST OF CONCRETE

FIBRE REINFORCED CONCRETE (COCONUT FIBRE)

% of fibre w/w	Sample no.	Weight (Lbs)	Strength (Psi)	Average Strength (Psi)	% deviation $\frac{F.S - P.S}{P.S.} \times 100$
0.1	1	26.67	213	214	+ 2.88
	2	26.66	215		
	3	26.87	214		
0.2	1	26.62	219		
	2	26.31	223	221	+ 6.25
	3	26.54	221		
0.3	1	26.05	234		
	2	25.72	236	235	+12.98
	3	25.83	235		
0.4	1	26.78	245	248	
	2	26.78	251		
	3	26.60	248		
0.5	1	27.14	230		
	2	27.14	232	231	+11.05
	3	27.22	231		
0.6	1	25.10	204		
	2	25.55	207	205	- 1.44
	3	25.80	205		
0.7	1	26.25	199		
	2	26.00	197	198	- 4.80
	3	26.10	198		
0.8	1	25.30	180		
	2	26.40	189	185	-11.05
	3	26.24	185		
0.9	1	26.85	166		
	2	26.00	162	164	-21.15
	3	26.60	164		

TABLE - 11

COMPRESSION TEST OF CUBE

FIBRE REINFORCED CONCRETE(JUTE FIBRE)

% of fibre w/w	Sample no.	Weight (Lbs)	Strength (Psi)	Average strength (Psi)	% deviation $\frac{F.S. - P.C.}{P.C.} x100$
0.1	1	16.84	2116	2302	- 32.69
	2	17.24	2488		
	3	17.10	2302		
0.2	1	17.14	2484		
	2	16.94	2402	2443	- 28.56
	3	17.50	2443		
0.3	1	16.48	2625		
	2	16.58	2526	2526	- 26.14
	3	17.20	2526		
0.4	1	16.80	2560		
	2	16.60	2572	2566	- 24.97
	3	16.95	2568		
0.5	1	16.20	2240		
	2	16.30	2488	2364	- 30.88
	3	16.40	2364		
0.6	1	16.60	2613		
	2	16.35	1991	2302	- 32.69
	3	16.55	2302		
0.7	1	16.10	2208		
	2	16.15	2064	2127	- 37.80
	3	16.22	2127		
0.8	1	16.20	1991		
	2	16.60	1865	1928	- 43.62
	3	16.80	1928		
0.9	1	17.10	1918		
	2	17.10	1884	1901	- 44.40
	3	17.30	1901		

TABLE - 12

COMPRESSION TEST OF CUBE

FIBRE REINFORCED CONCRETE(COCONUT FIBRE)

% of fibre w/w	Sample no.	Weight (Lbs)	Strength (Psi)	Average strength (Psi)	% deviation $\frac{P.S. - P.S.}{P.S.}$ x100
0.1	1	17.65	2262	2251	- 34.18
	2	17.70	2240		
	3	17.58	2251		
0.2	1	17.40	2342		
	2	17.90	2476	2409	- 29.56
	3	17.80	2409		
0.3	1	17.80	2526		
	2	17.60	2676	2601	- 23.95
	3	17.70	2602		
0.4	1	17.30	2489		
	2	17.60	2555	2522	- 26.25
	3	17.50	2522		
0.5	1	18.10	2338		
	2	18.25	2282	2310	- 32.46
	3	18.20	2310		
0.6	1	16.30	2058		
	2	16.60	2240	2149	- 37.16
	3	16.40	2149		
0.7	1	16.10	2064		
	2	16.40	1912	1988	- 41.87
	3	16.30	1988		
0.8	1	16.30	1928		
	2	16.50	1936	1932	- 4350
	3	16.20	1932		
0.9	1	16.30	1854		
	2	15.90	1818	1836	- 46.31
	3	15.95	1836		

Proceedings of the International Symposium on Fibre Reinforced Concrete, December 16-19, 1987, Madras, India

FLEXURE BEHAVIOUR OF NOTCHED COIR REINFORCED CONCRETE BEAM UNDER CYCLIC LOADING

R. N. SINGH
Associate Professor,Department of Civil Engg.,
Bihar Institute of Technology,Sindri,Dhanbad
Pin - 828 123 INDIA

SYNOPSIS

This paper presents three point bend test on unnotched and notched randomely oriented Fibre (coir) Reinforced Concrete (FRC) beam. The concrete mix proportion is cement:Sand:coarse aggregate as 1:2:4 with Water-cement ratio as 0.55 and weight fraction of coir as 3 percent by weight of cement. The dimensions of the beam are 80X160X750 mm. The non-dimensional notch depth (a/w) are 1/16,3/32 and 1/8 where 'a' and 'w' are crack length and depth of beam respectively.

The beams were tested under cyclic loading. The cross head speed was maintained at 0.2 mm/minute during loading. The fracture load was applied in 3-4 cycles. The load at interval of 0.5 kN and corresponding transeverse deflection of the central point for various load cycles were recorded.

Flexure strength has been computed ignoring slow crack growth. Stress concentration effect has been obtained. Load **deflection curves** for various load cycles have been given. Coir reinforcement in conjuction with load cycling enhances useful life of concrete. Initial crack fracture toughness (K_{ICI}) has been computed. The scatter in the fracture toughness value is within 15 percent.

INTRODUCTION

Fibre reinforced metals and alloys are in extensive use for high strength light weight requirements. Concrete also has been reinforced with bamboo, steel fibre and polymer. But the use of coir as construction material for reinforcing concrete is at hypothetical stage. Singh et al[1] have attempted to determine the strength and other mechanical properties of coir and reinforced mortar beams. But the mix proportioning,mechanical properties of the FRC are yet to be standardised. Present investigation is an attempt in this direction.

Application of linear elastic fracture mechanics was attempted by Kaplan [2] for fracture toughness determination of mortar and concrete. James and Naus [3] showed influence of paste,mortar and concrete on critical stress intensity factor. Shah and Mc. Garry [4] studied notch sensitivity of mortar and concrete. Walsh [5] concluded notch insensitivity of shallow concrete beams. Kim et al [6] have confirmed notch sensitivity of shallow concrete beams and have developed uncertainty over existence of fracture toughness for concrete. Visalvanich and Naaman [7] evaluated fracture properties of asbestos cement, plain mortar and steel fibre reinforced mortar.Wecharatana and shah [8] studied crack growth resistance in mortar and steel fibre reinforced mortar with 1% volume fraction of 6.35 mm smooth steel fibres. Singh et al [1] have confirmed existence of stress concentration effect and attempted fracture toughness determination in shallow coir reinforced mortar beams.

The author is not aware of any work on flexural and fracture behaviour of coir reinforced concrete. It is therefore necessary to investigate the behaviour of coir reinforced concrete beam under monotonic and cyclic loading. It is also necessary to investigate the notch sensitivity effect and fracture toughness of such FRC. This paper presents an attempt in these directions. The author is also contemplating to simulate fracture toughness of FRC by using a two dimensional general purpose finite element programme developed for simulation of elastic plastic fracture mechanics [Singh and Ramakrishnan [9]].

THEORY

The stress field in the vicinity of a crack tip in polar coordinate is defined as

$$\sigma_{ij} = \frac{K_I}{\sqrt{2\pi r}} f_{ij}(\theta) \qquad (1)$$

where,

$K_I = \sigma\sqrt{\pi a}$

a = crack length

r = distance of point (i,j) from crack tip

σ = uniform stress in the material for from the crack tip

θ = angle with respect to crack/extended crack plane

K_I = Stress intensity factor in mode I

As the point (i,j) under consideration approaches crack tip, $r \to 0$ and $\sigma_{ij} \to \infty$, that is, high stress concentration occurs at the crack tip and material fails at relatively low far field applied stress. But before catastrophic fracture of material takes place, the initial crack grows and reaches a critical value. The crack extension occurs when the stress at the crack tip reaches a critical value and K_I reaches the critical value K_c. This value is called fracture toughness and in plain strain case is written as $K_{IC} = \sigma_c \sqrt{\pi a}$.

In case of three point bend test σ_c is obtained from the characteristic dimensions of the specimen and maximum fracture bending moment. Of course, the crack length at this level should be the sum of the initial crack length and the slow crack growth that occurs before catastrophic fracture. Measurement of slow crack growth in case of concrete is extremely difficult but somehow could be obtained by compliance method. Brown and Srawley [10] have developed following expression for computation of fracture toughness of high strength metallic materials for beam span depth ratio of 4.

$$K_{IC} = \frac{6 M_c a^{1/2}}{b w^2} \left[1.93 - 3.07 \left(\frac{a}{w}\right) + 14.53 \left(\frac{a}{w}\right)^2 - 25.11 \left(\frac{a}{w}\right)^3 + 25.80 \left(\frac{a}{w}\right)^4 \right] \quad (2)$$

where M_c = Critical bending moment induced at the centre of beam and other terms are as already defined.

Kim et al [6] subsequently used this expression to find K_{ICI} for concrete beam. This investigation uses this expression for computation of K_{ICI} for FRC.

EXPERIMENTAL PROGRAM

Materials

Commercially available coir was cleaned,entire foreign particles (dust, lump etc.) removed and the fibres were tried to be separated. The coirs of varying natural length 80 mm down and varying diameter were used for the purpose. Tests on average diameter of fibre as o.5 mm and gauge length 50 mm exhibited breaking strength of 40.5×10^3 - 98.4×10^3 kN/m^2 and strain of 10-32%. The initial modulus of elasticity was found to be 1.9×10^6 kN/m^2[1]. Coarse aggregate passing IS 20 mm and retained on IS 6 mm, Barakar sand passing IS 2 mm down and local fresh natural portland cement from A.C.C. were used.

Test Specimen

Beam of dimensions 80 X 160 X 750 mm were decided to be cast in wooden mould. Groove was made on inner side of the wooden wall to insert m.s. plate piece which intruded into the beam section providing notch of appropriate depth while casting. Eight specimens of notched beams with initial notches of 10 mm, 15 mm and 20 mm on the tension side and two unnotched beams were cast. The configuration of the single edge notch bend specimen is shown in fig.1. It was decided to test cylinders of 150 X 300 mm for determination of modulus of elasticity.

The specimen depth was chosen to examine whether stress concentration effect exists in the shallow beams which have smaller depths than that of Walsh [5] minimum criterion (225 mm) and if it existed, to compare the K_{IC} values obtained for different notch depths. The notched crack length-beam depth ratio of the specimens, a/w, varied from 0.0625-0.1250.

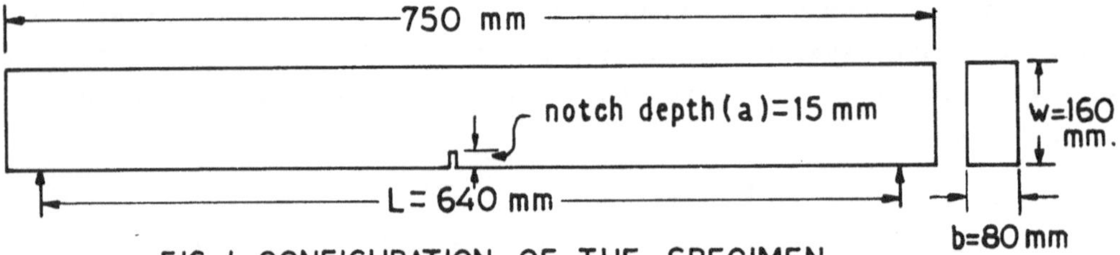

FIG. I. CONFIGURATION OF THE SPECIMEN

Mix proportioning

The use of coir as reinforcing material in concrete is at a hypothetical stage. The procedure of mixing coir in concrete is cumbersome. Obviously, the procedure is yet to be standardised. At the present stage of development, the proportion of cement:sand:coarse aggregate(ca) was taken as 1:2:4 with water cement ratio(w/c) of 0.55 and weight fraction (w_f) of fibre as 3 percent by weight of cement.

Mixing and Casting

Hand mixing was used for convenient handling of coir. Sand and cement were mixe d dry and kept separately. Then coarse aggregate, fibres and dry mix of cement and sand were kept in three layers and mixed thoroughly. Appropriate quantity of water was sprinkled on the dry mix in layers. Every layer together with the underlying ones(for thesecond and third layers) was thoroughly mixed after sprinkling water. The wet mixing continued till uniform colour of mix was obtained. While mixing separation of fibres and removal of lumps of coir was done. Mixing procedure was felt to be extremely tedious. Formation of very small lumps of fibres could not be avoided. The fibres were randomely oriented in the mix.

Strength test specimens were cast as per standard procedure. Beams were cast in wooden moulds in three layers. Great care had to be taken in maintaining the position of the mild steel plate piece fixed in groove on wooden side wall for making notch of appropriate depth. Each layer was compacted with 80 blows with 16 mm dia. and 600 mm long mild steel rod. Fibres created finishing problems at the top surface. Specimens were cured in fresh water after 24 hours of casting till they were taken out of the curing tank 48 hours before testing.

TEST PROCEDURE

The beams were removed from the curing tank 2 days before the day of testing. They were dried out for one day and then gauge points were fixed for measuring longitudinal extension/shortening. The surface of the beam above simple support and under load line were filed and cleaned to bring the same in horizontal planes for desired loading. The prenotch was cleaned by means of fine saw to ensure uniform depth.

The specimens were placed in position in a universal testing machine for three point bend test. Instrumentation for measurement of vertical deflection at centre of beam was made. The test set up is shown in fig.2. It was decided to load the specimen under cyclic loading. Load was applied at constant cross head displacement rate of 0.2 mm/minute.While the loading rate was fixed by the displacement rate of the cross head, the unloading rate could not be ascertained at a fixed rate. However, by means of stopwatch, the loading and unloading time for a particular load-cycle was ensured to be approximately same.

The load at interval of 0.5 kN and corresponding transverse and longitudinal displacement of the beam were recorded while both loading and unloading. The maximum load was applied in 3-4 cycles. Load cycling was done after maximum load also. Measurement of slow crack growth was attempted by cathetometer microscope but it could not be successful. Crack opening also could not be measured. The specimens were tested at varying ages of 29,31,33,37 and 78 days. As expected the age of specimen would affect the various parameters.

RESULTS AND DISCUSSIONS

Load-Deflection Behaviour

Typical load-deflection curves for two beams are shown in figs.3 and 4. As would be expected the loci of the peak of various cycles resemble somewhat to the load-deflection curve for monotonic loading. However,in absence of the monotonic load-deflection curve the quantitative variation, if any, in the maximum load at various deflection cannot be commented. Maximum deflection in both the two beams is 0.3175 mm. As would be

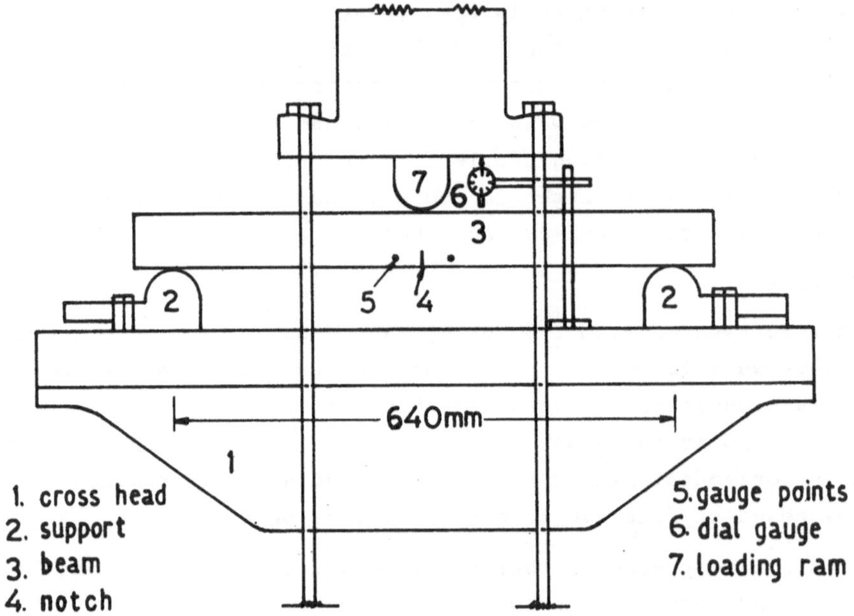

FIG.2. TEST SETUP

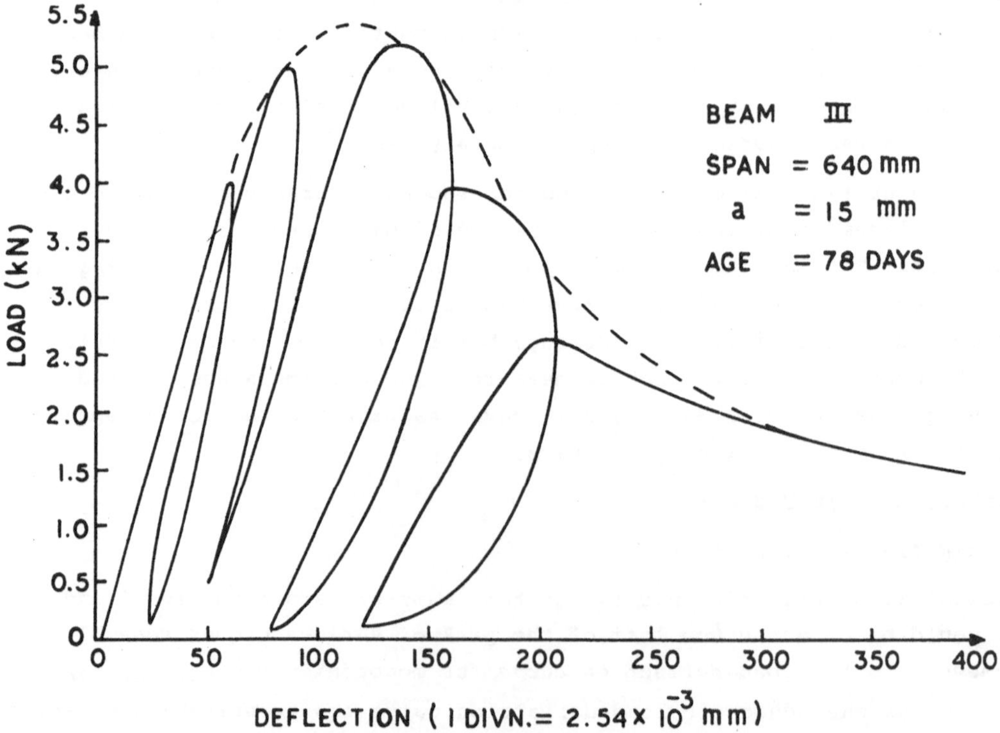

FIG.3. LOAD DEFLECTION CURVE

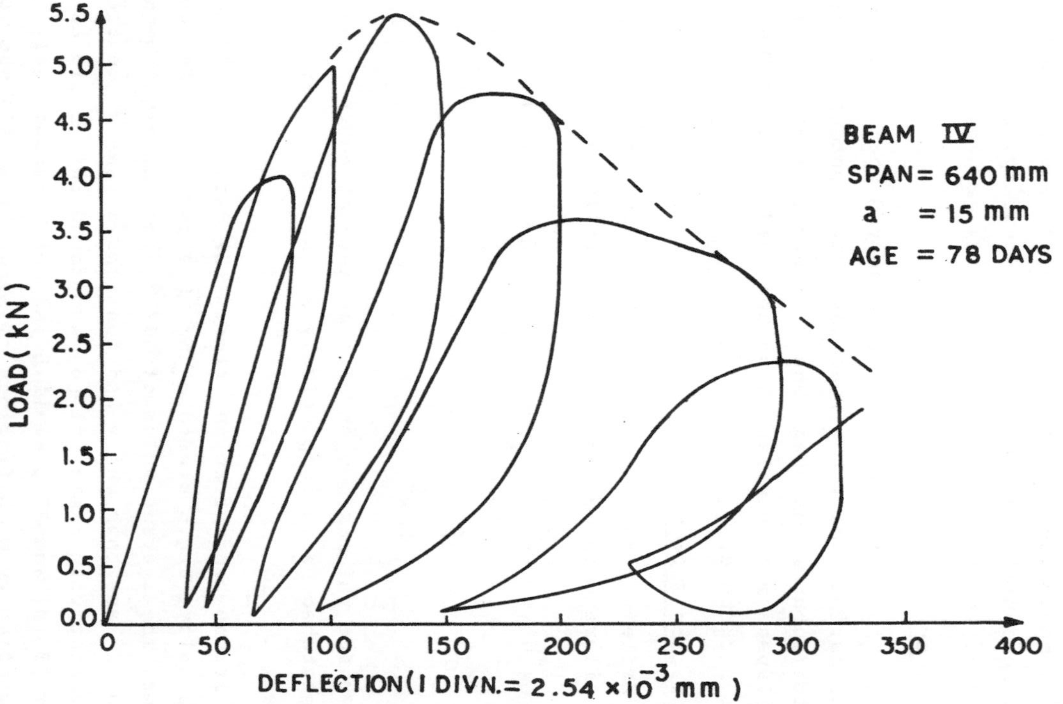

FIG. 4. LOAD DEFLECTION CURVE

expected this deflection is more than that in the case of plain concrete beam. It was also observed that no ingredient of the beam fell down under such a large deflection unlike plain and steel reinforced concrete beam. Though the full depth of the beams was cracked the two sides remained intact. Thus the coir reinforcement imparts ductility to the beam. The widening of crack could not be measured but it was felt that the crack width was much small and hence the fibres contribute crack arresting property.

Measurement of crack length at the peaks of various load cycles would have helped in computation of compliance and regression analysis but the same could not be done to desired degree of accuracy, because of experimental dificulties in measurement of slow crack growth.

Load-Time Behaviour

Figure 5 presents load-time curve. It is observed that in the second reloading after maximum load, the load was kept constant at 68% of the maximum load for about 10 minutes. Under monotonic loading the beam would have collapsed much earlier. Thus the load cycling enhances useful life of structure.

Stress Concentration Effect

The nominal tensile stress developed in the notched and unnotched specimens at failure were evaluated from the following equations.

Notched, $$\sigma_N = \frac{6\ Mn}{b(w-a)^2} \qquad (3)$$

Unnotched, $$\sigma_U = \frac{6\ Mu}{b\ w^2} \qquad (4)$$

where Mn and Mu are the maximum bending moments at centre span of notched and unnotched beams respectively at failure. Other terms are as defined earlier.

In absence of stress concentration effect, the ratio σ_N / σ_U could be expected to be unity. Fig.6 presents σ_N / σ_U versus a/w relation for the 8 notched beams. The average value of σ_N / σ_U for the eight beams comes to 0.825 which is less than unity and confirms stress concentration effect in agreement with Kim et al [6] who found this ratio as 0.73 and 0.76 for 75 mm (3 in) and 37.5 mm (1.5 in) beams respectively. It is also in confirmity with findings of Singh et al [1] who found the average value of 30 mortar and reinforced mortar beams less than unity. This investigation invalidates argument of Walsh [5], who concluded that there was no stress concentration effect in less than 225 mm deep beams,

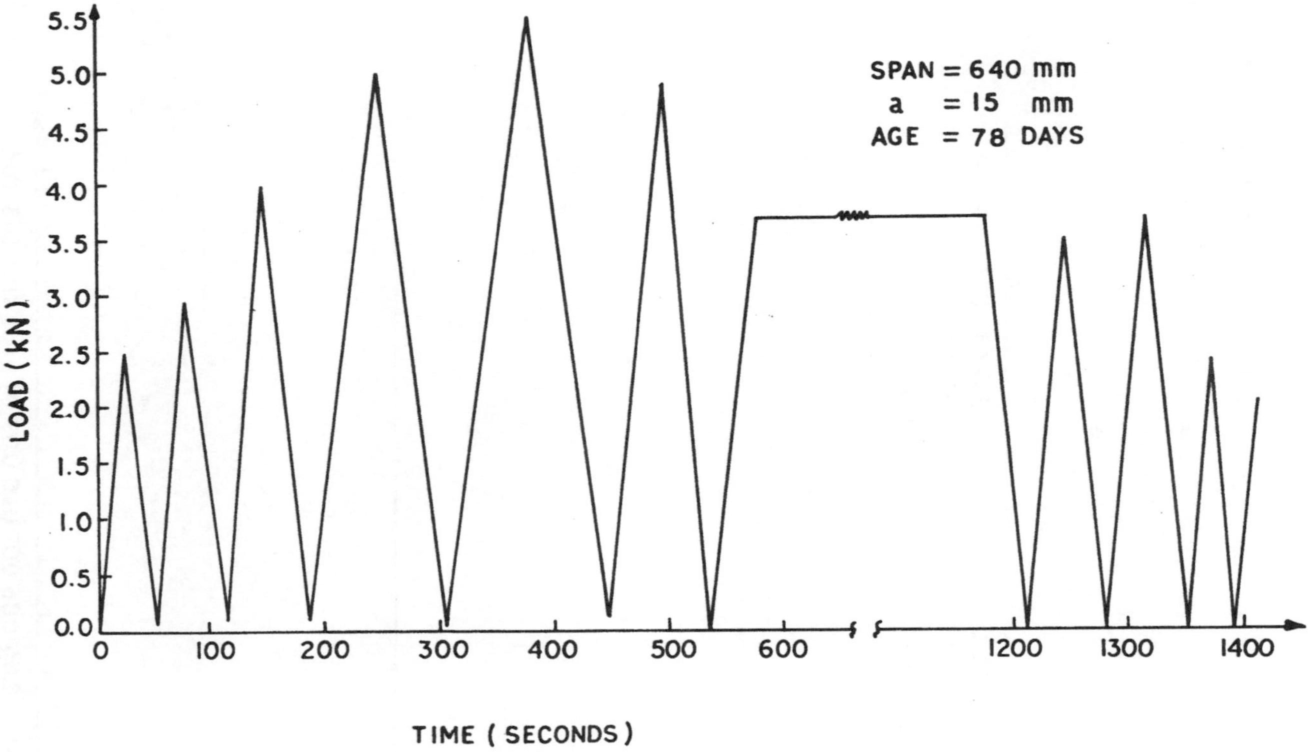

FIG. 5 LOAD - TIME CURVE

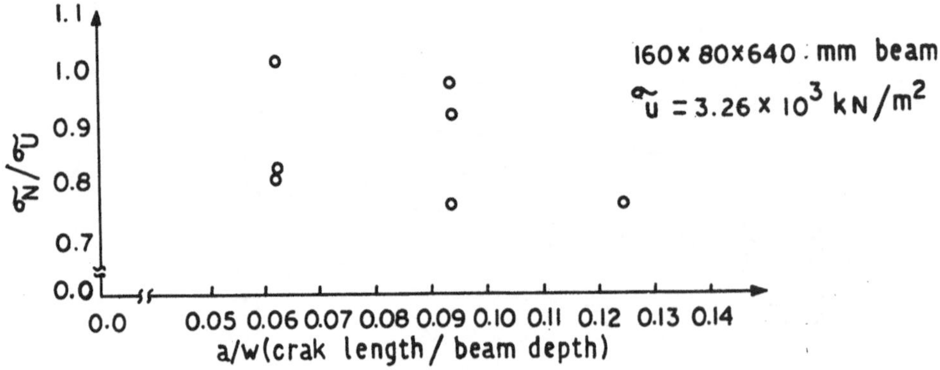

FIG. 6. σ_N/σ_u vs. a/w

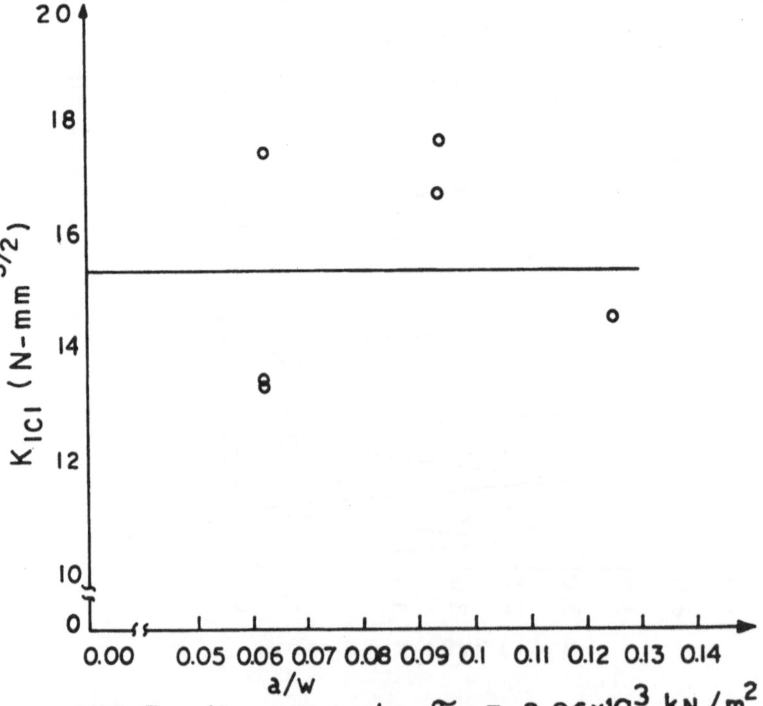

FIG. 7. K_{IC1} vs. a/w, $\sigma_u = 3.26 \times 10^3$ kN/m^2

and confirms stress concentration effect in shallow FRC beams.

Initial Crack Fracture Toughness (K_{ICI})

The K_{ICI} value have been computed from the characteristic specimen dimensions, fracture load and initial crack length using equation (2). Fig.7 presents K_{ICI} vs.a/w relation. The average value of K_{ICI} is 15.33 $N\text{-}mm^{-3/2}$. Maximum scatter in K_{ICI} value is found to be 15 percent. This scatter is attributed partly to the variation in age of specimens, and indicates the existence of fracture toughness value. Kim et al [6] report maximum scatter in K_{ICI} values of 21 and 27 percent from average in 75 mm (3 in) and 37.5 mm (1.5 in) deep cement, sand, gravel concrete beams respectively. Singh et al [1] report this scatter of 4 percent to 27 percent in mortar and reinforced mortar beams. In light of such large scatter the existence of a critical value of fracture toughness at which the specimens fracture cannot be ensured unless extensive experimentation is done. It also indicates the need to control the dimensions of the test specimens in terms of beam thickness- depth ratio (b/w) and beam depth-span length ratio (w/L) in order to produce a consistent fracture toughness or otherwise by testing different sizes of specimens.

CONCLUSIONS AND RECOMMENDATIONS

Following are the original contributions of the investigation.

1. Mix proportioning and method of mixing coir for reinforcing concrete is presented but it is yet to be standardised.
2. Coir provides ductility to concrete. It allows the beam to deflect more. It also contributes crack arresting property to concrete. It holds the ingredients from falling down even after failure and saves property and life from damage.
3. Useful life of FRC beam is increased because of cyclic loading.
4. Stress concentration effect exists in shallow FRC beams.
5. There is an indication of constant value of K_{ICI} though maximum scatter of 15% has been observed in K_{ICI} value.

The recommendations of the investigation are as follows:

1. Standard method of mix proportioning for FRC should be established by varying weight fraction of fibres.
2. Extensive experimentation for fracture toughness determination with varying characteristic dimensions of notched beams and incorporating slow crack growth should be performed.

REFERENCES

[1] Singh,R.N.,Jha,J.and Singh,C.K.,Strength and Fracture Behaviour of

Fibre Reinforced concrete,Sixth International Conference on Fracture, New Delhi, India,Dec.4-10.1984,P.S.Vol,pp 184-186.

[2] Kaplan,M.F.,Crack Progation and the fracture of concrete,Journal of the American Concrete Institute,Proceeding Vol.58,No.5,Nov.1961, pp.591-610

[3] James,L.and Naus,D.J.,Fracture Toughness of Portland Cement Concrete, Journal of the American Concrete Institute, Proceedings,Vol.66, No.6, June 1969, pp 481-489.

[4] Shah,S.P. and Mc.Garry,F.J.,Griffith Fracture criterion and Concrete, Proceedings, ASCE,Vol.97 EM 6,Dec.1971,pp 1663-1676.

[5] Walsh,P.F.,Fracture of Plain Concrete, The Indian Concrete Journal, Vol.46,No.11,Nov.1972,pp 467,470 and 476.

[6] Kim,M.M.,Ko,H.Y. and Gerstle,K.H.,Determination of Fracture Toughness of concrete, Fracture in concrete, proceedings at the ASCE National convention,Hollywood,Florida,October 27-31, 1980 pp 1-14.

[7] Visalvanich, K.and Naaman, A.E., Evaluation of Fracture Techniques in cementitious composites, Fracture in Concrete, Proceedings at the ASCE National Convention, Hollywood,Florida,October 27-31,1980,pp.65-81.

[8] Wecharatana,M.and Shah,S.P.,Resistance to Crack Growth in Portland cement composites, Fracture in concrete, Proceedings at the ASCE National Convention,Hollywood,Florida,October 27-31,1980 pp-82-105.

[9] Singh R.N. and Ramakrishnan,C.V.,Simulation of Elastic-Plastic Fracture Mechanics Using viscoplastic Modelling with Isoparametric Finite Elements, Sixth International Conference on Fracture(ICF6) New Delhi,India-Dec.4-10,1984,Vol VI, pp 3989-3996.

[10]Brown,W.F. and Srawley J.E., Plain Strain Toughness Testing of High strength Metallic Materials, ASTM,STP No.410,1969.

Proceedings of the International Symposium on Fibre Reinforced Concrete, December 16-19, 1987, Madras, India

PROPERTIES OF GLASS FIBRE REINFORCED CEMENT COMPOSITES WITH AND WITHOUT POLYMER IMPREGNATION

M. Neelamegam and B. Venkateswarlu
Scientists, Structural Engineering Research Centre, Madras - 600 113, India

ABSTRACT

Reinforced cement mortar is increasingly used for thin structural and non-structural components in civil engineering constructions. Unreinforced cement composites are highly brittle and hence they show sudden failure without any ductility. Fibre reinforced composites possess higher ductility and toughness than the conventional cement composites, and so they have been recently introduced in many structural applications. Glass fibre is a versatile reinforcement for thin microconcrete members as it imparts better workability than steel fibres and it does not pose any corrosion problem. Experiments have been conducted at Structural Engineering Research Centre, Madras, on glass fibre reinforced cement composites with and without polymer impregnation to study mechanical and durability properties. Polymer impregnation does not give scope for ingression of external elements for any chemical reaction inside the composite material and so non-alkali resistant glass fibres were used in these experiments.

Using two-phase composite material approach of continuous fibres for the randomly distributed fibre reinforced mortars, the ratio of tensile strength of the fibre reinforced composite, σ_{ct}, to the tensile strength of the matrix, σ_{mt}, is derived as

$$\frac{\sigma_{ct}}{\sigma_{mt}} = (1+\alpha P_f)$$

where P_f, is the volume fraction of fibres with respect to the matrix volume, and α is the ratio of bond strength of fibres in the matrix to the tensile strength of the mortar. The values of α are evaluated from the experimental results and presented in this article. Mechanical and durability properties of glass fibre reinforced cement composites with and without polymer impregnation are also reported in this paper.

INTRODUCTION

Cement composites without well distributed reinforcement are very brittle. To improve their ductile behaviour and other mechanical properties, such as tensile, flexural, and impact strength, steel and asbestos fibres were used in many components made of cement composites. Use of asbestos fibres in cement composites is discouraged as its exposure is detrimental to the health of human beings. Although steel fibres improve ductility, flexural strength and toughness, they result in poor workability of mix due to clumping effect while mixing and casting. Corrosion damage and increased weight density are the other drawbacks of steel fibres.

Glass fibre reinforced mortar composites (GFRMC) were, therefore, attempted first in Russia [1]. These studies showed deterioration of glass fibres due to alkali attack of cement paste. Hence high zirconia alkali resistant glass fibres have been tried in U.K. and U.S.A. [2,3,4]. These fibres were found to be very expensive and so their use to reinforce cement composites exposed to normal environment conditions was questioned [5,6]. Moreover, the recent experiments showed that no reaction takes place between E-glass fibres and cement under conditions, such as low alkalinity in cement [7]. Hence experiments have been conducted on specimens made of cement mortar reinforced with commerically available E-glass fibres. In order to improve the strength properties and also to prevent penetration of moisture and air into the voids in the matrix, polymer impregnation was also adopted in some of the specimens. Results of these tests on GFRMC specimens with and without polymer impregnation were discussed in this paper.

MATERIALS

Ordinary Portland cement [8] and river sand conforming to zone III grade (Fig. 1) of IS specifications [9] were used to prepare the test specimens. E-glass fibres of 6 mm and 12 mm length were selected to reinforce the mortar. Physical properties of these fibres are given in Table 1 [10]. Methyl methacrylate (MMA) monomer with benzoyl peroxide as a catalyst was used for impregnation of the GFRMC specimens.

Table 1: Physical and Chemical properties of glass fibres [4,10]

Physical Properties		Chemical analysis	
Diameter of filament	9μ - 12.7μ	Oxide	% by wt.
Number of filaments	200	SiO_2	52.4 - 54.0
Tensile strength of glass fibre	2850 - 3164 MPa	Al_2O_3	13.9 - 14.4
		B_2O_3	10.6 - 11.0
Young's modulus of glass fibre	74000 MPa	TiO	0.02
Specific gravity	2.69	MgO	3.4 - 4.6
Exterior surface area of strand	$0.62\ mm^2/mm$	CaO	17.3
		Na_2O	0.08 - 0.1
		K_2O	0.03
		Fe_2O_3	0.1

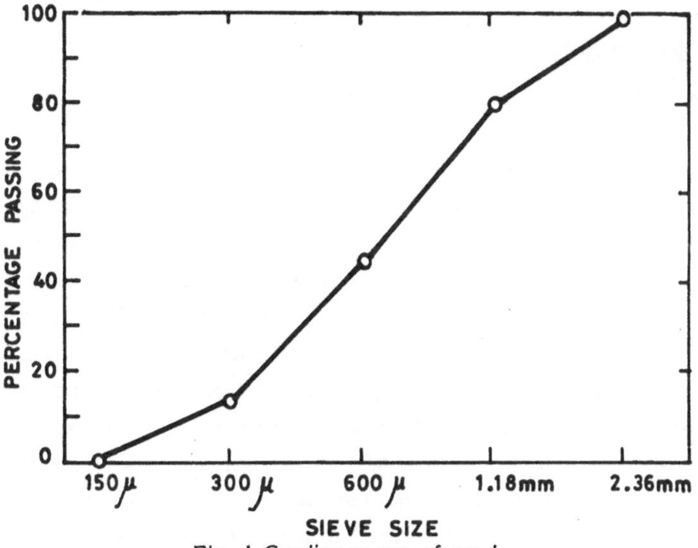

Fig. 1 Grading curve of sand

TEST SPECIMENS

Two trial mixes with cement and sand in proportions of 1:2 and 1:3 by weight were studied with different water-cement ratios. Of these mixes, 1:2 cement mortar and 0.45 water-cement ratio was found to give higher consistency and density. This mix was also found to give lower monomer loading. Hence it was selected for preparing GFRMC specimens containing glass fibres in different proportions upto 4% by weight of dry cement mortar. After mixing cement and dry sand, the required quantity of water was added and the mortar was thoroughly mixed. While adding the water, chopped glass fibres were distributed uniformly in the matrix. Hand mixing was used as the quantity of mortar in each batch of mix was small.

Cylinders of 75 mm diameter and 150 mm height were cast for testing split tensile and compressive strength of GFRMC. Dog-bone shaped specimens shown in Fig. 2 were cast for direct tension test. The moulds were filled with mix in a suitable number of layers and compacted on a vibrating table. After demoulding, the specimens were cured under water for sufficient period.

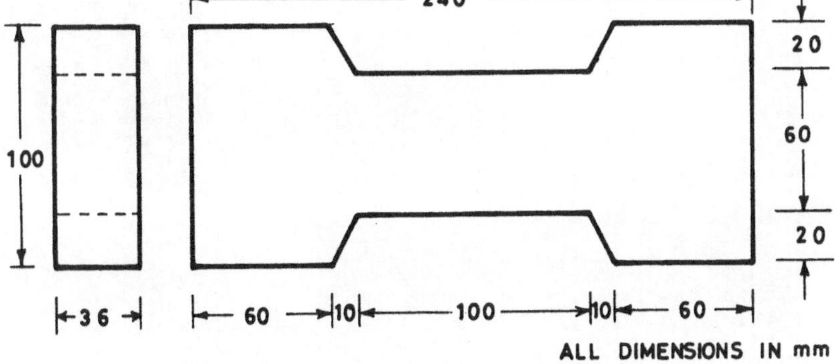

Fig. 2. Direct tension test specimen

POLYMER IMPREGNATION

The cured GFRMC specimens selected for polymer impregnation were weighed and kept in a hot air oven at a temperature of 110°C to 120°C. After removing moisture from the voids in the specimens, they were taken out from the oven and cooled to the room temperature. The dry specimens were weighed and kept in an impregnation chamber (Fig. 3). The chamber was then closed in an air-tight lid which has connections with valves for a pressure pump, a vacuum pump, and a monomer feed tank. After closing all the valves, except the connections for vacuum port, the vacuum pump was started and the air from the voids in the specimens was sucked out under a vacuum of 76 cm of Hg in the chamber. After maintaining the vacuum pressure in the chamber for one hour, the vacuum port was closed and the monomer port was opened. Methyl methacrylate mixed with benzoyl peroxide in the monomer feed tank was slowly transferred into the impregnation chamber. After filling the tank with the required quantity of monomer, the monomer inlet valve was closed. The air pressure valve was then opened and the tank was pressurised to one atmosphere in order to reduce the soaking period and to effect penetration of monomer to full volume of the specimens. The pressure was maintained for four hours so that all the voids in these specimens were filled with the catalysed monomer. The remaining monomer in the tank was collected and weighed. The monomer impregnated specimens were removed from the chamber and weighed. They were kept in a hot water bath at a temperature of 80°C to 90°C for eight hours to completely polymerise the impregnated monomer. They were removed and cooled to the room temperature. The final weights of the polymer impregnated GFRMC (PIGFRMC) specimens were noted to ascertain the quantity of polymer filled in the voids.

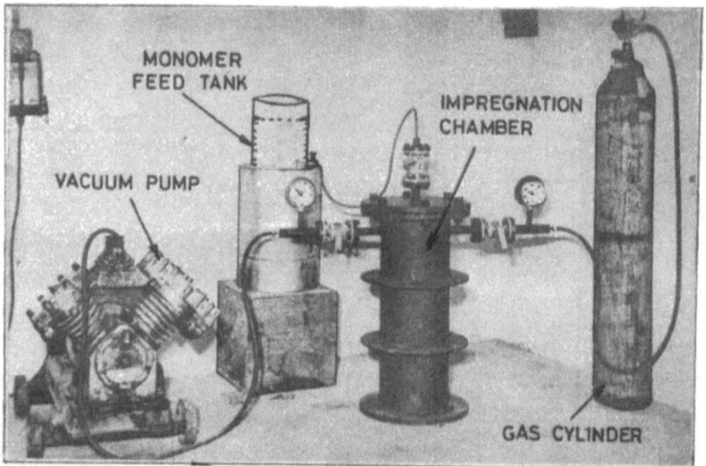

Fig. 3 Polymer impregnation set-up

TEST RESULTS AND DISCUSSIONS

Workability

The workability of fresh cement mortar with and without fibres was measured by the flow table test. The results of 1:2 cement mortar with water-cement ratio varying from 0.45 to 0.6 are shown in Fig. 4. These results show that the workability of cement mortar reduced with the addition of fibres. As seen from the test results, the percentage of flow of the cement mortar without fibres for water-cement ratios equal to and more than 0.5 was considerably more than in the case of the mix with water-cement ratio of 0.45. But this difference of flow for the mixes with glass fibres was negligible, particularly when the fibre content was more than 2%. Moreover, the loss in strength of the mix with water-cement ratio above 0.45 was very pronounced. As the workability of the mix with water-cement ratio of 0.45 was adequate, this mix was selected for studying the mechanical and durability properties of GFRMC. The workability of GFRMC with 6 mm long fibres was found to be more than that with 12 mm long fibres. This difference in workability was more pronounced with water-cement ratios more than 0.45. Hence 6 mm long fibres were selected for studying the mechanical and durability properties of GFRMC.

Tensile Strength

Tensile strength of cement composites is the most important property as it governs compressive strength and other properties. Fibres are added to cement mortar mainly to improve the tensile strength and so this property is discussed first.

Griffith's criterion was first adopted by Romouldi and Batson to predict the tensile strength of cement mortar [11]. Their experiments on cement mortar specimens reinforced with closely spaced continuous wires showed that the tensile strength of mortar was greatly influenced by the spacing of reinforcement. Realising the practical difficulties in arranging reinforcement with continuous wires at very close spacing Romouldi and Mandel carried out tests on cement mortar specimens reinforced with randomly distributed short steel fibres [12]. The results of these tests were found to be the same as in the case of mortar specimens reinforced with continuous wires.

It is, therefore, proposed to use two phase composite material approach of continuous fibres for the randomly distributed fibre reinforced mortars. The equilibrium of total forces in the composite material gives:

$$\sigma_{ct} \cdot A_c = \sigma_{mt} \cdot A_m + \sigma_{ft} \cdot A_f \qquad \ldots\ldots \quad (1)$$

where A_c, A_m, and A_f are the cross-sectional areas of the composite, cement matrix and the fibres normal to the applied force, respectively, σ_{ct}, σ_{mt} and σ_{ft} are the tensile strengths of the cement composite, matrix, and fibres, respectively.

Eq. 1 can be written as:

$$\frac{\sigma_{ct}}{\sigma_{mt}} = \left(1+\frac{\sigma_{ft}}{\sigma_{mt}} P_f\right) / \left(1+P_f\right) \qquad \ldots\ldots \quad (2)$$

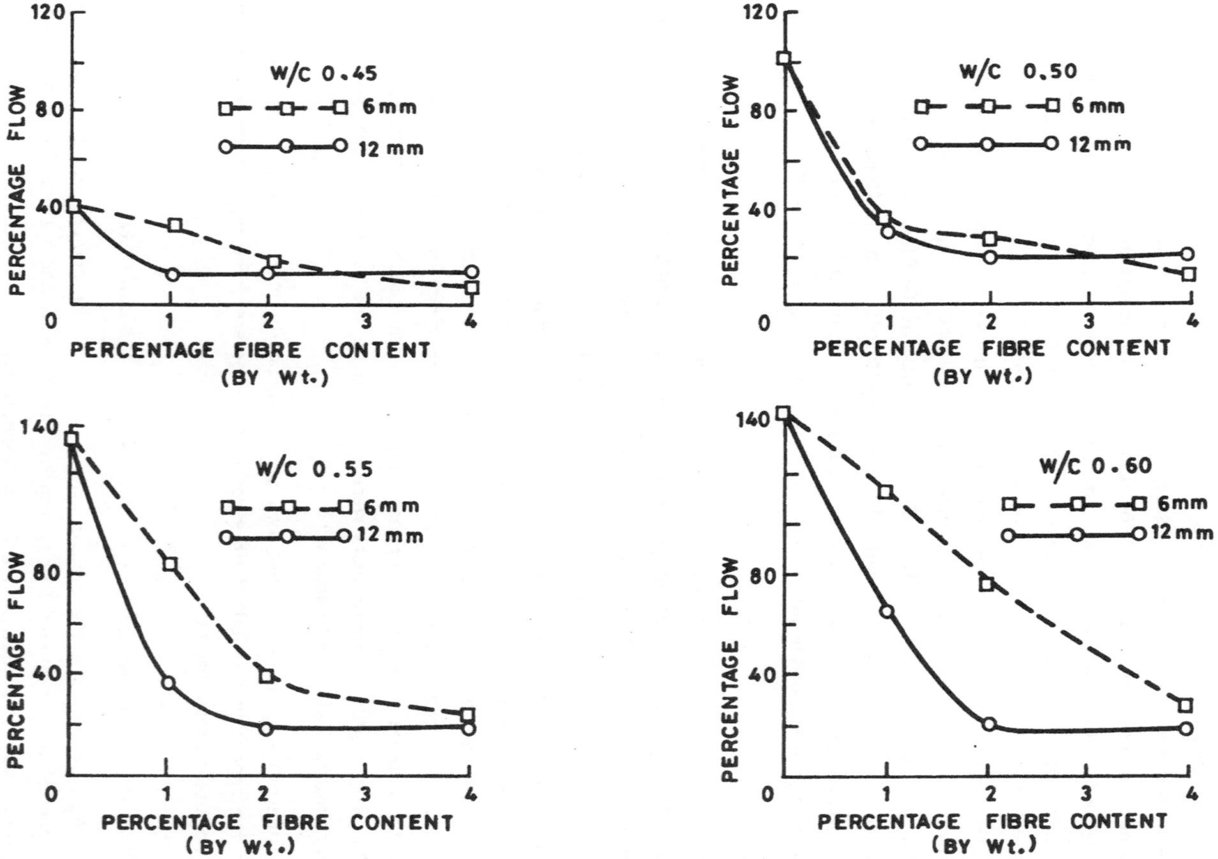

Fig. 4 Workability of glass fibre reinforced mortar composites

where P_f is the volume fraction of fibres with respect to the matrix volume. Eq. 2 shows that the ratio of the tensile strengths of mortar with and without fibres depends upon the volume fraction of fibres and the ratio of the tensile strengths of fibres and matrix. The tensile strength of glass fibres is very high compared to their bond strength in mortar and hence failure of the cement composites reinforced with glass fibres aligned with the loading depends upon the bond strength. In the case of cement mortar reinforced randomly with short fibres, the ultimate bond strength depends upon many factors, such as orientation and length of fibres, and strength of the matrix.

Therefore, Eq.2 is modified as:

$$\frac{\sigma_{ct}}{\sigma_{mt}} = (1+\alpha P_f)/(1+P_f) \qquad \text{..... (3)}$$

where α is the ratio of the bond strength of the fibres in the matrix to the tensile strength of the mortar.

Since $P_f \ll 1$

$$\frac{\sigma_{ct}}{\sigma_{mt}} = (1+\alpha P_f)(1-P_f) \qquad \text{..... (4)}$$

The value of α is determined from the tensile strength tests. Among the different types of tests for tensile strength of cement composites, the split cylinder strength test is the most popular and convenient one, and is recommended in the codes of practice for design and construction of concrete structures. The variation of split tensile strength with the percentage content of 6 mm fibres is shown in Fig. 5. The values of α for these observations were found to vary from 2.6 to 5. Even though polymer impregnation of cement mortar without fibres improved the split tensile strength by about three times, addition of fibres did not give appreciable increase in the tensile strength of PIGFRMC.

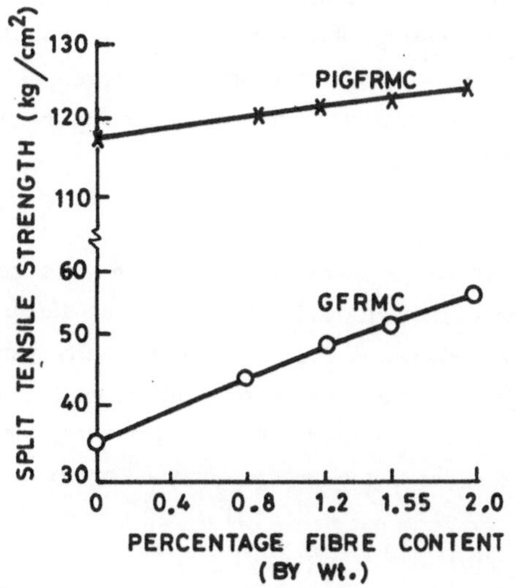

Fig. 5 Tensile strength of GFRMC and PIGFRMC

Compressive Strength

When the cement composite specimen is tested under uniaxial compression, it fails due to split tension on the principal plane. Therefore, the compressive stgrength is related to the tensile strength. The most commonly used relationship between tensile and compressive strength of cement composite is as follows:

$$\sigma_{ct} = RF_{cu}^{0.5} \qquad ..... (5)$$

where F_{cu} is the compressive strength of cement composite, R is a coefficient, and σ_{ct} is the same as defined previously. Therefore, the strength ratios of cement composites with and without fibres may be represented as follows:

$$\frac{F_{cu}}{F_{mu}} = \left(\frac{\sigma_{ct}}{\sigma_{mt}}\right)^2 \qquad ..... (6)$$

where F_{mu} is the compressive strength of cement mortar without fibres and all the other terms are the same as defined in the previous equation. Kukreja et al found that for steel fibre reinforced concrete, the compressive strength increases with fibre contact as follows [13]:

$$\frac{F_{cu}}{F_{mu}} = 1+KP_f \qquad ..... (7)$$

where F_{cu}, F_{mu}, and P_f are the same as defined previously and K is a coefficient. From the experiments in steel fibre reinforced concrete, the value of K was determined as 12.3.

From Eqs. (4), (6), and (7)

$$\alpha = 1+0.5\ K \qquad ..... (8)$$

This value gives $\alpha = 7.15$. For glass fibre reinforced composites, there was practically no increase in compressive strength. The density and the compressive strength of GFRMC were found to decrease with increasing fibre content as shown in Fig. 6. This decrease was more pronounced in the case of polymer impregnated specimens. As in the case of split tensile strength, polymer impregnation increased the compressive strength of cement mortar without fibres by about three times.

Flexure

Tests on beam strips of 10 cm x 3.6 cm x 50 cm size were conducted to study the flexural strength and toughness of GFRMC. The results of flexural tests with central and one third point loading are shown in Fig. 7. These results show marked increase in flexural compressive strength with fibre content. The deflection curves shown in Fig.8 also indicate increased ductility with fibre content. However, the polymer impregnated specimens exhibited brittle failure as in the case of uniaxial compression.

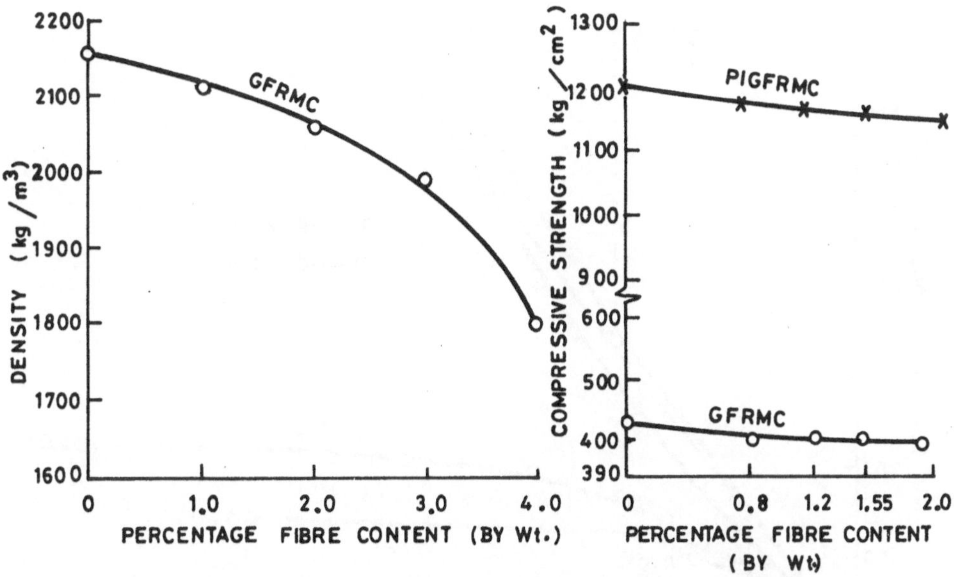

Fig. 6 Density and Compressive Strength of GFRMC and PIGFRMC

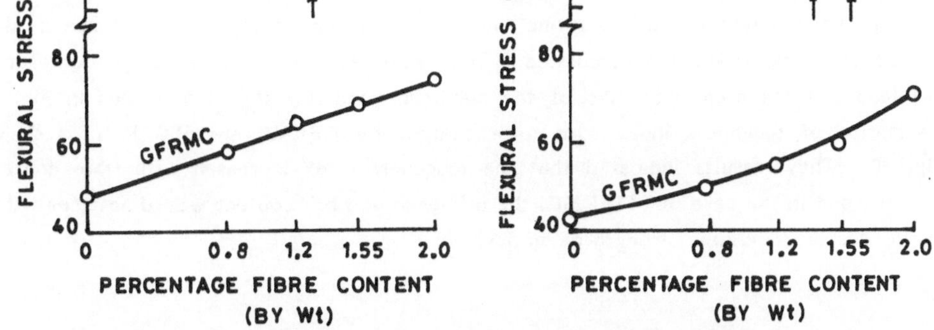

Fig. 7 Flexural strength of GFRMC and PIGFRMC

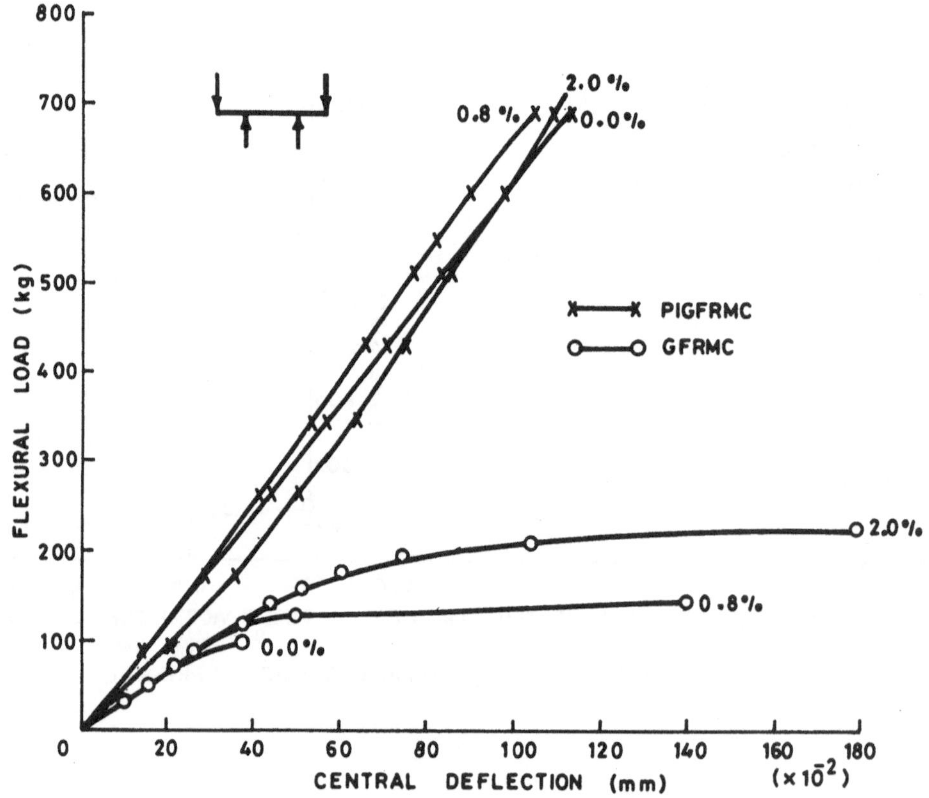

Fig. 8 Deflection of GFRMC and PIGFRMC specimens

Toughness

Toughness of GFRMC is measured by the energy absorbed prior to complete failure of the specimen. The toughness is more conveniently expressed as an index which is given by the ratio between the total area under the load deflection curve upto failure load and the area of the same curve upto cracking load. In the case of mortar specimens, particularly when they are impregnated, the cracking load cannot be easily identified. Hence in this study, the toughness index was calculated by the ratio between the area of the curve upto failure load and the area upto 70% of the maximum load (0.7 P_{max}) as shown in Fig. 9. The variation of toughness index with fibre content for GFRMC and PIGFRMC is shown in Fig. 10. These results indicated that the toughness index increased with fibre content in GFRMC, and in the case of PIGFRMC, the influence of fibre content was almost negligible due to low ductility and high strength of matrix.

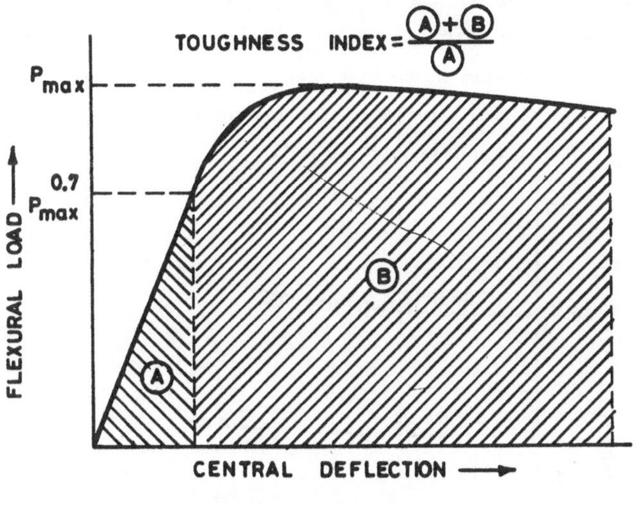

Fig. 9 A typical load-deflection curve for toughness index

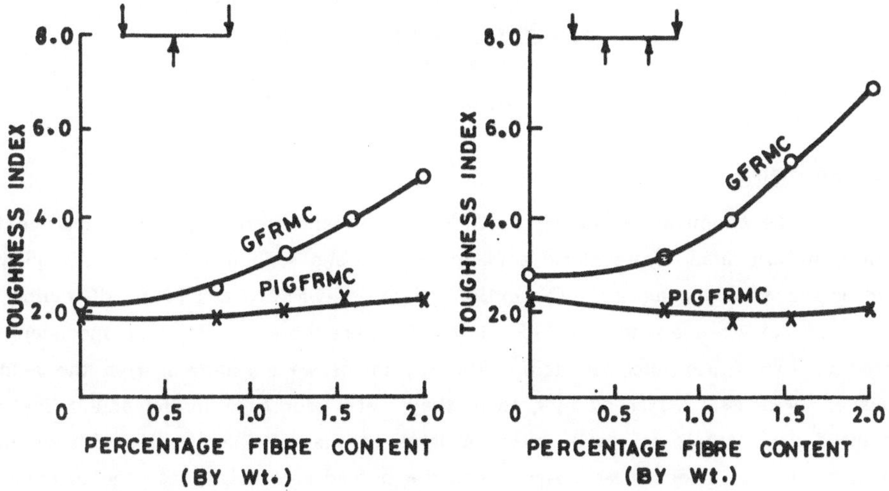

Fig. 10 Variation of toughness index with fibre content

Abrasion Resistance

The abrasion resistance of a material is determined from the wearing resistance recommended in the Indian Standards [14] . The wearing resistance of a specimen is measured by the loss in thickness due to abrasion and is calculated by the following formula:

$$t = \frac{10\,(W_1 - W_2)}{W_1} \cdot \frac{V_1}{A} \qquad \text{..... (9)}$$

where W_1, and W_2 are the initial and final weights, respectively, of the specimen subjected to abrasion, V_1 and A are the original volume in cm^3 and area in cm^2 of the specimen, respectively, and t is the loss in thickness of the specimen in mm due to abrasion. The results of abrasion of GFRMC and PIGFRMC are shown in Fig. 11. These results showed that the loss in thickness of GFRMC and PIGFRMC increased with fibre content unlike in the case of steel fibres. However, the loss in thickness of PIGFRMC specimens was lower than in the case of unimpregnated GFRMC.

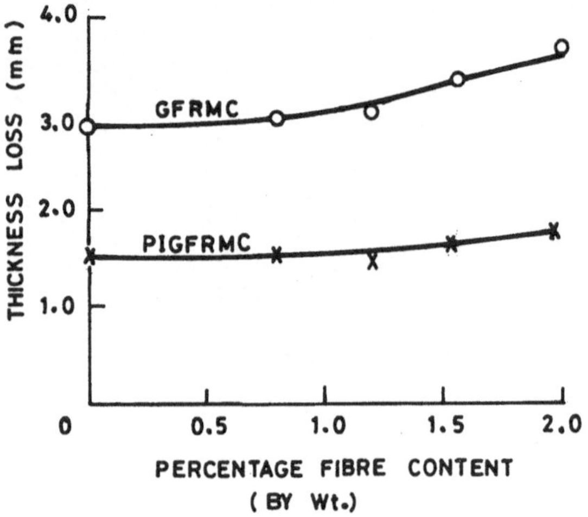

Fig. 11 Abrasion loss of GFRMC and PIGFRMC

Chemical Resistance

To study the chemical resistance of GFRMC with and without polymer impregnation 15% dilute sulfuric acid, hydrochloric acid, nitric acid and sodium hydroxide solution were selected as the attacking agents. The initial weights of GFRMC and PIGFRMC cylindrical specimens of 7.5 cm diameter and 15 cm height were noted. The test specimen was immersed in 15% dilute sulphuric acid. The specimens were removed from the acid at intervals of 24 hours. After washing them they were kept back in the acid. The acid was changed at intervals of seven days. Weights of the specimen at intervals of about 3 days were noted. From these observations the percentage weight loss due to chemical attack was computed. Similar tests were conducted separately using 15% dilute HCl, HNO_3,and NaOH. The percentage loss in weight due to chemical attack is shown in Fig. 12. The curves in Fig. 12 indicate that GFRMC was severely attacked by 15% dilute H_2SO_4. However this loss was considerably less in the case of PIGFRMC. The GFRMC specimens lost weight upto about 80% in about 40 days period where as the PIGFRMC lost only upto about 20% during the same period. This loss in both cases was found to slightly increase with addition of fibres. The effect of 15% dilute HCl and HNO_3 was very little in GFRMC and PIGFRMC. The maximum loss due to attack by HCl was only about 25% in GFRMC and about 10% in cement mortar. These were further reduced in the case of HNO_3. The attack of 15% dilute NaOH on cement mortar with and without glass fibres was practically nil. In fact the test specimens showed increased weight due to

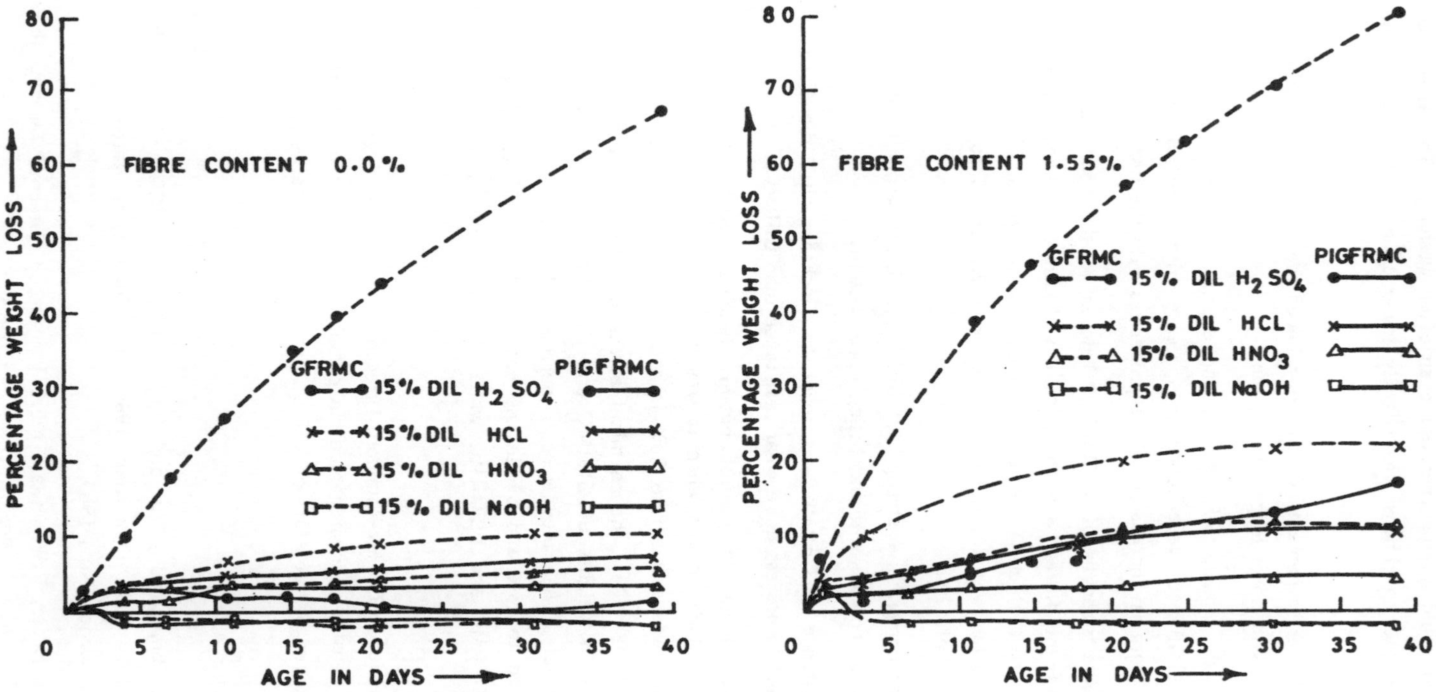

Fig. 12 Effect of chemicals on GFRMC and PIGFRMC

deposition of salts which could not be removed even after washing. The attack of these acids and base on PIGFRMC specimens was slightly lower than in the case of GFRMC.

CONCLUSIONS

1. Addition of glass fibres to cement composites reduces the workability. This reduction is more pronounced with fibres longer than 6 mm.

2. Split tensile strength, modulus of rupture, and ductility which is measured by the toughness index of GFRMC were found to increase with fibre content. However, PIGFRMC showed marginal increase of these properties with the fibre content.

3. The ratio of tensile strength of GFRMC is related to the bond strength and the volume fraction of the fibres, and the tensile strength of cement matrix as given by Eq.4. The value of α which is the ratio of the bond strength of fibres to the tensile strength of the mortar was found to vary between 2.5 and 5 for volume fraction of 6 mm long fibres upto 2%.

4. The density and the compressive strength of GFRMC with and without polymer impregnation were found to decrease with increasing fibre content.

5. The abrasion resistance of GFRMC was found to increase with the fibre content. PIGFRMC was found to possess more wear resistance than GFRMC.

6. The PIGFRMC was found to possess better chemical resistance than GFRMC. The GFRMC showed 80% of loss when it was immersed in 15% dilute H_2SO_4 for 40 days. This loss was reduced to 20% in the case of PIGFRMC. In the case of 15% dilute HCI and HNO_3 the chemical attack on GFRMC was not so severe. The maximum loss in weights of GFRMC and PIGFRMC were found to be 25% and 15%, respectively, when they were immersed in 15% dilute HCI for 40 days. These were reduced to 15% and 8% in the case of dilute HNO_3. There was no deterioration of GFRMC and PIFGRMC when they were immersed in 15% dil NaOH.

ACKNOWLEDGEMENT

The authors are thankful to Dr. M. Ramaiah, Director of Structural Engineering Research Centre, Madras, for his constant encouragement in pursuing the studies on concrete composites. The services rendered by the staff of SERC in preparing the manuscript are acknowledged with thanks.

REFERENCES

1. Biryukovich, K.L., Biryukovich, Yu, and Biryukovich, D.L., "Glass Fibre Reinforced Cement," Published by Budivel'nik Kiev, 1964, Civil Engineering Research Association Translation No. 12, London, Nov. 1965.

2. Majumdar, A.J., and Nurse, R.W., "Glass Fibre Reinforced Cement," Building Research Establishment Current Paper, CP94/75, U.K., Oct. 1975.

3. Ali, M.A., Majumdar, A.J., and Singh, B., "Properties of Glass Fibre Cement - The

Effect of Fibre Length and Content," Building Research Establishment Current Paper CP38/76, U.K., June 1976.

4. Marsh, Jr. Henry, N., and Clarke, Jr. Lawrence, L., "Glass Fibre Reinforced Cement Base Materials," International Symposium: Fibre Reinforced Concrete, ACI Publication SP-44, American Concrete Institute, 1974.

5. Bigen, J., "Glass Fibre Reinforced Cement: Improvements by Polymer Additions," Symposium, Advances in Cement Matrix Composites, Materials Research Society, University Park, Pennsylvania, Nov. 1980.

7. Goeman, F., "Performance of E-Glass Fibres as Reinforcement for Portland Cement Mortar," An International Symposium: Fibre Reinforced Concrete, ACI Publication, Sp-44, American Concrete Institute, 1974.

8. Indian Standards, "IS:269 - 1967, Indian Standard Code of Practice on Specification for Ordinary, Rapid Hardening, and Low Heat Portland Cement," Indian Standards Institution, New Delhi, 1967.

9. Indian Standards, "IS:383-1970, Specifications for Coarse and Fine Aggregates from Natural Sources for Concrete," Indian Standards Institution, New Delhi, 1970.

10. Dastur, S.M., "Quality Aspects of Fibre Glass Reinforcements," Symposium on FRP - A Versatile Construction Material for Chemical Industry, Indian Plastics Institute, Madras, 1985.

11. Romouldi, J.P., and Batson, G.B., "Behaviour of Reinforced Concrete Beams with Closely Spaced Reinforcement," ACI Journal, Proceedings V.60, No.6, June 1963.

12. Romouldi, J.P., and Mandel, J.A., "Tensile Strength of Concrete Affected by Uniformly Distributed and Closely Spaced Short Lengths of Wire Reinforcement," ACI Journal, Proceedings V.61, No.6, June 1964.

13. Kukreja, C.B., Kaushik, S.K., Kanchi, M.B., and Jain, O.P., "Effect of Steel Fibres on Compressive Strength and Compressive Strain of Concrete," Journal of the Institution of Engineers (India), Civil Engg. Div., V.62, Pl. C12, September 1981.

14. Indian Standards, "IS:1237-1980, Cement Concrete Flooring Tiles," Indian Standards Institution, New Delhi, 1980.

Proceedings of the International Symposium on Fibre Reinforced Concrete, December 16–19, 1987, Madras, India

FIBRE REINFORCED CONCRETE WITH NON-METALLIC FIBRES- AN EXPERIMENTAL STUDY

1. R. Nagaraja
Director, I.C.S. J.N.Technological University, Hyderabad (A.P)

2. B.L.P. Swami
Assistant Professor, Civil Engineering Department,
J.N.T.U. College of Engineering, Anantapur (A.P.)

SYNOPSIS

Concrete properties especially like resilience and energy absorption capacity could be significantly improved by the addition of non-metalic fibres. In the present investigation the strength properties of concrete reinforced with short pieces of random fibres of nylon, bamboo and coir have been studied. Tests were performed on specimens to determine the compressive strength, split tensile strength, flexural strength, energy absorption and modulus of elasticity and the results are compared with those of plain concrete.

It is concluded that the addition of non-metallic fibres do not contribute to the improvement of compressive strength compared to plain concrete. The tensile load carrying capacity of FRC is fairly increased over that of plain concrete. The stress-strain curve in the initial elastic range is more stable in the case of FRC. The plastic rotation characteristics are improved with the addition of fibres. The non-metallic fibres also contribute towards increased resistance to impact and blast.

1. INTRODUCTION

The weakness of concrete in tension is well known. Its high compressive strength and capacity to deform significantly under stresses near failure stress is well established. However, under shock loads concretes tend to be quite brittle and under effects of blast tend to splinter out causing a large extent of secondary damage to the surroundings. Specially in such structures where more than the high compressive strength, the energy obsorption or ductility characteristics of concrete is called into play any resilient fibre metallic or non-metallic with a reasonably good tensile strength would be useful as reinforcement. The principal requirement for such a composite of concrete and non-metallic fibres is a reasonably good bond between the fibre and the concrete.

R.N. Swamy [14] studied the mechanical properties of steel fibre reinforced concrete. Laboratory investigation carried out by Sridhara [12] indicated varying degrees of improvement in the impact strength of concrete due to the provision of fibres. He employed coir, jute, glass fibre and stretched nylon as the fibres and concluded that the impact strength at 90 days of the fibre concrete had increased three fold for 3% coir and 3% jute fibres. This increase for nylong yarn was nearly 17 times. Siraskar and Kumar [11] conducted tests on fibre concrete pill boxes. Goldfein [2] reported that among all the fibres tried, nylon fibre had the unique advantage of imparting resilience to concrete.

2. DETAILS OF INVESTIGATION

In the present investigation the strength properties of concrete reinforced with short pieces of random fibres of nylon, bamboo and coir are studied in tension, compression and bending. The aspect ratio of the fibres and the concrete mix were maintained constant throughout the test series. The details of fibres, test specimens and the tests performed are described as follows.

2.1 Details of Fibres

The volume percent of fibres adopted was 0.5 and 1 for nylon , 2 and 3 for bamboo and 1.5 and 2 for coir. The aspect ratio of the fibre is decided by the case of mixing as a fibre length having large aspect ratio is likely to lump inside the mixer.

2.2 Details of Specimens

A large number of 15cm standard cubes and 15cm X 15cm X 70cm long beams were cast. Nominal concrete mix of 1:2:4 by weight with a water cement ratio of 0.65 with medium workability was used for casting the specimens. For each percentage of fibre 5 specimens were used. All the specimens were cast and cured as per standard procedures and tested at the age of 28 days.

2.3 Testing

Tests were performed on the specimens to determine 1) Compressive strength 2) Split tensile strength 3) Flexural strength 4) Energy absorption and 5) Modulus of elasticity.

3. RESULTS AND CONCLUSIONS

3.1 Compressive Strength

From the value of the average cube compressive strength (Table 1), it is observed that the inclusion of nylon fibre has no effect on the crushing strength. All the plain concrete specimens exhibited severe spalling at failure whereas the nylon FRC specimens exhibited very mild spalling. In the case of specimens with bamboo fibres the average cube compressive strength is decreased. This is probably due to the lower specific gravity of the bamboo fibre. However, spalling has been arrested appreciably. The behaviour of the specimens with coir fibres is more or less similar to those with bamboo fibres.

3.2 Tensile Strength

The split tension test results are shown in Table 2. The addition of nylon fibres by 0.5% and 1% indicated an increase in the average tensile strength by 25.94% and 26.03% respectively. In the case of specimens with bamboo fibres by 2% and 3% by volume this increase is 15.65% and 20.72% respectively. With coir fibres, there is no improvement in the tensile strength of concrete. However, specimens with all the above fibres avoided the brittle nature of failure which is common with plain concrete specimens.

3.3 Flexural Strength

It can be observed (Table 3) that the addition of nylon fibres by 0.5% and 1% has increased the flexural strength by 7.3% and 24.4% respectively. There is no improvement in the flexural strength with bamboo fibres. With coir fibres there is some reduction in the flexural strength but the tensile crack propogation in the specimens was considerably reduced.

3.4 Flexural Load Vs Deflection

The load-deflection curves (Figs. 1 to 3) show that the transfer of stress from the concrete matrix to the fibres takes place when the matrix starts cracking and this is indicated by the change in the slope of the curve from linearity. With further loading the fibres start either pulling out or debonding with the matrix until it reaches its ultimate failure load. With higher percentage of nylon fibres there is considerably increase in the flexural strength of concrete.

Modulus of elasticity as computed from the observed deflection is found to be considerably decreasing in the case of FRC compared to plain concrete. This enables the FRC more ductile.

3.5 Moment Vs Rotation

From the typical graphs plotted between moment and rotation (Figs 3 to 6), It can be observed that the presence of fibres in concrete facilitated the beams to develop higher rotation characteristics towards failure. The rotation capacity is more with nylon fibres, less with bamboo fibres and still lesser with coir fibres whereas plain concrete specimens exhibit pure brittle failure.

3.6 Resilience and Ductility Characteristics

The computed values of strain energy of the FRC specimens are shown with those of plain concrete in Table 4. With the addition of nylon and bamboo fibres there is improvement in the strain absorbing capacity by various degrees whereas there is no improvement with coir fibres. In general, it can be concluded that the FRC is more resilient and ductile compared to plain concrete.

3.7 Shear Strength

The ultimate shear strength of FRC is worked out from the compressive and tensile strengths by making use of the relation given by Murashev et al [9]. The results are shown in Table 5. It is seen that the ultimate shear strength is fairly increasing due to the addition of nylon fibres and this increase is less with bamboo fibres. But with coir fibres the ultimate shear strength is decreasing. This may be attributed to the stretching property of coir fibre in wet condition.

3.8 Cracking Characteristics

It is observed from the tested specimens that the addition of fibres is effective in arresting the growth as well as propogation of cracks. This can be due to the dissipation of part of the energy in destroying the bond between the fibres and matrix, thus the energy available for developing crack surface is considerably reduced. It has also been found that FRC beams have failed with a single crack traversing almost straight up to the neutral axis at the cracking load in addition to the numerous micro cracks that have formed in the tensile zone of the beam. The plain concrete beams have failed suddenly into two pieces without any warning.

Based on the appreciable resistance to the growth and propogation of cracks in FRC, it can also be concluded that the resistance of FRC to impact and fatigue loadings is fairly greater than plain concrete.

REFERENCES

1. Dinesh Mohan and Rehsi, S.S., "Steel fibre reinforced concrete units", Indian Concrete Jl., V.51, Dec. 1977, pp 370-373.
2. Goldfein, S., "Plastic fibrous reinforcement for portland cement", Technical Report No.1757, U.S. Army (Engrs) R and D Labs, Fort Belvoir, Virginia, Oct.1968.
3. Grimer, F.J. and Ali, N.A., "The strength of cements reinforced with glass fibres", Magazine of concrete Research, March 1969, V.21, pp 23-30.
4. Hannant, D.G., "Steel fibres and light weight beams" , Concrete, Dec. 1973.
5. Hollister, G.S. and Thomas, C., "Fibre reinforced materials", Elsevier Publishing Co. Ltd., 1966.
6. Hughes, B.P. and Fattuhi, N.I., "Load-deflection curves for fibre reinforced concrete beams in flexure", Magazine of Concrete Research, V.29, Dec.1977, pp 199-206.
7. Lankard, D.R., "Fibre concrete application ", Rilem Symp. 1975, pp 3-17.
8. Mc. Currich, L.H. and Adams, M.A.J., "Fibres in cement concrete", Concrete,

April 1973, pp 51-53.

9. Murashev, V., Siglow, E. and Baikov, V., "Design of reinforced concrete structures", MIR Publishers, Moscow, 1968.

10. Parameswaran, V.S. and Rajagopalan, K., "Strength of concrete beams with aligned or random steel fibre micro reinforcement", Rilem Symp. 1975, pp 95-103.

11. Siraskar, K.A. and Kumar, S., "Fibre reinforcement for shatter proofness of concrete", Proc. of Modern Trends in Civil Engineering, Roorkee, 1972.

12. Sridhara, S., Kumar, S. and Sinare, M. ., "Fibre reinforced concrete", The Indian Concrete Jl., V.45, No.10, October 1971 pp 428-442.

13. Swamy, R.N., "Delft conf. on fibre concrete materials", Concrete, Dec. 1973, P 44

14. Swamy, R.N. and Ali Noori, K., "Bond strength of steel fibre reinforced concrete", Concrete, August 1974, pp 36-37.

TABLE 1

DIRECT COMPRESSION TEST

(15cm X 15cm X 15cm cubes)

Sl. No.	Fibre	Vol. % of Fibre	Specimen No.	Weight in Kgs.	Ult.Comp. load in Kgs.	Ult. Comp. stress Kg/Cm²	Average Comp. stress Kg/Cm²
1.	Plain Concrete	0.0	JP/8	8.20	62,000	275.55	
			JP/1	8.10	62,800	279.11	276.18
			JP/4	8.15	61,600	273.88	
2.	Nylon	0.5	JNI/6	7.80	61,400	280.00	
			JNI/5	8.05	65,600	291.56	286.96
			JNI/4	8.10	65,100	289.33	
3.	Nylon	1.0	JN2/10	8.00	63,100	280.44	
			JN2/2	8.05	64,000	284.09	282.67
			JN3/8	8.25	63,600	282.69	
4.	Bamboo	2.0	JB2/2	7.90	55,800	248.00	
			JB2/10	7.65	54,600	242.67	243.11
			JB2/6	8.00	53,700	238.67	
5.	Bamboo	3.0	JB3/7	7.80	48,800	216.89	
			JB3/6	7.80	56,600	251.56	236.45
			JB3/5	7.80	54,200	240.99	
6.	Coir	1.5	JC1/6	7.90	44,500	197.78	
			JC1/7	7.90	47,800	212.44	196.44
			JC1/8	7.90	40,300	179.11	
7.	Coir	2.0	JC2/3	7.25	33,000	146.67	
			JC2/4	7.45	33,000	146.67	143.41
			JC2/10	7.15	30,800	136.89	

TABLE 2

SPLIT TENSION TEST

(15cm X 15cm X 15cm cubes)

Sl. No.	Fibre	Vol. % of Fibre	Specimen No.	Weight in Kgs.	Ult.Tensile load in Kgs.	Ult. Tensile stress Kg/Cm²	Average Tensile stress Kg/Cm²
1.	Plain Concrete	0.0	JP/10	8.20	7225	20.44	
			JP/9	8.20	7050	19.94	21.28
			JP/7	8.30	8300	23.47	
2.	Nylon	0.5	JN1/8	8.10	9625	27.22	
			JN1/2	8.10	9750	27.57	26.80
			JN1/10	8.15	9050	25.60	
3.	Nylon	1.0	JN2/6	8.00	9850	27.86	
			JN2/1	8.15	9350	26.44	26.82
			JN2/9	8.05	9250	26.16	
4.	Bamboo	2.00	JB2/9	8.00	8100	22.91	
			JB2/1	7.90	8500	24.04	24.61
			JB2/3	8.00	9500	26.87	
5.	Bamboo	3.00	JB3/7	7.70	8760	24.77	
			JB3/8	7.85	9840	27.83	25.70
			JB3/3	7.85	8660	24.59	
6.	Coir	1.5	JC1/1	7.95	7240	20.48	
			JC1/2	7.85	7000	19.80	20.56
			JC1/3	7.90	7570	21.41	
7.	Coir	2.0	JC2/5	7.55	7850	22.20	
			JC2/1	7.20	7000	19.80	20.55
			JC2/2	7.60	6950	19.66	

TABLE 3

FLEXURE TEST

(15cm X 15cm X 70cm beams)

Sl. No.	Fibre	Vol. % of Fibre	Speci-men No.	Weight in Kgs.	Ult.Flexural load in Kgs.	Average load in Kgs.	Strength ratio
1.	Plain Concrete	0.0	JP/7	37.23	2800	1913	1.000
			JP/4	37.40	2100		
			JP/2	37.05	1840		
2.	Nylon	0.5	JN1/2	37.25	2000		
			JN1/4	37.47	2000	2053.33	1.073
			JN1/3	37.55	2160		
3.	Nylon	1.0	JN2/5	37.60	2200		
			JN2/4	37.76	2900	2380	1.241
			JN2/2	37.53	2040		
4.	Bamboo	2.0	JB2/5	36.60	1860		
			JB2/4	36.65	2200	2020	1.031
			JB2/2	36.62	2000		
5.	Bamboo	3.0	JB3/4	35.70	1700		
			JB3/3	36.00	1800	1746.67	0.913
			JB3/2	36.10	1740		
6.	Coir	1.5	JC1/2	35.05	1480		
			JC1/3	35.00	1400	1426.67	0.746
			JC1/5	35.01	1400		
7.	Coir	2.0	JC2/4	34.90	1580		
			JC2/3	34.80	1400	1460	0.763
			JC2/2	34.85	1400		

TABLE 4

COMPARISON OF STRAIN ENERGIES OF PLAIN CONCRETE AND FRC

(Young's Modulus of Concrete = 2×10^5 Kg/Cm^2)

Fibre	Specimen No.	% of fibre	Ult. load	Moment	Rotation	Strain energy	Average energy	Energy Improvement factor
Plain	JP/1	---	1800	-	-	6.40		
	JP/4	---	2100	-	-	8.70	7.26	
	JP/2	---	1840	-	-	6.69		
Nylon	JN1/3	0.5	2160	21600	$12.80/10^4$	13.82		
	JN1/5	0.5	2000	20000	$10.20/10^4$	10.20	12.01	1.65
Bomboo	JB2/2	2.0	2000	20000	$11.60/10^4$	11.60		
	JB2/3	2.0	2000	20000	$14.60/10^4$	14.60	12.60	1.74
	JB2/1	2.0	2000	20000	$11.60/10^4$	11.60		
Coir	JC1/5	3.0	1600	16000	$9.60/10^4$			
Coir	JC1/5	1.5	1400	14000	$13.20/10^4$	9.24	7.15	0.98
	JC1/1	1.5	1100	11000	$9.20/10^4$	5.06		

TABLE 5

ULTIMATE SHEAR STRENGTH OF FRC

Sl. No.	Fibre	Vol. % of fibre	Cube Comp. strength Kg/Cm^2	Tensile strength Kg/Cm^2	Shear strength Kg/Cm^2	Average Ult.shear strength Kg/Cm^2
1.	Plain	---	275.55	20.44	43.95	
		---	279.11	19.94	43.69	44.86
		---	273.88	23.47	46.95	
2.	Nylon	0.5	280.00	27.22	51.13	
			291.56	27.51	52.51	51.35
			289.33	25.60	50.40	
3.	Bamboo	2.0	248.00	22.91	44.15	
			242.67	24.04	44.73	45.26
			238.67	26.87	46.90	
4.	Coir	1.5	197.78	20.48	37.27	
			212.44	19.80	37.98	37.17
			179.11	21.40	36.27	

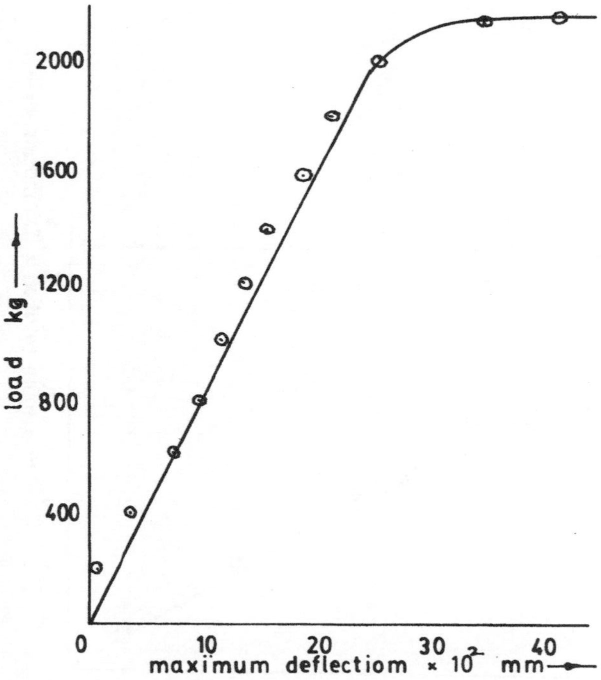

FIG.1. LOAD- DEFLCTION CURVE WITH 2 PERCENT NYLON FIBRE.

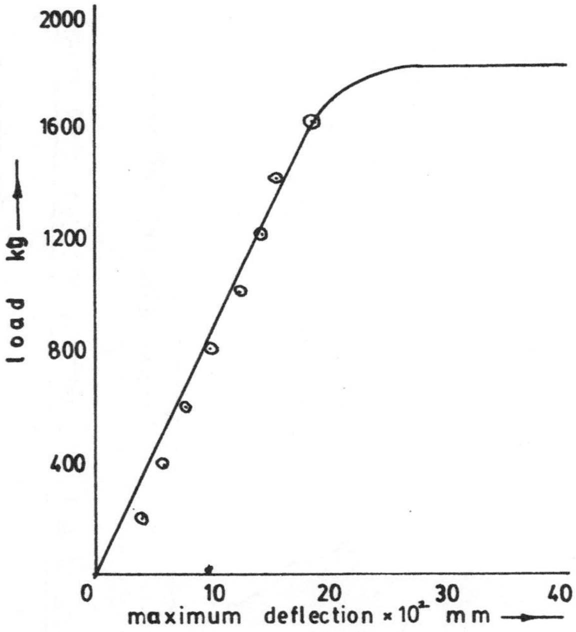

FIG.2. LOAD- DEFLECTION CURVE WITH 2 PERCENT BAMBOO FIBRE.

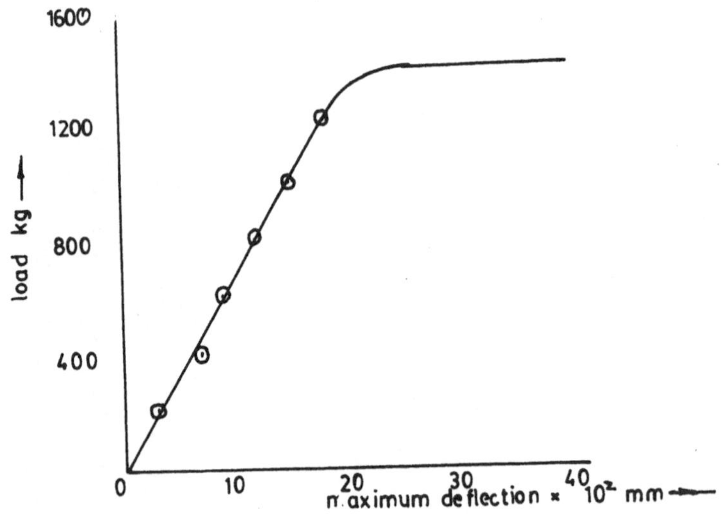

FIG.3. LOAD-DEFLECTION CURVE WITH 2 PERCENT COIR FIBRE.

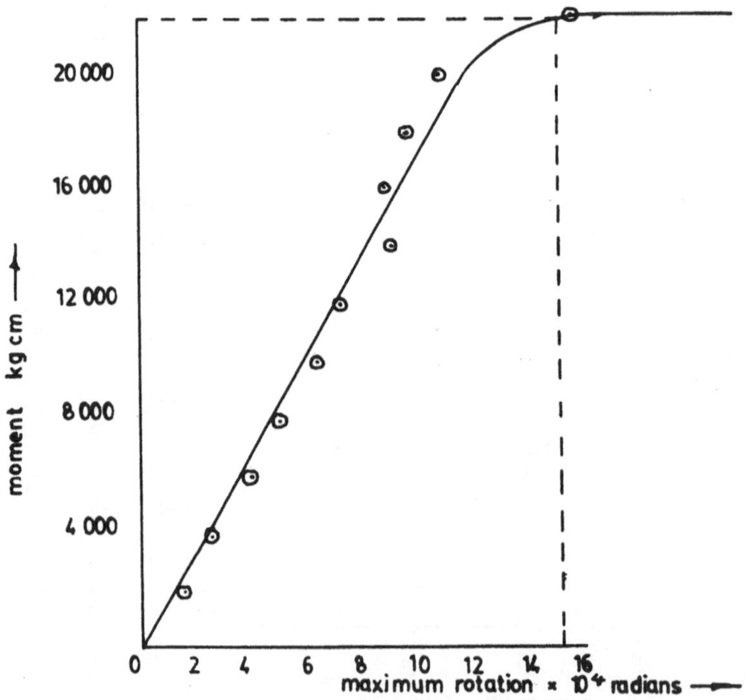

FIG.4. MOMENT-ROTATION CURVE WITH 2 PERCENT NYLON FIBRE.

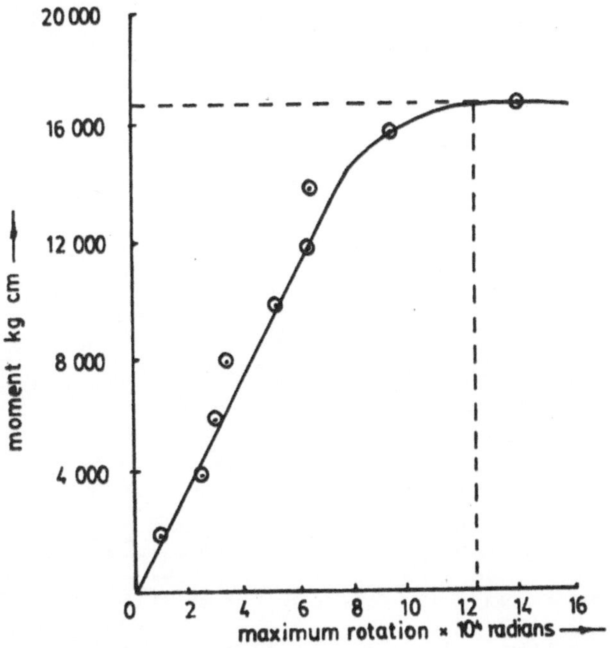

FIG.5. MOMENT-ROTATION CURVE WITH 2 PERCENT BAMBOO FIBRE.

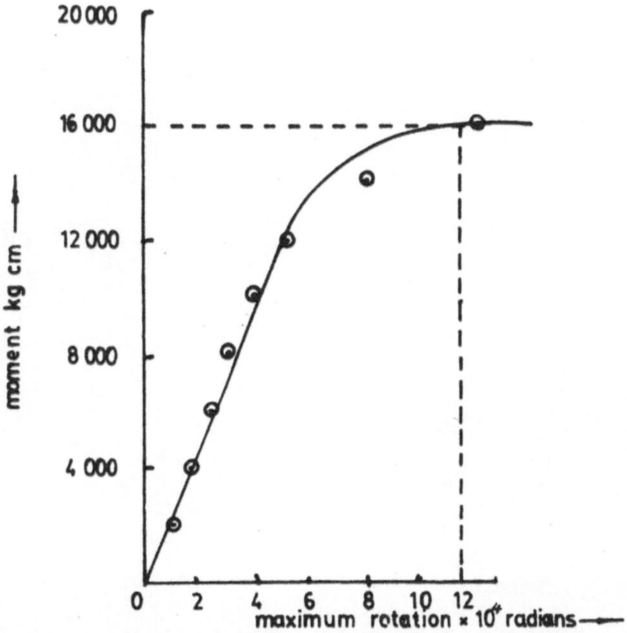

FIG.6. MOMENT-ROTATION CURVE WITH 2 PERCENT COIR FIBRE.

AUTHOR INDEX